Vol. 32. **Determination of Organic Compounds: Methods and Procedures.** By Frederick T. Weiss

Vol. 33. **Masking and Demasking of Chemical Reactions.** By D. D. Perrin

Vol. 34. **Neutron Activation Analysis.** By D. De Soete, R. Gijbels, and J. Hoste

Vol. 35. **Laser Raman Spectroscopy.** By Marvin C. Tobin

Vol. 36. **Emission Spectrochemical Analysis.** By Morris Slavin

Vol. 37. **Analytical Chemistry of Phosphorus Compounds.** Edited by M. Halmann

Vol. 38. **Luminescence Spectrometry in Analytical Chemistry.** By J. D. Winefordner, S. G. Schulman and T. C. O'Haver

Vol. 39. **Activation Analysis with Neutron Generators.** By Sam S. Nargolwalla and Edwin P. Przybylowicz

Vol. 40. **Determination of Gaseous Elements in Metals.** Edited by Lynn L. Lewis, Laben M. Melnick, and Ben D. Holt

Vol. 41. **Analysis of Silicones.** Edited by A. Lee Smith

Vol. 42. **Foundations of Ultracentrifugal Analysis.** By H. Fujita

Vol. 43. **Chemical Infrared Fourier Transform Spectroscopy.** By Peter R. Griffiths

Vol. 44. **Microscale Manipulations in Chemistry.** By T. S. Ma and V. Horak

Vol. 45. **Thermometric Titrations.** By J. Barthel

Vol. 46. **Trace Analysis: Spectroscopic Methods for Elements.** Edited by J. D. Winefordner

Vol. 47. **Contamination Control in Trace Element Analysis.** By Morris Zief and James W. Mitchell

Vol. 48. **Analytical Applications of NMR.** By D. E. Leyden and R. H. Cox

Vol. 49. **Measurement of Dissolved Oxygen.** By Michael L. Hitchman

Vol. 50. **Analytical Laser Spectroscopy.** Edited by Nicolo Omenetto

Vol. 51. **Trace Element Analysis of Geological Materials.** By Roger D. Reeves and Robert R. Brooks

Vol. 52. **Chemical Analysis by Microwave Rotational Spectroscopy.** By Ravi Varma and Lawrence W. Hrubesh

Vol. 53. **Information Theory As Applied to Chemical Analysis.** By Karel Eckschlager and Vladimír Štěpánek

Vol. 54. **Applied Infrared Spectroscopy: Fundamentals, Techniques, and Analytical Problem-solving.** By A. Lee Smith

Vol. 55. **Archaeological Chemistry.** By Zvi Goffer

Vol. 56. **Immobilized Enzymes in Analytical and Clinical Chemistry.** By P. W. Carr and L. D. Bowers

Vol. 57. **Photoacoustics and Photoacoustic Spectroscopy.** By Allan Rosencwaig

Vol. 58. **Analysis of Pesticide Residues.** Edited by H. Anson Moye

Vol. 59. **Affinity Chromatography.** By William H. Scouten

Vol. 60. **Quality Control in Analytical Chemistry.** By G. Kateman and F. W. Pijpers

Vol. 61. **Direct Characterization of Fineparticles.** By Brian H. Kaye

Vol. 62. **Flow Injection Analysis.** By J. Ruzicka and E. H. Hansen

Vol. 63. **Applied Electron Spectroscopy for Chemical Analysis.** Edited by Hassan Windawi and Floyd Ho

Vol. 64. **Analytical Aspects of Environmental Chemistry.** Edited by David F. S. Natusch and Philip K. Hopke

Vol. 65. **The Interpretation of Analytical Chemical Data by the Use of Cluster Analysis.** By D. Luc Massart

Vol. 66. **Solid Phase Bi** ... **cts.** Edited by William H. Scouten

(*continued on back*)

5/9/89

Inductively Coupled Plasma Emission Spectroscopy

Part 2

CHEMICAL ANALYSIS

A SERIES OF MONOGRAPHS ON
ANALYTICAL CHEMISTRY AND ITS APPLICATIONS

Editors
P. J. ELVING, J. D. WINEFORDNER
Editor Emeritus: **I. M. KOLTHOFF**

VOLUME 90

A WILEY-INTERSCIENCE PUBLICATION

JOHN WILEY & SONS

New York / Chichester / Brisbane / Toronto / Singapore

Inductively Coupled Plasma Emission Spectroscopy

Part 2

APPLICATIONS AND FUNDAMENTALS

Edited by

P. W. J. M. Boumans

Philips Research Laboratories
Eindhoven, The Netherlands

A WILEY-INTERSCIENCE PUBLICATION

JOHN WILEY & SONS

New York / Chichester / Brisbane / Toronto / Singapore

Library of Congress Cataloging in Publication Data:

Inductively coupled plasma emission spectroscopy.

(Chemical analysis, ISSN 0069-2883; v. 90)
"A Wiley-Interscience publication."
Includes index.
Contents: pt. 1. Methodology, instrumentation, and performance—pt. 2. Applications and fundamentals.
1. Plasma spectroscopy. I. Boumans, P. W. J. M. (Paul Willy Josephy Maria) II. Series.

QD96.P62I4 1987 543'.0873 86-18984
ISBN 0-471-85378-X (v. 2)

Printed in the United States of America

10 9 8 7 6 5 4 3 2

CONTRIBUTORS

Ramon M. Barnes, Department of Chemistry, University of Massachusetts, Amherst, Massachusetts

M. W. Blades, Department of Chemistry, University of British Columbia, Vancouver, British Columbia, Canada

Andrew W. Boorn, Sciex, Thornhill, Canada

Maher I. Boulos, Department of Chemical Engineering, University of Sherbrooke, Sherbrooke, Québec, Canada

J. A. C. Broekaert, Institut für Spektrochemie und angewandte Spektroskopie, Dortmund, Federal Republic of Germany

Richard F. Browner, School of Chemistry, Georgia Institute of Technology, Atlanta, Georgia

Jan-Ola Burman, Svenskt Stål AB, Division Metallurgi, Luleå, Sweden

K. Fuwa, Department of Chemistry, Faculty of Science, University of Tokyo, Bunkyo, Tokyo, Japan

Hiroki Haraguchi, Department of Chemistry, Faculty of Science, University of Tokyo, Bunkyo, Tokyo, Japan

F. J. M. J. Maessen, Laboratorium voor Analytische Scheikunde, Universiteit van Amsterdam, Amsterdam, The Netherlands

J. W. McLaren, Analytical Chemistry Section, Chemistry Division, National Research Council of Canada, Ottawa, Canada

J. M. Mermet, Laboratoire des Sciences Analytiques, Université Claude Bernard—Lyon I, Villeurbanne, France

K. Ohls, Hoesch Hüttenwerke, A.G., Dortmund, Federal Republic of Germany

B. L. Sharp, The Macaulay Institute for Soil Research, Craigiebuckler, Aberdeen, Scotland

D. Sommer, Hoesch Hüttenwerke, A.G., Dortmund, Federal Republic of Germany

G. Tölg, Laboratorium für Reinststoffanalytick, Max-Planck-Institut für Metallforschung (Stuttgart), Dortmund, Federal Republic of Germany

Jon C. Van Loon, Departments of Geology and Chemistry and Institute for Environmental Studies, University of Toronto, Toronto, Ontario, Canada

PREFACE

This book is the second part of the two-volume treatise *Inductively Coupled Plasma Emission Spectroscopy* and addresses applications and fundamentals. Part 1 deals with methodology, instrumentation, and performance. The original intention was to cover the complete subject in a single volume. However, the explosive growth of inductively coupled plasma–atomic emission spectroscopy (ICP–AES) in recent years and the scarcity of adequate and modern textbooks on emission spectroscopy indicated the need for a more comprehensive treatment. Therefore, this work was set up in its present form, with a fourfold aim:

1. To fill an essential gap in the AES literature.
2. To provide a critical and tutorial survey of more than 20 years of research, development, and application in the field of ICPs and related plasma sources.
3. To act as a handbook and textbook for the novice and the expert.
4. To serve as an aide-memoire and major source of reference for broad groups of analytical spectroscopists, analytical chemists, physical chemists, and physicists, including researchers, technicians, and applied analysts.

Part 1 comprises nine chapters. Five of them were written by myself: Introduction to AES (Chapter 1), Plasma Sources other than ICPs (Chapter 2), ICPs (Chapter 3), Basic Concepts and Characteristics of ICP–AES (Chapter 4), and Line Selection and Spectral Interferences (Chapter 7). I shared the authorship with G. M. Hieftje in a chapter on Torches (Chapter 5) and with J. A. C. Broekaert in a chapter on Sample Introduction (Chapter 6). The topic Spectrometers was entrusted to J. W. Olesik (Chapter 8) and Detection and Measurement to H. Bubert and W.-D. Hagenah (Chapter 9).

The heavy seal that my authorship set on Part 1 emphasizes my concern as editor to produce a textbook that treats the basic principles of emission spectroscopy in a tutorial, systematic, and consistent way. I have been conscious, however, that a consistent treatment of a subject by a single author entails the hazards of bias and shortsightedness. Therefore, I employed the scholarship, experience, and specific expertise of a group of authors to cover those topics that can be easily split off the main line without risking loss of consistency and coherence. These authors have been allotted only a relatively modest place in Part 1, but had the opportunity to set their stamp fully in Part 2, where my task was confined to that of a rather critical editor and text revisor.

The difference in expertise and scientific background among the authors has resulted in a balanced distribution of diverse, though consistent, views and opinions throughout the work as a whole. The participation of authors from various countries, mainly in Western Europe and North America, has further contributed to an equilibrated treatment of the various topics. I have made efforts to formulate their assignments strictly and to inspect their texts carefully in order to prevent undue overlap. Only in those cases where the expression of different, but complementary visions appeared beneficial have I tolerated some overlap.

The thematic arrangement of the topics has resulted in a compilation of chapters that are more or less self-existent and may be consulted independently of each other. In the treatment of the topics, either a tutorial approach or a review character predominates, depending on the subject matter. In all instances, however, the literature has been covered comprehensively.

For practical reasons the complete material was split into two parts, published as separate volumes, each being self-consistent within certain limits. Part 1 covers the basis of ICP–AES as an analytical method and deals with: (1) the fundamental analytical concepts, namely, detection limits, precision, accuracy, dynamic range, multielement capability, spectral interferences, and line selection; (2) the principles of the instrumentation, such as, sample introduction devices, torches, spectrometers, and devices for detection and measurement. For obvious reasons a treatment of the principles of generators and a detailed survey of commercial instruments have been omitted.

Part 2 comprises 12 chapters dealing with the following subjects. The first six chapters review applications of ICP–AES according to field: Metals and Industrial Materials by K. Ohls and D. Sommer (Chapter 1), Geological by J.-O. Burman (Chapter 2), Environmental by J. W. McLaren (Chapter 3), Agriculture and Food by B. L. Sharp (Chapter 4), Biological–Clinical by F. J. M. J. Maessen (Chapter 5), and Organics by A. W. Boorn and R. F. Browner. The basis of this subdivision may be disputed, since similar sample types may be encountered in different fields. Therefore some overlap in the discussion of the analysis problems and their solutions is inevitable. On the other hand, the boundaries within which the authors had to stay gave them ample opportunities for in-depth discussions. Crucial is that different authors have laid different emphases and have expressed different visions. Therefore these chapters on applications reflect, in their entirety, the fruits of broad experience and profound knowledge. Let the reader benefit from seeing reviews and views of this caliber assembled in a single volume, and, in particular, by also considering chapters that at first sight might seem to be beyond his field of interest. Interaction is always beneficial.

Chapter 7, Direct Elemental Analysis of Solids by ICP–AES, written by J. C. Van Loon, H. Haraguchi, and K. Fuwa, links up with the general treatment of sample introduction in Chapter 6 of Part 1 and concludes the part of this book dealing with applications.

The subsequent text covers the fundamentals of ICP–AES in four chapters: Fundamental Aspects of Aerosol Generation and Transport by R. F. Browner (Chapter 8), Plasma Modeling and Computer Simulation by M. I. Boulos and R. M. Barnes (Chapter 9), Spectroscopic Diagnostics by J. M. Mermet (Chapter 10), and Excitation Mechanisms and Discharge Characteristics—Recent Developments by M. L. Blades (Chapter 11). These chapters are tutorial reviews in which classical notions and concepts, as well as recently gained insights, are treated.

As a further stimulus to beneficial interactions, Chapter 12, contributed by J. A. C. Broekaert and G. Toelg, deals with the Status and Trends of Development of Atomic Spectrometric Methods for Elemental Trace Determinations. This chapter provides an assessment that may stimulate the reader to look also beyond the strict boundaries of ICP–AES and, more generally, beyond the limits of emission spectroscopy.

The work as a whole is primarily intended for spectroscopists and analytical chemists in the field of ICP–AES. However, the way in which the subjects are addressed makes the information also of major interest to workers in related fields: AES using plasma sources other than ICPs, ICP atomic fluorescence spectrometry (AFS), and ICP mass spectrometry (MS). More generally, anyone wishing to become acquainted with the essentials of emission spectroscopy will either find much of the desired information directly in this work or will be guided by it to more specialized literature.

I hope that this treatise will be a useful source of reference to many workers in the fields of AES and ICP–AES concerned with research, development, or applications in either universities, industrial organizations, or governmental institutes. Let the insights gained by consulting this work stimulate the further progress of emission spectroscopy.

I want to thank the authors for their contributions, their willingness to accept my suggestions, and their perseverance to carry this project appropriately to its end. I also want to thank the Series Editor, Professor J. D. Winefordner, for his patience and encouragement and for his understanding of my position as editor.

P. W. J. M. BOUMANS

Eindhoven, The Netherlands
June 1987

CONTENTS

CHAPTER 1. APPLICATIONS: METALS AND INDUSTRIAL MATERIALS 1
K. Ohls and D. Sommer

CHAPTER 2. APPLICATIONS: GEOLOGICAL 27
J.-O. Burman

CHAPTER 3. APPLICATIONS: ENVIRONMENTAL 48
J. W. McLaren

CHAPTER 4. APPLICATIONS: AGRICULTURE AND FOOD 65
B. L. Sharp

CHAPTER 5. APPLICATIONS: BIOLOGICAL–CLINICAL 100
F. J. M. J. Maessen

CHAPTER 6. APPLICATIONS: ORGANICS 151
A. W. Boorn and R. F. Browner

CHAPTER 7. DIRECT ELEMENTAL ANALYSIS OF SOLIDS BY INDUCTIVELY COUPLED PLASMA EMISSION SPECTROMETRY 217
J. C. Van Loon, H. Haraguchi, and K. Fuwa

CHAPTER 8. FUNDAMENTAL ASPECTS OF AEROSOL GENERATION AND TRANSPORT 244
R. F. Browner

CHAPTER 9. PLASMA MODELING AND COMPUTER SIMULATION 289
M. I. Boulos and R. M. Barnes

CHAPTER 10. SPECTROSCOPIC DIAGNOSTICS: BASIC CONCEPTS 353
J. M. Mermet

CHAPTER 11. EXCITATION MECHANISMS AND DISCHARGE CHARACTERISTICS—RECENT DEVELOPMENTS 387
M. W. Blades

CHAPTER 12. STATUS AND TRENDS OF DEVELOPMENT OF ATOMIC SPECTROMETRIC METHODS FOR ELEMENTAL TRACE DETERMINATIONS 421
J. A. C. Broekaert and G. Tölg

INDEX 459

Inductively Coupled Plasma Emission Spectroscopy
Part 2

CHAPTER

1

APPLICATIONS: METALS AND INDUSTRIAL MATERIALS

K. OHLS and D. SOMMER

Hoesch Hüttenwerke, A.G.
Dortmund, Federal Republic of Germany

1.1. Introduction
- 1.1.1. Position of Inductively Coupled Plasma–Atomic Emission Spectrometry (ICP–AES) in Industrial Laboratories
- 1.1.2. Instrumentation Used in Industrial Laboratories
- 1.1.3. Trace Analysis: Sample Preparation
- 1.1.4. Examples of Line Sets Used in Polychromators

1.2. Analysis Methods Using Solutions
- 1.2.1. Iron and Ferrous Alloys
- 1.2.2. Metals and Nonferrous Alloys
- 1.2.3. Oxidic Materials
- 1.2.4. Metals in Organics

1.3. Trace Analysis by ICP–AES
- 1.3.1. Injection Technique after Extraction
- 1.3.2. Sample Elevator Technique

1.4. Direct Analysis of Solids by ICP–AES

1.5. Hydride Generation Technique

1.6. ICP As a Detector for Gas Chromatography

References

1.1. INTRODUCTION

1.1.1. Position of Inductively Coupled Plasma–Atomic Emission Spectrometry (ICP–AES) in Industrial Laboratories

Analysis in industrial laboratories is concerned with two types of problems:

1. Analytical control of the production processes using high-speed spectroscopic methods, such as spark AES and X-ray fluorescence spectrometry (XRFS).

2. Analytical control of the various raw materials and the eventual products, including analyses upon special demands from the production line.

Analyses in category 1 are based on empirical calibrations, the correctness of which has to be checked continuously and monitored via accurate analyses to maintain the high level of accuracy required. Analyses of this type and those of category 2 require highly qualified and skilled analysts, because customers offer new and even more difficult analysis problems, originating, for example, from the effects of traces of elements on the properties of the products. Because labor is expensive, these methods have to be used so that the work can be done with a minimum of manpower. Therefore atomic absorption spectrometry (AAS) has been substituted in recent years for many chemical analysis techniques, such as gravimetric, volumetric, electrochemical, and spectrophotometric methods. An alternative of AAS is ICP–AES, particularly if ICP is combined with a polychromator, thus permitting simultaneous multielement analyses.

For trace analysis, ICP–AES is a useful supplement to flame AAS, electrothermal atomization (ETA) AAS, and inverse voltammetry. Addition of ICP–AES to the array of methods requires only the development of some special procedures of sample preparation and sample introduction.

Applications have been worked out in connection with the development of ICP instrumentation, such as the analysis of steels [1], the platinum-group metals and gold [2] or a Sm–Co alloy, and some elements in metallic thin layers [3].

The ICP operating conditions used for steel and metal analysis with low-power Ar ICPs do not differ essentially: 1.2–2-kW input power, 10–18-L/min outer (+ intermediate) gas flows, about 1-L/min carrier flow (at the same time serving as the nebulization gas of a concentric glass or cross-flow nebulizer), a solution uptake rate of 1–3 mL/min, and an observation height of 15–18 mm above the coil. Salt concentrations of about 5 mg/mL are common, and in dissolution one uses hydrochloric or nitric acid. A striking feature in the various reports is that such elements as B, Ce, La, Nb, P, Pb, Pr, Sn, Ta, V, and Zr did not produce the well-known difficulties encountered in AAS.

The analysis sample of a metal is normally prepared by acid dissolution, resulting in a relatively low salt content of the solution. Sample solutions of oxidic materials must be prepared by a fusion technique using sodium tetraborate, for example, usually followed by acid dissolution of the melt. The resulting higher salt content may affect the nebulizer characteristics (uptake rate, efficiency, aerosol transport rate). This fact has to be borne in mind when a blank solution is adopted for the analytical procedure.

Pure metals or low-alloyed steels contain a single matrix only, but the element concentrations to be determined are relatively low. Reference and blank solutions hardly cause problems except for the purity of the chemicals used, which is often unknown and must therefore be checked.

Alloys and oxidic materials normally contain two or more matrices and the concentration level of many concomitants is higher. Finally, the traces in oxidic samples have become important for a better control of the recycling of metal containing slags and for coping with environmental problems of bulk materials. Using ICP–AES for the detection and determination of such traces one meets many difficulties primarily caused by spectral interferences (see Part 1, Chapter 7).

Nitrogen–argon ICPs operating at a 3-kW power level have the advantages that they tolerate the introduction of a higher amount of sample per unit time and that they easily accept sample types with widely differing characteristics, such as aqueous solutions, emulsions, and organics. Really unambiguous and objective comparisons between the advantages and disadvantages of low-power Ar and high-power N_2–Ar ICPs, as to their use in the daily practice of an analytical service laboratory, do not exist. It is chiefly a question of personal preference, confidence, and "faith."

1.1.2. Instrumentation Used in Industrial Laboratories

An important requirement is that the instrumental system should operate stably over a long period of time. It is the task of instrument manufacturers to design systems that fulfill this requirement. As has been discussed in Part 1, Chapters 1 and 8, equipment for ICP–AES consists of the ICP source and the spectrometer, the latter being either a polychromator or a monochromator. The choice between polychromator and monochromator depends on whether many elements have to be simultaneously determined in well-known samples or relatively few elements must be determined in samples of varying composition and origin. Often both types of instruments will be needed in a laboratory to cope adequately with all analytical problems offered.

Which ICP–AES equipment has to be chosen in a particular situation depends on the analytical problems, the available budget, and the facilities offered by the various instrument manufacturers. It is beyond the scope of this chapter to discuss these points.

In our case, we started in Europe relatively early with the introduction of ICP–AES in a steelworks laboratory, so that, more or less, we had to develop the equipment ourselves. This explains, for example, our preference for a high-power radio frequency (rf) generator (up to 3.5 kW output), since it has proved to be a versatile instrument that can be used to operate both a low-power Ar ICP and a high-power N_2–Ar ICP. However, at present many highly developed ICP systems are on the market, so judgments about the preferable choice are hardly possible. This holds even more so because the prime task of industrial laboratories is to analyze samples, not to make systematic comparisons between instruments of different types, and to publish the results. Therefore, we must limit the further discussion to the recounting of some experiences that still appear of general interest.

It is well recognized that a crucial component of an ICP system is the aerosol generation unit, which consists of a nebulizer and a spray chamber. The long-term stability of the ICP system depends to a considerable extent on the type of nebulizer, the stability of the nebulizer gas flow, and the type of solutions aspirated. In general it is recommended that separate supplies be used for the various gases involved in ICP generation and solution nebulization (outer, intermediate, and carrier gas). In a routine laboratory, the gas supplies should be designed so that each gas reservoir can be replaced without interrupting the operation of the ICP. This point has been mentioned only once in the literature [4].

The gas conduit pipes and tubes, sample tubes, and complete nebulizer system must be kept clean. The drain from the spray chamber should be controlled so that the pressure in the spray chamber does not fluctuate (see Part 1, Chapter 6). The analyst should be well aware that the operation of the nebulizer and the quality and quantity of aerosol entering the plasma depend appreciably on the type of solution aspirated. Therefore, standardized sample preparation techniques are indispensable, since changes in viscosity, salt, or acid content of the solutions all effect the aerosol generation and therefore the analyte and background signals. Also the effects of chemical reactions between constituents in the solutions leading to losses, especially at trace levels, should be borne in mind. Additional effects may manifest themselves if aerosol desolvation is used [5].

1.1.3. Trace Analysis: Sample Preparation

If we define trace analysis in metals and oxides as concentration determinations at less than 50-μg/g (ppm) levels, this means that the analysis refers to concentrations of less than 500 ng/mL in the solution, when the latter contains 1 mg/mL (1% m/m) of the original sample. Trace analysis in this sense often requires special sample preparation techniques including enrichment procedures. Raising the solute concentration in the solution does not necessarily lead to better detection limits and easily gives rise to nebulizer problems, such as clogging and salt deposits.

Extraction procedures are very useful in trace analysis. There are two approaches:

1. Extractive separation of the matrix to reduce the spectral interference from lines of the matrix on lines of the trace elements. Extractions of this type need not be quantitative as to the removal of the matrix. However, the traces should not be partly entrained with the matrix.

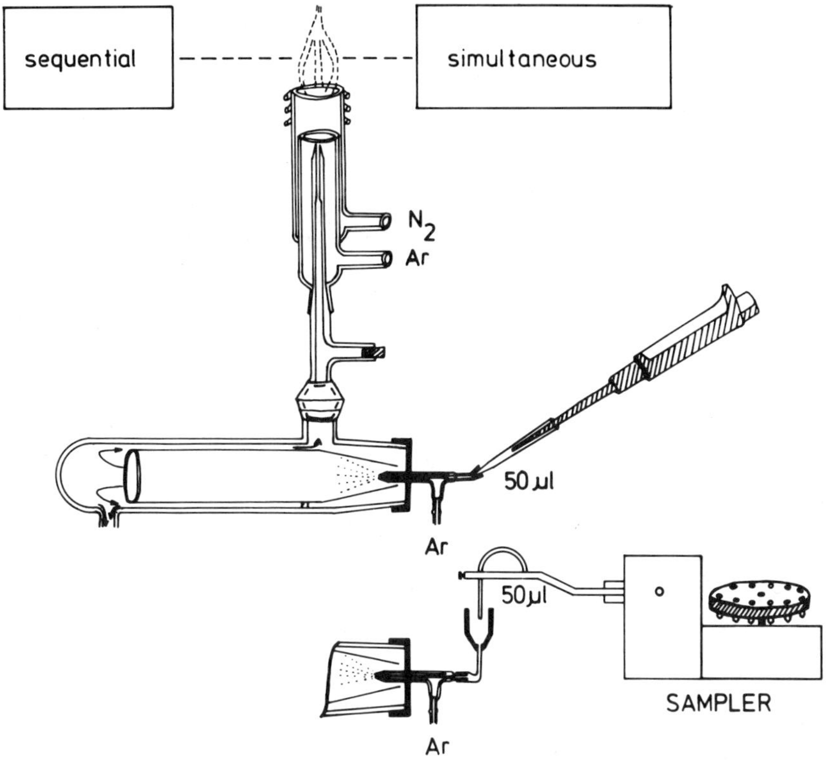

Figure 1.1. Schematic representation of the manual and automated technique.

2. Specific extraction of an analyte using a specific reagent. This extraction should be quantitative.

Sample solutions resulting from extraction often have a small volume and a high salt content. The direct injection into the nebulizer tube with a pipette or via an automatic device can be advantageously used (see Fig. 1.1 and Part 1, Chapter 6). In the authors' laboratory 50-μL samples are customary. The salt content of sample solutions can be reduced by pressure bomb dissolution. However, only small amounts of sample can be used then, which requires the availability of homogeneous material or necessitates many repeated determinations.

Another approach used in this laboratory is to separate the elements of interest by means of gas chromatography after converting them into metallo-organic compounds with complexing agents such as dithiocarbamate. The ICP is then used as an element-specific detector, which is sometimes more sensitive than common nonspecific detectors such as the flame ionization detector [6].

Hydride generation is a further alternative for separating the relevant elements (As, Sb, Pb, Se, or Sn) from the original sample matrix [7] (see also Part 1, Chapter 6).

In general, trace analysis using ICP–AES has not yet superseded ETA–AAS, but future developments might tip the balance. This situation contrasts with the use of ICP–AES for determining major and minor constituents in (solutions of) reference samples. In this area, ICP–AES has become more economical then conventional flame AAS, because of the simultaneous multielement capability of ICP–AES.

1.1.4. Examples of Line Sets Used in Polychromators

A polychromator should be provided with such an array of exit slits that the most important sample types can be covered. As an example, Table 1.1 gives a survey of the composition of some materials important in the steel industry.

The choice of appropriate spectral lines is dictated by the requirement that line coincidences must be avoided. Further limitations arise from the spectral range of the instrument and from mechanical constraints, such as the minimum distance between two exit slits. Therefore compromises are required, which often preclude the use of the best lines.

Typical line coincidence problems are encountered with the following lines:

B 249.68 – W 249.66 nm
Al 396.15 – Mo 396.15 nm
Ti 334.94 – Cr 334.93 nm
Mn 257.61 – W 257.62 nm
Cu 324.75 – Mo 324.76 nm

As an example of a suitable line set, Table 1.1 shows the lines available on a polychromator in this laboratory. The table also lists:

1. The detection limits for an Ar ICP operated at 1.5 kW.
2. The lowest concentrations that can be determined.
3. The best analysis line of the relevant element [8–10].
4. In the case that not the best line was adopted, the reason for this.

Table 1.1 also contains experimental values of the lowest concentrations in the original solid sample that can be determined with a given precision, in this case a 10% relative standard deviation (RSD). These concentrations are generally a factor of 500–1000 (700 on average) above the detection limits in the solution, where a factor of 100 arises from the conversion of concentration in

Table 1.1. Analysis Lines Used for Simultaneous ICP Spectrometry and Detection Limits for Argon ICP (1.5 kW)

Common Concentration Level of Samples in Steel Laboratories							
Steels % (m/m)	Oxidic Materials % (m/m)	Element	Line (nm)	Detection Limit (μg/mL)	Lowest Concentration (μg/g)	Best Lines (nm)	Reason of Discrepancy
<0.001	<0.01	Ag(I)	224.641	0.2		328.068	
—	50–0.01	Al(I)	308.215	0.02	10	309.278	V 309.311
1.0–0.0001	—	Al(I)	396.152	0.02	10	396.152	(Ca)
—	—	Ar(I)	404.442	—		415.859	Reference
0.1–0.0005	0.1–0.0001	As(I)	197.197	0.1	10	197.197	—
0.01–0.00005	—	B(I)	208.959	0.01		208.959	—
—	10–0.001	B(I)	249.678	0.03			(Fe, W)
<0.001	5–0.0001	Ba(II)	233.527	0.005		455.403	≤430 nm
0.5–0.001	<0.01	Bi(I)	223.061	0.05		195.389	Hg 194.227
0.01–0.0001	40–0.01	Ca(II)	393.366	0.005	2	393.366	—
<0.00001	0.01–0.00001	Cd(I)	228.802	0.003		228.802	— (As)
0.1–0.001	<0.1	Ce(II)	418.660	0.03		395.254	Al 396.152
20–0.001	<0.01	Co(I)	345.350	0.02		228.616	Cd 228.802
25–0.1	20–0.01	Cr(I)	425.435	0.1	50	205.552	Sb 206.833
2–0.001	0.1–0.001	Cr(II)	267.716	0.005	2		
1–0.001	0.1–0.0001	Cu(I)	324.754	0.0005	1	324.754	—
99.5–50	60–0.01	Fe(II)	259.940	0.002		259.940	—
<0.0001	<0.01	Hg(II)	194.227	0.0005		253.652	P 253.565
0.1–0.001	<0.1	La(II)	394.910	0.02		394.910	(Ar)
0.5–0.001	60–0.01	Mg(II)	279.553	0.0005		279.553	—
5.0–0.001	1–0.01	Mn(II)	257.610	0.001	1	257.610	—
10–0.001	<0.5	Mo(II)	281.615	0.01	5	202.030	Nb 202.932
5–0.001	<0.5	Nb(II)	202.932	0.1	50	309.418	Al 309.278

Table 1.1. (*Continued*)

Common Concentration Level of Samples in Steel Laboratories							
Steels % (m/m)	Oxidic Materials % (m/m)	Element	Line (nm)	Detection Limit (μg/mL)	Lowest Concentration (μg/g)	Best Lines (nm)	Reason of Discrepancy
0.01–0.0001	<0.01	Nd(II)	405.996	0.02		430.358	≤430 nm
20–0.001	0.1–0.0001	Ni(II)	231.604	0.01	10	231.604	—
0.1–0.0001	15–0.001	P(I)	253.565	1		213.618	Zn 213.856
0.5–0.0001	<0.05	Pb(II)	220.353	0.5	50	405.783	Nd 405.996
0.01–0.0001	<0.01	Pr(II)	422.535	0.0001		417.939	Ce 418.660
0.01–0.0001	<0.01	Sb(I)	206.833	0.1	10	206.833	—
0.1–0.0001	<0.1	Se(I)	196.026	0.05		196.026	—
5–0.001	50–0.01	Si(I)	288.158	0.01	10	251.611	—
0.01–0.0001	<0.01	Sn(II)	189.980	0.1	10	189.980	
<0.001	<0.01	Sr(II)	407.771	0.00005		407.771	—
1–0.0001	<0.1	Ta(II)	240.063	0.01		301.254	—
0.1–0.0001	<0.1	Te(I)	238.578	0.1	50	214.281	Zn 213.856
1–0.001	0.5–0.001	Ti(II)	334.941	0.0005	1	334.941	—
5–0.001	0.5–0.001	V(II)	310.230	0.005	1	309.311	Al 309.278
20–0.001	0.5–0.001	W(I)	400.883	0.05		207.911	B 208.959
<0.001	10–0.01	Zn(I)	213.856	0.005		206.200	Sb 206.833
1–0.0001	0.5–0.001	Zr(II)	343.823	0.005		343.823	—

mass/mass to mass/volume (dilution factor) under the assumption that the solution contains 1% (m/v) of the solid sample. The remaining factor of 5–10 (7 on average) links up well with the factor of 5 derived theoretically in Part 1, Section 4.2, where a concentration equal to five times the detection limit was defined as the limit of determination, that is, the concentration at which the RSD in the net line signal is 10%. Spectral interferences may cause the limit of determination to become substantially larger than five times the conventional detection limit (see Part 1, Section 7.7.7).

In connection with Table 1.1, we further note interferences from W, Mo, and Cr on various lines when the concentrations of these concomitants exceed 1% (m/m) in the solid sample. Also the interference of Pb 220.35 nm by Ni 220.35, Co 220.34, and W 220.38 nm should be noted. Difficulties are encountered when the concentration of any of the concomitants is above 5% (m/m).

The determination of the important elements S and P in steel requires a vacuum spectrometer because the best lines are located in the vacuum UV (VUV) region: S 180.73, S 182.04, and P 178.28 nm. The optical path between source and spectrometer must be flushed with argon. Nevertheless, the determination of S in steel meets with difficulties because the concentration range of interest is 1–50 μg/g, which corresponds to 0.01–0.5 μg/mL in a 1% (m/v) solution of the sample. One way to deal with this problem is the use of chemical reactions leading to the evolution of hydrogen sulfide, as has been described for a dc plasma [11].

The determination of P also meets with difficulties. Only one line (P 178.28 nm) is free from interference in the presence of Fe. Instruments can be used that offer the possibility to measure this P line in the third order. In that case, the limit of determination is 1 μg/g (ppm) P in steel [12]. Since the dilution factor will be often larger than 100 (even up to 1000) to prevent problems with high salt contents, the limits of determination in the solid will be frequently less favorable than in Table 1.1.

In this laboratory favorable experience was gained with a N_2–Ar ICP operating at 3-kW of input power. A comparison of the detection limits achieved with this source and those with an Ar ICP operated at 1.5 kW is shown in Table 1.2. Neither systematic analysis of the data nor a precise breakdown of the causes of the differences has been attempted.

1.2. ANALYSIS METHODS USING SOLUTIONS

This section provides a survey of application studies described in the literature. For the reader's convenience we have included with the analytes the spectral lines used. These are given in parentheses (nm) behind the element symbols.

Table 1.2. Comparison of Detection Limits Achieved with an Ar–ICP and a N_2–Ar ICP under Optimum Conditions and Related to the Same Polychormator

		Detection Limits (μg/L)	
Element	Line (nm)	Argon/Argon 1.5 kW	Nitrogen/Argon 3.0 kW
Sn(II)	189.980	100	7
Hg(II)	194.227	0.5	7
Se(I)	196.026	50	1
As(I)	197.197	100	10
Nb(II)	202.932	100	4
Sb(I)	206.833	100	4
B(I)	208.959	10	4
Zn(I)	213.856	5	1
Pb(II)	220.353	500	5
Bi(I)	223.061	50	5
Ag(I)	224.641	200	6
Cd(I)	228.802	3	4
Ni(II)	231.604	10	6
Ba(II)	233.527	5	4
Te(I)	238.578	100	10
Ta(II)	240.063	10	9
B(I)	249.678	30	7
P(I)	253.565	1000	6
Mn(II)	257.610	1	0.3
Fe(II)	259.940	2	4
Mg(II)	279.553	0.5	0.5
Mo(II)	281.615	10	9
Si(I)	288.158	10	3
Al(I)	308.215	20	6
V(II)	310.230	5	2
Cu(I)	324.754	0.5	3
Ti(II)	334.941	0.5	0.7
Zr(II)	343.823	5	4
Co(I)	345.350	20	4
Ca(II)	393.366	5	0.4
La(II)	394.910	20	3
Al(I)	396.152	20	4
W(I)	400.883	50	2
Nd(II)	405.996	20	8
Sr(II)	407.771	0.05	0.2
Ce(II)	418.660	30	3
Pr(II)	422.535	0.1	2
Cr(I)	425.435	100	10

1.2.1. Iron and Ferrous Alloys

Butler, Fassel, and Kniseley were the first to report analytical ICP work for iron, steels, and related alloys [13]. They used a low-power Ar ICP and a high-performance polychromator (1-m Paschen–Runge mounting, 2160 lines/mm grating, 0.46 nm/mm reciprocal linear dispersion). About 1 g of steel was dissolved in a mixture of HCl and HNO_3 and diluted to a concentration of 5 mg/mL of sample. The elements determined were Al (396.1), Ce (456.2), Cr (357.8), Cu (324.7), La (398.8), Mn (403.0), Ni (351.5), Nb (405.8), Pb (405.7), Pr (417.9), W (400.8), and Zr (349.6). The paper gives the detection limits for aqueous solution containing 0.5 mass % of Fe.

The determination of As (228.81), B (249.68), Bi (306.77), Ce (413.76), La (408.67), P (253.40), Sn (242.17), and Ta (295.19) in stainless and low-alloyed steels or Cr (302.43) and Ni (300.36) has been described [14]. A 40-MHz ICP operating at 1.5–2 kW [15] was used and pneumatic and ultrasonic nebulization were compared for the introduction of solutions of high salt content. The selection of lines, interferences caused by line broadening, and the choice of the operating parameters were discussed with a view toward minimizing the effects.

A 27-MHz ICP was applied [16] to the determination of Al (394.40), Ce (418.66), Co (238.89), Cr (360.53), Cu (324.75), La (398.85), Mn (257.61), Mo (317.03), Nb (405.89), Ni (231.60), Ti (363.55), V (311.07), W (400.88), Y (371.03), and Zr (343.82) in steels and alloys. The results were compared with those obtained by AAS and spectrophotometry. Trace analysis in alloys was also performed using different acid dissolution techniques. The elements Zn (202.55), Sb (206.84), P (213.62), Pb (220.35), Cd (226.50), Co (238.89), Ta (265.32), and Sn (284.00) were covered.

Another system [17] used comparable operating conditions, but involved lines typically used with polychromators. For example, the following lines were used for an iron matrix: Al (308.22), As (193.70), B (249.77), Co (228.62), Cr (267.72), Cu (324.75), Fe (238.86, internal standard), Mn (257.61), Mo (202.03), Nb (313.08), Ni (231.60), P (214.91), Si (288.16), Ta (240.06), Ti (334.94), V (292.40), W (207.91), and Zr (339.20).

A 27-MHz Ar ICP [18] was used for the determination of Al (396.15), Nb (309.42), Ti (334.94), V (309.31), and Zr (343.82) in steels after line selection studies. The sample solution, prepared by acid dissolution, contained 0.2 g/100 mL of sample.

In recent years applicational work has been done for steels and related materials by four working groups sponsored by the Commission of the European Communities. A line coincidence table for 40 elements that do not show any spectral interference from the matrix up to a concentration of 5 mg/mL of Fe has been worked out by Wittman [20]. A 50-MHz Ar ICP [19] was used in this work in combination with either pneumatic or ultrasonic nebulization.

The development of analytical procedures for steel samples was studied by Hughes [21], who used a high-power N_2-Ar plasma generated with a free-running, 10-kW generator (27 MHz), a Greenfield torch, a pneumatic glass nebulizer (Meinhard), and a high-performance polychromator with 35 lines. The double grating instrument used permits the coverage of two spectral ranges: directly in the range 158–403 nm (dispersion, 0.56 nm/mm) and via a periscope mounting in the range 383–619 nm (0.54 nm/mm). The following lines were set up in the former range: P (178.28), S (180.73), As (193.70), Sb (206.84), Zn (213.86), Cd (226,50), Co (228.62), Ni (231.60), Te (238.58), Ta (240.06), B (249.77), Si (251.61), Hg (253.65), Mn (257.61), Fe (259.94), Cr (267.71), Mg (279.55), Sn (284.00), Mo (287.15), Al (308.22), V (311.07), Ca (315.89), Cu (327.40), Ag (328.07), Ti (334.94), Zr (339.20), Ce (395.25), whereas Mg (383.83), Ca (393.37), W (400.88), Pb (405.78), Sr (407.77), La (408.67), Ba (455.40), and Na (589.00) were accommodated in the latter.

The analysis of steels was investigated by Ohls [22] using both types of ICP sources referred to in Section 1.1. Several rare earth metals are important in steel, if desoxidation has been achieved by addition of "Mischmetal." These are Ce (0.02–1.3% m/m), La (0.01–0.5% m/m), Nd (0.01–0.4% m/m), and Pr (0.003–0.14% m/m). The best line for the B determination is B 182.60 nm, which requires a vacuum instrument or apparatus flushed with an inert gas [23].

Hancart [24] developed an analytical procedure for some ferrous alloys. An internal standard was used to permit the accurate determination of both high and low concentrations, for example, Ni in the case of FeSi. The procedure was developed for a 50-MHz Ar ICP in combination with a 1.5-m polychromator. Thus, it was possible to determine up to 50% (m/m) of Fe (271.49 nm) and up to 80% (m/m) of Si (288.16), but also 0.1–0.3% (m/m) of Mn.

The analytical procedures, in general, do not differ drastically, since even different ICPs require approximately similar conditions, in particular, the salt content of the solution and the amount of sample that can be fed into the plasma in unit time. This rather favorable circumstance permits the development of universal sample preparation procedures for different groups of samples.

As an example, Fig. 1.2 shows a schematic representation of the dissolution procedure for steel samples used in this laboratory. This approach is used for the determination of most elements in steel, except for C, P, and S. If a residue is left after the acid dissolution, the residue is fused with sodium tetraborate, sodium carbonate, or potassium carbonate. The last two fluxes are used for determining B. The dissolution technique represented in Fig. 1.2 has been checked with many standard reference materials. No difficulties are encountered in the concentration range above 0.005% m/m of most elements present in steel. Evidently, appropriate blank solutions must be incorporated.

Polychromators are generally used for multielement analyses of routine samples. As discussed in Part 1, Chapter 8, polychromators of various size and type

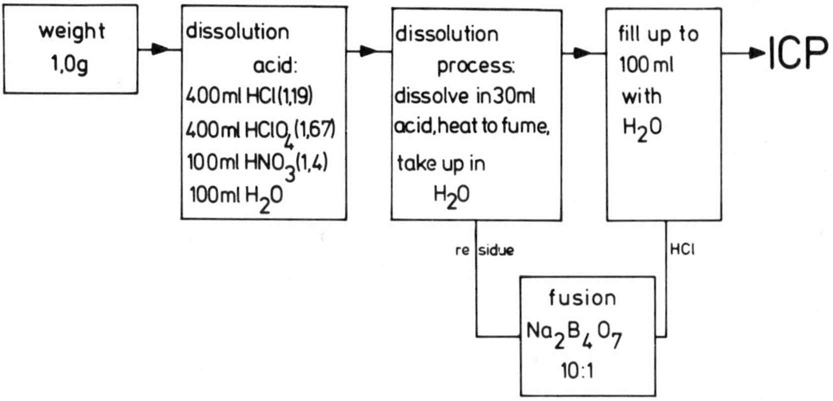

Figure 1.2. Scheme of a general method for steel analysis.

are commercially available; they may incorporate such features as automatic background correction, which is achieved by moving the entrance slit or a refractor plate behind the slit. Modern instruments are also provided with such software that corrections for (spectral) interferences can be applied (see Part 1, Sections 1.6 and 7.6).

Special analyses involving new sample types and/or analytes require the availability of instruments that permit a flexible choice of both the ICP conditions and, most importantly, the wavelengths for analysis. For this purpose, computerized monochromators of various size and type are commercially available. They permit both single-element analysis and sequential multielement analysis (cf. Part 1, Chapter 8).

1.2.2. Metals and Nonferrous Alloys

Pure Al can be analyzed after dissolving Al metal (0.5 g) in NaOH (10 mL of a 30% solution). Then, concentrated HNO_3 (30 mL) is added and the solution is filled up with H_2O to 100 mL. All impurities such as Cr (0.01–0.2% m/m), Fe (0.2–1.2), Mg (0.05–1.3), Ni (0.02–0.5), Cu (0.05–0.8), Pb (0.01–0.1), Sb (0.04–1.3), Sn (0.01–0.15), Ti (0.01–0.15), and Zn (0.1–1.2) can be determined simultaneously. A cross-flow nebulizer or a concentric glass nebulizer operated with water-saturated Ar can be used with such a high salt content of mainly soluble $NaNO_3$.

A combination of a low-power Ar plasma source and a direct reading spectrometer has been used [17] to analyze aluminum- and copper-based alloys by means of a line set (nm) fixed for (1) aluminum—Al (308.22, internal standard), Cr (267.72), Cu (324.75), Fe (259.94), Mg (383.23), Mn (257.61), Ni (231.60), Pb (220.35), Sb (217.58), Sn (189.99), Ti (334.94), and Zn (213.86);

and (2) copper—Ag (328.07), Al (308.22), As (193.70), Cd (226.50), Cu (213.60, internal standard), Fe (259.94), Mn (257.61), Ni (231.60), P (253.57), Pb (220.35), Sb (217.58), Si (288.16), Sn (189.99), and Zn (206.20).

The analysis of Al–Ti alloys [25] was performed by means of a N_2–Ar ICP generated with a 6-kW generator of the free-running type. The application of ICP–AES to the analysis of nickel-based superalloys has also been studied [26], and another team determined the alloying constituents (Sn, Mo, Nb) and the minor impurities of Zn alloys [27] by using sequential ICP instrumentation. A simultaneous determination of the platinum-group metals, Au and Ag has been reported [28] to be unsatisfactory without a separation procedure to remove the elements from the matrix. A N_2–Ar ICP was used here.

Impurities of high-purity Co, Mn, and Ni have been successfully determined by using a high-power (4–5 kW, 27 MHz) N_2–Ar ICP in combination with an extraction procedure [29]. The elements Ag, Al, Be, Bi, Co, Cr, Cu, Fe, Mn, Mo, Ni, Pb, Sb, Sn, Ti, V, and Zr were separated from the matrix by organic solvents containing various chelating agents. Sc and Au were added as internal standards.

1.2.3. Oxidic Materials

Oxides have to be analyzed for the main components, for accompanying elements in low concentration, and for traces if there is any connection with environmental effects. Most oxidic materials cannot be completely dissolved by acids. Therefore, the normal way to prepare the sample solution is fusion with fluxing agents such as sodium tetraborate or carbonate.

Only one simple method for slag analysis using acid dissolution has been reported [30]: 1 g of slag was dissolved in 20 mL of HCl (1 + 1) and oxidized by adding 5 mL of concentrated HNO_3; then 20 mL of $HClO_4$ was added and the solution was heated to fuming. After cooling, the residual salts were dissolved in 100 mL of water, and further diluted 10 times just before the measurement. The range of RSD for the results of CaO, MgO, Al_2O_3, TiO_2, MnO, V_2O_5, Cr_2O_3, and total Fe was between 0.4 and 2.3%.

In general, fusion must be used to prepare analytes from oxidic materials. Many fusion techniques are well known and have been described in the literature, but it seems impossible to standardize such procedures. ICP–AES requires this, because of possible interference effects caused by easily ionizable elements [31] and stray light and line broadening effects, especially associated with the presence of Ca and Mg [14,32] (see Part 1, Chapters 7 and 8).

Oxidic materials can be analyzed successfully by means of standardized analytical methods. In this laboratory such a method has been developed (Fig. 1.3) using sodium tetraborate as fluxing agent: 0.1 g of oxide is mixed with 1 g of the flux to get a concentration of nearly 1 g/100 mL. Calibration and blank

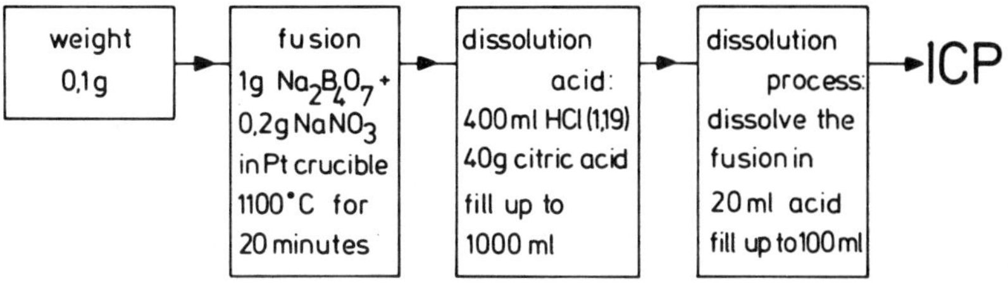

Figure 1.3. Scheme of a general method for oxidic sample analysis.

procedures involve fusion of the same amount of sodium tetraborate without sample, with standard reference samples, or with synthetic mixtures of pure oxides. This technique can be used for microanalysis, too. Oxidic residues of different kinds were analyzed by reducing the weight (0.1–1 mg) and adding the chemicals in proportionally smaller amounts [33]. Using a microfusion technique, a sample volume of 1–5 mL can be obtained depending on the initial weight.

The precision of ICP–AES results obtained from oxidic samples depends largely on the mass content of the components that normally varies more than in metals. In the mean range of 1% (m/m), an RSD of 1–3% can be achieved in most cases.

The possibility of determining major constituents has been reported for iron ores [34]. Ores were fused with a mixture of sodium carbonate and tetraborate (1 + 1), and the melt was dissolved in HNO_3, diluted, and analyzed by a simultaneous technique. The resultant precision was comparable to that obtained by conventional methods of chemical analysis. Correction factors for overlap of spectral lines were given.

Iron ores have also been analyzed using solutions prepared with various fusing agents [24] such as a mixture of sodium tetraborate, carbonate, and potassium disulfate (1 + 1 + 1) or a mixture of lithium carbonate and boron oxide. The analyte concentration was about 2 g/100 mL. $NiCl_2$ was added as internal standard. Standard deviations of 0.25% for Fe (65–25% m/m), 0.19% for CaO (25–0.1%), 0.24% for SiO_2 (25–2%), 0.1% for Al_2O_3 (11–1%), 0.05% for MgO (2–0.05%), and 0.009% for TiO_2 (0.5–0.01%) were achieved.

Sample preparation and analysis methods were also developed for coal, fly ash, and slags [35] to determine toxic elements like Pd, Cd, Ni, Mg, Be, Bi, Sb, As, and Sn; essential trace elements like Cr, Mn, Co, Cu, Zn, Se, Mo, B, and V; and major constituents like Si, Al, Fe, Ca, Na, and K. Results obtained on corresponding standard reference materials were compared with those found by neutron activation analysis or AAS. The precision obtained was about 5% RSD.

Further analytical techniques for oxidic materials have been developed in the related fields of analysis of geological and environmental materials (see Chapters 2 and 3).

1.2.4. Metals in Organics

Industrial laboratories are interested in the analysis of two different types of organic solvents containing metals or metallic compounds. The first type involves organic solvents like ethers and ketones used for extraction procedures in combination with chelating agents to separate the metallic ions from the matrix elements. The other category of organics encompasses lubricating or engine oils containing (wear) metals in the form of chemical compounds or metallic and oxidic particles.

For the determination of the wear metals Al, Cr, Cu, Fe, Ti, Ni, Mg, and Mn in engine oil, a method has been described using a high-power N_2–Ar ICP [36]. Another method [37] for wear metals in lubricating oil used a low-power Ar ICP and 4-methyl-2-pentanone (MIBK) for dilution of the oil. Al, Ag, Ca, Cr, Cu, Fe, Mg, Mn, Ni, Pb, Si, Sn, Ti, V, and Zn were determined. The detection limits were in the range of 0.1–0.01 $\mu g/g$ of oil except for Ca and Cu (0.0004 and 0.007). The precision obtained by means of reference samples was about 5% RSD on average, except for low contents of Ag and Al (RSD, 10–30%). Alternative methods using low-power Ar ICPs for wear metal determinations have been reported [38,39].

The direct analysis of undiluted oil is difficult to calibrate because of the influence of the viscosity [40], which can differ by more than 100 cSt in practical work. The effects of viscosity differences are eliminated by a 1 : 10 dilution with xylene.

In our laboratory the high-power N_2–Ar ICP is used for wear metal analysis of lubricants diluted by xylene [41]. In principle, there is no significant difference with the detection limits reported for a low-power Ar ICP [37]. For Mg, Al, Si, Ca, Ti, Cr, Mn, Fe, Ni, Cu, Zn, and Cd, a precision in the range of 0.3–0.5% RSD could be achieved at the 30 $\mu g/g$ concentration level. For B the RSD was higher (1.9%).

The accuracy of wear metal analysis depends largely on the sampling method and the particle size. The sample solution can be stabilized, depending on the type of particles, by ultrasonic treatment. Often, the nebulizer may block, especially if used motor oils are analyzed. Therefore, the results of routine analysis must be compared with those obtained with more common methods such as spark-AES using a rotating disk electrode [22,41] (cf. Part 1, Section 1.7) or ETA–AAS [42]. Reference analyses are required either to detect influences from molecular absorption in ETA–AAS or generally to account for unknown effects from added synthetic organics to lubricating oils.

On the other hand, the choice of the method for the determination of wear metals in oils depends on the size and structure of the particles contained in the oil samples. If the particles can be destroyed, for example, by ultrasonic treatment, ICP–AES or rotating disk AES can be used, while the analysis can be done by ETA-AAS irrespective of the particle form and size [42]. ICP–AES is limited by the nebulizer characteristics and the rotrode technique does not work well if the particles cannot be completely transported to the electrode gap.

In summary, ICP–AES can be a good technique for the analysis of metals in undigested organics, as well as for aqueous solutions, if several important points are taken into consideration in the analytical procedure, such as the effects of sample and solvent on the background signal, the viscosity of sample and blank solution, the selection and preparation of standards by addition of metalloorganic compounds soluble in the solvent used, and the behavior of sample and solvent in relation to the aerosol production.

1.3. TRACE ANALYSIS BY ICP–AES

Trace analysis in metals, oxides, or industrial chemicals is defined here as analysis involving element concentrations up to 50 $\mu g/g$ (ppm) of sample. This means, that the real concentration in the solution will be less than 0.5 $\mu g/mL$, the solution containing less than 1% (m/m) of solid sample. For many elements this concentration is rather close to the detection limit of ICP-AES. Therefore, several trace enrichment or matrix separation techniques have been developed [43–47]. These two approaches differ basically. Enrichment methods should be quantitative or give a reproducible yield. Enrichments with a yield below 90% should not be used. For example, traces can be enriched on activated carbon after reacting with chelating agents. After the carbon collector has been treated with nitric acid, the concentrated elements can be easily determined [48].

Another method for determining very low concentrations of Hg, Ag, Cu, Pb, Cd, Sn, As, Se, Te, Zn, Co, and Ni by a preconcentration method uses precipitate exchange on thin sulfide layers [49]. A very simple technique applicable to both preconcentration and separation is cation or anion exchange [50]. For the determination of Cd, Zn, Pb, among others, preconcentration by means of solvent extraction has been described [51]. Solvent extraction allows both enrichment of the trace elements and/or separation from the matrix elements. Most of these techniques have been developed for AAS analysis. A few were specifically designed for ICP–AES [29,52–57]. Preconcentration of traces or separation from the matrix is often required, not only because the detection limits are not low enough but also because analysis lines experience interference from lines of the matrix, as happens in the cases of Fe, W, Mo, for example [58].

1.3.1. Injection Technique after Extraction

Greenfield and Smith [36] developed an "injection technique" in connection with the construction of useful spray chambers. They worked with a medical injection syringe to determine several elements in water and organic solutions by injecting 1–25 μL of solution containing 0.1–1 μg/mL of analyte. The RSD was about 5%. The present authors modified this technique to inject small sample volumes of high salt contents, for example, by means of microliter pipettes into a flame for AAS [59]. A special advantage was the elimination of blocking of the burner slit by crystallization of solids during continuous nebulization. Much additional work has since been done [60], but the technique seemed to be forgotten in ICP–AES, possibly because low-power Ar ICPs could not tolerate sudden air inlet or sudden sample injection [61]. Recent developments are described in Part 1, Chapter 6.

A method has been developed in this laboratory [33] for determining very low trace concentrations in pure iron after matrix extraction [62]. A high-power N_2-Ar ICP and a conventional Meinhard nebulizer are used. The end of the nebulizer is shaped into a small funnel for the direct injection of 50-μL samples into the inner capillary tube. The emission signal obtained with a 50-μL sample is about 50% of the signal obtained during continuous sample uptake, but the injected volume is 1/30 of that needed in the continuous nebulization mode. This technique permits the determination of up to 40 elements in only one sample solution of less than 0.5 mL; even various determinations can be done. To determine traces in pure Fe, for example, the Fe is extracted by di-isopropyl-ether and the traces are enriched by boiling down the residue to 0.2 mL (Fig. 1.4). Thus, several elements can be determined in iron at a concentration level of 1–10 μg/g with an RSD of 5% on average. Using an automatic sample injection device (Fig. 1.2) designed for AAS, the RSD could be improved to 2–3%.

1.3.2. Sample Elevator Technique

This trace analysis method and its further development, described in detail in Chapter 7 and in Part 1, Chapter 6, is suitable for both solution and solid sample analysis. It is a typical microsample technique working with 10-μL sample vol-

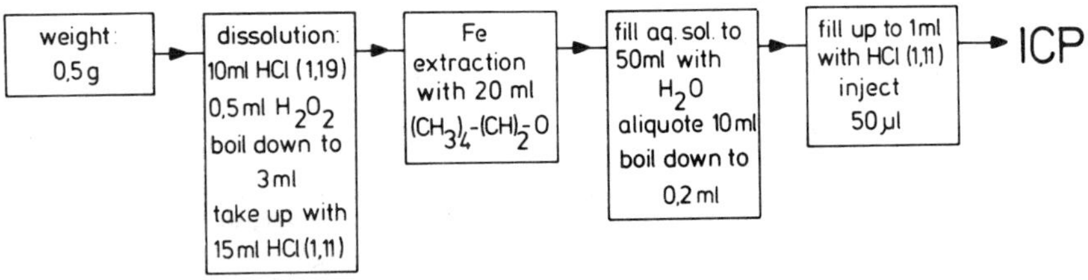

Figure 1.4. Scheme for a trace analysis method using extraction and concentration followed by the injection technique.

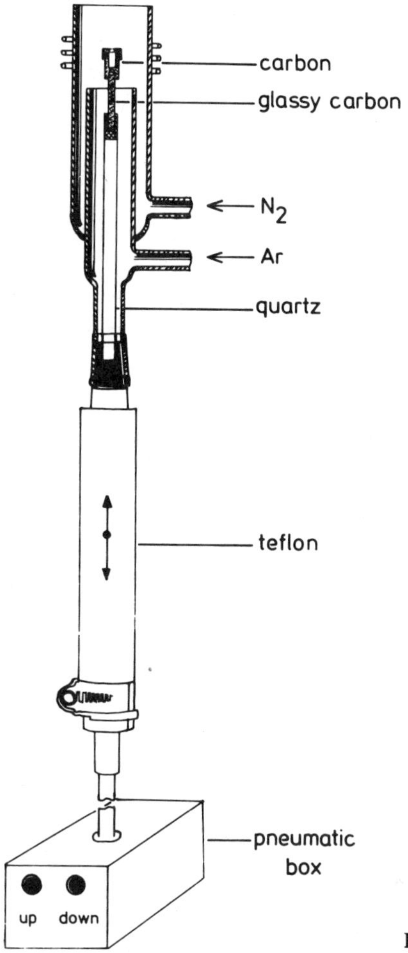

Figure 1.5. Device for sample elevator technique.

umes applied to a graphite crucible. This sample cup is directly inserted into the plasma by an elevator system (Fig. 1.5). Enrichment of liquid samples can be achieved by repeatedly drying 10-mL samples in the same cup using a surface evaporator. A 10- or 20-fold enrichment is possible.

1.4. DIRECT ANALYSIS OF SOLIDS BY ICP–AES

The direct analysis of solids by ICP-AES is specifically treated in Chapter 7 and is also covered in part in Part 1, Chapter 6. Therefore, only a few points will be mentioned here.

The most promising method for industrial use is the combination of a spark (or interrupted arc) for sample ablation and an ICP for the excitation of the vapor formed in the spark and subsequently transported to the ICP. We have applied this technique [63] to the determination of As, Cu, Cr, Mn, Ni, and V in steel at a concentration level of 100 $\mu g/g$. The RSD was about 1–5%. Recently, a systematic study of direct analysis of solid samples using spark ablation combined with excitation in an ICP, was made by Aziz et al. [63a]. The present authors [63] applied the technique also to oxidic materials, which were mixed with copper powder and briquetted to discs of about 30 mm in diameter.

The solid sampling technique is suitable for rapid analysis involving element concentrations down to 0.01% (m/m) and up to 50–60% (m/m) using a single sample only. Often, only very small portions of oxidic material are available, for example, in the case of oxidic-phase residues after dissolution of steel or in the case of filter residues resulting from environmental sampling. Then the use of a small cup electrode can be helpful. The sample is loaded in the cup together with some conducting powder.

Finally, we should point to the use of devices for insertion of solids into the ICP via a carbon or tungsten cup. This approach is discussed extensively in Chapter 7 and in Part 1, Chapter 6. One version of this technique was developed in this laboratory [64] as the sample elevator technique (SET). In this set-up it is combined with a high-power N_2-Ar ICP. As Fig. 1.5 indicates, the sample holder is moved pneumatically up and down to permit raising the filled cup to the appropriate position in the ICP and subsequently changing it for a cup loaded with the next sample. With solid samples it is often advantageous to cover the cup with a lid so as to prevent an explosion-like escape of the sample (vapor). Solutions can be also handled, but must be first dried by keeping the cup just below the coil before raising it into the center of the ICP. Table 1.3 lists detec-

Table 1.3. Comparison of Detection Limits Obtained by the Injection Technique or the Sample Elevator Technique (SET)

		Detection Limit (μg/g)	
Element	Line (nm)	Injection Technique (50 μL)	SET (30 μg)
Al(I)	308.215	74	1.9
Co(II)	228.616	13	2.5
Cu(I)	324.754	30	0.5
Mn(II)	257.610	0.75	0.03
Ni(II)	231.604	29	2.5
Si(I)	288.158	45	1.1

tion limits achieved with either the injection technique (50-μL samples) or the SET (30-μg samples).

1.5. HYDRIDE GENERATION TECHNIQUE

In the last decade some elements have become important that were not normally determined in steel laboratories, namely As, Se, Cd, Sn, and Tl. Their determination is now requested at the trace concentration level in steels, slags, and galvanic slurry. Although for most of these elements the detection limits of ICP–AES are lower than those of flame AAS, they are not always good enough. Therefore, traces of As, Se, and Sn must be determined by the hydride technique to meet the demand of the very low detection limits required.

The first publication on hydride generation for determining As using flame AAS was published in 1969 [65]. Much work on the hydride techniques for all hydride-forming elements (As, Se, Bi, Sn, Ge, Pb, Sb, and Te) has since been done, in particular in combination with AAS. More recently, hydride generation has been also combined with ICP–AES, and this topic is treated in Part 1, Chapter 6. For the steel matrix rather comprehensive work has been done by Fleming and Ide [66], and by Broekaert and Leis [67], the latter with particular reference to AAS.

One of the problems of hydride generation is the possible occurrence of a manifold of interference effects (see, for example, [68–71]), which can occur in the reaction cell, the vapor phase, and/or the detector (ICP). Sommer and Ohls [7] determined Sn in low-alloyed steel, in pure aluminum, and in yellow brass. A commercially available hydride generator was connected to a high-power N_2-Ar ICP. In contrast to As and Se, which could be determined in a simple way in steel and slags, many parameters had to be optimized for the determination of Sn, such as the acid concentration, the speed of magnetic stirring, the concentration of $NaHB_4$ to 0.8% (m/v), and the pressure of the supporting gas. An interfering effect of Fe(III) could be removed by adding ascorbic acid. Whereas the elements Al, Ca, Cr, Mn, and V did not cause any interference effect, Co, Cu, Ni, Mo, Si, and W lowered the emission signal significantly. The Sn determination is consequently not feasible in alloyed steels or nonferrous metals without special chemical operations. For example, Co and Cu at low concentrations must be complexed by EDTA. At higher concentrations the Sn traces must be coprecipitated on manganese hydroxide [72].

To get higher Sn emission signals at the extremely low contents in steels or alloys, it is useful to add thioglycolic acid to the sample solution. Detailed descriptions of the procedures are given in [7]. The detection limit of Hg is reasonable, namely 0.04 μg/mL, if the line Hg 253.65 nm can be used, but in case of line interference (Fe 253.66 nm) the cold vapor technique is indispensable. It can be applied in ICP–AES in a similar way as in ETA–AAS [73,74].

1.6. ICP AS A DETECTOR FOR GAS CHROMATOGRAPHY

Gas chromatography (GC) is a very sensitive technique for the separation of easily vaporizable organic solvents. Unfortunately, the usual GC detectors are not very specific for metal atoms. Therefore, GC systems have been applied with AAS [75,76] and all three types of plasma sources (MIP, DCP, and ICP: Part 1, Chapter 2) [77–81]. The first paper reporting on a GC–MIP combination appeared in 1965 [77]. The method was mainly used to detect traces of Cl, Br, I, S, and P contained in organics.

Compared with GC–MIP interfacing relatively little has been reported with regard to GC–DCP [82,83] and GC–ICP combinations [80,81]. While Windsor and Denton [81] have determined nonmetallic elements by means of the low-power Ar ICP at 500 W, working with pure organic compounds like thiophene for the detection of S and tetravinylsilane for Si, Sommer and Ohls [86] have used a real matrix. The Pb content in different types of gasoline was determined, for example. For easy handling, a commercial GC output was connected to a special gas inlet of the Greenfield torch. The common ICP parameter setting used with solutions was employed.

The purpose of applying a GC–ICP detector in steel analysis is to determine

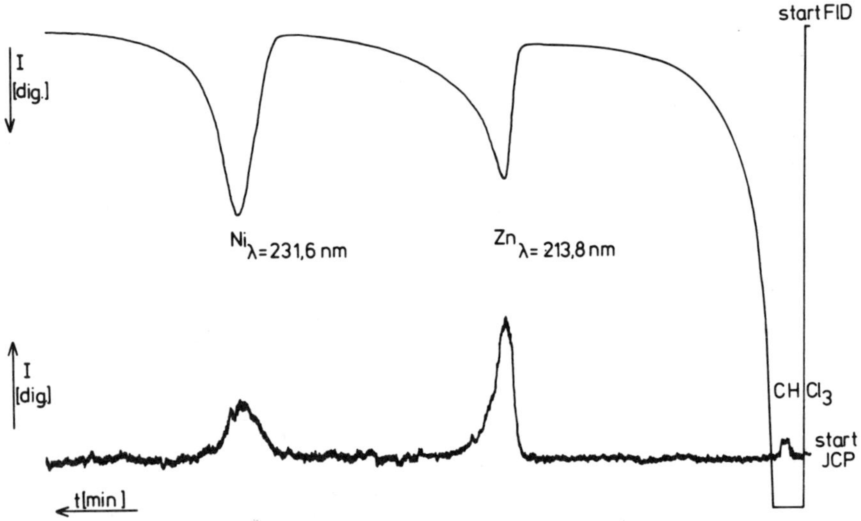

Figure 1.6. FID and ICP peaks after GC separation of Ni and Zn dithiocarbamate (GC/FID, top; GC/ICP, bottom).

very low trace concentrations of accompanying elements in pure iron or pig iron and in pure metals. The trace metals can be complexed by chelating agents, like dithiocarbamate, separated by GC, and simultaneously determined by ICP–AES. The unspecific signals for Ni and Zn observed with a flame ionization detector (FID) are compared in Fig. 1.6 with element-specific signals of the ICP recorded at the corresponding analysis lines.

The identification of ferrocene, organoarsenic, organomercury, and organolead compounds by interfacing of high-performance liquid chromatography (HPLC) and an ICP was achieved by Uden [84]. Relatively little has been reported on the use of inorganic or organometallic derivatization reactions in conjunction with ICP detection [85,86], although this interfacing combines a selective separation technique with a sensitive multielement detector capability.

These methods can be expected to become useful for the determination of inorganic constituents in rolling mill oils or emulsions, for example, and for detecting metallo-organic impurities in industrial organic solvents and chemicals.

REFERENCES

1. K. Ohls, K. H. Koch, and H. Grote, *Fresenius Z. Anal. Chem.* **284,** 177 (1977).
2. R. B. Wemyss and R. H. Scott, *Anal. Chem.* **50,** 1694 (1978).
3. P. W. J. M. Boumans, L. C. Bastings, F. J. de Boer, and L. W. J. van Kollenburg, *Fresenius Z. Anal. Chem.* **291,** 19 (1978).
4. R. N. Kniseley, H. Amenson, CC. Butler, and V. A. Fassel, *Appl. Spectrosc.* **28,** 285 (1974).
5. G. R. Kornblum and L. de Galan, *Spectrochim. Acta* **32B,** 455 (1977).
6. D. Sommer and K. Ohls, *Fresenius Z. Anal. Chem.* **295,** 337 (1979).
7. D. Sommer, K. Ohls, and A. Koch, *Fresenius Z. Anal. Chem.* **306,** 372 (1981).
8. P. W. J. M. Boumans, *Line Coincidence Tables for Inductively Coupled Plasma Atomic Emission Spectrometry.* Pergamon Press Oxford (1980, 1984).
9. P. W. J. M. Boumans, A. W. Witmer, F. J. de Boer, and M. Bosveld, *ICP Information Newslett.* **3,** 213 (1977).
10. R. K. Winge, V. J. Peterson, and V. A. Fassel, *Appl. Spectrosc.* **33,** 206 (1979).
11. P. D. Swaim and S. R. Ellebracht, *Anal. Chem.* **51,** 1605 (1979).
12. K. H. Koch, Radex-Rundschau, p. 780 (1982).
13. C. C. Butler, V. A. Fassel, and R. N. Kniseley, *Anal. Chem.* **47,** 825 (1975).
14. R. Diemiaszonek, J.-L. Mouton, and C. Trassy, *Analusis* **7,** 96 (1979).
15. J.-C. Souilliart and J. Robin, *Analusis* **1,** 427 (1972).
16. V. Rett and I. Hlavacek, *Hutn. Listy* **34,** 428 (1979).
17. A. F. Ward and L. F. Marciello, *Anal. Chem.* **51,** 2264 (1979).

18. A. Gomez Coedo, M. T. Dorado Lopez, and J. L. Jimenez Seco, *Rev. Metal. CENIM* **15,** 97 (1979).
19. M. H. Abdallah, R. Diemiaszonek, J. Jarosz, J. M. Mermet, J. Robin, and C. Trassy, *Anal. Chim. Acta* **84,** 271 (1976).
20. A. Wittmann, *Final Report of the Institut de Recherches de la Siderurgie Project*, Metz (1981).
21. H. Hughes, *Final Report of the British Steel Corporation Project*, Middlesbrough (1981).
22. K. Ohls, *Final Report of the Hoesch Project*, Dortmund (1980).
23. G. F. Wallace, 7th FACSS Meeting, Philadelphia, (1980), Abstr. 86.
24. J. Hancart, *Final Report of the Centre Recherche Metallurgique Project*, Liége (1981).
25. L. Ebdon, M. R. Cave, and D. J. Mowthorpe, *ICP Information Newslett.* **5,** 146 (1979).
26. D. E. Fornwalt, R. Yungk, and R. Cone, 7th FACSS Meeting, Philadelphia (1980), Abstr. 93.
27. J. Jarbo, M. T. Hurteau, and J. P. Mislan, Chemical Institute of Canada's 63rd Chemical Conference, Ottawa (1980), Abstr. 1.
28. G. L. Everett, *ICP Information Newslett.* **5,** 145 (1979).
29. G. E. E. Balaes and A. E. Watson, Report No. 1992, National Institute of Metallurgy (NIM), Randburg, South Africa (1978).
30. K. Sato, I. Tanaka, M. Akiyama, and T. Otsuki, 31st Pittsburgh Conference, Atlantic City (1980), Abstr. 57.
31. G. Horlick and M. W. Blades, *Appl. Spectrosc.* **34,** 229 (1980).
32. G. F. Larson, V. A. Fassel, R. K. Winge, and R. N. Kniseley, *Appl. Spectrosc.* **30,** 384 (1976).
33. A. Wittmann, J. Hancart, W. Hughes, and K. Ohls, in R. M. Barnes, ed., *Developments in Atomic Plasma Spectrochemical Analysis*, Heyden, Philadelphia (1981) p. 550.
34. Y. Endo and N. Sakao, *Tetsu-to-Hagané* **66,** 119 (1980).
35. D. Yates, T. Clevenger, D. James, and E. Hinderberger, 22nd Rocky Mountains Conference, Denver (1980), Abstr. 223.
36. S. Greenfield and P. B. Smith, *Anal. Chim. Acta* **59,** 341 (1972).
37. V. A. Fassel, C. A. Peterson, F. N. Abercrombie, and R. N. Kniseley, *Anal. Chem.* **48,** 516 (1976).
38. J. Soudière and P. Juvanon du Vachat, Applied Research Laboratories Analysis Report AQ/10403, Ecublens (1976).
39. R. Fischer, *Erdöl und Kohle* **31,** 200 (1978).
40. M. S. Cresser and R. R. Browner, *Anal. Chim. Acta* **113,** 33 (1980).
41. K. Ohls, *Erdöl und Kohle*, Compendium, (1977/78) p. 194.
42. D. Sommer and K. Ohls, *Fresenius Z. Anal. Chem.* **298,** 123 (1979).

43. R. F. Sanzolone, T. T. Chao, and G. L. Crenshaw, *Anal. Chim. Acta* **105,** 247 (1979).

44. A. M. T. C. Horta and A. J. Curtius, *Anal. Chim. Acta* **96,** 207 (1978).

45. A. Dornemann and H. Kleist, *Fresenius Z. Anal. Chem.* **291,** 349 (1978).

46. K. R. Sperling, *Fresenius Z. Anal. Chem.* **292,** 113 (1978).

47. T. Uchida, M. Nagase, I. Kojima, and C. Ida, *Anal. Chim. Acta* **94,** 275 (1977).

48. E. Jackwerth and H. Berndt, *Anal. Chim. Acta* **74,** 299 (1975).

49. A. Disam, P. Tschöpel, and G. Tölg, *Fresenius Z. Anal. Chem.* **295,** 97 (1979).

50. R. Kuroda and T. Seki, *Fresenius Z. Anal. Chem.* **296,** 146 (1979).

51. R. E. Sturgeon, S. S. Berman, A. Desaulniers, and D. S. Russell, *Talanta* **27,** 85 (1980).

52. E. Grallath, P. Tschöpel, G. Kölblin, U. Stix, and G. Tölg, *Fresenius Z. Anal. Chem.* **302,** 40 (1980).

53. R. M. Barnes and J. S. Genna, *Anal. Chem.* **51,** 1065 (1979).

54. A. Sugimae, *Bunseki Kagaku* **29,** 184 (1980).

55. S. S. Berman, J. W. McLaren, and D. S. Russell, *ICP Information Newslett.* **6,** 173 (1980).

56. G. Knapp, S. Raptis, and B. Schreiber, *ICP Information Newslett.* **6,** 194 (1980).

57. Y. Endo and N. Sakao, *A. u. R.* **17,** 105 (1979).

58. K. Ohls and D. Sommer, *Erdöl u. Kohle* **37,** 177 (1984).

59. E. Sebastiani, K. Ohls, and G. Riemer, *Z. Anal. Chem.* **264,** 105 (1973).

60. H. Berndt and E. Jackwerth, *J. Clin. Chem. Clin. Biochem.* **17,** 71 and 489 (1979).

61. J. A. C. Broekaert and F. Leis, *Anal. Chim. Acta* **109,** 73 (1979).

62. K. Ohls and D. Sommer, Kontron Seminar, Munich (1977).

63. K. Ohls and D. Sommer, *Fresenius Z. Anal. Chem.* **296,** 241 (1979) and *ICP Information Newslett.* **4,** 247 (1978.)

63a. A. Aziz, J. A. C. Broekaert, K. Laqua, and F. Leis, *Spectrochim. Acta* **39B,** 1091 (1984).

64. D. Sommer and K. Ohls, *Fresenius Z. Anal. Chem.* **304,** 97 (1980).

65. W. Holak, *Anal. Chem.* **41,** 1972 (1973).

66. H. D. Fleming and R. G. Ide, *Anal. Chim. Acta* **83,** 67 (1976).

67. J. A. C. Broekaert and F. Leis, *Fresenius Z. Anal. Chem.* **300,** 22 (1980).

68. G. Mausbach, *GIT Fachz. Lab.* **23,** 898 (1979).

69. F. D. Pierce and H. R. Brown, *Anal. Chem.* **49,** 1417 (1977).

70. J. Guimont, M. Pichette, and N. Rhéaume, *At. Absorption Newslett.* **16,** 53 (1977).

71. P. N. Vijan and D. Leung, *Anal. Chim. Acta* **120,** 141 (1980).

72. O. G. Koch and G. A. Koch-Dedic, *Handbuch der Spurenanalyse*, Springer Verlag, Berlin (1964) p. 892.

73. I. Kunert, J. Komarek, and L. Sommer, *Anal. Chim. Acta* **106,** 285 (1979).

74. J. F. Chapman and L. S. Dale, *Anal. Chim. Acta* **101,** 203 (1978).
75. G. E. Paris, W. R. Blair, and F. E. Brickman, *Anal. Chem.* **49,** 378 (1977).
76. W. de Jonghe, D. Chakraborti, and F. Adams, *Anal. Chim. Acta* **115,** 89 (1980).
77. A. J. McCormack, S. C. Tong, and W. D. Cooke, *Anal. Chem.* **37,** 1471 (1965).
78. Y. Talmi, *Anal. Chim. Acta* **74,** 107 (1975).
79. P. C. Uden, B. D. Quimby, R. M. Barnes, and W. G. Elliott, *Anal. Chim. Acta* **101,** 99 (1978).
80. M. S. Black and R. F. Browner, *Anal. Chem.* **53,** 249 (1981).
81. D. L. Windsor and M. B. Denton, *Appl. Spectrosc.* **32,** 366 (1978).
82. R. J. Lloyd, R. M. Barnes, P. C. Uden, and W. G. Elliott, *Anal. Chem.* **50,** 2025 (1978).
83. P. C. Uden, R. M. Barnes, and F. P. di Sanzo, *Anal. Chem.* **50,** 852 (1978).
84. P. C. Uden, in R. M. Barnes, ed., *Developments in Atomic Plasma Spectrochemical Analysis*, Heyden, Philadelphia (1981) p. 302.
85. C. Cast, J. C. Kraak, H. Poppe, and F. J. M. J. Maessen, *J. Chromatogr.* **185,** 549 (1979).
86. D. M. Fraley, D. Yates, and S. E. Manahan, *Anal. Chem.* **51,** 2225 (1979).

CHAPTER

2

APPLICATIONS: GEOLOGICAL

JAN-OLA BURMAN*

Department of Geology
University of Luleå
Luleå, Sweden

2.1. Introduction
2.2. Sample Dissolution
2.2.1. Introduction
2.2.2. Fusion Methods
2.2.3. Direct Acid Dissolution Methods: Open Systems
2.2.4. Direct Acid Dissolution Methods: Closed Digestion Systems
2.2.5. Selective Sample Preparation
2.3. Nebulization
2.4. ICP as an Analytical Tool for Geological Samples
2.4.1. Introduction
2.4.2. ICP–AES Compared with AAS
2.4.3. ICP Versus Spark AES
2.4.4. ICP Versus Arc AES
2.4.5. ICP–AES Versus Capacitively Coupled Microwave Plasma–AES
2.4.6. ICP–AES Versus XRFS
2.5. Obtained Versus Needed Accuracy and Vice Versa
2.5.1. Introduction
2.5.2. Comparison of Accuracy Among Different Methods
2.6. Special Applications of ICP–AES in Geology
2.6.1. Introduction
2.6.2. Determination of Rare Earth Elements
2.6.3. Marine Geochemistry

The analytical and geochemical research have been supported financially by the National Swedish Board for Technical Development (STU) and the Swedish Natural Science Research Council (NFR) under grant Prof. Kurt Boström. The author also thanks K. Boström for the critical reading and valuable comments on the manuscript. I wish to sincerely thank Dr. P. W. J. M. Boumans for his help during the preparation of the manuscript of this chapter.

* Present address: Svenskt Stål AB, Division Metallurgi, Luleå, Sweden

2.6.4. Biogeochemical Prospecting
2.6.5. Analysis of Fluid Inclusions in Geological Samples
2.7. Conclusions
References

2.1. INTRODUCTION

The large variety in the composition of geological samples and the refractory properties of many elements are some of the reasons why multielement analyses of geological samples are considered both ticklish and time-consuming. The demand for numerous, fast, and cheap and yet accurate multielement analyses in geological prospecting, for example, necessitated the development of new instrumental methods to replace several classical methods as described by Washington [1]. The instrumental methods included colorimetric procedures [2, 3], atomic emission spectrometry using arc and spark excitation [4, 5], atomic absorption [6] and X-ray fluorescence spectrometry (XRFS) [7–10], and various trace analysis techniques compiled by Reeves and Brooks [11]. The rapid development of inductively coupled plasma–atomic emission spectrometry (ICP–AES) instruments and methods in recent years, was not yet covered in that compilation, which was published in 1978. At that time ICP–AES was just beginning to enjoy widespread acceptance, particularly in the geological field, as is evident from the recent publication of a monograph by Thompson and Walsh [12].

One of the first ICP papers dealing with geological applications was published by Scott and Kokot [13], who compared ICP–AES and flame atomic absorption spectrometry (FAAS) for the determination of Cu, Zn, Ni, Co, and Pb. Watson and Russell [14, 15] and Watson and Steele [16] started to use ICP–AES for routine analysis of geological samples. Silicate analyses were described by Govindaraju et al. [17, 18] and Burman et al. [19, 20] using a fusion method. Uchida et al. [21] discussed a HCl/HF procedure in a sealed Teflon vessel. Both Brenner et al. [22] and Walsh and Howie [23] have documented good analytical results for major and minor element determinations performed under routine conditions and have claimed ICP–AES to be an outstanding general analytical procedure. Blast furnace and steel converter slag have been successfully analyzed [24] by a procedure developed for geological samples [19].

2.2. SAMPLE DISSOLUTION

2.2.1. Introduction

The sample dissolution procedure is critical in at least two respects: first, samples should be completely dissolved to get reproducible and accurate results, and second, nebulizer problems should be avoided. Owing to their refractory properties, many minerals can withstand extremely hard chemical treatments; the residues are hardly or not detectable with the naked eye. Typical trouble-makers are elements such as Ti, Zr, W, and Cr. Furthermore, the nebulization is very critical. Thus, several years of experience in routine operation with the ICP–AES systems at the University of Luleå have shown that the nebulizer and sample transport system caused most problems.

The sample dissolution procedure and the nebulizer characteristics must match each other to achieve good performance, since heavy salt loads tend to give drift caused by clogging of the nebulizer. ICP nebulizers are generally more prone to salt deposits than nebulizers used in FAAS, mainly because the aerosol carrier gas flow used in 1- to 2-kW ICPs is lower than in FAAS (cf. Part 1, Chapter 6). An irony is that many fluxes happen to cause more nebulizer problems than other salts. The maximum salt load therefore is also a function of the type of salt being aspirated.

2.2.2. Fusion Methods

The analysis of silica-rich materials requires awkward digestion procedures, involving, for instance, lithium metaborate or sodium peroxide. Solution techniques such as colorimetry and AAS generally require a large excess of $LiBO_2$; thus, Shur and Ingamells [25] recommend a sample-to-flux ratio of 1:5 and Ingamells [26] a ratio of 1:7 for complete dissolution of silica rocks. Govindaraju et al. [18] also used an excess of flux (1:5) for AES analysis of major and minor constituents of silicate rocks using a capacitively coupled microwave plasma (CMP). In this laboratory we initially used a CMP and employed the fusion method outlined by Govindaraju et al. [18] with some minor changes [20] to dissolve the borate bead in toto instead of taking only a fraction of it [18] (see Table 2.1). The modified procedure encompasses fewer steps and is less time-consuming.

The fusion method applicable to CMP analyses does not work properly for the ICP [19, 20, 27]. Metaborate-fused samples caused a continuous drift in the readings because the nebulizer (a concentric glass type, Meinhard T-200-A4) clogged and became inoperable within 20–30 min after nebulization had started. Inspection by microscope revealed clogging of the gas capillary; such a deposit changes the gas flow which leads to drift and irreproducible readings. The difficulties can be overcome by changing the sample-to-flux ratio to decrease the salt load [19] on the nebulizer and also by saturating the carrier gas by means of a wetting system (see Section 2.3 and [27]). It is indeed possible to fuse silicate rocks and slags by a sample to $LiBO_2$ ratio of 1:1 [19, 27]. Samples prepared in this way cannot be analyzed by FAAS, since the Si results show very large scattering. It is plausible that the silica exists as dissolved macro-complexes that are not decomposed in a relatively cold flame but are completely destroyed in a hot plasma. Ingamells [26] reported that polymerization of silicate and aluminate proceeds in the solutions, even if no visible precipitation

Table 2.1. Conditions for Fusion Methods Generally Used for Silicate Analyses

Type of Flux	Sample : Flux (g : g)	Final Volume (mL)	Nebulizer	Reference	Method
$LiBO_2$	0.1 : 0.5	50	AA-type	[25]	AAs
$LiBO_2$	0.2 : 1	100	AA-type	[26]	Colorimetry
$LiCO_3/H_3BO_3$	0.4 : (0.5 : 1)	200[a]	CMP-type	[18]	CMP
$LiBO_2$	0.05 : 0.35	100	CMP-type	[20]	CMP
$LiBO_2$	0.5 : 1.5	250	Cross-flow	[23]	ICP
Na_2O_2	0.2 : 2	200[b]	Cross-flow	[22]	ICP
$LiBO_2$	0.2 : 2	200[b]	Cross-flow	[22]	ICP
$LiBO_2$	0.1 : 0.1	100	Concentric	[19]	ICP

[a]0.2 g of the total (0.4 + 0.5 + 1 = 1.9 g) fused product diluted to 200 mL.
[b]High-power ICP, 11 kW, gives less clogging due to higher central gas flow.

occurs, which leads to low results if the colorimetric analyses are delayed more than a few hours. Such a polymerization due to aging obviously has no harmful effect on the results obtained by ICP methods [19, 21–24, 27, 28].

Walsh and Howie [23] use Pt crucibles to fuse the rocks with $LiBO_2$. Graphite crucibles are preferred by several workers [18, 19, 20, 22] because they are easier to use and cheaper when running large series. An electrical muffle furnace can easily be loaded with 8–10 samples and there is no need for continuous supervision by the operator. Govindaraju et al. [18] use an automatic tunnel furnace with a capacity of 60 samples/h. Sets of 200 crucibles filled with sample and flux are treated in each run. The samples are transported on a train, which moves through the tunnel at a controlled rate. This well-designed fusion step is followed by automatic acid dissolution, which still further minimizes the tedious and time-consuming work associated with the dissolution of rock samples. The automated production of liquid samples makes it possible to do sample preparation with a small staff, the amount of work being almost on the same level as for solid sample methods such as X-ray fluorescence spectrometry (XRFS).

The fusion methods discussed previously are summarized in Table 2.1. These methods are not sensitive enough for trace analyses down to the 0.5–1 ppm level. The load limitation of the nebulizer makes it impossible to increase drastically the amount of sample and flux per unit volume.

Govindaraju [28] developed a preconcentration technique by using small ion-exchange columns made of Tygon tubings and containing about 100 mg of an acid cation exchanger. This approach allows the separation of trace elements such as Sn, W, U, Th, Au, Pt, Be, Ga, Mo, Nb, As, Sb, and Bi, which often are difficult to determine in geological samples. The method is based on the $LiBO_2$ fusion described in [18].

Barnes et al. [29, 30] extended the earlier reported applications, using polydithiocarbamate also for preconcentration of trace elements of $LiBO_2$-fused geological samples. Ion exchanges that do not complex alkali and alkaline earth elements, in particular Li, K, Na, Mg, and Ca borates, are very advantageous for preconcentration techniques; they can be used not only to separate the traces from the flux components, but also from Ca and Mg, which are known to cause large background enhancements as a result of line-broadening and scattering [31, 32] (cf. Part 1, Chapters 7 and 8).

2.2.3. Direct Acid Dissolution Methods: Open Systems

HF–$HClO_4$ dissolution is still more frequently used than ion exchange for low level trace element determinations. The salts obtained by this method are easier to nebulize than borates. A 1-g sample is normally diluted to a final volume of 100 mL, or equal proportions. The HF-$HClO_4$ method described by Riley [2] is well established, and many laboratories use procedures close to this for FAAS analysis. This method is more difficult to automate than the fusion method because different rock types react in different ways upon the same acid attack. Even after repeated treatment there might remain residues of refractory minerals containing elements such as Al, Ti, Zr, and Cr. The only fast way to make these residues accessible for solution analysis is by a fusion step. There is a risk that refractory oxides will again form if the HF-$HClO_4$ mixture is evaporated to dryness or near dryness with too much heat. This can be avoided by using a water bath or an infrared heater. Silica is lost by evaporation of gaseous SiF_4 when an open system is used. This is a serious drawback for geologists because silica is one of the most important major elements used for the classification of rocks. Therefore, two different solutions are necessary to obtain complete information on major and trace elements. The disadvantage of Si loss can be avoided by using closed systems, but trace determinations with these methods are not free from objections either because of nebulizer limitations. However, loss of Si can in some circumstances also be an advantage, since it reduces the salt load considerably; many geological samples contain more than 50% SiO_2.

2.2.4. Direct Acid Dissolution Methods: Closed Digestion Systems

Plastic containers resist HF and are therefore used for the decomposition of silicatic materials. Uchida et al. [33] use a combination of Teflon and polypropylene vessels to decompose rocks with HF and HCl at room temperature. Boric acid is added to complex the excess HF, which makes it possible to transport the sample through a glass nebulizer and a silica torch without harmful etching by HF. Silica is also complexed and remains in the solution. Odegard [34] uses

Table 2.2. Use of H_3BO_3 as an Agent for Complexing HF in Dissolution Methods for Closed Systems Used for ICP–AES

Sample : H_3B_3 Ratio (g : g)	Final Volume (mL)	Nebulizer	Reference
0.2 : 2.25	2000	Cross-flow	[34]
0.05 : 0.24	100	Cross-flow	[33]

the same reagents as [33], but polyethylene vessels, which are first sealed and kept at room temperature, followed by a two-step treatment on a water bath. These two methods are limited by the maximum salt load that the nebulizer can tolerate. The salt contribution from the boric acid dominates over those from sample and acids (see Table 2.2). Dissolution in closed systems therefore has similar limitations as fusion methods because of the need for a complexing agent such as H_3BO_3. Commercially available and more complicated Teflon-coated pressure vessels, for example, autoclaves from Perkin-Elmer and Parr, can also be used, but preferably only for small sample series for both practical and economic reasons. Closed systems offer good results, but are not ideal for rapid determinations.

2.2.5. Selective Sample Preparation

Geological samples are frequently dissolved by a selective procedure. Some sediments can be rich in a carbonate phase that is easily dissolved in cold or moderately warm dilute acid; by such an approach, carbonates can be separated from the rest of the sample. Another selective technique involves the use of moderately warm HCl and a reducing compound like hydroxyl-amin-hydrochloride or citric acid; such procedures preferentially dissolve metal hydroxide phases, whereas silicates are only slightly attacked. Even more sophisticated methods of this type have been proposed [35]. These methods have proved to be very effective for geochemical prospecting. Despite the selective attack of the samples, anomalies are generally detected by this technique.

2.3. NEBULIZATION

The cross-flow nebulizer is generally less prone to salt deposits than the concentric glass nebulizer (cf. Part 1, Chapter 6). The solid contents of the solutions associated with the procedures summarized in Tables 2.1 and 2.2 are about the maximum allowed. The performance of the concentric nebulizer is considerably improved by saturating the sample carrier gas with water by means of a wetting device described by Burman [27]. The wetting system (Fig. 2.1) was

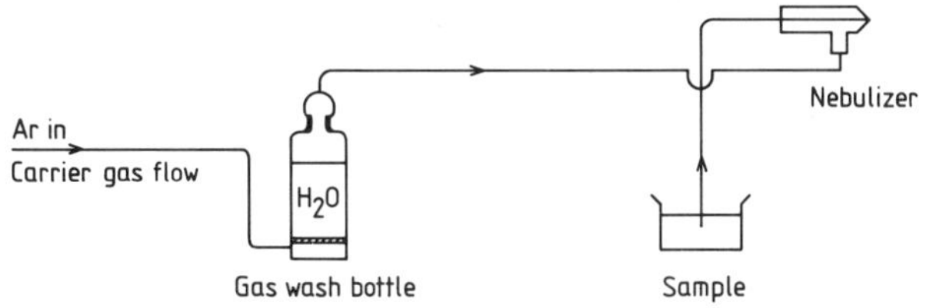

Figure 2.1. Schematic diagram of device for wetting the carrier gas. The carrier gas is saturated with water by leading the gas through an ordinary gas bottle prior to passing to the nebulizer. [J.-O. Burman, in R. M. Barnes, ed., *Developments in Atomic Plasma Spectrochemical Analysis* (1981) p. 15. Copyright (1981), Heyden & Son, Ltd. Reprinted by permission of John Wiley & Sons, Ltd.]

tested by running the same $LiBO_2$-fused sample at several dilutions with and without saturated carrier gas (see Figs. 2.2a,b). The clogging tendencies and resulting drift, as well as the wear of the capillary tips, drastically decreased by the use of the wetting device in combination with a Meinhard nebulizer. Similar tests with a cross-flow nebulizer resulted in detrimental effects on the stability. The cross-flow nebulizer without wetter still tolerated higher salt load than the concentric nebulizer with wetter.

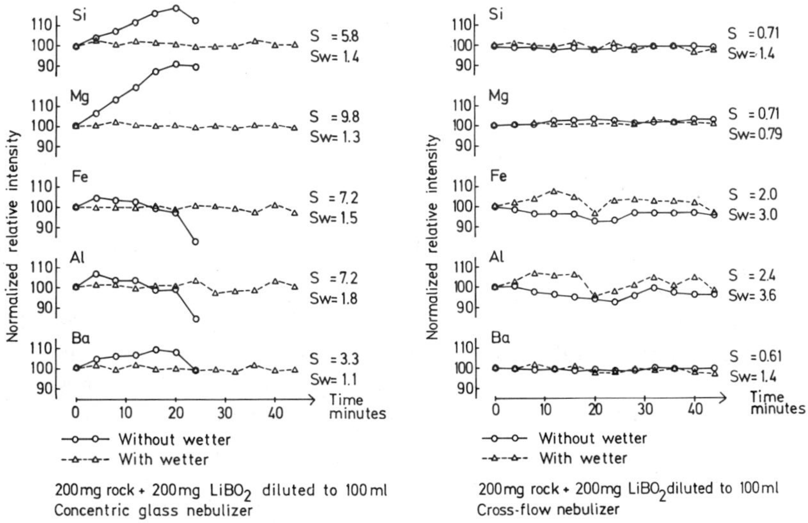

Figure 2.2. The wetter device improves the performance of the concentric glass nebulizer (*left*), whereas the stability of the cross-flow nebulizer (*right*) is worsened. S, Standard deviation without wetter device; Sw, standard deviation using wetter device. [J.-O. Burman, in R. M. Barnes, ed., *Developments in Atomic Plasma Spectrochemical Analysis* (1981), p. 15. Copyright (1981) Heyden & Son, Ltd. Reprinted by permission of John Wiley & Sons, Ltd.]

2.4. ICP AS AN ANALYTICAL TOOL FOR GEOLOGICAL SAMPLES

2.4.1. Introduction

The critical question to be answered when a new system for geological analysis has to be chosen is: Which elements have to be determined and how many analyses per year are to be made? Generally speaking one has to distinguish four cases:

1. Many analyses involving major and trace elements.
2. Few analyses involving major and trace elements.
3. Many analyses involving major elements.
4. Few analyses involving major elements.

The need for trace element information alone is often restricted among geologists, since also the concentration of major constituents is of main interest in geological research and prospecting. For all four alternatives listed above ICP–AES can be used with preference for the simultaneous mode of measurement (polychromator) for alternatives 1 and 3 and a sequential mode [32, 36, 37] for alternatives 2 and 4; however, other analytical techniques may also offer systems that can be adequately applied.

There has been a tendency to market new spectroscopic techniques such as FAAS, electrothermal atomization (ETA) AAS, and plasma source AES methods completely free from interferences. The experience gained by analysts using, for example, ETA–AAS, has not confirmed such claims and hence one may be suspicious when ICP–AES is stated to be nearly free from interferences. Indeed, several studies show that ICP–AES suffers from some interference, but it is a technique with few and restricted limitations compared with other methods. ICP–AES is not a miracle method answering all analytical needs, but it is a technique well suited for routine analysis of most geological materials including silicate-rich samples.

FAAS had a period of strong growth during the 1960s and ETA–AAS grew during the subsequent decade. It appears reasonable to assume that ICP–AES will have its period of growth during the 1980's. The excellent characteristics of this method (cf. Part 1, Chapter 4) seem to make this a rather modest prediction.

2.4.2. ICP–AES Compared With AAS

FAAS has successfully been used for analyses of geological samples for a long time, mainly as a complement to or replacement of already established solid sample techniques such as arc AES or XRFS. The geologists' demand for numerous multielement analyses makes AAS a less suitable choice because it is a single element method, even though AAS might be the ultimate choice for the laboratory running small series. Today, sequential slew-scan monochromators with ICP and the top-of-the-line FAAS spectrometers are competitors because they have analysis capacities of the same order of magnitude, for alternatives 2 and 4 mentioned in Section 2.4.1. A closer look reveals that the ICP system has a higher capacity but is also more expensive. Simultaneous ICP spectrometers and FAAS instruments cannot be considered as competitors because they are not intended for equal work loads. Analytically, ICP–AES has some fundamental advantages over FAAS:

1. ICP covers approximately 70 elements and is generally more sensitive than FAAS.
2. ICP–AES has far better sensitivity for B, P, C, S, and the rare earth elements.
3. Chemical interferences for refractory elements such as Al, Ti, W, Zr, and Cr are almost nonexistent in the high temperature and inert atmosphere of the ICP; this contrasts with the relatively cold and chemically complex environment of the flame.
4. ICP–AES has a linear dynamic range of 4–6 orders of magnitude and will therefore require far less frequent dilutions of samples than AAS.

However, one cannot bypass the problem of spectral interferences encountered in any AES method using a high-temperature source. This severe problem may be encountered in trace analysis. Adequate measures are indispensable to cope with them (see Part 1, Chapter 7).

ETA–AAS is only complimentary to ICP–AES owing to its low analysis rate. However, it is a much more sensitive micromethod applicable to sample sizes of $< 100\ \mu L$ and is capable of trace and ultratrace determinations in geological samples. Interferences are the rule rather than the exception, making the analyses ticklish and still more time-consuming and expensive. Many applications are better and easier solved with preconcentration procedures followed by FAAS or even multielement ICP detection.

Large series of geological samples are much better analyzed by ICP–AES, whereas FAAS is more suited for small series with a small number of elements. The ideal instrumental configuration is a combination of several techniques, for example, AAS and ICP–AES, but this is unfortunately beyond the budget of many smaller laboratories.

2.4.3. ICP Versus Spark AES

Spark AES is most extensively used for production and process control in the metallurgical industry where very rapid analyses are required. Spark AES has been used by Danielsson et al. [38, 39] for geological samples. In this approach the ground sample was fed in powder form on a tape that passes a spark gap at constant speed. The method had good sensitivity but the reproducibility for trace elements was relatively poor because of the very small amounts of samples actually analyzed and consequent repercussions from sample inhomogeneities. However, the technique has given good results for major and minor constituents. ICP gives somewhat better accuracy at the cost of a more complicated sample preparation procedure, whereas the demand for skilled instrument operators is higher for the tape machine.

2.4.4. ICP Versus Arc AES

Arc AES has been a tool for routine geoanalyses during more than five decades and is still used in many laboratories, as evidenced by work of Watson and Russell [40]. The increased precision, accuracy, and detection limits, despite the dilution of the sample, are arguments favoring ICP–AES. The greatest advantage of arc AES is the simplicity of sample preparation, which involves only grinding and mixing with graphite and possibly a buffer. Thus, Maessen et al. [41] showed that fusion of the sample with $LiBO_2$ gives better accuracy and precision due to removal of mineralogical effects, which, however, entails more time-consuming sample preparation and a large risk of contamination.

2.4.5. ICP–AES Versus Capacitively Coupled Microwave Plasma

The capacitively coupled microwave plasma (CMP) (cf. Part 1, Chapter 2) was the first commercial plasma source AES system in Europe. ICP and CMP have been compared by Boumans et al. [42], who found that the ICP yields better detection limits and less interference. Nevertheless with proper buffering by strontium, Govindaraju et al. [18] succeeded in obtaining excellent results for major and minor constituents using CMP. Burman and Boström [43] emphasized the application of geological samples by running four series with both ICP and CMP. The series was prepared to resemble different calibration procedures frequently used, namely (1) synthetic solutions based on a mixture of metal salts, (2) synthetic solutions with the addition of a corresponding amount of $LiBO_2$ flux, (3) HF–$HClO_4$ dissolved standard rocks, and (4) $LiBO_2$-fused standard rocks. Representative results are shown in Fig. 2.3.

The four calibration series are plotted together to facilitate direct comparison for the same element. Evidently the CMP readings scatter much more than the ICP results. The slopes of the calibration curves for CMP differ widely owing to the influences of the matrix. Therefore, buffering of the samples and standard solutions is indispensable. ICP excitation is far superior because of the relative freedom from matrix effects thus making it easy to prepare standard solutions for ICP–AES. Even if solutions from different dissolution methods are mixed,

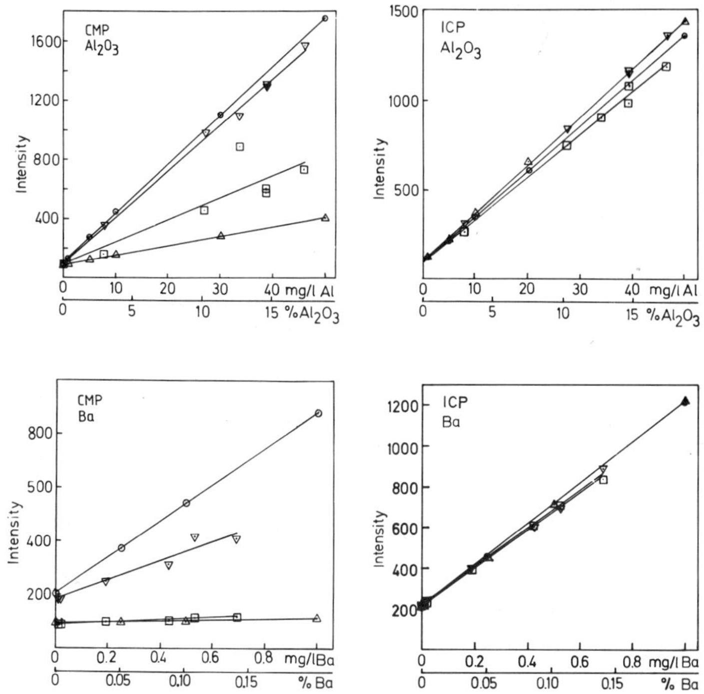

Figure 2.3. Calibration curves obtained with CMP and ICP. Series 1 (△), Synthetic calibration solutions containing metal salt diluted to 2% (v/v) HNO_3. Series 2 (○), Synthetic calibration solutions prepared as series 1 but with addition of 350 mg of $LiBO_2$ per 100 mL. Series 3 (□), Calibration solutions made by $HF/HClO_4$ of six certified rocks, 50 mg to a final volume of 100 mL 2% (v/v) HNO_3. Series 4 (▽), Calibration solutions made by $LiBO_2$ fusion of certified rocks, 50 mg of rock and 350 mg of $LiBO_2$ of a final volume of 100 mL 2% (v/v) HNO_3. Al and Ba excitation in four matrices by CMP and ICP. The slope for the CMP may differ by as much as a factor of 32. The concentration axis shows both the concentration in the solution (mg/L) and the concentration in the solid rock (%). [Reprinted with permission from J.-O. Burman and K. Boström, "Comparison of Different Plasma Excitation and Calibration Methods in the Analysis of Geological Materials by Optical Emission Spectrometry," *Anal. Chem.* **51**, 516 (1979). Copyright (1979), American Chemical Society.]

it is still possible to make accurate analyses, but the contamination from reagents can cause troubles if the same amounts are not used in all samples. Therefore, the samples should be uniformly prepared to minimize the effect of contamination from reagents. In summary, no real difficulties are encountered in making synthetic calibration solutions for trace element analysis since the absence of the main elements of a geological matrix causes no problems. This makes it easy to prepare calibration solutions for the ICP–AES determinations of trace elements for which certified values are lacking.

2.4.6. ICP–AES Versus XRFS

ICP and XRFS are very good multielement techniques and both are perfectly suited for analysis of large series of samples. Sample preparation is much easier for XRFS; it comprises direct briquetting of ground samples or various fusion methods. After the more time-consuming dissolution with a dilution factor of 1 : 100, ICP–AES still has superior detecting power compared with XRFS, with the exception of S, P, and the halogens, of which F, Cl, and Br are even beyond the capability of ICP–AES. Light elements (Li, B, Be), however, cannot be determined with XRFS. The shortcomings and advantages of the two methods must, as always, be considered in the light of the specific analytical service needed by geologists. XRFS is obviously insufficient for some trace analyses, but it might still be sensitive enough for prospecting, since only concentrations significantly higher than the background in the rocks or soils are of interest. Such anomalies can often be detected by XRFS. The naturally occurring trace levels are often accessible by ICP–AES but not XRFS, and this is of importance in several fields of geological research. Multielement analyses comprising major and trace constituent determinations are essential to bring new understanding of fundamental geological processes. Several novel ICP systems have been specifically tailored to exploit the ICP capabilities in specialized geological research programs. The incorporation of an ICP in these systesm [44–47] is so essential that a comparison with systems based on other analytical techniques is impossible (see also Section 2.6).

Generally elements present at high concentrations can be easier and better determined by XRFS than by ICP–AES, again because of the relatively simple and cheap sample preparation procedures for XRFS. The accuracy of XRFS is equal to or better than that of ICP–AES, when sophisticated computer software for matrix corrections in XRFS is used. The application of software corrections is limited to geological samples of relatively uniform composition. If, for instance, iron ores or manganese nodules with high Fe and Mn concentrations are analyzed by using ordinary geological XRFS program, the accuracy will be drastically degraded, whereas even substantial changes in the sample composition have only negligible effects on the accuracy of ICP–AES.

2.5. OBTAINED VERSUS NEEDED ACCURACY AND VICE VERSA

2.5.1. Introduction

High accuracy is almost always expensive. The relation between accuracy and price is far from linear, which should induce the geologist or any other customer of analyses to be modest in his demand for accuracy. Too often analysts are asked to do analyses with the best possible accuracy, when a more relevant demand should be: Do the analysis with the worst possible accuracy. The results must be accurate enough to fulfill the aim of the investigation, but need not be better; all extravagant efforts to raise the accuracy represent waste of time and money. Too many customers of analytical service are too ambitious in their demand for accuracy. The term "best possible" is perhaps used in a routine way, without any thought about its meaning. It is very important therefore to have a continuous debate about the interrelation between quantity, quality, and price, not at least in the geological field.

The lack of relevant self-criticism among analysts is another important point in the discussion about obtained versus needed accuracy. Many of us sincerely hope and wish to deliver results of the quality demanded by the customer, which may result in too optimistic a judgment of the quality of the analyses. Accuracy and precision must be distinguished carefully. The repeatability of the method is not a measure of the overall accuracy [48].

At the end of the 1940s Fairbairn and co-workers at the U.S. Geological Survey started an investigation of the reliability of chemical rock analyses by distributing two rocks, G-1 and W-1, to 34 distinguished analysts [49, 50]. After compilation of the results, they concluded: "First results show that rock analyses are generally not as precise as has been assumed." Certification of Geological Reference Samples (GRS) have been most valuable in improving the analytical methods (see Flanagan [51–54], Geischer [55, 56], and Govindaraju [57, 58]). GRSs are used to check the results of, for example, AAS and AES techniques and, furthermore, are the basis for calibration and matrix correction procedures in XRFS. This leads to the rather frustrating starting point: "In order to do the XRFS analysis you must know the analysis of samples." The rock reference samples are generally certified through careful analysis performed by many laboratories, where only laboratories skilled in the art are allowed to participate. Nevertheless rather large discrepancies are found between the reported results. Participation in the certification of GRSs is advantageous in various respects. First, it gives information about the status of the methods used at one's own laboratory; second, it is of great value for obtaining access to all information reported from other laboratories using different analytical techniques. Geologists and perhaps also some analysts become very surprised when realizing the scatter about the certified values for trace elements.

Geostandards Newsletter (edited by K. Govindaraju [59]) is a journal devoted to the study and promotion of geological reference samples. It is the only one in the field. The publication of interlaboratory analyses of new GRSs and also the continuous updating of data on older GRSs make this journal most valuable for both analysts and geologists.

2.5.2. Comparison of Accuracy Among Different Methods

Fair comparison between the accuracy obtained by different methods is difficult. It rarely happens that two methods are compared under exactly equal circumstances; different samples may be used, different persons with varying skills do the work, and different procedures are employed involving, for example, a different number of dissolutions or readings per dissolution or different degrees of sophistication of applied computer-controlled matrix corrections. It must be also remembered that analysts writing articles present a subjective picture of the relative importance of the different properties of the methods being compared.

The analyses of the three new GRSs during 1980 gave an excellent opportunity to compare methods. The data were presented in *Geostandards Newsletter*, compiled by Govindaraju [58]. The same samples were analyzed in many independent ways, but the large number of participants made it possible to divide the results into groups, one for each main method used. In all, 122 laboratories and 243 analysts belonging to the International Working Group "Analytical Standards of Minerals, Ores and Rocks" took part in this collaborative study of three samples: anorthosite AN-G, Basalt BE-N, and granite MA-N. All the interesting results from [58] cannot be reproduced here due to space limitations, since more than 6000 units of analytical data on major, minor, and 53 trace elements were reported.

Table 2.3. Results of Major and Minor Element Determinations Obtained during Certification of Three French Geological Reference Samples[a]

	AN-G			BE-N			MA-N		
	n	$\bar{x}$	s	n	$\bar{x}$	s	n	$\bar{x}$	s
SiO_2 %									
Total	112	46.33	0.69	113	38.43	0.62	113	66.66	0.79
AAS	15	46.08	0.91	15	38.35	0.79	16	66.80	1.05
Wet chemical	47	46.35	0.5	46	38.41	0.56	48	66.51	0.58
XRFS	41	46.32	0.74	42	38.51	0.68	40	66.70	0.85
Plasma AES	9	46.67	0.61	10	38.34	0.40	9	67.04	0.40
Recommended		46.30			38.20			66.60	
Al_2O_3 %									
Total	115	29.82	0.94	115	10.16	0.41	111	17.68	0.61
AAS	31	29.77	1.36	31	10.14	0.38	28	17.46	0.81
Wet chemical	27	29.81	0.90	25	10.26	0.55	28	17.69	0.61
XRFS	42	29.80	0.59	43	10.12	0.36	42	17.83	0.37
Plasma AES	12	30.05	0.61	13	10.18	0.29	11	17.54	0.62
Recommended		29.80			10.12			17.62	
TiO_2 %									
Total	111	0.22	0.04	117	2.65	0.14	86	0.04	0.06
AAS	14	0.23	0.04	15	2.61	0.17	7	0.05	0.03
Wet chemical	42	0.22	0.06	43	2.68	0.19	32	0.05	0.08
XRFS	43	0.22	0.02	43	2.63	0.07	40	0.03	0.03
Plasma AES	12	0.23	0.02	14	2.63	0.09	6	0.02	0.01
Recommended		0.22			2.62			0.01	

[a] Data from Govindaraju [58].

Si, Al, and Ti are representative of the major and minor elements (see Table 2.3). Wet chemical methods, including gravimetry, colorimetry, and volumetry, constitute one group of methods. Instrumental methods such as AAS, XRFS, and AES make up three other groups. The mean and standard deviation for all observations are presented under "Total." The term "plasma AES" is used throughout the text in this section because one of the laboratories uses a CMP [18] while all other laboratories used ICP.

The following comments are for Si and Al: The spread about the mean value is highest for AAS and lowest for wet chemical methods. The XRFS mean values are closer to the proposed values than those for plasma AES, but the spread is generally lower for plasma AES than for XRFS despite the relatively small number of plasma AES observations. The order in the "accuracy race" then becomes as follows: Wet chemical methods are the superior ones, XRFS and plasma AES share the second place, and the AAS methods end up slightly behind. The result is well known to analysts: Wet chemical methods for Si and Al are more accurate than instrumental methods, but the former methods are labor intensive and slow.

Comments for Ti include: The results for Ti contrast to those for Si and Al in that the wet chemical methods do not perform as well as the instrumental methods. Plasma AES gives the best results, very closely followed by XRFS. There is a small advantage for AAS over wet chemical methods. The low Ti content of 0.02% in MA-N is definitely the best determined by plasma AES.

Instrumental methods dominate in trace element analyses; wet chemical methods are consistently omitted in Table 2.4. The number of trace analyses

Table 2.4. Results of Trace Element Determinations Obtained during Certification of Three French Geological Reference Samples[a]

	AN-G			BE-N			MA-N		
	n	$\bar{x}$	s	n	$\bar{x}$	s	n	$\bar{x}$	s
Ba (ppm)									
Total	36	37.6	11	46	1001	141	31	43	11.7
AAS	6	49.1	9.2	7	1007	117	7	42.4	8.4
XRFS	13	34.9	11	19	1069	107	10	38.1	12
Plasma AES	7	35.1	9.2	9	990	112	6	45.7	9.1
Recommended		34			1025			42	
Cu (ppm)									
Total	40	18.8	4.9	42	73	11.3	43	137	23
AAS	13	19.5	2.8	15	72	6	18	141	13
XRFS	14	18.9	6.2	13	73	8.4	13	136	26
Plasma AES	5	17.8	4.8	7	74	15	5	145	19
Recommended		19			72			140	

[a] Data from Govindaraju [58].

for plasma AES does not exceed nine, making the mean and standard deviations less significant. Comments for Ba and Cu include: AAS and XRFS and plasma AES give almost the same performance in the region of 40–1000 ppm of Ba. In the interval of 20–150 ppm of Cu, the spread is lower for AAS and higher or of the same magnitude for plasma AES and XRFS.

The results in Tables 2.3 and 2.4 give an indication of the capabilities of ICP–AES. The interested reader should consult the original source [58] for complete information. To conclude, the result of this independent and objective assessment of methods in connection with the certification of GRSs is that ICP–AES generally can give an accuracy that is the same as or better than that achievable with other methods.

2.6. SPECIAL APPLICATIONS ON ICP–AES IN GEOLOGY

2.6.1. Introduction

Rock analysis for prospecting purposes is probably the most used ICP–AES application in geology. Other special applications that take advantage of the ICP–AES characteristics have already been developed, despite the fact that ICP–AES is a relatively new technique. The following sections summarize some of these applications.

2.6.2. Determination of Rare Earth Elements

The pattern of rare earth elements (REE) often gives valuable information in geological research, for example, as a basis for proposing models for the origin of rocks. Earlier REE determinations have encountered considerable analytical difficulties owing to the need for complicated and costly methods, such as neutron activation analysis and spark source mass spectrometry. The high sensitivity of emission lines of REE make ICP–AES an excellent alternative to the methods used earlier. Broekaert et al. [60] and Walsh [61] described methods for REE analysis of geological samples. Ion-exchange separation gave very good recoveries and a good precision for La, Ce, Pr, Nd, Sm, Eu, Gd, Dy, Er, Yb, and Lu [61]. The major constituents were first eluted, whereas the REEs were held quantitatively on the resin, which thereby gave advantageous separation between REEs and stray light-generating elements like Ca and Mg. The determination of REEs by ICP–AES is also very advantageous compared with the techniques most used at present in geological research and exploration like AAS and XRFS.

2.6.3. Marine Geochemistry

ICP–AES is now used at numerous scientific institutions, for example, the University of Luleå, for research in marine geochemistry. Sediments are analyzed following the procedures of rock analysis [19]. To improve the knowledge about geochemical cycling of the elements in the marine environment, it was necessary to include plankton [62], which represents the dominant part of the biomass in the sea, and also suspended matter in rivers draining to the sea as well as suspended matter sustained in the sea. Plankton resists acid attacks surprisingly well, making it unavoidable to use pressure autoclaves for HNO_3 dissolution. Plankton holds high Na, Ca, and Mg concentrations, making it difficult to determine some trace elements because of the line broadening and stray-light effects from Ca and Mg [31]. Suspended matter in river and sea water is analyzed after filtration at 6-atm pressure through 0.45-μm Millipore filters. Sample volumes of 10–100 L are used, depending on the load of suspended matter. The procedure follows the following sequence: sampling, filtration in field, acid dissolution, evaporation and ashing, $LiBO_2$ fusion, and multielement ICP–AES determination. ICP–AES has proved to be an excellent multielement technique for this application with a sample size of 1–5 mg, which, owing to the fusion step, also includes refractory elements like Si, Al, Ti, and Zr [63–67]. The conservative behavior of Al, Ti, and Zr in nature makes these refractory elements important references in different geochemical studies. It is hardly possible to find one analytical technique except for ICP–AES, which is capable of coping with the broad variety of applications in marine geochemical research.

2.6.4. Biogeochemical Prospecting

Mineralizations may be found by analyzing soil and vegetation from the ground standing over a potential ore, because high metal concentrations in the ground and the groundwater may be reflected in the vegetation. Furthermore, it is known that particulate matter from vegetation surfaces enters the air in some aerosol-like form called ''exudates,'' which, in turn, might reflect an anomalous composition in the chain from ground to vegetation. Two methods for sampling of this aerosol followed by multielement ICP–AES have been described. They are known as Airtrace and Surtrace [44, 46, 47].

Sampling by the Airtrace method is achieved from airplane flying at low magnitude. The particles are collected on an adhesive tape placed at the end of a large funnel mounted on the aircraft. The sample on the tape is vaporized by a CO_2-laser and flushed into an ICP–AES by an argon gas stream. Dissolution and nebulization are avoided, allowing trace determinations, even if the total amount of sample is less than 1 mg. Surtrace (surface trace) is based on the same principle, but the sampling equipment is carried by humans, resulting in a more local reconnaissance. These methods are exciting and highly specialized ICP–AES applications. ICP–AES is also used for biogeochemical prospecting in a more traditional way by sampling of vegetation followed by dissolution and multielement analysis.

2.6.5. Analysis of Fluid Inclusions in Geological Samples

ICP–AES has been used to investigate the composition of fluid inclusions in sample of pegmatite and carbonatite [45]. Material ejected from the sample during decrepitation at elevated temperatures is transferred to the plasma by a stream of Ar. Up to 40 elements can be determined simultaneously in the decrepitate. The method is rapid and permits the determination of a number of important elements which in conventional fluid inclusions are present in concentrations so low that they can be detected only with difficulty, even with very sensitive methods.

2.7. CONCLUSIONS

There has always been a need for general multipurpose analytical methods that answer as much as possible the analysis demands of geologists. Sensitive and accurate methods exist for almost all elements and species in geological samples, but for economic reasons the trade-off between quantity, quality, and price makes it impossible to use many of the methods on a routine basis.

Neither ICP–AES nor any other method is a "black box method," giving good analytical results by just pushing the button, but ICP–AES is very well suited for many analytical tasks in a geological laboratory if compared with other established methods. The general properties of ICP–AES permit the use of the technique in even relatively small research institutions or laboratories, because of the favorable cost/profit ratio. Complementations and replacements in the instrumental outfit of geological laboratories can no longer be done without careful consideration of the potential of ICP–AES. The short history of ICP–AES has already proved that a very significant part of the ICP users are working in the geological field.

REFERENCES

1. H. S. Washington, *The Chemical Analysis of Rocks*, Wiley, New York (1930).
2. J. P. Riley, *Anal. Chim. Acta* **19,** 413 (1958).
3. E. B. Sandell, *Colorimetric Determinations of Traces of Metals*, Wiley-Interscience, New York (1959).

4. T. Török, J. Mika, and E. Gegus, *Emission Spectrochemical Analysis*, Adam Hilger, Bristol (1978).
5. P. W. J. M. Boumans, *Theory of Spectrochemical Excitation*. Hilger & Watts, London/Plenum Press, New York (1966).
6. E. E. Angino and G. K. Billings, *Atomic Absorption Spectrometry in Geology*, Elsevier, Amsterdam (1972).
7. R. O. Müller, *Spectrochemical Analysis by X-ray Fluorescence*, Plenum Press, New York (1972).
8. B. E. Leake, G. L. Hendry, A. Kemp, A. G. Plant, P. K. Harvey, J. R. Wilson, J. S. Coats, J. W. Aucott, T. Lünel, and R. J. Howarth, *Chem. Geol.* **5,** 7 (1969).
9. J. Kikkert, *Spectrochim. Acta* **38B,** 809 (1983).
10. J. Kikkert, *Spectrochim. Acta* **38B,** 1497 (1983).
11. R. D. Reeves and R. R. Brooks, *Trace Element Analysis of Geological Materials, Chemical Analysis Series*, Vol. 51, Wiley-Interscience, New York (1978).
12. M. Thompson and J. N. Walsh, *A Handbook of Inductively Coupled Plasma Spectrometry*, Blackie, Glasgow and London (1983).
13. R. H. Scott and M. L. Kokot, *Anal. Chim. Acta* **75,** 257 (1975).
14. A. E. Watson and G. M. Russell, *ICP Information Newslett.* **3,** 273 (1977).
15. A. E. Watson and G. M. Russell, *ICP Information Newslett.* **3,** 409 (1978).
16. A. E. Watson and T. M. Steele, *ICP Information Newslett.* **5,** 409 (1980).
17. K. Govindaraju and G. Mevelle, *Spectrochim. Acta* **38B,** 1447 (1983).
18. K. Govindaraju, G. Mevelle, and C. Chourad, *Anal. Chem.* **48,** 1325 (1976).
19. J.-O. Burman, C. Pontér, and K. Boström, *Anal. Chem.* **50,** 679 (1978).
20. J.-O. Burman, B. Boström, and K. Boström, *Geol. Foeren. Stockholm Foerh.* **99,** 102 (1977).
21. H. Uchida, T. Uchida, and C. Iida, *Anal. Chim. Acta* **108,** 87 (1979).
22. I. B. Brenner, A. E. Watson, G. M. Russell, and M. Concalves, *Chem. Geol.* **28,** 321 (1980).
23. J. N. Walsh and R. A. Howie, *Mineral. Mag.* **43,** 967 (1980).
24. J.-O. Burman, in R. M. Barnes, ed., *Developments in Atomic Plasma Spectrochemical Analysis*, Heyden, Philadelphia (1981), p. 564.
25. N. H. Shur and C. O. Ingamells, *Anal. Chem.* **38,** 730 (1966).
26. C. O. Ingamells, *Anal. Chem* **38,** 1228 (1966).
27. J.-O. Burman, in R. M. Barnes, ed., *Applications of Plasma Emission Spectrochemistry*, Heyden, Philadelphia (1979), p. 15.
28. K. Govindaraju, Int. Winter Conf. Developments in Plasma Spectrochemical Analyses, San Juan, Puerto Rico (1980).
29. R. M. Barnes and J. S. Genna, *Anal. Chem.* **51,** 1065 (1979).
30. A. Miyazaki and R. Barnes, *Anal. Chem.* **51,** 299 and 364 (1981).
31. G. F. Larsson, V. A. Fassel, R. K. Winge, and R. N. Kniseley, *Appl. Spectrosc.* **30,** 384 (1976).

32. J.-O. Burman, B. Johannson, B. Morefält, K. H. Närfeldt, and L. Olsson, *Anal. Chim. Acta* **133,** 379 (1981).

33. H. Uchida, T. Uchida, and C. Iida, *Anal. Chim. Acta* **108,** 87 (1979).

34. M. Odegard, *Jarrell-Ash Newslett.* **2,** 4 (1979).

35. R. Chester and M. J. Hughes, *Chem. Geol.* **2,** 249 (1967).

36. M. A. Floyd, V. A. Fassel, R. K. Winge, J. M. Katzenberger, and A. P. D. Silva, *Anal. Chem.* **52,** 431 (1980).

37. T. Catterick and D. A. Hickman, *Analyst* **104,** 516 (1979).

38. A. Danielsson, F. Lundgren, and G. Sundqvist, *Spectrochim. Acta* **15,** 122 (1959).

39. A. Danielsson, and G. Sundqvist *Spectrochim. Acta* **15,** 125, 134 (1959).

40. A. E. Watson and G. M. Russell, *Spectrochim Acta* **33B,** 143 (1978).

41. F. J. M. J. Maessen, J. W. Elgersma, and P. W. J. M. Boumans, *Spectrochim. Acta* **31B,** 179 (1976).

42. P. W. J. M. Boumans, F. J. de Boer, F. J. Dahmen, H. Hölzel, and A. Meier, *Spectrochim. Acta* **30B,** 449 (1975).

43. J.-O. Burman and K. Boström, *Anal. Chem.* **51,** 516 (1979).

44. D. N. H. Horler, J. Barber, and A. R. Barringer, *Geochem. Expl.* **13,** 41 (1980).

45. M. Thompson, A. H. Rankin, S. J. Walton, C. Halls, and B. N. Foo, *Chem. Geol.* **30,** 121 (1980).

46. A. R. Barringer, AIRTRACE—An Airborne Geochemical Technique, 1st William T. Pecora Memorial Symp., Oct. 1975, Sioux Falls, South Dakota.

47. A. R. Barringer, J. H. Davis, and L. Dauber, SURTRACE—An Airborne Geochemical System, Proc. 12th Int. Symp. Remote Sens. Environ., Manila, Philippines, April 1978.

48. K. Boström, Lecture notes for geochemistry course Mn 10, Univ. of Luleå.

49. H. W. Fairbairn, W. G. Schlecht, R. E. Stevens, W. H. Dennen, L.-H. Ahrens, and F. Chayes, *U.S. Geol. Surv.*, Bull. 980 (1951).

50. H. W. Fairbairn, *Geochim. Cosmochim. Acta* **4,** 143 (1953).

51. F. J. Flanagan, *U.S. Geol. Surv. Bull.* **1113,** 113 (1960).

52. F. J. Flanagan, *Geochim. Cosmochim. Acta* **33,** 81 (1960).

53. F. J. Flanagan, *Geochim. Cosmochim. Acta* **38,** 1731 (1974).

54. F. J. Flanagan, *U.S. Geol. Surv. Prof.* Pap. 840 (1976).

55. M. Fleischer, *Geochim. Cosmochim. Acta* **29,** 1263 (1965).

56. M. Fleischer, *Geochim. Cosmochim. Acta* **33,** 65 (1969).

57. K. Govindaraju, *Geostandards Newslett.* **3,** 3 (1979).

58. K. Govindaraju, *Geostandards Newslett.* **4,** 49 (1980).

59. *Geostandards Newslett.*, K. Govindaraju, ed., C.R.P.G., 15, Notre Dame des Pauvres, B.P. 20, 54501 Vandoevre-lès Nancy, France.

60. J. A. C. Broekaert, F. Leis, and K. Laqua, *Spectrochim. Acta* **34B,** 73 (1981).

61. J. N. Walsh, F. Buckley, and J. Bakker, *Chem. Geol.* **33,** 141 (1981).

62. K. Boström and B. Boström, Unpublished results.

63. K. Boström, J.-O. Burman, B. Boström, C. Pontér, and S. Brandlöf, *Finn. Mar. Res.* **244,** 8 (1978).

64. K. Boström, L. Wiborg, and J. Ingri, *Mar. Geol.* **50,** 1 (1982).

65. K. Boström, J.-O. Burman, C. Pontér, and J. Ingri, *Mar. Chem* **10,** 335 (1981).

66. J.-O. Burman. *Ecol. Bull. Stockholm* **35,** 99 (1983).

67. K. Boström, J.-O. Burman and J. Ingri, *Ecol. Bull. Stockholm* **35,** 39 (1983).

CHAPTER

3

APPLICATIONS: ENVIRONMENTAL

J. W. McLAREN

Analytical Chemistry Section, Chemistry Division,
National Research Council of Canada,
Ottawa, Canada

3.1. Introduction
3.2. Dissolution of Solid Samples
3.3. Calibration Procedures
3.4. Types of Spectroscopic Interferences Encountered with Environmental Samples
3.5. Analysis of Samples Related to Water Quality Monitoring
3.5.1. Water Samples
3.5.1.1. Fresh Waters
3.5.1.2. Seawater
3.5.1.3. Waste Waters
3.5.1.4. Preconcentration Techniques for Water Analysis
3.5.2. Water Quality Monitor Organisms
3.5.3. Freshwater and Marine Sediments
3.6. Analysis of Samples Related to Air Quality Monitoring
3.6.1. Airborne Particulate Matter
3.6.2. Coal and Coal Fly Ash
References

3.1. INTRODUCTION

The application of inductively coupled plasma–atomic emission spectrometry (ICP–AES) to the analysis of environmental materials such as natural and waste waters, sediments, biological tissues, and air particulates is now widespread. The rapid acceptance of this relatively new technique by environmental scientists can be attributed to a number of its features which are particularly advantageous in this field.

1. ICP–AES is a rapid multielement technique that permits simultaneous determination of major, minor, and trace constituents.

2. Although the sensitivity of ICP–AES for many of the commonly determined trace elements (for example, Cu, Cd, Ni, Pb, Zn) is not greatly different from that of a popular alternative technique, flame atomic absorption spectrometry (AAS), it is much superior for the more refractory metals (for example, V, Be) as well as for some nonmetals such as boron and phosphorus.
3. The remarkably low susceptibility of ICP–AES to interelement interferences, other than spectroscopic interferences, makes calibration for a variety of environmental materials very much easier than it is for alternative atomic emission sources such as the dc arc.
4. The compatibility of the ICP with devices for generating the hydrides of elements such as As, Sb, and Se permits the determination of these elements, often simultaneously, at a much higher sensitivity than is possible with normal sample introduction by pneumatic nebulization.

This chapter presents a review of applications of ICP–AES to the analysis of environmental samples. In this first section, methods of sample dissolution are reviewed and calibration procedures, as well as possible interferences, are discussed. In the subsequent sections, applications to natural waters (including fresh waters and seawater) and waste waters, water quality monitor organisms (for example, fish, molluscs, algae), and freshwater and marine sediments are described. Finally, applications to airborne particulate matter, coal ash and coal fly ash, are reviewed. It is neither possible nor desirable, in a chapter of this length, to provide a comprehensive review, but an attempt has been made to provide an overview of the wide variety of samples that can be analyzed. As the analysis of many solid environmental samples (for example, sediments, airborne particulates) is rather similar to geological applications discussed in Chapter 2 and the analysis of many others (for example, fish, molluscs) is similar to applications described in Chapters 4 and 5, water analysis has been given greater emphasis in this chapter.

3.2. DISSOLUTION OF SOLID SAMPLES

At present at least, accurate ICP–AES analysis of environmental materials other than natural or waste waters requires dissolution of the samples. The choice of an appropriate procedure is complicated by the temptation, on the one hand, to dilute the original sample as little as possible so that the maximum number of trace elements can be determined, and, on the other hand, the desire to avoid dissolved solids concentrations higher than 1–2% in the solution presented to the ICP [1]. Fortunately, the sensitivity of ICP–AES permits the direct determination of many trace elements in environmental materials, even when the dissolution procedure involves a dilution of 100-fold or more.

The dissolution of environmental materials can be accomplished by a variety of techniques that fall into two major categories: acid digestion and fusion [2, 3]. The choice of an optimum method for subsequent multielement analysis is governed by a number of considerations. The method should effect a complete dissolution of all the elements of interest without loss of any as volatile species. Few, if any, of the standard procedures will dissolve all of the possible mineral species in sediments, for example, although the fusion procedures are generally more comprehensive than the acid digestions. The loss of certain elements (for example, Si, B, As, Be, and Sb) as volatile fluorides is a well-known limitation of some acid digestion techniques that use hydrofluoric acid. However, the use of fluxes to effect the dissolution may raise the dissolved solids concentrations in the solutions to inconveniently high levels for the subsequent instrumental determination, necessitating a large dilution. In addition, the fluxes may introduce significant quantities of impurities that raise the blank levels (and hence, the detection limits) for the trace elements. In these latter respects, the acid digestion procedures have definite advantages over the fusion methods, because the fluxes cannot be readily obtained as pure as the acids, and also because excess acids can often be easily removed from the samples by evaporation.

The development of a single procedure that is appropriate for the preparation of a wide variety of environmental materials for subsequent ICP–AES analysis is an even greater challenge. Two such procedures that have been reported exemplify the two general approaches to sample digestion. Representative of acid digestion procedures is a method proposed by McQuaker et al. [4] for the preparation of waters, soils, tissues, and airborne particulates. An open-vessel digestion with either nitric or perchloric acids, or nitric, perchloric, and hydrofluoric acids (depending on the sample type) followed by evaporation to fumes of perchloric acid, dilution, and filtration (if necessary) results in solutions of uniform acid concentration (3.5% $HClO_4$). The presence of a constant known acid concentration in the sample solutions greatly simplifies the calibration procedure [5–7]. Representative of fusion procedures is a method proposed by Floyd and co-workers [8] in which a 0.2-g sample is fused with 2 g of sodium hydroxide in a graphite crucible, after which the melt is dissolved in dilute hydrochloric acid. In this case, it is necessary to add to the calibrating reference solutions an amount of sodium equivalent to that added to the samples as sodium hydroxide. Other variations of these two basic approaches will be encountered in the following sections.

3.3. CALIBRATION PROCEDURES

A feature of the ICP–AES calibrations that greatly facilitates simultaneous determination of major, minor, and trace elements is their linearity over a very wide concentration range (generally five to six orders of magnitude, starting from the detection limit). Thus, the intermediate dilutions typical of sequential multielement analysis by AAS are unnecessary. But the efficient exploitation of this advantage requires careful planning of the preparation of a series of multielement reference solutions for calibration purposes. Also, a more elaborate mathematical analysis of emission intensity data, which normally span several orders of magnitude of concentration, is necessary to arrive at a calibration that is accurate over the entire range.

A number of complications arise in the preparation of a series of multielement reference solutions. The chemical compatibility of the various forms of the elements of interest in a particular acidic medium must be considered to avoid precipitation in the mixed solution. The introduction of significant amounts of the trace analytes as impurities in the major elements must be avoided. The absence of any spectroscopic interferences between the various elements in each reference solution is also very desirable. Problems of this type have been discussed by McQuaker et al. [9] in a report describing a calibration scheme for some 30 elements in environmental materials. It is generally convenient to divide the elements of interest into groups that are chemically compatible and at similar concentrations. For example, in the calibration scheme cited previously, the 30 elements were divided into five groups containing a total of 20 reference solutions. The number of reference solutions for any particular element ranged from five to 15 depending on the necessary concentration range. It was not necessary to run all 20 reference solutions for each calibration; six of them (of which one was the blank) were designated as "normalization standards" for recalibration purposes.

The process of determining an accurate calibration equation from the emission intensity data has been addressed in two publications by Maessen and Balke [10, 11], who pointed out that the apparent analytical advantage contributed by the wide linear dynamic range of ICP–AES is to some extent offset by the fact that the variance of the emission response is not constant over the entire range. This constancy of variance is a fundamental prerequisite for application of least-squares procedures in determining calibration equations, unless weighting factors are used. Typically, the standard deviation of the ICP emission response is constant for the first two to three orders of magnitude of concentration (starting at the detection limit), after which it increases in proportion to the signal intensity (cf. Part 1, Section 4.2.2). In this latter regime, a logarithmic transformation of the calibration data renders the variance constant. Maessen and Balke showed that use of a linear scale over a concentration range spanning four orders of magnitude resulted in large inaccuracies at low concentrations. A logarithmic scale over the entire range gave somewhat better results overall, but best results were obtained by dividing the concentration range in half, with a linear scale for the lower two orders and a logarithmic scale for the upper two.

Thus, the most appropriate calibration procedure depends upon the concentration range that must be spanned, despite the wide linear dynamic range of ICP–AES.

3.4. TYPES OF SPECTROSCOPIC INTERFERENCES ENCOUNTERED WITH ENVIRONMENTAL SAMPLES

Perhaps the greatest contrast between ICP–AES and AAS lies in the domain of interelement interferences; the types of interferences most commonly encountered in each of the two techniques are virtually mutually exclusive. In AAS, interelement interferences are mostly chemical in nature, whereas spectroscopic interferences such as line overlaps are virtually nonexistent. In ICP–AES, just the opposite is true. A series of publications has shown that, with an appropriate choice of plasma operating conditions, chemical interferences are reduced to negligible proportions, at least for trace analysis [1, 12–18], but in turn there is a continuously growing literature describing the nature of spectroscopic interferences in ICP–AES and appropriate procedures for elimination of, or correction for, these interferences [19–28].

In view of the discussion of spectroscopic interferences in Part 1, Chapters 7 and 8, further detailed discussion here would be redundant. Analysis of a variety of complex environmental materials ensures that all the possible types will be encountered sooner or later. The most difficult samples, from this point of view, are the geological materials such as sediments [29, 30]. These materials normally contain high concentrations of aluminum, calcium, and magnesium, which cause significant background shifts in many regions of the spectrum, and also high concentrations of iron, and to a lesser extent, titanium, which can give rise to serious line overlap interferences in the determination of trace elements.

The characterization of, and correction for, various spectroscopic interferences encountered in a wide variety of marine samples have been discussed by McLaren and co-workers [28–30], and their findings are applicable to most other types of environmental materials. A means of scanning across a short region (0.1–0.2 nm) of the spectrum in the vicinity of a proposed analyte line is an invaluable tool in the identification of interferences. Efficient and accurate correction for these interferences normally requires a means of measurement of the background intensity adjacent to the analyte wavelength (preferably on both sides), as well as a means of applying previously determined correction factors to compensate for line overlap interferences [31, 32] (cf. Part 1, Sections 7.5 and 7.6).

3.5. ANALYSIS OF SAMPLES RELATED TO WATER QUALITY MONITORING

3.5.1. Water Samples

3.5.1.1. Fresh Waters

The application of ICP–AES to multielement analysis of fresh waters has been quick to gain acceptance for several reasons:

1. The samples can be introduced directly to the ICP.
2. Spectroscopic interferences are minimal.
3. ICP–AES detection limits permit the direct analysis for many elements of environmental interest at criteria levels for drinking water and irrigation water set by regulatory agencies.

A review of the literature indicates three areas in which difficulties have been encountered:

1. Spectroscopic interferences due to stray light from intense calcium and magnesium lines, particularly in early work with hard water samples.
2. Particulate clogging of certain types of pneumatic nebulizers by unfiltered water samples.
3. Insufficient detection power for the direct determination of some metals of interest.

Winge et al. [33] described water analysis using one of the earliest commercially available ICP spectrometers and also with a custom instrument comprising an ICP source and a 1-m vacuum polychromator. Stray light induced background shifts, arising from scatter from very intense calcium and magnesium lines, were measured for 28 wavelengths. A more recent report [34] describes modifications to a very similar polychromator to reduce stray light interferences. The mechanically ruled grating was replaced with a holographic grating, and interference filters were installed immediately in front of the photomultiplier tube detectors for all wavelengths less than 250 nm. The authors concluded that the installation of the holographic grating was the primary reason for the much improved stray light characteristics of the spectrometer.

It should be noted that the importance of sources of background shifts other than stray light was not fully appreciated before the publication of a landmark paper by Larson and Fassel in 1979 [25] in which line-broadening and radiative recombination background interferences in ICP–AES were described. Of particular interest in the context of water analysis is the very considerable broadening of the wings of intense calcium and magnesium resonance lines such that significant background shifts are observed as much as 10 nm from the peak centers, and the presence of a series of broad Mg triplets in the 257- to 274-nm region, as well as a recombination continuum below about 255 nm. Thus, even with improvements in spectrometer design, including the use of holographic gratings, interference filters, and redesigned baffling, significant background shifts due to true spectroscopic phenomena must be addressed. Various background correction strategies based on rotatable quartz refractor plates [26–28] or movable entrance slits [7] have been developed (cf. Part 1, Chapters 7 and 8).

Problems with accidental clogging of conventional pneumatic nebulizers by suspended particulate matter have led to considerable efforts to produce other types of nebulizers. A number of designs utilizing the Babington principle have been reported [35–42] (cf. Part 1, Chapter 6). The main advantage of such an approach is that the sample does not flow through a small easily clogged orifice. Sensitivity, precision, and detection limits attainable with Babington-type nebulizers appear to be comparable to those obtained with concentric and cross-flow designs. Garbarino and Taylor [41] described the development and optimization of a Babington nebulizer intended specifically for freshwater analysis. These nebulizers also have a greater tolerance for high dissolved solids than cross-flow or concentric nebulizers.

Although ICP–AES detection limits with conventional pneumatic nebulization are in many cases adequate for regulatory purposes, they are not low enough to permit the actual direct determination of many metals of environmental interest in most freshwater samples. A recent report [43] described an evaluation of an ultrasonic nebulization system, with an associated desolvation apparatus, as an alternative to a concentric nebulizer for introduction of freshwater samples to the ICP. The nebulizer used was a modified version of a continuous-type ultrasonic nebulizer developed by Olson et al. [44]. As expected, detection limits attained with this system were five to 10 times lower. Unfortunately, an order of magnitude improvement in detection limits is not always enough. The adequacy of this approach depends on whether it is necessary actually to determine the concentrations of trace metals in the water, or merely to ensure that they lie below certain criteria levels. If an actual determination is required, it will in many cases be necessary to preconcentrate the metals of interest prior to their introduction to the ICP. Goulden and Anthony [45] reported direct determination of 14 trace elements in freshwater samples. After a 10-fold preconcentration in the digestion process, the samples were introduced to the ICP with conventional nebulization, but with a heated spray chamber.

3.5.1.2. Seawater

Despite the rapid acceptance and current popularity of ICP–AES for multielement analysis of freshwater, applications of the technique to seawater analysis have been limited. The very low concentrations of most of the metals of interest in all but heavily polluted samples preclude their direct determination. In addition, the 3.5% dissolved solids concentration typical of seawater is problematic for many of the sample introduction systems and torch configurations used in ICP–AES. Thus, a chemical separation of the elements of interest from the seawater matrix, which also serves to preconcentrate these elements, is normally required. Recent data on the concentrations of elements such as Cu, Cd, Pb, and Zn in uncontaminated seawater indicate that concentration factors of about 500 are necessary to permit efficient multielement trace analysis of seawater by ICP–AES, at least with conventional pneumatic nebulization [46–48]. The use of ultrasonic nebulization with aerosol desolvation, which is known to improve detection limits by approximately an order of magnitude [44], will lower the required concentration factor to about 50. The challenge, then, in applying ICP–AES to seawater analysis is to develop a preconcentration procedure that is sufficiently rapid to avoid offsetting the speed advantage of ICP–AES compared with slower, but more sensitive techniques such as graphite furnace atomic absorption spectrometry and anodic stripping voltammetry.

Preconcentration of trace metals from seawater or artificial seawater for subsequent ICP–AES determinations has been accomplished by liquid–liquid extraction, coprecipitation, ion-exchange, and adsorption procedures [46–52]. Liquid–liquid extractions are attractive because of their speed and simplicity, but the maximum concentration factors possible are often insufficient. Sugimae [49] reported determination of Fe, Mn, Cu, Ni, Zn, and V in polluted seawater following a preconcentration of 20- to 40-fold achieved by complexation of the metals with diethyldithiocarbamate (DDTC) followed by extraction into chloroform. The combined chloroform extracts were evaporated, and the residue was wet-ashed with a mixture of nitric and hydrochloric acids, and then dissolved in dilute nitric acid. A similar procedure reported by McLeod et al. [46] achieved higher concentration factors (250–500) by redissolution of the residue in a much smaller volume of acid. Simultaneous determination of Fe, Cd, Cu, Mo, Ni, V, and Zn in open sea samples was accomplished. This higher concentration factor is more appropriate for typical seawater samples, but is achieved only at the expense of a time-consuming evaporation of the chloroform extracts to dryness.

Coprecipitation of Cr(III), Mn, Co, Ni, Cu, Cd, and Pb from seawater samples with indium hydroxide was suggested by Hiraide et al. [50] as an effective preconcentration technique for subsequent ICP–AES analysis. For 1.2-L samples, the procedure afforded a concentration factor of 240. Essentially quantitative recovery of the seven metals from artificial seawater, albeit at concentrations considerably above normal levels, was demonstrated. Significant quantities of alkali metals and alkaline earths were also precipitated, but these were reported not to hinder the ICP–AES determination of the trace metals.

The simultaneous determination of Fe, Mn, Cu, Ni, and Zn in coastal seawater by ICP–AES after preconcentration on a chelating ion-exchange resin (Chelex-100) was demonstrated by Berman et al. [47]. Concentration factors of 25 proved to be sufficient for these elements when ultrasonic nebulization with aerosol desolvation was employed, but, even with concentration factors of up to 100, Cd, Co, and Pb could not be determined. Although the Chelex-100 separation allows higher concentration factors than does liquid–liquid extraction without a subsequent evaporation, the method is rather time-consuming (sample throughput rate is 1–2 $mL \cdot min^{-1}$) and the removal of the calcium and magnesium from the resin prior to elution of the trace metals requires careful washing procedures.

Another method of preconcentration that shows great potential for seawater analysis is the adsorption of the metals, or complexes of the metals, onto an inert stationary phase from which they can subsequently be readily removed. Watanabe et al. [51] demonstrated this approach by concentration of several trace metals from coastal seawater by complexation with 8-hydroxyquinoline followed by adsorption on C_{18}-bonded silica gel. After passage of the seawater sample, and washing of the column, the metal complexes were eluted with methanol. Following evaporation of the methanol and wet-ashing of the residue, the metals were dissolved in 0.1 *M* nitric acid. A concentration factor of 200 was obtained for a 1-L seawater sample, permitting the determination of several metals by ICP–AES with pneumatic nebulization. A more detailed investigation of this technique was reported by Sturgeon et al. [52], who subsequently developed a modification that eliminates the need to add 8-hydroxyquinoline to the seawater initially and permits the elution of the metals from the column with an aqueous acid mixture [48]. The seawater, after pH adjustment, is passed over a column of silica-immobilized 8-hydroxyquinoline at a flow rate of 15–20 $mL\ min^{-1}$. After washing of the column, the metals are eluted with a mixture of 1 *M* HCl/0.1 *M* HNO_3. Concentration factors of 200 or more are feasible for 1- to 2-L seawater samples.

Although considerable progress has been made in the analysis of seawater by ICP–AES, inadequate detection power continues to be a problem for many samples. As no "breakthrough" that would improve detection limits by an order of magnitude appears imminent, the emphasis on future work will probably be on improved preconcentration techniques.

3.5.1.3. *Waste Waters*

The elevated levels of metals and other pollutants in many industrial discharge waters and municipal waste waters permits their direct determination by ICP–AES without a prior concentration. In addition, the determination of boron and phosphorus may be possible. In some cases, however, the high dissolved solids concentration may necessitate a prior dilution.

Broekaert and Leis [53] reported the direct determination of boron and eight metals in a variety of industrial waste waters. Deleterious effects of high salt concentrations (up to 2%) in some of the samples were circumvented by substitution of a rudimentary flow-injection technique for the normal continuous solution aspiration. Sample aliquots of up to 500 μL were injected into the nebulizer uptake tube by means of a peristaltic pump. Despite a twofold to 10-fold deterioration of detection limits compared with those for continuous nebulization, multielement analysis of effluents from steel plants and zinc-plating plants could be achieved.

Ishizuka et al. [54] reported the determination of phosphorus in municipal waste waters by ICP–AES as an attractive alternative to colorimetric or indirect atomic absorption procedures. It was demonstrated that differences in emission response for various inorganic and organic forms of phosphorus are small. A detection limit of 20 $\mu g \cdot L^{-1}$ was attained with the most sensitive line (213.618 nm), but it was noted that a serious spectroscopic interference by Cu rendered this line unusable for some samples.

3.5.1.4. *Preconcentration Techniques for Water Analysis*

A review of all the metal preconcentration techniques that are potentially compatible with ICP–AES for water analysis is certainly beyond the scope of this section, but several recently reported methods involving a combination with ICP–AES will be reviewed briefly.

Siggia and co-workers [55,56] have synthesized several chelating ion-exchange resins for trace metal separation from urine, and fresh and salt waters. The concentration of a number of trace metals in urine for subsequent ICP–AES determination, using a poly(dithiocarbamate) resin was demonstrated by Barnes and Genna [57]. A concentration factor of 125 was possible for 250-mL samples, at a sample throughput rate of 2.5 $mL \cdot min^{-1}$. Complete dissolution of the resin in 1:1 nitric acid/sulfuric acid was necessary for efficient metal recovery. Exclusion of the alkali and alkaline earth metals, but also iron and manganese, was noted. Colella et al. [56] demonstrated preconcentration of trace metals from natural waters using a poly(acrylamidoxime) resin. In this case, a 24-h batch equilibration of 1 L of water with 0.5 g of the resin was followed by recovery of the metals by 3 h immersion in 6 *M* hydrochloric acid. Concentration of Fe, Cu, Cd, Pb, and Zn from seawater and pondwater samples was described, although the metal concentrations in the former were certainly far above normal values, even for a coastal sample [47].

A convenient means of preconcentration for elements such as As, Sb, Bi, Se, and Te is available via generation of their gaseous hydrides, usually formed by reaction of an acidified sample solution with sodium borohydride (see Part 1, Section 6.5). This technique was developed in the early 1970s for use with AAS. Its adaptation to ICP–AES was pioneered by Thompson and co-workers, starting in 1978 [58–61]. The greatly improved detection limits available for the hydride-forming elements via this route permit their otherwise impossible determination in a wide variety of environmental materials, including water samples [61–67] (cf. Part 1, Section 4.1.7.6).

Goulden et al. [66] reported a method by which simultaneous determination of As, Se, Sb, Sn, and Bi in river and lake waters is possible. The technique involves an initial fourfold preconcentration and persulfate oxidation of the water sample followed by hydride generation in an automated system. Reported detection limits for As, Se, Sb, Sn, and Bi were 0.02, 0.03, 0.02, 0.05, and 0.05 $\mu g \cdot L^{-1}$ respectively. The accuracy of the technique for arsenic and selenium determinations in reference soils and sediments was demonstrated, but no data (other than the detection limits) were given for water analysis. In a recent publication, Nygaard and Lowry [67] describe a procedure for simultaneous determination of arsenic, antimony, and selenium in a variety of environmental materials; data were presented for U.S. National Bureau of Standards and U.S. Environmental Protection Agency reference water samples, as well as for a variety of waste waters.

3.5.2. Water Quality Monitor Organisms

In view of the very low concentrations of many trace metals of environmental interest in natural waters, and the difficulties associated with their accurate determination, it is fortunate that organisms such as fish, molluscs, and phytoplankton, which tend to concentrate these metals by various physical and biochemical processes, can often be used as indirect indicators of water quality.

The preparation of biological tissues for ICP–AES analysis involves either a dry-ashing at 400–500°C or a wet-ashing with one or more mineral acids or hydrogen peroxide [2]. Wet-ashing can be performed in either open or closed vessels. McQuaker et al. [4] recommended the use of a mixture of nitric and perchloric acids for the digestion of tissue samples, with the addition of hydrofluoric acid only when destruction of a silicate matrix is necessary. Similar procedures were used by McLaren and co-workers [28,29] for the dissolution of various freeze-dried marine tissues, and by McKinney and Schlict [68] for analysis of fish tissues.

Once samples of this type have been dissolved, their analysis by ICP–AES is usually quite straightforward, provided that concentrations of the metals of interest are not too low. Dissolved solids concentrations are normally low, as the bulk of the sample is lost as carbon dioxide during the digestion, and spectroscopic interferences are minimal. For some elements (for example, Pb, As, Sb, Se) a preconcentration step may be necessary. Hydride generation procedures, referred to in Section 3.5.1.4, can readily be adapted to determination of As and Se in fish [66].

3.5.3. Freshwater and Marine Sediments

The trace metal concentrations of freshwater and marine sediments can also be used as indicators of water quality. Even with a dilution factor of 100-fold or more in the dissolution procedure, most trace elements of interest can be directly determined by ICP–AES, provided that proper attention is paid to a variety of spectroscopic interferences that arise when trace elements are determined in the presence of much higher concentrations of iron, aluminum, calcium, magnesium, and titanium.

As noted in Section 3.2, either acid digestion or fusion techniques can be used for dissolution of sediments. For determination of trace metals of environmental interest, acid digestion procedures are strongly favored [69–72]. A procedure developed by McLaren and co-workers [30] permitted the direct determination of six major and minor elements and eight trace elements in near-shore marine sediments by ICP–AES. Dissolution of the samples was achieved with a mixture of nitric, perchloric, and hydrofluoric acids by heating in sealed Teflon vessels. The method was not suitable for determination of As, Cd, and Mo in these materials because of inadequate sensitivity, nor for Cr and Ti because of incomplete dissolution of these metals.

Hydride generation procedures (discussed in Section 3.5.1.4 and Part 1, Section 6.5) can be applied to analysis of sediments. Pahlavanpour et al. [61] reported the simultaneous determination of As, Sb, and Bi in soils and sediments by hydride generation after digestion with concentrated hydrochloric acid at 150° in sealed tubes. Detection limits of about 0.1 $\mu g/g$ for all three analytes were obtained. This procedure did not permit the simultaneous determination of Se, for which a separate method was developed [62]. In contrast, a report by Nygaard and Lowry [67] described a method permitting simultaneous determination of As, Sb, and Se in environmental materials with solution detection limits of 1, 3, and 1 $\mu g \cdot L^{-1}$, respectively.

3.6. ANALYSIS OF SAMPLES RELATED TO AIR QUALITY MONITORING

3.6.1. Airborne Particulate Matter

The application of ICP–AES to air quality monitoring has been principally in the area of airborne particulates. Samples are normally collected by passage of a known volume of air through a suitable filter. Variations of this technique are used both in industrial workplaces to safeguard occupational health, and also outdoors to monitor atmospheric discharges of smelters and coal-fired electric generating stations. This latter type of monitoring has become especially important in recent years with renewed interest in coal as a relatively cheap and abundant fuel source.

Two detailed reports, describing collection and ICP–AES analysis of airborne particulate matter in Japan and the United States, respectively, give an excellent overview of this area and indicate the difficulties encountered [73,74].

Sugimae [73] described the state of the art for air particulate monitoring in Japan. All aspects of the analytical procedure, including filter selection, air sampling, filter digestion, and ICP–AES analyses, were reviewed. Samples were taken with both high-volume (1500 L $\cdot$ min^{-1}) and low-volume (20 L $\cdot$ min^{-1}) equipment. For the former, 8 × 10-in. glass fiber filters were employed, whereas, for the latter, circular filters of 47 or 110 mm were used. Glass fiber filters were preferred to cellulose or polycarbonate filters, even though it was realized that the relatively high and variable levels of metallic impurities in glass filters is the limiting factor for determining detection limits.

Various procedures for removing the particulate metals from the glass filters were evaluated by Sugimae. These included several mixed acid digestions, nitric acid/hydrogen peroxide digestion after low-temperature ashing, and sodium carbonate fusion after pretreatment of the filter with a mixture of hydrofluoric and sulfuric acids in a platinum crucible. Only the latter procedure resulted in total dissolution of the filter, but this also liberated acid-insoluble metallic impurities, thus raising the blanks. Ultimately, a nitric acid/hydrogen peroxide digestion was chosen as the standard technique, although it was recognized that this method did not achieve complete dissolution of two of the nine metals studied (beryllium and chromium).

Very few interferences were encountered in the subsequent ICP–AES analyses, the exceptions being spectroscopic interferences by iron on the Be(I) 234.861 and Cr(II) 283.563-nm lines, and by magnesium on the Zn(II) 202.548-nm line.

An equally detailed report by Olson et al. [74], describing procedures developed at the U.S. Department of Energy, Ames Laboratory, provides some interesting contrasts to the Japanese methodology. Samples were collected on polycarbonate filters (Nuclepore Corporation, 0.8-μm pore size, 25-mm diam.). These filters are well known in environmental circles for their relatively low and constant blank levels. The digestion procedure, which involves wet-ashing with a mixture of sulfuric acid and 50% hydrogen peroxide, results in total dissolution of the filter and its contents. The dissolved and diluted particulates were analyzed for nine trace metals by ICP–AES with ultrasonic nebulization and aerosol desolvation. The report also contains a description of an ingenious procedure for preparing "reference" or "standard" filters with known loadings of the metals of interest.

Two other publications link ICP–AES directly to environmental air monitoring. Lynch and co-workers [75] described the use of ICP-AES analysis of air particulates collected on glass fiber filters to monitor emissions from a lead–zinc smelter. A brief report by Zimmerman et al. [76] describes analysis of airborne particulate matter for industrial hygiene monitoring.

In summary, it appears that ICP–AES can be readily applied to analysis of airborne particulate matter and that the principal problems that must be overcome (representative sampling, and control of the analytical blank) are not strongly dependent on the method used for the measurement of the metal concentrations.

3.6.2. Coal and Coal Fly Ash

Since coal-fired electric generating stations represent a considerable source of atmospheric pollution, there has been growing interest in recent years in the composition of coal and coal fly ash.

A detailed description of a multitechnique, multielemental analysis of coal and fly ash has been reported by Nadkarni [77]. ICP–AES plays a major role in this scheme, as it is used for determination of eight major and 20 trace elements. The dissolution of coal samples is a two-step process. First the coal is reduced to an ash by either high-temperature ashing or low-temperature oxygen plasma ashing. The coal ash (or fly ash) is then dissolved by a lithium tetraborate fusion for determination of the major elements or by a mixed acid (aqua regia/hydrofluoric acid) digestion in a sealed Teflon vessel ("Parr" bomb) at 110°C for determination of the trace elements. In the latter case, it is occasionally necessary to filter off some undissolved carbon particles prior to dilution to volume.

The instrument used for the subsequent multielement determinations by ICP–AES has been described by Botto [31]. More recently Botto has described the addition of a second polychromator with higher resolving power and a more flexible background correction system to the same instrument [78]. This modification very significantly reduced the number and magnitude of serious spectroscopic interferences and permitted the accurate determination of several additional trace elements of environmental importance (cf. Part 1, Section 7.7.7).

REFERENCES

1. P. W. J. M. Boumans and F. J. de Boer, *Spectrochim. Acta* **32B,** 365 (1977).
2. T. T. Gorsuch, *The Destruction of Organic Matter*, Pergamon Press, Oxford (1970).
3. W. M. Johnson and J. A. Maxwell, *Rock and Mineral Analysis*, *Chemical Analysis Series*, Vol. 27, Chap. 4. Interscience, New York (1981).
4. N. R. McQuaker, D. F. Brown, and P. D. Kluckner, *Anal. Chem.* **51,** 1082 (1979).
5. S. Greenfield, H. McD. McGeachin, and P. B. Smith, *Anal. Chim. Acta* **84,** 67 (1976).
6. S. Greenfield, H. McD. McGeachin, and F. A. Chambers, *ICP Information Newslett.* **3,** 117 (1977).
7. R. L. Dahlquist and J. W. Knoll, *Appl. Spectrosc.* **32,** 1 (1978).
8. M. A. Floyd, V. A. Fassel, and A. P. D'Silva, *Anal. Chem.* **52,** 2168 (1980).
9. N. R. McQuaker, P. D. Kluckner, and G. N. Chang, *Anal. Chem.* **51,** 888 (1979).
10. F. J. M. J. Maessen and J. Balke, in R. M. Barnes, ed., *Developments in Atomic Plasma Spectrochemical Analysis*, Heyden, Philadelphia (1981) p. 128.
11. F. J. M. J. Maessen and J. Balke, *Spectrochim. Acta* **37B,** 37 (1982).
12. G. F. Larson, V. A. Fassel, R. H. Scott, and R. N. Kniseley, *Anal. Chem.* **47,** 238 (1975).
13. G. F. Larson and V. A. Fassel, *Anal. Chem.* **48,** 1161 (1976).
14. D. J. Kalnicky, V. A. Fassel, and R. N. Kniseley, *Appl. Spectrosc.* **31,** 137 (1977).
15. M. H. Abdallah, J. M. Mermet, and C. Trassy, *Anal. Chim. Acta* **87,** 329 (1976).
16. H. Kawaguchi, T. Ito, K. Ota, and A. Mizuike, *Spectrochim. Acta* **35B,** 199 (1980).
17. S. R. Koirtyohann, J. S. Jones, C. P. Jester, and D. A. Yates, *Spectrochim. Acta* **36B,** 49 (1981).
18. M. W. Blades and G. Horlick, *Spectrochim. Acta* **36B,** 881 (1981).
19. P. W. J. M. Boumans, *Spectrochim. Acta* **35B,** 57 (1980).
20. J. M. Mermet and C. Trassy, *Spectrochim. Acta* **36B,** 269 (1981).
21. A. Batal and J. M. Mermet, *Spectrochim. Acta* **36B,** 993 (1981).
22. R. W. Winge, V. A. Fassel, V. J. Peterson, and M. A. Floyd, *Appl. Spectrosc.* **36,** 210 (1982).
23. G. F. Larson, V. A. Fassel, R. K. Winge, and R. N. Kniseley, *Appl. Spectrosc.* **30,** 384 (1976).
24. V. A. Fassel, J. M. Katzenberger, and R. K. Winge, *Appl. Spectrosc.* **33,** 1 (1979).
25. G. F. Larson and V. A. Fassel, *Appl. Spectrosc.* **33,** 592 (1979).

26. R. K. Skogerboe, P. J. Lamothe, G. J. Bastiaans, S. J. Freeland, and G. N. Coleman, *Appl. Spectrosc.* **30,** 495 (1976).
27. S. R. Koirtyohann, E. D. Glass, E. J. Hinderberger, and F. E. Lichte, *Anal. Chem.* **49,** 1121 (1977).
28. J. W. McLaren and S. S. Berman, *Appl. Spectrosc.* **35,** 403 (1981).
29. S. S. Berman, J. W. McLaren, and D. S. Russell, in R. M. Barnes, ed., *Developments in Atomic Plasma Spectrochemical Analysis*, Heyden, Philadelphia (1981) p. 586.
30. J. W. McLaren, S. S. Berman, V. J. Boyko, and D. S. Russell, *Anal. Chem.* **53,** 1802 (1981).
31. R. I. Botto, in R. M. Barnes, ed., *Developments in Atomic Plasma Spectrochemical Analysis*, Heyden, Philadelphia (1981) p. 141.
32. R. I. Botto, *Anal. Chem.* **54,** 1654 (1982).
33. R. K. Winge, V. A. Fassel, R. N. Kniseley, E. De Kalb, and W. J. Haas, Jr., *Spectrochim. Acta* **32B,** 327 (1977).
34. C. E. Taylor and T. L. Floyd, *Appl. Spectrosc.* **34,** 472 (1980).
35. P. J. McKinnon, K. C. Giess, and T. V. Knight, in R. M. Barnes, ed., *Developments in Atomic Plasma Spectrochemical Analysis*, Heyden, Philadelphia (1981) p. 287.
36. R. C. Fry and M. B. Denton, *Anal. Chem.* **49,** 1413 (1977).
37. R. C. Fry and M. B. Denton, *Appl. Spectrosc.* **33,** 393 (1979).
38. J. F. Wolcott and C. B. Sobel, *Appl. Spectrosc.* **32,** 591 (1978).
39. R. F. Suddendorf and K. W. Boyer, *Anal. Chem.* **50,** 1769 (1978).
40. B. Thelin, *Analyst* **106,** 54 (1981).
41. J. R. Garbarino and H. E. Taylor, *Appl. Spectrosc.* **34,** 584 (1980).
42. L. R. Layman and F. E. Lichte, *Anal. Chem.* **54,** 638 (1982).
43. C. E. Taylor and T. L. Floyd, *Appl. Spectrosc.* **35,** 408 (1981).
44. K. W. Olson, W. J. Haas, Jr., and V. A. Fassel, *Anal. Chem.* **49,** 632 (1977).
45. P. D. Goulden and D. H. J. Anthony, *Anal. Chem.* **54,** 1678 (1982).
46. C. W. McLeod, A. Otsuki, K. Okamoto, H. Haraguchi, and K. Fuwa, *Analyst* **106,** 419 (1981).
47. S. S. Berman, J. W. McLaren, and S. N. Willie, *Anal. Chem.* **52,** 488 (1980).
48. R. E. Sturgeon, S. S. Berman, S. N. Willie, and J. A. H. Desaulniers, *Anal. Chem.* **53,** 2337 (1981).
49. A. Sugimae, *Anal. Chim. Acta* **121,** 331 (1980).
50. M. Hiraide, T. Ito, M. Baba, H. Kawaguchi, and A. Mizuike, *Anal. Chem.* **52,** 804 (1980).
51. H. Watanabe, K. Goto, S. Taguchi, J. W. McLaren, S. S. Berman, and D. S. Russell, *Anal. Chem.* **53,** 738 (1981).
52. R. E. Sturgeon, S. S. Berman, and S. N. Willie, *Talanta* **29,** 167 (1982).
53. J. A. C. Broekaert and F. Leis, *Anal. Chim. Acta* **109,** 73 (1979).

54. I. Ishizuka, K. Nakajima, and H. Sunahara, *Anal. Chim. Acta* **121,** 197 (1980).
55. D. S. Hackett and S. Siggia, in *Environmental Analysis*, Academic Press, New York (1977) p. 253.
56. M. B. Colella, S. Siggia, and R. M. Barnes, *Anal. Chem.* **52,** 2347 (1980).
57. R. M. Barnes and J. S. Genna, *Anal. Chem.* **51,** 1065 (1979).
58. M. Thompson, B. Pahlavanpour, S. J. Walton, and G. F. Kirkbright, *Analyst* **103,** 568 (1978).
59. M. Thompson, B. Pahlavanpour, S. J. Walton, and G. F. Kirkbright, *Analyst* **103,** 705 (1978).
60. M. Thompson and B. Pahlavanpour, *Anal. Chim. Acta* **109,** 251 (1979).
61. B. Pahlavanpour, M. Thompson, and L. Thorne, *Analyst* **105,** 756 (1980).
62. B. Pahlavanpour, J. H. Pullen, and M. Thompson, *Analyst* **105,** 274 (1980).
63. M. Ikeda, J. Nishibe, S. Hamada, and R. Tujino, *Anal. Chim. Acta* **125,** 109 (1981).
64. T. Nakahara, *Anal. Chim. Acta* **131,** 73 (1981).
65. J. A. C. Broekaert and F. Leis, *Fresenius Z. Anal. Chem.* **300,** 22 (1980).
66. P. D. Goulden, D. H. J. Anthony, and K. D. Austen, *Anal. Chem.* **53,** 2027 (1981).
67. D. D. Nygaard and J. H. Lowry, *Anal. Chem.* **54,** 803 (1982).
68. G. L. McKinney and G. R. Schlict, *Jarrell-Ash Plasma Newslett.* **3** (4), 4 (1980).
69. R. T. T. Rantala and D. H. Loring, *At. Absorption Newslett.* **12,** 97 (1973).
70. R. T. T. Rantala and D. H. Loring, *At. Absorption Newslett.* **14,** 117 (1975).
71. H. Agemian and A. S. Y. Chau, *Anal. Chim. Acta* **80,** 61 (1975).
72. S. A. Sinex, A. Y. Cantillo, and G. R. Helz, *Anal. Chem.* **52,** 2342 (1980).
73. A. Sugimae, *ICP Information Newslett.* **6,** 619 (1981).
74. K. W. Olson, W. J. Haas, Jr., and V. A. Fassel, in *Analyses of Airborne Particulates and Human Urine by Inductively Coupled Plasma-Atomic Emission Spectrometry*, NIOSH Publication No. 79-110 (1978).
75. A. J. Lynch, N. R. McQuaker, and D. F. Brown, *J. Air Pollut. Control Assoc.* **30,** 257 (1980).
76. R. Zimmerman, T. Burk, and O. Hoitomt, *Jarrell-Ash Plasma Newslett.* **2** (3), 6 (1979).
77. R. A. Nadkarni, *Anal. Chem.* **52,** 929 (1980).
78. R. I. Botto, *Spectrochim. Acta* **38B,** 129 (1983).

CHAPTER

4

APPLICATIONS: AGRICULTURE AND FOOD

B. L. SHARP

The Macaulay Institute for Soil Research
Aberdeen, Scotland

4.1. Introduction

4.2. Sample Types and Typical Concentration Ranges

- 4.2.1. Soils
- 4.2.2. Plant Materials
- 4.2.3. Fertilizers
- 4.2.4. Foods and Feedstuffs
- 4.2.5. Animal Tissues and Body Fluids

4.3. Sample Preparation Procedures

- 4.3.1. Introduction
- 4.3.2. Sampling and Sample Handling
- 4.3.3. Dry-Ashing/Dissolution
- 4.3.4. Wet-Ashing
- 4.3.5. Fusion/Dissolution
- 4.3.6. Extractable Soil Contents
- 4.3.7. Preconcentration Techniques
  - *4.3.7.1. Extraction Methods*
  - *4.3.7.2. Ion Exchange*
  - *4.3.7.3. Hydride Generation*

4.4. Speciation

4.5. Instrumentation and Technique Capability

- 4.5.1. Power of Detection
- 4.5.2. Comparative Merits of Single- and Multichannel Spectrometers
- 4.5.3. Nebulizers and Sample Introduction
- 4.5.4. Interference and Optimization

4.6. Conclusions

References

4.1. INTRODUCTION

Agriculture is an excellent example of an area in which scientific advances can be shown historically to have followed the availability of analytical techniques.

Today, however, the solution of agricultural problems is much less limited by analytical inadequacies because the advent of atomic absorption spectrometry (AAS) has solved many of the problems of elemental analysis. Satisfactory methods exist for the majority of metals. The problems arise from the ever-increasing demand for analytical data. Hitherto, samples were presented with a request for the determination of usually < 5 elements. The same samples today often require analysis for 10 or more elements. This reflects the greater understanding of the role of elemental content in biological systems, although it remains doubtful if all the information requested can yet be put to good use. The application of spectroscopic techniques to speciation studies is an important advance, because the active components of the system are determined by this approach and this will undoubtedly change the nature of the analytical task.

The introduction of the inductively coupled plasma (ICP) to agricultural analysis will not, for the reasons given, have the same revolutionary impact as the introduction of AAS. It will, however, ease the burden on the analyst and thereby enable an improved analytical service to be offered. Much of the basic work necessary for the implementation of the ICP as a routine analytical technique has already been carried out and satisfactory commercial systems are available. Realization of the full potential of the technique, however, awaits the development of universal and reliable sample preparation and sample introduction procedures. There is a vast, and still increasing, literature on these topics in relation to AAS, and it is a sign of the growing maturity of the ICP technique that the recent trend in the literature has been to concentrate on these subjects. The analysis of agricultural materials for elemental content embraces most sample matrix types, with the exception of those derived from metallurgical products and industrial chemicals. The number of elements of interest is fortunately limited to those that have known biological functions or are of importance because of their toxicity to plant or animal life. Therefore, it is appropriate to begin an assessment of the ICP for agricultural analysis by a discussion of the various sample types and an approximate indication of their elemental content.

4.2. SAMPLE TYPES AND TYPICAL CONCENTRATION RANGES

The most common sample types encountered in agricultural elemental analysis are soils, plant materials, fertilizers, food, animal feedstuffs, animal tissues, and body fluids. Normally, determinations are made of the total contents of small groups of elements within each material. Soils are also analyzed for their extractable or available content, as this is indicative of the ability of a soil to supply nutrients for plant growth.

4.2.1. Soils

The inorganic or mineral content of soil is produced by weathering and erosion of the parent rocks. The principal rock types are of igneous and metamorphic origin, composed largely of silicates and oxides, or of sedimentary origin. The latter includes sandstones consisting mainly of quartz or other resistant silicates or of limestones composed of carbonate material. A few metamorphic rocks such as marble, formed by the alteration of limestones, are low in silicate. In the process of soil formation, secondary minerals are formed, the most important of which are the oxides of aluminum and iron and the layer-silicate clay minerals.

The organic matter content of the soil is produced by the decomposition of plant and animal tissues. The important processes involved are humification, nitrogen transformation, and the translocation of material by the various soil organisms. Compared with the processes responsible for mineral formation, biological transformation is rapid and usually occurs on a time scale of 1–10 y. The organic fraction of the soil is far less well characterized than the mineral fraction, not surprisingly, in view of its amorphous character and the complexity of the compounds contained therein. Humus represents the "final" state attained by organic matter in the soil and this has been shown to be comprised of heterogeneous polymers that are formed from carbohydrates, polyphenols, amino acids, and other substances. The principal components, other than C, H, and O, of the organic fraction are Ca, K, Mg, Na, N, P, and S. Few data are available on the total elemental content of the organic fraction as it is virtually impossible to differentiate the mineral and organic components definitively. In any event, most determinations are carried out on the extractable content, which is more relevant for agricultural purposes. Nevertheless, for a soil composed almost entirely of organic matter, that is, peat, typical concentrations of the principal components are [1]: Ca, 0.1–4.5%; K, 0.008–0.039%; Mg, 0.058–0.36%; N, 1.0–2.65%; Na, 0.023–0.099%; and P, 0.011–0.59%, all expressed as percent totals of oven dry weight.

The total elemental content of soils can be determined directly on the solid by d.c. arc emission spectrometry, X-ray fluorescence, and spark source mass spectrometry or by a solution technique following dissolution of the matrix by a fusion or acid digestion. The determination of available content is carried out following extraction with a reagent for which there has been demonstrated a correlation between the concentration in the extract and the plant uptake of a particular element. A variety of extractants is used, depending on local soil type and conditions and on the element involved. Common extractants are EDTA, acetic acid, water, and ammonium acetate. Table 4.1 summarizes the total contents of various elements in soils and the contents extractable with EDTA and acetic acid [2].

Table 4.1. Contents of Various Elements in Soils[a]

Element	Total Content: Soil: Typical Concentration $\mu g \cdot g^{-1}$	Total Content: Analysis[b] solution ($\mu g \cdot mL^{-1}$)	Extractable Content[c] (normal range; $\mu g \cdot mL^{-1}$ in extract: 0.05 *M* EDTA	Extractable Content[c]: 0.5 *M* Acetic Acid
Ag	0.1	0.001		
As	6	0.006		
B	10	0.01		
B				0.005–5[d]
Ba	1000	1		
Be	6	0.006		
Cd	0.5	0.0005		
Cd			<0.002–0.06	<0.0003–0.008
Co	15	0.015		
Co			<0.01–0.8	<0.001–0.05
Cr	100	0.1		
Cr			0.02–0.8	<0.0003–0.03
Cu	20	0.02		
Cu			0.06–2	<0.001–0.08
Ga	25	0.025		
Ge	1	0.001		
Hg	0.1	0.0001		
Li	50	0.05		
Mn	800	0.8		
Mn			1–20	0.1–2.5
Mo	2	0.002		
Mo			<0.006–0.2	<0.0001–0.0008
Ni	50	0.05		
Ni			0.04–1.0	0.003–0.1
Pb	20	0.02		
Pb			0.2–2	<0.00005–0.1
Rb	100	0.1		
Sc	8	0.008		
Se	0.5	0.0005		
Sn	3	0.003		
Sn			<0.004–0.2	<0.0005–0.005
Sr	300	0.3		
Sr				0.005–0.25
Ti	4000	4		
Ti			0.1–2	<0.003–0.03
V	100	0.1		
V			0.04–1	<0.001–0.03
Y	40	0.04		
Zn	50	0.05		
Zn			<0.6–4	<0.05–0.75
Zr	400	0.4		

[a]Table extracted from data tabulated by Ure [2]. (Data reproduced by permission of Dekker, New York.)

[b]$LiBO_2$ fusion HNO_3 dissolution: 100 mg soil, 100 ml solution, i.e., 1000× dilution.

[c]EDTA extract: 15 g soil, 75 mL extract, i.e., 5× dilution. Acetic acid extract: 20 g soil, 800 mL extract, i.e., 40× dilution.

[d]Hot water extraction: 20 g soil, 40 mL extract, i.e., 2× dilution.

4.2.2. Plant Materials

Sixteen elements have been positively identified as being essential for plant growth; these include the macroelements C, Ca, H, K, Mg, N, O, P, and S and the microelements B, Cl, Cu, Fe, Mn, Mo, and Zn. Plant tissues are comprised of carbohydrates, fats, proteins, nucleoproteins, and the enzymes that enable the plant system to function. C, H, N, O, P, and S are used in the tissue structures; Ca, Cl, K, and Mg are used principally for the electrolyte supply; and the micro- or trace elements are used for building the enzymes. Many other elements are found in plants, although their presence is regarded as a consequence of the growth environment rather than as essentials to physiological functioning.

Plants are grown for food and, therefore, to the list of endogenous plant elements that it is necessary to determine must be added those that are essential for supporting animal life and those that represent a toxicity hazard. Fourteen trace elements have been identified as being essential to support higher forms of animal life, namely Co, Cr, Cu, F, Fe, I, Mn, Mo, Ni, Se, Si, Sn, V, and Zn. It has been suggested that As, Br, Cd, Ga, and Ge might also be necessary for some forms of life [3]. Finally, Cd, Hg, Pb, and Tl must be added to the list because of their potential toxicity and their widespread distribution in the environment. Although the presence of many trace elements is essential to plant and animal welfare, it is also the case that many can only be tolerated in a narrow concentration range. For example, in the case of B, sugar beet should contain a minimum of 20 $\mu g \cdot g^{-1}$ (dry matter content) whereas amounts greater than 200 $\mu g \cdot g^{-1}$ are toxic in cereals [4].

Table 4.2 shows the common concentration ranges of various elements in plant materials (agricultural crops) expressed as $\mu g\ g^{-1}$ in dry matter.

4.2.3. Fertilizers

Modern inorganic fertilizers are largely manufactured products of carefully controlled composition used for supplying N, P and K, Ca, and Mg and also for supplementing the trace element supply. Assay of most of these materials is straightforward, using recommended or official methods. The trace element composition of inorganic fertilizers varies considerably depending on the source of the raw material and the manufacturing process.

Organic manures such as farmyard manure, liquid manure, seaweed, fish meal, composts, sewage sludges, and so forth, tend to be used mostly for small-scale intensive production where the greater costs and handling difficulties are offset by the higher unit crop values. Trace element composition is generally similar to that of the plant materials or animal excrement from which the manures are derived. Sewage sludges and municipal composts are an important ex-

Table 4.2. Contents of Various Elements in Plant Materials

Element	Common Level[a] ($\mu g \cdot g^{-1}$)	Reference
Ag	0.02–0.6	5
Al	10–1000	4
As	0.01–1.5	5
B	1–100	4
Ba	10–220	6
Ca	200–3000	4
Cd	0.05–0.9	5
Co	0.02–0.5	4
Cr	0.1–0.5	6
Cu	1–25	4
Fe	20–500	4
Hg	0.013–0.17	5
I	0.07–10	5
K	1000–5000	4
Li	0.8–1.3	5
Mg	500–10000	4
Mn	5–500	4
Mo	0.2–10	4
Na	100–10000	4
Ni	0.1–20	4
P	200–17900	7
Pb	0.1–5	4
S	200–19800	7
Se	0.001–0.5	5
Si	200–8000	5
Sn	0.2–6.8	5
Sr	4–20	6
Ti	0.1–20	4
V	0.3–0.7	6
W	0.01–0.06	5
Zn	5–200	4

[a]Content expressed as $\mu g \cdot g^{-1}$ in dry matter.

ception, however, because of their contamination with toxic elements from industrial effluent and urban wastes.

Waste materials or industrial by-products, including metal slags, frits, coal ash, fly ash, soot, and sulfur, are also used as fertilizers. As with sewage sludges, such materials may be heavily contaminated and trace element screening should be a prerequisite for their use.

The materials used as fertilizers are so diverse that tabulating trace element content is not possible in the context of a single table; therefore, for further information the reader is referred to the communication by Swaine [8] or the publications of bodies such as the Association of Official Analytical Chemists (USA).

4.2.4. Foods and Feedstuffs

Food materials are almost exclusively organic in nature, although the trace elements contained in them can be present in inorganic forms added as dietary supplements or present as a by-product of processing or packaging. As with fertilizers, the base materials are extremely diverse and defining typical concentration ranges for trace elements is difficult. An excellent guide to the composition of food materials is afforded by the elemental content of the American Food and Drug Administration's "Total Diet Composites" [9], which consist of groups of food items that have been cooked or prepared as they might be in the kitchen. They can, therefore, be considered to represent typical food samples. Table 4.3 lists the elemental content of the composites as determined by neutron activation analysis by Tanner and Friedman [10].

4.2.5. Animal Tissues and Body Fluids

The principal sample types encountered in this grouping include mammalian soft tissues, skeletal tissues, blood, milk, urine, and faeces. The range of soft and skeletal tissues each tend to have similar elemental compositions, although some tissues are known to act as accumulators for certain elements. These are listed in Table 4.4; the data are for human tissues.

The range of elements with known biological functions was listed in the section on plant materials and, as with plants, animal tissues contain a large number of additional elements that are the product of the environment. These include Al, Ag, Au, Cd, Ge, Hg, Pb, Rb, Sb, Ti, Zr, and many others. There are now several major compilations of the elemental content of human tissues [3, 5, 11, 14], but there are not such comprehensive compilations for animals. Generally, livestock animals are found to have similar elemental content to humans, although there are some minor differences. The literature values for the concentrations of many elements, notably Ag, Al, As, B, Ba, Cd, Cr, and Mo, vary by an order of magnitude or more and analytical data often show a continual downward trend with time, probably indicating improved control of contamination during sampling and analysis.

Table 4.3. Elemental Content of Total Diet Composites[a,b]

		Concentration in ppm ($\mu g/g$) Wet Weight							
Element	Range	Dairy Products 1	Meat, Fish, Poultry 2	Grain and Cereal Products 3	Potatoes 4	Leafy Vegetables 5	Legume Vegetables 6	Root Vegetables 7	Garden Fruits 8
Ca		1143–2370	<200–1646	322–1122	79–227	346–1019	248–961	241–578	114–282
	Average	1540	<300*	606	<160*	600	450	400	<200*
Mn		<1–<6	<4–<6	4–12	1.3–9	1–8.1	1–6	1.1–4.5	<2–<5
	Average	<2*	<5*	<6*	2.9*	2.3*	<4*	2.4*	<3*
Mg		99–284	213–487	240–669	166–494	103–402	128–477	98–269	96–272
	Average	164	<340*	360*	340	200*	300	170	<200*
Al		<0.5–21	<1–5.6	5.6–92	2.6–22	3.8–13.5	0.7–13.7	0.5–4.6	1.8–35
	Average	2.9*	2.4*	32.4	10.4	8.6	3.4	1.5	9.9
Na		852–1755	3363–5082	3479–6000	158–1000	174–1626	1826–7396	404–1552	1878–4022
	Average	1240	4150	4700	430	780	3200	760	3000
K		815–3131	1848–5619	<2000–5958	4400–10321	1225–3400	<1100–4308	1239–3487	960–2830
	Average	1540*	2670*	<2600*	6240	2191	1623*	2452	1626*
Cu		<3–7	<4–19	8–12	<3–7	<2–5	2.7–8	<2–<4	2.4–<8
	Average	3*	<10*	<10*	4*	<3*	<6*	<3*	<5*
Cl		1358–3154	4218–6623	4912–7845	616–2647	424–2550	2421–10191	410–2152	2209–5844
	Average	2040	5080	6160	1180	1160	4300	960	4100
Sb		<0.002–0.02	<0.004–0.015	0.006–0.05	<0.003–0.03	0.001–0.027	<0.002–0.014	<0.002–<0.03	0.002–0.011
	Average	<0.004*	0.008*	<0.01*	<0.008*	<0.006*	0.008*	<0.004*	<0.006*
Hg		<0.001–0.003	0.004–0.03	<0.002–0.013	<0.001–0.013	<0.001–0.009	<0.001–0.002	<0.001–0.002	<0.001–0.004
	Average	<0.001*	0.017	<0.003*	<0.002*	<0.001*	<0.001*	<0.001*	<0.001*
Co		0.003–0.031	0.006–0.132	0.011–0.107	0.014–0.084	0.008–0.041	0.013–0.056	0.005–0.033	0.009–0.040
	Average	0.011	0.030	0.029	0.049	0.019	0.027	0.014	0.018
Fe		1.5–16	18–39	21–49	9–27	5.6–21	10.3–43	5.1–9	5.8–14
	Average	3*	24	36	15	12	21	7	<9
Zn		3.4–7.3	20–58	5.1–8.2	1.6–7.8	1.9–7.5	4.7–16.8	1.6–4.2	1.9–7.5
	Average	5.1	30	7.4	4	3	8.1	2.6	3.4
Se		0.009–0.05	0.11–0.24	0.11–0.30	0.007–0.05	0.005–<0.06	0.005–<0.06	0.004–0.039	0.006–0.008
	Average	0.02*	0.17	0.17	<0.02*	<0.01*	<0.016*	<0.02*	0.016*

[a]*, Medians.

[b]Data taken from Tanner and Friedman [10]. (Table reproduced by permission of Elsevier Scientific Publishing Company, Amsterdam.)

Table 4.4. Elemental Concentrations of Accumulator Tissues[a]

Element	Accumulator Tissue	Concentration ($\mu g \cdot g^{-1}$)	References
As	Hair	0.03–2.0	5, 11, 12
	Nails	0.2–3.0	5, 11
Ba	Bone	3–70	5, 11, 13, 14
Ca	Bone	170,000	5, 11
Cd	Kidney (increasing with age)	18–310	5, 11, 14
	Liver (up to 50 years)	2–22	5, 11, 14
Cs	Muscle	0.07–1.6	5, 11
F	Bone	2000–12000	5, 11
	Bone (from animals) unexposed	(300–600)	15, 16, 17
Fe	Bone marrow and erythrocytes	3–380	5, 11
	Liver	250–1400	5, 3
Hg	Kidney	0.3–12	5, 11, 18, 19
I	Thyroid	2000–5000	3
	Hair	0.1–15	5, 11
P	Bone	62000–71000	5, 11
Pb	Bone	3.6–30	5, 11, 20, 21, 22
S	Hair	42000–60000	5, 23, 24
Sb	Hair	0.09–3	5, 11
Sr	Bone	36–140	5, 11, 14
	Tooth	21–280	25, 26, 27
Zn	Prostate gland	186–1340	3, 28, 29
	Choroid of the eye	up to 138000	30

[a]Data extracted from H. J. M. Bowen [5]. (Reproduced by permission of Academic Press, Inc., London.)

4.3. SAMPLE PREPARATION PROCEDURES

4.3.1. Introduction

Inductively coupled plasma–atomic emission spectrometry (ICP–AES) is predominantly a solution analysis technique, whereas most agricultural samples are solids. This single fact imposes the greatest limitation on the application of the ICP for agricultural analysis. The requirement for a dissolution procedure imposes a dilution factor of usually 10- to 1000-fold, makes sample preparation the rate-determining step in the analysis, and may limit the accuracy of analyses through the introduction of contamination or the loss of analyte elements. These disadvantages, however, are largely offset by the ease of standardization that attends the requirement for dissolution. Like most other atomic spectrometric techniques, ICP–AES is a relative or comparative technique depending upon calibration, or to use preferred analytical terminology, standardization, for its quantitative use. Dissolution techniques, therefore, may be seen primarily as a means of obtaining from the solid sample a true solution, a colloidal solution, or an inhomogeneous suspension. In practice, the mineralization of the sample matrix, which is the usual method of dissolution, also makes it possible to calibrate the system using readily available, high-purity inorganic laboratory chemicals. The use of solutions for analysis has the further advantage that preconcentration techniques, ranging from simple reduction of volume by evaporation to complex solvent extraction procedures, hydride generation, or ion-exchange separation and preconcentration, can readily be employed.

Many of the procedures discussed in the following sections have been developed for AAS and are not yet fully tested for use with ICP-AES. However, in view of the reduced chemical interferences associated with ICP–AES, they should be readily adaptable, and indeed, some may be usable in simplified form. For example, Allain and Mauras [31] determined Al in blood, urine, and water following simple dilution and Kirkbright et al. [32] determined P in milk powders after mixing with glacial acetic acid and dilution. A useful tabulation of the extensive literature on sample preparation covering the last 15 years may be found in the Annual Reports on Analytical Atomic Spectroscopy and its successor, the Atomic Spectrometry Updates in the Journal of Analytical Atomic Spectrometry published by the Royal Society of Chemistry (UK). A further useful source is Bock [33].

4.3.2. Sampling and Sample Handling

Obtaining a representative sample is obviously of crucial importance, and yet this is often extremely difficult. The sampling of soils from fields is an excellent example of a difficult sampling problem and has been the subject of discussion by several authors, including Jackson [34], Scott, Mitchell, and Purves [4], and Peck and Melsted [35]. Obtaining homogeneous and representative subsamples for analysis from larger entities such as plants may be accomplished by milling, which produces a thorough mixing of the various components.

Before acquiring a sample, it is necessary to assess the contamination risks from the tools to be used and to select materials accordingly. This is particularly important when dealing with blood and tissue samples where hypodermic needles can be the source of contamination as can metal forceps, knives, and so forth. Sample containers are a further potential source of contamination [36]. For soils, polythene bags are satisfactory; brown paper bags can also be used although these can be a source of B contamination. For laboratory ware, Pyrex, quartz, low-density polythene, polypropylene, polycarbonate, and polytetrafluorethylene (PTFE) have proved to be acceptable, and this is in agreement with the findings of Fisher et al. [37]. Container stoppers of rubber contain Zn and are a common source of contamination; labels may also contaminate if placed in the container with the sample. Reagent purity may be assured by techniques such as subboiling distillation [38] and electrolysis. Many laboratories, including this one, have installed laminar air-flow hoods to provide clean air (Class 100) work stations.

The key to correct sampling and control of contamination is attention to detail, and it is essential that the analyst concerns himself with these aspects rather than accepting samples without knowledge of their history. The problems of contamination of biological samples have been discussed by Anand et al. [39] and by Zief and Mitchell [40].

Most agricultural samples require drying in preparation for weighing and dissolution and the methods include air-drying, oven-drying, vacuum-drying, and freeze-drying. Air-drying at $<25°C$ is the preferred method for soils. For plant materials, a maximum temperature of 80°C is desirable [4]. Most biological samples are best dried at temperatures not exceeding 80–100°C. Volatility losses are negligible for most elements, with the exception of a few such as Hg and Se, where small but significant losses can occur with some types of biological materials. For these elements, vacuum-drying at room temperature or freeze-drying should be considered, although freeze-drying losses of organomercurials have been reported [41]. For materials of animal origin, losses have been reported for Cd, Pb, and Se from oyster tissue dried at 120°C [42] and for Ce, Co, Mn, Pa, Ru, and Zn from mollusc samples dried at 100°C [43]. However, Koirtyohann and Hopkins [44] found no losses of Cd, Cr, Fe, and Zn during drying at 110°C.

4.3.3. Dry-Ashing/Dissolution

Dry-ashing/dissolution is one of the most widely used procedures and although it is slow, sometimes requiring up to 24 h per sample, it has the advantage of being readily adaptable for large batch processing and does not require constant operator attention. Most published applications of ICP–AES to the analysis of biological materials have employed dry-ashing for sample preparation, notably for plant materials [45, 46], fruit juices [47], tissues [48], and faeces and urine [49] although in this procedure a fluxing agent, sodium carbonate, was also used. In this laboratory, plant materials are ashed by the following procedure: 1–5 g oven-dry plant material is ashed overnight at 450°C in silica, the ash is moistened and evaporated to dryness twice with 2-mL portions of 6 *M* HCl on a steam bath, the residue is then treated with 2 *M* HCl, and after warming, the extract is filtered through a Whatman No. 30 paper and diluted to volume (50–250 mL) with water. The dilution factor for dry-ashing and dissolution generally falls within the range of 10–50, which is less than with most other dissolution methods. It should be noted however that for a typical plant, the mineral content is only one-tenth of the total, so that the analyte dilution may be much higher.

The principal disadvantage of dry-ashing is the potential loss of analyte that can occur through volatilization of elements such as As, Cd, Hg, Pb, and Se or by retention on the crucible or insoluble residue of Al, Ca, Cr, Cu, Fe, Ni, and V, although it should be stressed that the losses depend on the nature of the substances being ashed and on the ashing or dissolution technique employed. Thus, Munter et al. [50] reported losses of Cd and Pb by volatilization, Fe, Al, and Cr through retention on the ash residue, and Cu through retention on the silica crucible, after ashing at 457–500°C. Dahlquist and Knoll [51] found losses of Fe, Ni, V, Al, and Cr due to retention on insoluble silica residues, although these were not observed when nitric acid was used as an ashing aid. Koirtyohann and Hopkins [44] demonstrated that the losses of Cd, Cr, Fe, and Zn from tissue samples ashed at below 600°C were also attributable to retention on the insoluble residue. The normal method of overcoming loss of analyte on a silica residue is to include treatment with hydrofluoric acid in the dissolution step [52, 53]. The literature on losses during ashing shows apparent contradictions, for example, Prasad and Spiers [54] preferred dry-ashing at 475°C prior to the determination of Ca, Fe, and Zn in plant materials, whereas Etherington and Davies [55] have presented evidence that Fe is lost from plant material as iron pentacarbonyl during dry-ashing at 500°C or even during acid digestion.

Normally dry-ashing is not considered appropriate for preparation of samples prior to the determination of As and Se. Thompson et al. [56] (see also [33] and [57]), however, reported that the addition of magnesium nitrate, as an ashing aid, to herbage enabled the retention of these elements during ashing and that the resultant ash dissolved readily in 1 + 1 hydrochloric acid which was used as the medium for generation of the hydrides. Similarly, addition of sulfuric acid to food samples may prevent losses of Cd, Cu, or Pb at ash temperatures up to 750°C [58].

In recent years, low-temperature RF plasma ashing has been used; although it operates at temperatures below 100°C, losses of volatile elements such as Hg [59] and even Cr [60, 61] have been reported. Since oxidation only proceeds at the sample surface, repeated mixing of the sample is essential, thus, large or granular samples can be difficult to ash.

The recovery for most elements is quantitative, and although large samples may take days to ash, unattended batch processing, as for normal ashing, is easily carried out [62–64]. Reports on this technique include its application to reference materials [65], urine [66], serum and tissues [67, 68], sugar [69], and cellulose [59]. The addition of gases other than oxygen may enhance the rate of ashing; thus, Carter and Yeoman [70] used an equimolar mixture of CF_4 with O_2 for the rapid oxidation of 10-μL volumes of whole blood.

4.3.4. Wet-Ashing

Wet-ashing is the most versatile of the available dissolution procedures. It is generally quite rapid, but some procedures can be hazardous and necessitate skilled operator attention. The dilution factor is higher than for dry-ashing, and usually is in the range 50–200. In addition, reagent impurities may raise the minimum level of analyte determinable.

For organic materials, the most satisfactory and universal digestion procedure is that based on a mixture of nitric and perchloric acids [71]. The correct and safe application of this procedure depends on achieving a smooth increase in the oxidation potential of the solution, thereby ensuring that the most labile compounds are oxidized by the mild action of the nitric acid and only the most resistant compounds remain when the residue is exposed to the very high oxidation potential of the fuming perchloric acid at the final stage of the digestion. Usually the ratio of the nitric to perchloric acids used is in the range 1 : 1 to 3 : 1 and about 20 mL of acid per gram of sample is required. Slow heating (in a Kjeldahl flask) is used at first to ensure maximum oxidation by the nitric acid before it is distilled off, and then the temperature is gradually raised until the vigorous perchloric acid reaction has subsided. Because the oxidation reaction can be exothermic, the temperature may rise too quickly and cause charring. If this occurs, heating should be stopped and the reaction flask cooled by immersion in cold water. More nitric acid, in 0.5-mL increments, is then added until the darkening clears, and then the flask can be safely reheated. Samples containing high fat or oil content require extra attention to prevent charring, which is made more likely because of the difficulty of maintaining an adequate surface area of the sample in contact with the acid. Such samples require regular shaking to disperse them in the oxidizing medium. No attempt should be made to use nitric acid with any material that might release glycerol because there is a danger of the formation of explosive nitroglycerine. Only properly trained personnel should carry out digestions involving perchloric acid and the reader is referred to the recommendations of the Analytical Methods Committee (Royal Society of Chemistry, UK) [72] on the safe handling of this reagent.

In a modification of the nitric/perchloric procedure, a small quantity of sulfuric acid [73, 74] is added to the mixture prior to commencing oxidation. This adds an additional element of safety because the residue is kept moist by the sulfuric acid until the perchloric acid has evaporated. The time required, however, for the digestion is much longer when sulfuric acid is included, the terminal stage being reached when fumes of sulfur trioxide are evolved. Addition of sulfuric acid was found to be necessary by Jones et al. [73] when determining As, Sb, and Se by hydride generation ICP-AES. It was suggested that the sulfuric acid prevented the formation of resistant organic materials that inhibited generation of the hydrides. It has been commonly supposed that losses of Cr during nitric/perchloric digestions occur through volatilization of chromium oxychloride; however, Scott [75] has shown that the Cr is lost through retention of Cr on the silica residue. Recently, this has been confirmed by Chao and Pickett [76] who found that the presence of sulfuric acid at the final stage of the oxidation facilitated the addition of hydrofluoric acid to dissolve the silica.

The high dilution factors associated with perchloric acid digests arise from the relative insolubility of the alkali metal perchlorates, and substitution by sulfuric acid produces alternative problems with insoluble sulfates such as those of Ca, Ba, and Pb.

Digestion procedures for determinations of the volatile elements require precautions to avoid volatilization losses. For Se, Clinton [77] digested 1 g of blood or forage sample with a 5:2 nitric/perchloric mixture in a Pyrex tube held at 210°C in a drilled aluminum block placed on a hot-plate. Two drops of kerosene were added to prevent frothing. More commonly, procedures based on nitric/sulfuric have been preferred, for example, that reported for As in soils and plants by Thompson and Thoresby [78]. For Hg determination, oxidation is performed with 4–6% permanganate following digestion with concentrated sulfuric acid [74, 79, 80] or nitric/sulfuric acids [81, 82].

Samples containing high oil, fat, or sugar content may not, for the reasons given previously, be amenable to nitric/perchloric digestion and here procedures based on sulfuric/50% (160 volume) hydrogen peroxide can be employed. Once again great caution must be exercised, and it is advisable to follow well-tested procedures such as those given by the Analytical Methods Committee (Royal Society of Chemistry, UK) [83].

The problems of digesting inorganic materials such as soil minerals are those associated with dissolution of the matrix. Most procedures for the determination of the total content of soils involve hydrofluoric acid to dissolve the silica and the method can be modified to retain the silica or remove it. Jackson [34] describes the following procedure: 0.1 g of finely ground soil is placed in a Pt crucible, wetted with water, and then 0.5 mL of perchloric acid plus 5 mL of 48% hydrofluoric acid added. The crucible, with the lid covering most of the top, is placed on a sand bath at a temperature of 200–225°C and the acids evaporated to dryness. If dark organic matter collects on the side or lid of the crucible, it is removed by heating briefly with a Meker burner. The residue is then treated by boiling it with 5 mL of 6 *M* HCl diluted with 15 mL of H_2O. High-alumina samples may require repeated treatments to obtain dissolution whereas a few minerals such as chromite and zircon do not dissolve and a fusion is necessary. Although small sample quantities are used in this procedure, there may still be difficulties in exposing soils of high organic content to hot perchloric acid. A modified procedure, incorporating nitric acid, as described by Scott and Thomas [84] might then be preferred.

The disposal of sewage sludges to the land involves agricultural laboratories in extensive monitoring of the treated soils for the accumulation of heavy-metal pollutants. For this purpose, digestion with aqua regia is a convenient method of sample preparation [85]. Aqua regia does not give the true total contents because the silicate-bound metals are not removed; however, since few pollutants are added in silicate form this is unimportant. The procedure used in this laboratory takes 0.25 g of ignited sewage sludge or 1 g of unignited soil to dryness six times with 10 mL of aqua regia (redistilled nitric acid/6 *M* HCl, 1 + 3) on a water-bath, with the final residue again taken to dryness with 10 mL of 6*M* HCl and dissolved in 50 mL (sludge) or 25 mL (soil) of 0.06 *M* HCl. Recent evaluations of this procedure [86] indicate that for contaminated soils, 80–90% (mean 86%) of the total contents of Cd, Cu, Mn, Ni, Pb, and Zn are extracted (in comparison with the contents determined following bomb digestion), whereas for uncontaminated soils, an average of 76% of the Ca, 59% of the Mn, and 69% of the Zn are extracted. For sewage sludges, Van Loon et al. [87, 88] had previously reported that aqua regia removed greater than 90% of the metals As, Cd, Cr, Cu, Fe, Mn, Ni, Pb, V, and Zn, although was less effective for Ag. Aqua regia digestion of sludges can also be carried out by refluxing the sample with the binary acid. In an even simpler procedure, Thompson and Wagstaff [89] refluxed 0.5 g of dried sludges with nitric acid alone in 50-mL calibrated glass tubes for 15 min with subsequent dilution to the mark with distilled water. Satisfactory results were reported for Cd, Cr, Cu, Ni, Pb, and Zn.

For difficult matrices and where it is essential to avoid the loss of volatile elements, pressure dissolution can be an effective technique. Price and Whiteside [90] described a useful method in which samples were heated with nitric and hydrofluoric acids and then subsequently reheated with boric acid to convert the insoluble fluorides to fluoroborates. Pressure dissolution is unfortunately not conveniently adapted to handle large sample throughputs, since, although batch processing is possible, the PTFE-lined stainless steel bombs used are too expensive to employ in large numbers.

4.3.5. Fusion/Dissolution

Fusions are generally best avoided where the aim is to determine trace elements, because the purity of the solid fluxes is more difficult to control than that of solution reagents. A further difficulty arises from the resulting increase in the total dissolved solids content of the analysis solution, which may be a factor of 5–10 over the sample weight. This presents particular problems for ICP–AES because restriction of the sample flow to, or blockage of, the nebulizer may occur due to the increased viscosity and tendency to crystallization of the analyte solution (see also Section 4.5). In addition, the presence of the flux may be responsible for interference effects due to changes in the excitation mechanism [91]. A further problem with fusions is the analyte dilution factor, which may be as high as 1000. Nevertheless, they are necessary for difficult-to-dissolve substances such as the minerals chromite and zircon for which sodium peroxide (Na_2O_2) or potassium bisulfite ($K_2S_2O_7$) may be used. A wide choice of fusion reagents is available, including sodium carbonate/borate for high alumina materials, lithium carbonate or lithium tetraborate or metaborate for high silica and acidic matrices, and potassium hydrogen sulfate for basic samples. For siliceous materials, lithium metaborate is most widely used [92] and has the advantage over lithium tetraborate that it is a mobile liquid at 1000°C.

4.3.6. Extractable Soil Contents

The determination of extractable content is used to assess the ability of a soil to supply nutrients to plants. A wide range of extractants is employed [34, 35, 74, 93, 94], and the choice for a particular element is determined through demonstration by experiment of a correlation between the extracted level and the level of plant uptake. The types of soil vary considerably from place to place and hence different extractants are favored in different geographic areas. Common extractants, used particularly on noncalcareous soils, include: acetic acid for cationic species such as Ca, Co, Fe, K, Mg, Na, Ni, and Zn; neutral ammonium acetate for elements occurring in anionic form such as Mo and Mn; water for B; and EDTA or DTPA (diethylene-triaminepentaacetic acid) for Cu, which is largely bound in the form of organic complexes, and also for Mn and Zn. For the nonmetals P and S, calcium chloride or ammonium fluoride [35] and potassium dihydrogen phospate [95], respectively, have been employed. For N, the Kjeldahl digestion method is normally used, but Alder et al. [96] preferred reduction with sodium hypobromite prior to determining N from its NH band emission (336.0 nm) in the ICP. The detection limit was reported to be 0.1 $\mu g \cdot mL^{-1}$ N for a 5-mL aqueous solution. Many other extractants are used as aids to determining the nature of binding of a particular element in the soil, but the levels extracted by these do not necessarily correlate with plant availability.

A number of workers have analyzed soil extracts using ICP–AES [46, 51, 97–101]. Dahlquist and Knoll [51] determined elements directly in ammonium acetate and DTPA extracts; Hern [97] also determined elements directly in DTPA but found addition of surfactant (to prevent nebulizer blockage) to be necessary for ammonium acetate extracts; Hoult et al. [98] reported the direct determination of the major components Ca, K, Mg, and P in acidified ammonium acetate (pH 4.8). Jones [46, 99] commented that it was not possible to determine elements directly in neutral ammonium acetate, although he found "Double Acid" (0.05 M HCl in 0.025 N H_2SO_4) an acceptable medium (there have been improvements in nebulizers since this work was published). Similarly, Soltanpour et al. [100] encountered difficulties with extracts in 1 *M* NH_4HCO_3/0.005 *M* DTPA but overcame them by dilution of the extract 1 : 5 with 0.5 *M* nitric acid; Manzoori [101] preferred to remove the extractant before analysis. Although some nebulizers can tolerate dissolved solids up to 25% m/v, it is advisable, if possible, to restrict dissolved solids to not more than 1–2%. Experience in this laboratory suggests that the ability of the ICP to determine elements directly in soil extracts depends entirely on the type of soil, since this determines the total solids content of the analyte solution. For calcareous soils and those containing soluble salts, the direct extracts may be unacceptable as may, for example, ammonium acetate extracts of highly organic soils. In many cases, however, direct analysis will be possible and certainly considerable time savings over flame atomic absorption methods are readily achieved through avoiding dilutions and eliminating the necessity for complete oxidation of the organic content of the extract. Hislop and Cook [102] (see also [95]) described a method of determining soil phosphate status based on the use of an anion-exchange resin. Such procedures are eminently suitable for ICP application because they remove the matrix elements.

4.3.7. Preconcentration Techniques

Although ICP–AES has excellent detection limits, several biologically important elements cannot be determined directly in sample digests. The large number of preconcentration techniques developed for AAS are readily applicable to ICP analysis, but the multielement nature of ICP–AES imposes the additional requirement that they be capable of concentrating groups of elements simultaneously.

4.3.7.1. Extraction Methods

Extraction preconcentration methods have the disadvantage that the analyte is usually in the organic phase and whereas organic solvents can be introduced into the ICP [103, 104], detection limits may be poorer than those achieved with aqueous solutions. This decreases the effectiveness of the preconcentration. Experiments in this laboratory have shown that a low-power Ar plasma will tolerate organic solvents of low volatility such as alcohols, although increased power as compared with aqueous solvents is required, but cannot be used with volatile solvents such as acetone (see Part 1, Section 4.7.8 and Chapter 6, however). Standardization is also more difficult with organic solvents, although Hon et al. [105] described the use of isobutyric acid as a solvent for lubricating oils which enabled inorganic standards to be used for AAS. Adaptations of this kind may well prove useful for ICP–AES. In recent years, several groups of workers have noted the poor stability of Pb and other ammonium pyrrolidine dithiocarbamate (APDC) complexes in methyl isobutyl ketone (MIBK) and have recommended back-extraction into aqueous media [106, 107]. Again, extension of this approach has obvious merits in relation to analysis by ICP. Many extraction-based methods have been developed for AAS, and the ones most likely to prove useful for ICP are those that quantitatively extract a group of elements from a given matrix, and subsequently allow back-extraction into an aqueous phase.

David [74] has tabulated extraction concentration procedures used for trace metals in agricultural materials, and the data with further additions are reproduced in Table 4.5.

4.3.7.2. Ion Exchange

Jones et al. [73] have proposed a multielement preconcentration scheme based jointly on ion-exchange and hydride generation (air/H_2/N_2 flame atomizer) methods for the analysis of plant and animal tissues, and foods. Samples are digested in nitric/perchloric/sulfuric acids and 5% of the resulting solution is taken for the direct determination of the alkali, alkaline earth metals, phosphorus, and the transition metals such as Mn and Fe. The remainder of the solution is passed through a Chelex 100 resin column to strip elements such as Cd, Cu, Mo, Ni, V, and Zn from the matrix. As, Sb, and Se are not retained on the column and are determined in the initial column effluent using the hydride generation technique. This method is an adaptation and extension of those used for seawater by Kingston et al. [125] and Sturgeon et al. [126].

The resin columns, in 200-mm × 8-mm polypropylene tubes, with polyethylene frit resin supports, are prepared as follows: 10 mL of a magnetically stirred slurry of 30 g of wet Na-form resin plus 150 mL of deionized water (note organically bound copper may be present in undistilled water) are pipetted into the tube; the approximately 2 g of wet resin is treated successively with two 15-mL portions of deionized water, two 15-mL portions of 15% (v/v) nitric acid, two 15-mL portions of deionized water to rinse the acid from the column and one 10-mL volume of 15% (v/v) ammonium hydroxide to convert the resin to the functional ammonium form. Finally, the column is rinsed with two more 15-mL portions of deionized water to remove excess ammonium hydroxide. The digest solution (95 mL) is adjusted to pH 5.3 ± 0.2 using ammonium hydroxide and nitric acid and then buffered with 1 mL of 1 *M* ammonium acetate solution. The authors noted that at this point some digests of National Bureau of Standards (NBS) biological SRMs developed flocculent white precipitates, although these did not interfere with the separation. The sample is then allowed to pass through the column under gravity feed and the effluent collected for determination of As, Sb, and Se. Sequestered alkali and alkaline earth elements are removed from the column with 40 mL of 1 *M* am-

Table 4.5. Extraction Concentration Procedures Used for Trace Metals in Agricultural Materials[a]

Element/s	Material	Extractant[b]	Reference
Cd, Cr, Pd	Fish	APDC/MIBK	106
Pb	Soils, sediments	APDC//MIBK	107
Cd, Co, Cu, Mn, Ni	Soils	dithizone/CCl_4	108
Cd	Soils	APDC/MIBK	109
Mo	Fertilizers	Oxine/$CHCl_3$	110
B	Soils	2-Ethyl-1,3-hexanediol/MIBK	111
Co	Animal tissues, plants	1-Nitroso-2-naphthol/$CHCl_3$	112
Mo	Plants, soils, waters	APDC or oxine MAK	113
Cu	Plants	APDC/MIBK	114
Co	Plants	1-Nitroso-2-naphthol/iso-amylacetate	115
Cd, Co, Cu, Mn, Ni, Pb, Zn	Soils	HMA-HMDC/butyl acetate	116
Cu	Soils, plants, fertilizers	APDC/MIBK	117
Zn	Soils, plants, fertilizers	APDC/MIBK	118
Ag, Bi, Cd, Co, Cu, Ga, In, Fe, Pb, Mn, Hg, Mo, Ni, Pd	Soils	PDCA/$CHCl_3$	119
Mo	Soils	CNS/MIBK	120
Fe, Cu, Mn, Co, Ni	Soils, water	IDE-C	121
Pb	Plants, foods	EDTA–benzene	122
Cu, Fe, Mn, Zn	Plants	PDCA–$CHCl_3$	123
Cd	Plants, foods	Dithizone–$CHCl_3$	124

[a]Table extracted from data published by David [74]. Reprinted with permission from *Anal. Atomic Spectrosc.*, **1,** D. J. David, copyright (1978), Pergamon Journals.

[b]APDC, Ammonium pyrrolidine dithiocarbamate; MIBK, methyl isobutyl ketone; oxine, 8-hydroxyquinoline; MAK, methyl amyl ketone; HMA, hexamethyleneammonium; HMDC, hexamethylenedithiocarbamate; PDCA, pyrrolidene dithiocarbamic acid; CNS, thiocyanate; IDE-C, iminodiacetic acid-ethyl-cellulose.

monium acetate solution and the excess rinsed from the column with 10 mL of deionized water. The trace elements are then stripped from the column using 10 mL of 15% (v/v) nitric acid, residual nitric acid is rinsed from the column with 4 mL of deionized water, and the final volume made up to 15 mL. This 10% (v/v) nitric acid solution of the trace elements is then analyzed by ICP–AES against multielement standards in 10% (v/v) nitric acid.

This procedure is the most promising yet devised, although erratic results were obtained for the important elements Cr, Fe, and Mn and, in siliceous materials where hydrofluoric acid is needed to dissolve the residue, poor recoveries of Al and Pb were reported. Separations by this method are relatively slow, requiring 4–8 h. However, up to 40 columns can be prepared in 3–4 h so that batch processing can be readily carried out.

Haas et al. [127] determined trace elements in urine samples by ICP-AES, but found insufficient sensitivity for the elements Al, As, Cd, Co, Cr, Mn, Ni, Pb, and V. Barnes and Genna [128] attempted to overcome the lack of sensitivity by preconcentration of the trace elements on a polydithiocarbamate resin. The unusual aspect of their technique is that the recovery step is accomplished by digestion of the resin in nitric/sulfuric acids which yields a very high concentration factor of 125. Quantitative recoveries were obtained for the 10 elements studied, namely, Bi, Cd, Cu, Hg, Ni, Pb, Se, Sn, Te, and U.

4.3.7.3. Hydride Generation

The cold vapor [79, 129–132] and hydride generation [78, 133–136] (cf. Part 1, Sections 6.5 and 6.6) techniques as applied to AAS have reached such an advanced state of development that Godden and Thomerson [137] concluded that no major improvements are likely in the near future.

The position with regard to adaption of the hydride generation technique to ICP-AES is quite different because the plasma, at least at low power levels of (<2 kW), is unable to accept the H_2, CO_2, and H_2O vapor by-products of the reduction reaction. Nevertheless, Thompson et al. [56, 138–142] by careful control of the experimental conditions [138], by operating at higher powers (2.7 kW), and by employing continuous mixing of the sodium tetrahydroborate reductant at low concentration (1% m/v in 0.1 *M* sodium hydroxide) with the sample, have successfully applied the technique to the analysis of soils, sediments, and herbage. Continuous pumping and mixing of the sample and reductant ensures a low H_2 flux to the plasma, however, it also produces a low hydride flux so that some of the potential sensitivity of the technique is sacrificed. A study of interferences [139] for the elements As, Bi, Sb, Se, and Te showed that although mutual interferences are negligible, Cu severely interferes with the production of the hydrides of Bi, Se, and Te and that Fe has a similar effect with Se. The Cu interference was overcome by coprecipitation of the hydride-forming elements on lanthanum hydroxide, which was then redissolved in 5 *M* HCl. Similar studies were carried out for Ge and Sn [140] and, although the operating conditions were similar to those used for the other elements, the interferences from Cu, Co, Ni, and Fe were severe. However, replacing hydrochloric with tartaric acid as the working medium was found to be an effective means of controlling them. The first application described was the determination of Se in soils and sediments [141]. A sequential nitric/perchloric digestion (at 170°C) was used, followed by coprecipitation on lanthanum hydroxide and prior reduction of Se(VI) to Se(IV) with bromide. The detection limit was 1 ng $\cdot$ mL^{-1}. Subsequently, As, Bi, and Sb were determined also in soils and sediments [142], with digestions being carried out in screw-capped borosilicate test tubes with hydrochloric acid. Potassium iodide was used to reduce As(V) and Sb(V) to the (III) state and to reduce the interference of Cu on Bi. Detection limits for the three elements were reported to be of the order of 0.1 $\mu g \cdot g^{-1}$. Recently, the same elements have been determined in samples of herbage [56] following dry-ashing with magnesium nitrate. A modified hydride generator was used that enabled the analyte solution to be alternated rapidly with the blank reaction medium, producing a steady flow of H_2 to the plasma thus enabling more stable tuning of the generator. Detection limits for the three elements in solution were reported to be approximately 0.2 ng $\cdot$ mL^{-1}.

Fry et al. [143] proposed an alternative approach to implementation of the hydride generation technique for ICP–AES for As which, although more cumbersome to use than those described above, provides higher sensitivities. In this system, the water vapor and CO_2 produced in the reaction vessel are removed on columns of Drierite and dry NaOH, respectively, and the arsine and hydrogen separated by freezing out the arsine in a liquid Ar trap, with the hydrogen being allowed to vent. Subsequently the trap is removed from the liquid Ar and placed in lukewarm water and the arsine is flushed into the plasma by the injector flow, there producing a transient As emission signal. By means of this separation and concentration procedure, a detection limit of 0.03 ng $\cdot$ mL^{-1} for As was achieved using a low-powered plasma (1.2 kW) and low-resolution monochromator to isolate the As spectral lines (193.696 or 228.812 nm).

The introduction of volatile halides into the ICP has been reviewed by Sloat [144].

4.4. SPECIATION

The importance of trace elements in agriculture arises from their function in catalysts of enzyme reactions. Underwood [3, p. 8] describes their role as extending from weak ionic strength effects to the highly specific associations that occur in metalloenzymes. In these, the metal atom is firmly attached to the protein in a fixed stoichiometry. The use of the ICP as an element-specific detector for gas chromatography (GC) and high performance liquid chromatography (HPLC) for determination of the active forms of trace elements must surely be regarded as an important advance. The information provided by such determinations is far greater than element contents alone and the increasing use of such techniques may mitigate the ever-present danger of acquiring data simply because it is possible to do so.

Several authors have published descriptions of the use of ICPs as detectors for GC [145–147] and HPLC [148–150] (also see Chapter 6 and Part 1, Section 6.6). For GC, the advantages of using an ICP compared with, for example, a microwave induced plasma (see Part 1, Chapter 2), are those related to its ability to handle larger sample sizes and freedom from the plating out effects that are often observed in microwave plasma discharge tubes. These improvements in performance, however, must be set against the considerably greater cost of the ICP compared with the microwave plasma, or for that matter flame or electrothermal atomization detectors or electrochemical detectors.

The more important application, particularly in relation to agricultural studies, is as a detector for HPLC. This has the advantages over GC that it can be used to separate nonvolatile, polar, thermally stable, or high-molecular-weight compounds. It is fortuitous that the flow rates of the mobile phase used in HPLC are similar to, or may be adapted to, the nebulizer uptake rates (0.5–2 $mL \cdot min^{-1}$) normally used in ICP–AES. Manahan et al. [148] demonstrated the separation of Cu-chelates on an Aminex A-14 column using 0.05 *M* aqueous ammonium sulfate, at a flow rate of 2.0 $mL \cdot min^{-1}$, as the mobile phase. The column effluent was conducted to the nebulizer by a 30-mm length of $\frac{1}{16}$ in. od, narrow-bore, flexible PTFE tubing. This interface is critical since it must have a minimum dead volume to avoid peak-broadening. Comparison of the peaks obtained from the ICP detector with those obtained using a UV molecular absorption detector showed that peak-broadening occurred due to mixing in the nebulizer spray chamber, although the ICP exhibited better sensitivity, 6.9 $ng \cdot mL^{-1}$ compared with 18 $ng \cdot mL^{-1}$. Similar results were obtained by Gast et al. [149] who studied the separation of compounds of As, Fe, Hg, Pb, and Mo. These authors encountered difficulty with the more volatile solvents, for example, hexane, but reported acceptable levels of peak-broadening. Interestingly, it was noted that at mobile-phase flow rates in the range 0.8–2.0 $mL \cdot min^{-1}$, the ICP acted as a concentration-sensitive detector, whereas at lower flow rates, it behaved as a mass-flow sensitive detector. Obviously these observations relate to the effective time constant of the nebulizer/spray chamber.

The potential of the technique is perhaps best demonstrated by the work of Morita et al. [150]. Using a multichannel emission spectrometer with a cross-flow nebulizer and standard spray chamber as a detector for gel permeation chromatography, they obtained multielement chromatograms for vitamin B_{12}

and for a synthetic mixture of proteins of increasing molecular weight (cytochrome-*c*, chymotrypsinogen-A, albumin 45,000, albumin 67,000, adolase, catalase, and ferritin). In the case of vitamin B_{12}, C and Co chromatograms were obtained; for the protein composite, C, Cu, Fe, Mn, P, and Zn were studied. From the ratios of the emission intensities and by estimation of the areas of the chromatographic peaks, the atom number ratios of the vitamin B_{12} and of cytochrome-*c* were estimated and shown to be close to the theoretical values. Further development of this technique depends upon improvements in nebulizer and spray chamber design, to avoid peak-broadening and to extend the ability to handle the very low mobile-phase flow rates (5 $mL \cdot h^{-1}$) which are necessary for separations on soft gel stationary phases.

4.5. INSTRUMENTATION AND TECHNIQUE CAPABILITY

4.5.1. Power of Detection

The detection capabilities of ICP-AES in comparison with other techniques namely: flame AAS, electrothermal atomization AAS, dc arc spectrometry, neutron activation analysis, spark source mass spectrometry and X-ray fluorescence spectrometry, have been assessed by Koch et al. [151] (see also 152) and their data are reproduced in Table 4.6. The ICP data are typical of those that might be obtained under compromise operating conditions for multielement determinations in distilled water. The true power of detection of the ICP, referred to the original sample, will be degraded by the presence of a matrix, particularly one producing spectral interferences, and additionally a typical dilution factor of 50, incurred in sample dissolution, must be taken into account. It follows that for a number of biologically important elements, that is, As, Co, Cr, Hg, Ni, Pb, Se, Sn and V, preconcentration/separation techniques are necessary.

Although ICP–AES is a multielement technique, its performance is most often compared with that of AAS, probably because both techniques are used primarily to determine the concentration of elements in solution. In comparison with flame AAS, the limits of detection of ICP–AES are better, notably for those elements, for example, B, Mo, Ni, Sn, and V, which have low atomization efficiencies in the flame. Electrothermal atomization AAS, however, generally provides lower limits of detection than ICP–AES, although it is not nearly so convenient to use on large numbers of samples, particularly if the analyte is present at widely varying concentrations, in which case time-consuming dilutions are required. There are of course many factors that must be taken into account in selecting an analytical technique, but undoubtedly the well-established merits of multielement capability, extended dynamic range, relative freedom from interferences, ease of sample presentation, and, when automated, high sample throughput will ensure an increasing use of ICP–AES in agricultural laboratories throughout the world.

Table 4.6. Detection Limits of Analytical Techniques[a]

Element	Inverse Voltammetry ($\mu g \cdot mL^{-1}$)	AAS ($\mu g \cdot mL^{-1}$)	ETA–AAS ($\mu g \cdot mL^{-1}$)	DC Arc ES ($\mu g \cdot mL^{-1}$)	ICP–AES ($\mu g \cdot mL^{-1}$)	NAA ($\mu g \cdot mL^{-1}$)	SSMS ($\mu g \cdot mL^{-1}$)	XRF ($\mu g \cdot mL^{-1}$)
Major essentials								
Na	—	0.02	0.006	1	0.001	0.003	0.002	10
K	—	0.06	0.001	5	0.2	0.03	0.003	0.1
Mg	—	0.01	0.00008	0.05	0.0005	0.02	0.003	2
Ca	—	0.08	0.003	0.05	0.0005	0.5	0.003	0.3
P	—	500	—	30	0.2	0.2	0.003	2
S	—	—	—	—	0.03	300	0.003	0.1
Trace essentials								
B	—	45	0.2	0.5	0.01	—	0.001	1
Co	0.0001	0.2	0.004	0.3	0.01	0.03	0.005	0.1
Cr	—	0.2	0.002	0.7	0.005	7	0.005	0.5
Cu	0.00002	0.2	0.003	0.05	0.005	0.005	0.01	1
F	—	—	—	—	—	0.06	0.002	3
Fe	0.0002	0.2	0.002	2	0.02	140	0.005	1
I	—	—	—	—	—	0.0005	0.01	0.2
Mn	—	0.1	0.0008	0.03	0.002	0.0001	0.005	0.3
Mo	—	0.8	0.02	2	0.01	0.2	0.03	2

Ni	0.02	0.2	0.01	0.5	0.02	0.1	0.007	0.5
Se	—	0.3	0.006	100	0.2	0.01	0.01	1
Si	—	5	0.01	0.5	0.03	—	0.003	0.2
Sn	0.00004	0.4	0.006	0.7	0.1	0.5	0.03	0.04
V	0.002	2	0.04	0.5	0.2	0.0003	0.004	0.3
Zn	0.00003	0.02	0.00006	3	0.005	0.05	0.01	1
Trace toxic								
As	0.002	0.2	0.004	20	0.2	0.002	0.006	1
Ba	—	0.5	0.02	0.2	0.0005	0.01	0.02	0.3
Be	—	0.03	0.0004	0.05	0.0005	—	0.001	1
Bi	0.00001	1	0.004	0.3	0.3	5	0.02	2
Br	—	—	—	—	—	0.003	0.01	2
Cd	0.00001	0.05	0.0001	2	0.005	0.03	0.03	0.7
Cl	—	—	—	—	—	0.006	0.004	0.4
Hg	0.0005	15	0.2	20	0.2	0.01	0.06	1
Li	—	0.05	0.003	0.2	0.002	0.005	0.0006	1
Pb	0.00002	0.6	0.002	2	0.02	25	0.03	2
Sb	0.00004	1	0.004	3	0.5	0.01	0.02	0.08
Tl	0.00002	0.6	0.01	3	0.5	—	0.02	—

[a]Table extracted from data tabulated by Koch et al. [151].

4.5.2. Comparative Merits of Single- and Multichannel Spectrometers

Agricultural laboratories generally analyze large numbers of samples but, for a single laboratory, these are often derived from relatively few sample types. Under such circumstances, it is relatively easy to justify the high cost of multichannel instrumentation (that is, a polychromator) and to expend the necessary effort to effect an appropriate methodology and calibration procedure [153]. Where sample throughput is not the principal requirement, and/or relatively few elements (<8) are to be determined in each sample, single-channel instrumentation based on a scanning monochromator should be considered. This is acceptable for the ICP because unlike the dc arc, it is stable with respect to time. The case for single-channel spectrometers is strengthened when a wide variety of matrices have to be handled, since most of the time taken for the analysis is devoted to sample preparation and laboratories with this type of problem could well find a multichannel system grossly underutilized. Of course, there is little alternative to a multichannel system if several elements are to be determined on a small sample volume (<5 mL) or if multielement speciation is required, for example, HPLC detection.

The considerations discussed above cannot be quantified because they relate to factors such as need, work-load, staffing, economics, and so forth, and each laboratory has different circumstances. In one aspect, however, it is possible to assess semiquantitatively the difference between the two types of instrumentation. Because of the problems of optical stability, the multichannel spectrometer normally operates with a narrow entrance slit, for example, 20 μm, but with a much wider exit slit, usually about 50 μm. The scanning instrument is not so constrained and it is more usual to operate with equal entrance and exit slits, typically, of the order of 20 μm. For instruments having a reciprocal linear dispersion of 0.8 nm/mm (typical of a 1-m spectrometer having a 1200 groove/mm grating), the respective bandwidths are 0.04 nm (50-μm exit slit) and 0.016 nm (20 μm exit slit), respectively (cf. Part 1, Chapter 8). The optimum slit width, defined as that slit width at which the product of the resolving power and light flux through the exit slit is a maximum [154], is given by $W_{opt} = 1.3 f \lambda$, where f is the f-number of the collimating mirror and λ the wavelength. For an $f/9$ spectrometer operating at 300 nm the optimum slit width is 3.5 μm; however, instruments are normally operated above optimum slit width, which eases the problems of mechanical stability and provides greater light throughput.

The lower spectral bandwidth possible with the scanning system has two important consequences. First, for an isolated spectral line there will be a slight improvement in the limit of detection due to an increase in the line-to-background ratio, but more importantly, substantial improvements in freedom from spectral interference can be realized in certain circumstances (see in particular

Part 1, Sections 4.1.7.7, 7.7.6, and 7.7.7). Boumans [155] has distinguished three conditions of spectral interference: (1) line overlap in which the spectral lines are to an extent superimposed, (2) line interference in which a concomitant line intrudes on the measuring spectral window, and (3) background interference in which the wing of a neighboring spectral line intrudes on the measurement window. In cases of close line overlap, improvements in spectral bandwidth have minimal effect; for background interference the interference decreases almost linearly with decrease in spectral bandwidth, but for line interference substantial gains are possible and ultimately the interference may be eliminated. Elimination or reduction of spectral interference also decreases the limit of detection (see, in particular, Part 1, Section 7.7.5). These observations point to some further conclusions about the relative merits of single and multichannel instrumentation.

The multichannel spectrometer provides moderate limits of detection and is best suited to operating at concentration levels well above the detection limit, where the improved measurement statistics allow full advantage to be taken of its ability to provide rapid interelement corrections but, this may be limited by the availability of channels. Some modern multichannel spectrometers employ spectrum shifters to enable scanning of the spectral lines; however, the greater spectral bandwidth caused by the wider exit slits implies that automatic background corrections performed with such devices will be less effective than on equivalent single-channel instruments.

The single-channel instrument is the preferred choice when ultimate power of detection is required. This is particularly so if the instrument is sufficiently automated to permit optimization of the operating parameters for each spectral line selected. The compromise operating conditions employed for simultaneous multielement determinations may impose half an order of magnitude sacrifice in limits of detection for some elements. On the other hand, implementation of interelement corrections on a single-channel instrument requires a longer analysis time to measure the additional spectral lines and hence more sample is consumed. This disadvantage, however, is offset by the greater ability of the single-channel system to perform accurate automatic background correction.

It should be emphasized that blank subtraction without matrix correction is not a satisfactory method of determining net line intensities from an ICP because it does not take account of the differences that occur between the spectrum of the sample and the standard, which are not related directly to the analyte (cf. Part 1, Sections 4.3, 4.6, and 7.5).

4.5.3. Nebulizers and Sample Introduction

The nebulizer/spray chamber combination is the weakest link in the ICP instrumentation scheme for three reasons: (1) nebulizers are prone to blocking when nebulizing solutions containing particulates and/or a high dissolved-solids content; (2) the efficiency of the nebulizer/spray chamber combination is low, typically 2%; and (3) at high concentrations, analyte flicker noise, derived from the nebulizer, is dominant [156].

Most samples encountered in the agricultural laboratory have a relatively high dissolved-solids content, although in preparing samples for analysis by ICP it is usual to attempt to limit the dissolved solids to about 2%. Given that variations in nebulizer performance due to changes in viscosity have been dealt with by standardizing the acid and matrix content of the solutions or by feeding the nebulizer from a peristaltic pump, the requirement is for a nebulizer/spray chamber/ torch combination that can continue to work for long periods without salting up. Recent improvements in torch design [157], in which the injector and outer torch tubes are streamlined, have improved the performance with organic solvents and have partially solved the problem of salting-up of the injector tube. An alternative solution to this problem has been suggested by Trassy [158] and involves using an additional gas flow to sheath the sample carrier gas flow as it passes through the injector, thereby preventing salt deposition on the tube walls.

Variation in nebulizer performance usually causes problems when dealing with unfamiliar samples that may be close to the solubility limits for certain components, or that contain fibrous or particulate matter. Wohlers and Hoffman [159], for example, have shown that the closeness to saturation is more important than the absolute dissolved solids content in causing blockage. Blockage due to high salt loading usually occurs in the gas flow channel, particularly in concentric nebulizers, although recent improvements, notably recessing of the liquid capillary tip have, to a degree, resolved this problem. Blockage of the liquid capillary is usually caused by the presence of particulate matter and particularly fibres from filter papers and clothing.

The Babington nebulizer [160–162] or its modern "V" groove, and cone entrainment derivatives [163–165] (cf. Part 1, Section 6.2.3) have the capability of nebulizing high dissolved solids content solutions and even slurries containing particles up to 1 mm in size, although for complete atomization in the ICP only particles with sizes below 10 μm should be introduced. Mohamed and Fry [165] have used the slurry atomization technique to determine Cu, Mn, and Zn in fresh beef liver and beef steak by AAS and it is evident that this approach will be adopted for ICP–AES in the near future.

4.5.4. Interferences and Optimization

The nature and cause of interferences encountered in ICP–AES are extensively dealt with in other chapters in this work. Most types of interference will be encountered from the complex matrices in agricultural samples. The commonest are those due to spectral interference and stray light effects, but most difficult to deal with are those related to the presence of easily ionizable materials. It is well known that such interferences may be gross or minimal, depending upon the operating conditions. Therefore, optimization is of considerable importance, but most studies of optimization, and descriptions of optimization procedures such as the Simplex method, have concentrated on improving the signal-to-background ratio [167–170]. This is understandable, but often the conditions for maximum signal-to-background are quite different from those required for minimizing interferences (cf. Part 1, Section 4.7.9). For real samples, it will nearly always be the case that the conditions should be selected to minimize the interferences and within that constraint some fine-tuning may be possible to improve the sensitivity.

4.6. CONCLUSIONS

The ICP is now a technique with proven merit for agricultural analysis and is being used in laboratories throughout the world [171]. It is unlikely to replace AAS or conventional arc spectrometry in the agricultural laboratory; it will, however, complement them and do much to meet the ever increasing demand for elemental analysis.

It seems unlikely, even with substantial improvements in sample introduction equipment, particularly the nebulizer, that the ICP can ever achieve the limits of detection reported for electrothermal atomization AAS. Furthermore, multielement systems are now being developed for AAS. Nevertheless, for routine analysis, the greater dynamic range of the ICP ensures a more accessible multielement capability and it is this aspect that perhaps most ensures its long-term value for agricultural analysis.

REFERENCES

1. D. W. Futty and F. T. Dry, "The Soils of the Country Round Wick," in, *Memoirs of the Soil Survey of Great Britain*, Scotland, Department of Agriculture and Fisheries for Scotland (1976).
2. A. M. Ure, "Atomic Absorption and Flame Emission Spectrometry," in K. Smith, ed., *Soil Analysis—Modern Instrumental Techniques*, Dekker, New York (1981).
3. E. J. Underwood, *Trace Elements in Human and Animal Nutrition*, 4th ed., Academic Press, New York/London (1977) p. 1.
4. R. L. Mitchell, D. Purves, R. O. Scott, and R. C. Voss, *Spectrochemical Methods for the Analysis of Soils, Plants and Other Agricultural Materials*, The Macaulay Institute for Soil Research, Aberdeen (1971).

5. H. J. M. Bowen, *Environmental Chemistry of the Elements*, Academic, London (1979).
6. R. L. Mitchell, *Spectrochemical Analysis of Soils, Plants and Related Materials*, Technical Communication No. 44A of the Commonwealth Bureau of Soils, Harpenden (1964).
7. K. C. Beeson, *The Mineral Composition of Crops with Particular Reference to the Soils in Which They were Grown*, United States Department of Agriculture, Miscellaneous Publication No. 369 (1941).
8. D. J. Swaine, *The Trace Element Content of Fertilizers*. Technical Communication No. 52, Commonwealth Bureau of Soils, Harpenden (1962).
9. D. D. Manske and R. D. Johnson, *Pestic. Monit. J.* **10,** 134 (1977).
10. J. T. Tanner and M. H. Friedman, *J. Radioanal. Chem.* **37,** 529 (1977).
11. G. Iyengar, W. E. Kollmer, and H. J. M. Bowen, *Elemental Composition of Human Tissues and Body Fluids*, Springer-Verlag, Berlin/Heidelberg/New York (1978).
12. H. Smith, *J. Forensic Sci. Soc.* **4,** 192 (1964).
13. E. M. Sowden and S. R. Stitch, *Biochem. J.* **67,** 104 (1957).
14. E. I. Hamilton, M. J. Minski, and J. J. Cleary, *Sci. Total Environ.* **1,** 341 (1972/1973).
15. J. M. Harvey, *Queensl. J. Agric. Sci.* **9,** 47 (1952); **10,** 127 (1953).
16. J. W. Suttie, P. H. Phillips, and R. F. Miller, *J. Natr.* **65,** 293 (1958).
17. J. A. Weatherall and S. M. Weidmann, *J. Pathol. Bacteriol.* **78,** 233 (1959).
18. M. M. Joselow, L. J. Goldwater, and S. B. Weinberg, *Arch. Environ. Health* **15,** 64 (1967).
19. L. Kosta, A. R. Byrne, and V. Zelenko, *Nature* **254,** 238 (1975).
20. F. S. Hsu, L. Krook, W. G. Pond, and J. R. Duncan, *J. Natr.* **105,** 112 (1975).
21. G. W. Monier-Williams, *Trace Elements in Food*, Chapman and Hall, London (1949).
22. H. A. Schroeder and I. H. Tipton, *Arch. Environ. Health* **17,** 965 (1968).
23. J. Bacso, P. Kovacs, and S. Horvath, *Radiochem. Radioanal. Lett.* **33,** 273.
24. V. Valkovic, *Trace Elements in Human Hair*, Garland S.T.P.M.
25. F. Losee, T. W. Cutress, and R. Brown, *Trace Subst. Environ. Health—7*, Proc. Univ. Mo. Annu. Conf., 7th, 1973 (1974), p. 19.
26. H. Spencer, M. Li, J. Samachson, and D. Laszlo, *Metab. Clin. Exp.* **9,** 916 (1960).
27. N. Wolf, I. Gedalia, S. Yariv, and H. Zuckerman, *Arch. Oral Biol.* **18,** 233 (1973).
28. C. A. Mawson and M. I. Fischer, *Biochem. J.* **36,** 696 (1953).
29. C. A. Mawson and M. I. Fischer, *Can. J. Med. Sci.* **30,** 336 (1952).
30. G. Weitzel, F. J. Stecker, U. Roester, E. Buddecke, and A. M. Fretzdorff, *Hoppe-Seyler's Z. Physiol. Chem.*, **292,** 221 (1953).

31. P. Allain and Y. Mauras, *Anal. Chem.* **51,** 2089 (1979).
32. A. M. Gunn, G. F. Kirkbright, and L. N. Ophelm, *Anal. Chem.* **49,** 1492 (1977).
33. R. Bock, *Decomposition Methods in Analytical Chemistry* (translated by I. L. Marr) Blackie, Glasgow (1979).
34. M. L. Jackson, *Soil Chemical Analysis*, Prentice-Hall Inc., New Jersey (1958).
35. T. R. Peck and S. W. Melsted, in L. M. Walsh and J. D. Beaton, (Eds.), *Soil Testing and Plant Analysis*, Soil Science Society of America Inc. (1973).
36. R. O. Scott and A. M. Ure, *Proc. Analyt. Div. Chem. Soc.* **9,** 288 (1972).
37. G. L. Fisher, L. G. Davies, and L. S. Rosenblatt, in P. D. La Fleur (Ed.), *Accuracy in Trace Analysis*, Nat. Bur. Stand. (U.S.) Spec. Publ. **422,** 575 (1976).
38. E. C. Kuehner, R. Alvarez, P. J. Paulsen, and T. J. Murphy, *Anal. Chem.* **44,** 2050 (1972).
39. V. D. Anand, J. M. White, and H. V. Nino, *Clin. Chem.* **21,** 595 (1975).
40. M. Zief and J. W. Mitchell, *Contamination Control in Trace Element Analysis*, Wiley, New York (1976).
41. K. K. S. Pillay, C. C. Thomas, J. A. Sondel, and C. M. Hyche, *Anal. Chem.* **43,** 1419 (1971).
42. H. O. Fourie and M. Peisach, *Analyst* **102,** 193 (1977).
43. P. Strohal, S. Lulic, and O. Jelisavcic, *Analyst* **94,** 678 (1969).
44. S. R. Koirtyohann and C. A. Hopkins, *Analyst* **101,** 870 (1976).
45. R. H. Scott and A. Strasheim, *Anal. Chim. Acta* **76,** 71 (1975).
46. J. B. Jones, in R. M. Barnes, (Ed.) *Applications of Inductively Coupled Plasma to Emission Spectroscopy*, Eastern Analytical Symposium, The Franklin Institute Press, Philadelphia (1977).
47. J. A. McHard, S. J. Foulk, S. Nikdel, A. H. Ullman, B. D. Pollard, and J. D. Winefordner, *Anal. Chem.* **51,** 1613 (1979).
48. J. Locke, *Anal. Chim. Acta* **113,** 3 (1980).
49. F. E. Lichte, S. Hopper, and T. W. Osborn, *Anal. Chem.* **52,** 120 (1980).
50. R. C. Munter, R. A. Grande, and P. C. Ahn, *ICP Information Newslett.* **5,** 368 (1979).
51. R. L. Dahlquist and J. W. Knoll, *Appl. Spectrosc.* **32,** 1 (1978).
52. Method No. 3.008, *Official Methods of Analysis*, AOAC, 12th Edition (1975).
53. J. Davidson, *J. Assoc. Off. Agric. Chem.* **14,** 551 (1931).
54. M. Prasad and M. Spiers, *J. Agric. Food Chem.* **26,** 824 (1978).
55. J. R. Etherington and M. S. Davies, *Plant Soil* **50,** 497 (1978).
56. B. Pahlavanpour, M. Thompson, and L. Thorne, *Analyst* **106,** 467 (1981).
57. D. Hoede and H. A. Van der Sloot, *Anal. Chim. Acta* **111,** 321 (1979).
58. M. Feinberg and C. Ducauze, *Anal. Chem.* **52,** 207 (1980).
59. C. E. Mulford, *At. Absorption Newslett.* **5,** 135 (1966).
60. J. Kumpulainen, *Anal. Chim. Acta* **113,** 335 (1980).

61. K. Oikawa and Y. Ohyagi, *Bunseki Kagaku* **25,** 631 (1976).
62. C. E. Gleit and W. D. Holland, *Anal. Chem.* **34,** 1454 (1962).
63. B. B. Stafford, in D. D. Hemphill, (Ed.), *Trace Substances in Environmental Health*, University of Missouri, Missouri, USA (1968) p. 269.
64. G. J. Lutz, J. Stemple, and H. L. Rook, in *Proc. Modern Trends in Activation Analysis*, (1976), p. 1310.
65. Z. Grobenski, M. Melcher, and B. Welz, *Z. Anal. Chem.* **290,** 114 (1978).
66. G. Nise and O. Vesterberg, *Scand. J. Work. Environ. Health* **5,** 404 (1979).
67. D. Shapcott, *Dev. Nutr. Metab.* **2,** 43 (Cr Nutr. Metab.) (1979).
68. R. Djudzman, E. Van den Eckhout, and P. Moerloose, *Analyst* **102,** 688 (1977).
69. W. Wolf, W. Mertz, and R. Masironi, *J. Agric. Food Chem.* **22,** 1037 (1974).
70. G. F. Carter and W. B. Yeoman, *Analyst* **105,** 295 (1980).
71. S. G. Capar and J. H. Gould, *J. Assoc. Off. Anal. Chemists* **5,** 1054 (1979).
72. Analytical Methods Committee (Royal Society of Chemistry, UK), *Analyst* **84,** 214 (1959).
73. J. W. Jones, S. G. Capar, and T. C. O'Haver, *Analyst* 107, 353 (1982).
74. D. J. David, *Progr. Anal. Atomic Spectrosc.* **1,** 225 (1978).
75. K. Scott, *Analyst* **103,** 754 (1978).
76. S. S. Chao and E. E. Pickett, *Anal. Chem.* **52,** 335 (1980).
77. O. E. Clinton, *Analyst* **102,** 187 (1977).
78. A. J. Thompson and P. A. Thoresby, *Analyst* **102,** 9 (1977).
79. V. A. Thorpe, *J. Assoc. Offic. Anal. Chemists* **54,** 206 (1971).
80. J. F. Uthe, F. A. J. Armstrong, and M. P. Sainton, *J. Fish. Res. Board of Canada,* **27,** 805 (1970).
81. A. M. Ure and C. A. Shand, *Anal. Chim. Acta.* **72,** 63 (1974).
82. M. T. Jeffus, J. S. Elkins, and C. T. Kerner, *J. Assoc. Offic. Anal. Chemists* **53,** 1172 (1970).
83. Analytical Methods Committee (Royal Society of Chemistry, UK), *Analyst* **101,** 62 (1976).
84. D. Scott and J. H. Thomas, *Soil Sci.* **124,** 319 (1977).
85. A. M. Ure, *Proc. Anal. Div. Chem. Soc.* **17,** 409 (1980).
86. M. L. Berrow and W. M. Stein, *Analyst* **108,** 277 (1983).
87. J. C. Van Loon, J. Lichwa, D. Ruttan, and J. Kinrade, *Water, Air and Soil Poll.* **2,** 473 (1973).
88. J. C. Van Loon and J. Lichwa, *Environ. Letts.* **4,** 1 (1973).
89. K. C. Thompson and K. Wagstaff, *Analyst* **105,** 641 (1980).
90. W. J. Price and P. J. Whiteside, *Analyst* **102,** 664 (1977).
91. J. A. C. Broekaert and F. Leis, *Spectrochim. Acta* **34B,** 167 (1979).
92. A. A. Verbeek, M. C. Mitchell, and A. M. Ure, *Anal. Chim. Acta* **135,** 215 (1982).

93. C. S. Piper, *Soil and Plant Analysis*, The University of Adelaide, Adelaide (1942).

94. D. J. Hissink, *Soil Sci.* **15,** 269 (1923).

95. N. M. Scott, *J. Sci. Food Agric.* **32,** 193 (1981).

96. J. F. Alder, A. M. Gunn, and G. F. Kirkbright, *Anal. Chim. Acta* **92,** 43 (1977).

97. J. L. Hern, in R. M. Barnes (Ed.), *Applications of Plasma Emission Spectrochemistry*, Heyden, Philadelphia (1979), p. 23.

98. D. W. Hoult, M. M. Beaty, and G. F. Wallace, *At. Spectrosc.* **6,** 157 (1980).

99. J. B. Jones, *Comm. Soil Plant Anal.* **8,** 349 (1977).

100. P. N. Soltanpour, S. M. Workman, and H. P. Schwab, *Soil. Sci. Am. J.* **43,** 75 (1979).

101. J. L. Manzoori, *Talanta* **27,** 682 (1979).

102. J. Hislop and I. J. Cooke, *J. Soil Sci.* **105,** 8 (1968).

103. A. W. Boorn, M. S. Cresser, and R. F. Browner, *Spectrochim. Acta* **35B,** 823 (1980).

104. T. Ito, H. Kawaguchi, and A. Mizuike, *Bunseki Kagaku* **28,** 648 (1979).

105. P. K. Hon, O. W. Lau, and C. S. Mok, *Analyst* **105,** 919 (1980).

106. I. Ikuno, J. A. Whitehead, and R. E. White, *J. Assoc. Off. Anal. Chemists* **61,** 664 (1978).

107. W. Schmidt and F. Dietl, *Z. Anal. Chem.* **291,** 213 (1978).

108. C. H. Williams, D. J. David, and O. Iismaa, *Comm. Soil Sci. Plant Anal.* **3,** 399 (1972).

109. M. J. Dudas, *At. Absorption Newslett.* **13,** 109 (1974).

110. S. R. Koirtyohann and M. Hamilton, *J. Assoc. Offic. Anal. Chemists* **54,** 787 (1971).

111. J. R. Melton, W. L. Hoover. and P. A. Howard, *J. Assoc. Offic. Anal. Chemists* **52,** 950 (1969).

112. J. Jago, P. E. Wilson, and B. M. Lee, *Analyst* **96,** 349 (1971).

113. L. R. P. Butler and P. M. Mathews, *Anal. Chim. Acta* **36,** 319 (1966).

114. W. J. Simmons and J. F. Loneragon, *Anal. Chem.* **47,** 566 (1975).

115. W. J. Simmons, *Anal. Chem.* **47,** 2015, (1975).

116. I. I. Petrov, P. L. Tsalev, and A. I. Barsev, *At. Spectrosc.* **1,** 47 (1980).

117. J. E. Allen, *Spectrochim. Acta* **17,** 459 (1961).

118. J. E. Allen, *Analyst* **86,** 530 (1961).

119. E. Lakanen, *At. Absorption Newslett.* **5,** 17 (1966).

120. C. H. Kim, C. M. Owens, and L. E. Smyth, *Talanta* **21,** 445 (1975).

121. Zs. Horvath, K. Falb, and M. Varju, *At. Absorption Newslett.* **16,** 152 (1977).

122. G. R. Sirota and J. F. Uthe, *Anal. Chem.* **49,** 823 (1977).

123. A. B. Baker and R. L. Smith, *J. Agric. Food Chem.* **22,** 103 (1974).

124. J. Lener and B. Bibr, *J. Agric. Food Chem.* **19,** 1011 (1971).

125. H. M. Kingston, I. L. Barnes, T. C. Rains, and M. A. Champ, *Anal. Chem.* **50,** 2064 (1978).
126. R. E. Sturgeon, S. S. Berman, A. De Saulniers, and D. S. Russell, *Talanta* **27,** 85 (1980).
127. W. J. Haas, Jr., V. A. Fassel, F. Grabau IV, R. N. Kniseley, W. L. Sutherland, *Adv. Chem. Ser.* **172,** 91 (1979).
128. R. M. Barnes and J. S. Genna, *Anal. Chem.* **51,** 1065 (1979).
129. F. M. Teeny, *J. Assoc. Offic. Anal. Chemists.* **61,** 43 (1978).
130. S. D. Dassani, B. E. McClellan, and M. Gordon, *J. Agric. Food Chem.* **23,** 671 (1975).
131. J. L. Kacprzak and R. Chvojka, *J. Assoc. Offic. Anal. Chemists.* **59,** 153 (1976).
132. Analytical Methods Committee (Royal Society of Chemistry, UK), *Analyst* **102,** 769 (1977).
133. D. D. Siemer and L. Hageman, *Anal. Lett.* **8,** 323 (1975).
134. J. Aggett and A. C. Aspell, *Analyst* **101,** 341 (1976).
135. A. E. Smith, *Analyst* **100,** 300 (1975).
136. D. D. Siemer and P. Koteel, *Anal. Chem.* **49,** 1096 (1977).
137. R. G. Godden and D. R. Thomerson, *Analyst* **105,** 1137 (1980).
138. M. Thompson, B. Pahlavanpour, J. S. Walton, and G. F. Kirkbright, *Analyst* **103,** 568 (1978).
139. M. Thompson, B. Pahlavanpour, J. S. Walton, and G. F. Kirkbright, *Analyst.* **103,** 705 (1978).
140. M. Thompson and B. Pahlavanpour, *Anal. Chim. Acta* **109,** 251 (1979).
141. B. Pahlavanpour, J. H. Pullen, and M. Thompson, *Analyst* **105,** 274 (1980).
142. B. Pahlavanpour, M. Thompson, and L. Thorne, *Analyst* **105,** 756 (1980).
143. R. C. Fry, M. B. Denton, D. L. Windsor, and S. J. Northway, *Appl. Spectrosc.* **33,** 399 (1979).
144. S. S. Sloat, *Report IS-T-776,* Iowa State University, Iowa (1977).
145. D. L. Windsor and M. B. Denton, *Anal. Chem.* **51,** 1116 (1979).
146. D. L. Windsor and M. B. Denton, *J. Chromatogr. Sci.* **17,** 492 (1979).
147. D. Sommer and K. Ohls, *Z. Anal. Chem.* **295,** 337 (1979).
148. D. M. Fraley, D. Yates, and S. E. Manahan, *Anal. Chem.* **51,** 2225 (1979).
149. C. H Gast, J. C. Kraak, H. Poppe, and F. J. M. J. Maessen, *J. Chromatogr.* **185,** 54 (1979).
150. M. Morita, T. Uehiro, and K. Fuwa, *Anal. Chem.* **52,** 349 (1980).
151. O. G. Koch, P. D. LaFleur, G. H. Morrison, E. Jackworth, A. Townshand and G. Tölg, *Pure Appl. Chem.,* **54,** 1565 (1982).
152. T. W. Osborn, *J. Agric. Food Chem.* **25,** 229 (1977).
153. N. R. McQuaker, P. D. Kluckner, and G. N. Chang, *Anal. Chem.* **51,** 888 (1979).
154. H. W. Faust, in E. L. Grove, (Ed.), *Analytical Emission Spectrometry,* Vol. 1, Part 1, Marcel Dekker Inc., New York (1971), p. 185.

155. P. W. J. M. Boumans, *Line Coincidence Tables for Inductively Coupled Plasma Atomic Emission Spectrometry*, Vol. 1, Pergamon Press, Oxford (1980).

156. G. L. Walden, J. N. Bower, S. Nikdel, D. L. Bolton, and J. D. Winefordner, *Spectrochim. Acta* **35B,** 535 (1980).

157. P. W. J. M. Boumans and M. C. Lux-Steiner, *Spectrochim. Acta* **37B,** 97 (1982).

158. C. Trassy, *ICP Information Newslett.* **11,** 589 (1980).

159. C. C. Wohlers and C. J. Hoffman, *ICP Information Newslett.***9,** 502 (1981).

160. R. S. Babington, *U.S. Patents*, 3,421,692; 3,421,699; 3,425,058; 3,425,059; 3,504,859.

161. R. C. Fry and M. B. Denton, *Anal. Chem.* **49,** 1413 (1977).

162. R. C. Fry and M. B. Denton, *Appl. Spectrosc.* **33,** 393 (1979).

163. R. F. Suddendorf and K. W. Boyer, *Anal. Chem.* **50,** 1769 (1978).

164. J. F. Wolcott and C. B. Sobel, *Appl. Spectrosc.* **32,** 591 (1978).

165. B. L. Sharp, Patent: The Conespray Nebulizer–Method of and Apparatus for the Nebulization of Liquid and Liquid Suspensions, British Technology Group Assignment Number 8432338; UKPA 8531504.

166. N. Mohamed and R. C. Fry, *Anal. Chem.* **53,** 450 (1981).

167. S. Greenfield and D. T. Burns, *Anal. Chim. Acta* **113,** 205 (1980).

168. L. Ebdon, M. R. Cave, and D. J. Mowthorpe, *Anal. Chim. Acta* **115,** 179 (1980).

169. L. Ebdon, D. J. Mowthorpe, and M. R. Cave, *Anal. Chim. Acta* **115,** 171 (1980).

170. S. P. Terblanche, K. Visser, and P. B. Zeeman, *Spectrochim. Acta* **36B,** 293 (1981).

171. S. Yamasaki, *Kagaku no Ryoiki Zokan* **127,** 182 (1980).

CHAPTER

5

APPLICATIONS: BIOLOGICAL–CLINICAL

F. J. M. J. MAESSEN

Laboratorium voor Analytische Scheikunde
Universiteit van Amsterdam
Amsterdam, The Netherlands

5.1. Analytical Aims
5.2. Sampling
5.2.1. Introduction
5.2.2. Sample Collection
5.2.3. Sample Storage
5.2.4. Sample Homogenization
5.2.5. Sample Pretreatment
5.3. Sample Digestion
5.3.1. Introduction
5.3.2. Ashing Methods
5.3.2.1. General
5.3.2.2. Dry-Ashing
5.3.2.3. Wet-Ashing
5.3.3. Solubilization and Leaching Methods
5.3.3.1. General
5.3.3.2. Dilution with Water
5.3.3.3. Extraction with Aqueous Leachants
5.3.3.4. Solubilization
5.3.4. Comparison of Digestion Methods
5.3.5. Disgestion Followed by Analyte Preconcentration and Matrix Separation
5.3.5.1. Introduction
5.3.5.2. Chelating Ion-Exchange Procedures
5.3.5.3. Extraction Procedures
5.3.5.4. Digests of Biological Standard Reference Materials
5.3.5.5. Digests of Real Biological Materials
5.4. Analytical Practice
5.4.1. Sample Introduction
5.4.1.1. General
5.4.1.2. Hydride Technique
5.4.1.3. Volatile Metal Chelates
5.4.1.4. Electrothermal Vaporization (ETV)
5.4.2. Illustrative Examples of Analytical Techniques

5.4.3. Materials Analyzed and Elements Determined
5.4.3.1. Chronological Review
5.4.3.2. Determination of Molecular Species
5.4.3.3. Speciation
5.4.3.4. Merits of ICP–AES
5.4.3.5. Research
References

5.1. ANALYTICAL AIMS

The need for elemental analysis of biological materials in animal, medical, and environmental sciences has been well documented [1–4], whereas the number of trace elements of known and potential importance for mammalian organism is steadily increasing [5–8]. At present, more than 30 trace elements are considered to be clinically relevant. These elements can be classified into three categories:

1. Essential trace elements: Ca, Cr, Cu, F, Fe, I, Mn, Mo, Se, Si, Sn, V, and Zn.
2. Nonessential trace elements used therapeutically: Al, Au, B, Bi, Co, Ge, Li, and Pt.
3. Nonessential and toxic trace elements: Ag, As, Ba, Be, Cd, Hg, Ni, Pb, Sb, Te, and Tl.

This classification is not unambiguous, however, because toxic elements such as As and Sb are also used therapeutically. Furthermore the classification into essential and nonessential is questionable [9]. According to Schwarz [7] there is no functional relationship between toxicity and essential biological function. Trace elements that are toxic at relatively small dose levels may be highly essential and indispensable for normal body function in even lower amounts. Each essential trace element is toxic when given in excess. Obvious examples are Cu, Fe, and especially Se. Numerical values of detection limits of some 70 elements have been discussed in Part 1, Section 4.1.7. Tables 4.2 and 4.3, in particular, list detection limits for argon ICPs. Data for both pneumatic and ultrasonic nebulization are covered; they are mainly for compromise operating conditions. The periodic table in Fig. 4.13 of Part 1 indicates the detection limits attainable with the most prominent lines and at the same time reveals for each element the number of prominent lines that yield a detection limit within a factor of 3 from that of the best line. The data are for pure aqueous solutions and medium resolving power. For real samples, Table 4.2 and Figure 4.13 give only an indication of the detection power achievable.

It must be realized that for assessing the applicability of the ICP to a specified analytical problem, Tables 4.2 and 4.3 in Part 1 have only an indicative meaning because most types of biological materials require a sample digestion procedure, which, depending on the nature of the procedure applied, leads to considerable differences between the analyte concentration in the initial sample and that in the eventual digest solution (see Section 5.3). It should be further emphasized that detection limits lower than the lowest shown in Table 4.2 of Part 1 can be obtained by optimizing the experimental variables for a particular element [10–14] or using ultrasonic nebulization (see Table 4.3, Part 1). Finally it should be noted that the detection limits of hydride-forming elements such as As, Bi, Ge, Hg, Pb, Sb, Se, Sn, and Te, can be substantially improved by introducing them into the plasma in the form of their hydrides [15–18] (see Table 4.6, Part 1). However, pneumatic nebulization of the sample solution is, despite drawbacks, still the most widespread approach for routine simultaneous multielement analysis of biological materials [19]. For that reason, the main line in this chapter follows applications using pneumatic nebulization. Alternative sample introduction techniques will be discussed, however, in connection with applications to biological materials. Chapter 6 of Part 1 specifically deals with sample introduction methods and reviews the various techniques and their applications in ICP–AES.

In his excellent review paper on elemental trace analysis of biological materials, Morrison [8] notes that studying the role of trace elements in biochemical metabolism requires three types of information, namely, concentration, distribution, and speciation. So far, only modest progress has been made on the establishment of the distribution of trace elements over the various components of the sample, such as serum protein fractions. The same holds for the chemical form or valence state of the trace element under consideration. This is not so surprising because unambiguous associations and correlations can be established only if the concentrations are accurately known. Providing accurate results about the overall composition of a complex biological sample still requires a major effort, however. Therefore, this chapter gives special attention to factors affecting the accuracy of elemental trace analysis of biological materials. Further, it should be realized that in the analysis of many biological–clinical materials, more is required than reliable data concerning total contents of particular elements. Analyte speciation is often of utmost importance because toxicity, bioavailability, mobility, and bioaccumulation may depend critically on the state of bonding, the element species, organic binding partner, or physicochemical state of the analyte element.

5.2. SAMPLING

5.2.1. Introduction

Not exclusively in ICP–AES analysis, but in all methods employed for trace content determinations, sample handling often constitutes the weakest link of the analytical chain. This is so because the procedures encountered under the heading "sampling" are most frequently responsible for serious systematic errors. Apart from collecting and storing, sampling comprises sample homogenization and pretreatment procedures. The number and complexity of manipulations involved in the preparation of the ultimate sample solution chiefly depends on the nature of the material to be analyzed. Evidently, additional procedures are required for chemical speciation of the analyte elements or separation of the sample into components prior to analysis, for example, separation of blood serum into various groups of proteins. The present discussion of sampling will be confined, however, to procedures used when the overall contents of the analytes have to be determined in an aggregate of constituents, considered as the sample. Even with this restriction, a complete discussion of sampling remains difficult if not impossible simply because the number of different species is too large. Therefore, the discussion below can merely serve as an illustration of the nature of the problems encountered. For additional information, an extensive literature survey by Maienthal and Becker [20] and a review paper by Sansoni and Iyengar [21] are recommended.

5.2.2. Sample Collection

Except when sampling sweat, saliva, or urine, surgical instruments are always needed, for example, needles for venipuncture, blades for sampling skin or tissues as liver and kidney, scissors for hair or nail clippings, and tweezers and forceps for handling the sample [22]. An indication of the extent to which sample contamination can occur was given in 1973 by Versieck et al. [23]. They simulated sampling practice with neutron-activated instruments and measured the contamination introduced by their accompanying radioactivity. The authors startled analysts in the life sciences by reporting the results of their examination on contamination introduced during collection of liver biopsies using stainless steel needles and surgical blades. Analytical results showed that in small-needle biopsies of the liver a positive error of up to 30% can be found for Cu, Mn, and Zn, whereas Ag and Se concentrations can be doubled. Contaminations by cobalt appeared to exceed the concentration found in normal liver by a factor of 4 in some cases. For Ni, and especially for Cr, contaminations of more than 100 times the normal concentration were frequently found. Operative surgical liver biopsies obtained with surgical blades led to less severe contamination. No noticeable error was demonstrated for Ag, Co, Cu, Fe, Mn, Sn, and Zn. But for Au a contamination of up to 30% was found and it appeared that the concentration of Cr and Ni, which are major components of stainless steel, can be doubled as a result of contamination. Later Speecke et al. [22] investigated blood and liver sampling for contamination by Co, Cr, Mn, and Ni using disposable needles for blood and both Menghini biopsy needles and surgical blades for liver. Blood sampling was performed by successively drawing of four series of 20-ml portions. The first portions appeared to be the most strongly contaminated. The concentration levels of the following 20-ml fractions did not show great mutual differences. Comparison of the concentrations of the first 20 mL with the lowest concentration in a series showed that thoroughly preflushing the needle before taking the definite sample can reduce the concentration of the contaminating elements considered with about a factor of 10. Nevertheless, with the exception of Mn, also the less-contaminated 20-mL portions contained still too much Co, Cr, and Ni for making reliable (Ni) or even preliminary (Co, Cr) analyses for these elements. However, various metallic tools made of high-purity steels and implements made of other materials have been useful as controlled contaminants [21]. Evidently, the main disadvantage of using such instruments is that their constituents cannot be determined in the biological material handled. Thus, for venipuncture it is expected that needles made of Pt–Ir alloy of the highest purity would be convenient for the analysis of nonprecious metals in blood [22], whereas Mikac-Devic et al. [24] recommended the use of Teflon intravenous catheters to avoid contamination by the usually applied stainless steel needles. Webb et al. [25] used borosilicate "knives," formed by sheer cleaving of glass rods, and polypropylene forceps for dissection of pig heart tissue in which 13 trace elements were determined. In the author's laboratory aliquots of human spleen, liver, and placenta samples were taken using the titanium knives made available by the International Atomic Energy Agency, Vienna. Gills and McClendon [26] observed that the composition of a polyethylene excision implement makes it optimum as a sampling device but its practicability for hard tissue must be further tested. The use of adequate laser beams as described by Hislop and Parker [27] for cutting hard and soft tissue could offer interesting possibilities; some loss of trace elements on the surface may occur but would be negligible with regard to the entire sample [20]. As can be assumed from the examples discussed, serious hazards of contamination of chemical materials exist in the sample collection step. Further, it should be realized that sampling and sample storage, at least in case of medical sampling, are often carried out by persons not thoroughly familiar with the requirements of trace element analysis, for example, pathologists. Therefore, close interaction between physician and analyst is imperative to ensure that no significant changes in composition occur during sampling, storage, and transport.

5.2.3. Sample Storage

Since the sample container represents one of the earliest and potentially one of the largest sources of sample contamination, much of the analytical accuracy will depend upon the choice of the container materials and the efficiency of the cleaning procedures applied. An additional aspect connected with the analysis of biological materials is the limited availability of biological standards with reliable data for many trace elements of clinical interest. This makes the use of inorganic aqueous standard solutions often necessary. Hence, the accuracy and stability of such solutions should also be considered. From normal laboratory glassware, quartz, or synthetic materials, trace elements may be easily introduced, especially into liquid samples like body fluids. Because of the inertness and relative cleaniness of polyethylene, Teflon, and high-purity quartz, Kuehner et al. [28] and Mitchell [29] selected these materials as the most suitable ones for sample containers. When involved with large quantities of samples, polyethylene is principally used instead of Teflon and quartz because of economic considerations [30]. Moody and Lindstrom [31] examined the levels of impurities present in 12 different plastics, as well as the quantities of impurities leached from the plastics by acid cleaning. The various Teflons and conventional polyethylene bottles were found the least contaminating containers, once they had been cleaned. The authors proposed a cleaning procedure needing several weeks and consisting of an intense soaking with diluted hydrochloric acid followed by diluted nitric acid and the purest available distilled water. Similar treatments were applied in other laboratories [32–34]. Also losses may occur. Thus, in the author's laboratory, sorption losses were studied for Ag, As, Cd, Se, and Zn at a concentration level of 10^{-7} M in aqueous solutions during storage in containers made of borosilicate glass, high-pressure polyethylene, and polytetrafluoroethylene (PTFE) [35]. Besides pH and storage time, the effect of the ratio of inner container surface to sample volume was studied. It was shown that reduction of contact time and specific surface may be helpful in lowering sorption losses. However, the sorption behavior of trace elements depends on a variety of factors which, taken together, make sorption losses rather difficult to predict. The various factors involved may be classified into four categories. The first category is concerned with the analyte itself, for example, chemical form and concentration. The second category includes the characteristics of the solution, such as the presence of acids, dissolved material, complexing agents, dissolved gases (especially oxygen, which may influence the oxidation state), and suspended matter, which is competitive in the sorption process. The third category comprises container properties such as chemical composition, surface roughness, and cleanliness. The fourth category consists of external factors, like temperature, contact time, access of light, and sample agitation. All of these factors must be considered in assessing the hazard of sorption losses during a complete analysis. A long-term storage of, for example, urine for quality control and intercomparison purposes is often unavoidable. Urine samples frequently show precipitates upon standing and it was found by Golimowski et al. [36] that if urine is not acidified immediately after collection, these precipitates absorb up to 30% of the lead and 5% of the cadmium content of the urine, even if later acidified to pH = 2. A further study performed by Stoeppler et al. [37] showed that a distinct adsorption of nickel at the precipitates occurred over the whole pH range studied (pH $<$ 1 to pH = 6), but was remarkable only at higher pH values. Effects from container composition, storage time, and temperature on serum mineral levels were examined by Fisher et al. [38]. Containers of five different compositions (Pyrex, polypropylene, polycarbonate, Teflon, Vycor) containing the same volume of human serum were stored at three different temperatures (25, 8, −15°C). The elements Ca, Cu, K, Mg, Na, and Zn were determined at approximately a geometric progression of storage times to 50 days of storage. Container effects were highly significant for K, Na, and Zn, but were small with varying container composition. Essentially, no changes were observed in the refrigerated and frozen samples up to 50 days, but changes did occur in the samples stored at room temperature. The authors recommend that serum samples should be quickly frozen after harvesting in tightly capped containers with a minimum of space above the serum. Ibbott [39] recommends separating the serum from the clot as soon as possible to avoid cell leakage. Omang and Vellar [40] examined concentration gradients of Ca, Cu, and Na in serum, sweat, and urine during storage, freezing, and thawing. Even after 3 weeks at room temperature, no concentration gradient was found for Cu and Na. However, the main part of the Ca content appeared to be precipitated from the urine and the serum samples. The samples were also fractionated by sequential thawing and freezing. In this way top–bottom concentration gradients of up to a factor of 100 were produced, demonstrating the need for thorough homogenization of defrosted biological fluids prior to analysis. Freezing is a suitable long-term preservation method and should be done at −20°C [21]. However, because of denaturation of proteins and redistribution of elements due to rupture of cell walls by ice crystals, some irreversible processes may take place.

Owing to strongly reduced possibility of leaching or absorption, dry solid samples do not undergo important positive or negative changes in concentrations upon storage. Hence, the most practical method for storage is lyophilization in which, after a fast-freezing process, the water content is slowly eliminated. A constant weight is usually reached after a 24- to 48-h period with most materials [22]. Compared with body fluids, biological tissues have the advantage of lower mobility and chemical reactivity. Therefore, the hazards of contamination from container walls and changes in the mean composition are smaller than for fluids. A general disadvantage is the problem of homogeneity and the difficulty in getting truly representative samples.

5.2.4. Sample Homogenization

Few special methods have been described for homogenization of biological tissues for trace analysis. Blenders consisting of rotating knives in jars or across sieves are normally used, but such equipment is usually made of steels, which contain elements of clinical interest. Replacing the stainless steel parts by high-purity metals having satisfactory mechanical properties does not offer a practical solution here because of the limited availability of suitable metals and the associated high costs. Only titanium has been applied in rotating knife blenders [41]. Bowen and Gibbons [42] reported on experiences with a Perspex mill used for the homogenization of dry biological material. Donev [43] described a homogenizer consisting of a Teflon piston in a Plexiglass or quartz tube. However, these two techniques are only suitable for soft tissues not containing tough material such as blood vessels. An additional drawback of methods involving direct processing of wet samples is that the blended material is very susceptable to segregation upon standing [44]. An efficient and clean technique, in principle suitable for the homogenization of all types of biological tissues, was suggested by Iyengar [45] and Iyengar and Kasperek [46] under the name of "Brittle Fracture Technique." In this method, which was originally developed for hard tissues, the sample material is placed in a lidded Teflon container along with a Teflon-coated steel ball. The container is then cooled in liquid nitrogen and subsequently vibrated at some thousands cycles per minute for a preset period of time. In view of the promising potentialities of this technique, a systematic study of the optimum experimental conditions was carried out in the author's laboratory [47]. Human placenta was selected as a test sample because it contains tissues with markedly different degree of toughness. The study as a whole was directed toward ICP–AES trace metal analysis of solid biological materials. Consequently, only nonmetallic materials like Perspex, polyethylene, and Teflon were considered for construction of the equipment for homogenization. The stability of these materials to rigorous cleaning procedures was tested by prolonged exposure to nitric acid [30, 31] and to moderately high temperatures. Various refrigerants and grinding conditions were applied. Fragmentation of the sample was determined by fractionated sieving and the degree of homogenization of the resulting mixture by means of radiotracers. The best fragmentation of placental tissue was achieved by procedures comprising two alternate grinding periods in combination with strong refrigerants. The homogeneity appeared to vary from good to excellent depending on the effort put into the grinding procedure. Even a 10-mg subsample could be made representative of a 10-g sample.

Biological fluids do not present much of a heterogeneity problem if the sample is freshly taken, or the completely defrosted clear sample is vigorously shaken for a few minutes just before extracting the subsamples. However, liquids such as milk pose some problems due to the limited stability of emulsions and tend to segregate on standing for longer times.

5.2.5. Sample Pretreatment

Before subjecting the sample to a number of processes aimed at the transformation of the sample into its eventual form, which is suitable for analysis, some additional pretreatment procedures may be required. The type of such a pretreatment depends on the nature of the analytical problem. For instance, addition to blood of anticoagulants as heparine, citrate, or EDTA is common practice. Frequently chemical preservatives are added to biological tissues or whole organs. As a matter of course, the type, amount, and purity grade of the additives should be selected in such a way that element contamination or change of the initial sample composition is prevented.

The futility of trace element analysis of samples affected with surface contaminations, for example, hair, is generally recognized. However, washing procedures pose a fundamental problem because the contaminations from external sources that are deposited at the material surface must be completely removed while at the same time the trace metals present in the internal structure of the material should be left unaffected. Since hair is known to be a depository for trace elements in the body, and besides, it is easy to collect and store, numerous washing procedures for removing surface contamination have been examined. Salmela et al. [48] recently summarized a number of washing solutions that have been used: distilled water, ionic and nonionic detergents, combinations of aqueous detergents and organic solvents, chelating agents, and mineral acids. According to Mattera et al. [49] the extent to which metal-to-hair binding can be explained in terms of metals that are of endogenous or exogenous origin is an unsettled question and currently a subject of debate. Assarian and Oberleas [50] have even concluded that trace element determinations in human hair are too sensitive to pretreatment procedures to be a reliable source of clinically important information. However, Salmela and al. (48) pointed out that for a number of elements, a level exists below which the concentration cannot be reduced by further washings. The importance of a standardized washing procedure is widely recognized. Nevertheless, different laboratories use different procedures and there is no consensus on how the washing should be done. Consequently, additional work is required to define a standard washing procedure.

5.3. SAMPLE DIGESTION

5.3.1. Introduction

As was recently emphasized by Mertz [51], sample digestion is perhaps the least understood step in the analysis of biological materials. Yet, for ICP–AES routine analysis of most types of samples, this step is necessary to transform tissues, and mostly fluids, too, into solutions that are both adaptable to the nebulizer and suitable for analysis in the plasma. The most important methods to be considered for ICP–AES are based on dry- or wet-ashing, extraction, and solubilization.

Dry-ashing comprises procedures carried out at elevated temperatures usually employing muffle furnaces, ashing procedures at lower temperatures using activated oxygen, and some specific combustion procedures. Wet-ashing refers to the quantitive destruction of organic matter by oxidizing acids, whether or not in combination with oxidants like concentrated hydrogen peroxide. The wet oxidation ash concentrate is then the sample for analysis. Wet digestion can be carried out in open or closed systems; the latter is indicated as ''pressure digestion.'' Extraction procedures generally imply that the analytes are leached either directly from the original samples or from samples that are digested prior to the extraction. Diluted acids are frequently employed as leachants in the case of direct extraction. Solubilization of biological material can be accomplished by mixing the sample with aqueous or organic solutions of quarternary ammonium hydroxide and by enzymatic digestions. Also mere dilution with water, for example, for analysis of blood serum, is considered as a sample solubilization method.

5.3.2. Ashing Methods

5.3.2.1. General

Ashing generally leads to complete mineralization of the sample. An indication of the chemical composition of ashes resulting from biological materials of human origin may be obtained from Tables 5.1 and 5.2, which list the contents of relevant major elements and iron present in various tissues and body fluids. Although not a major element, iron is also included because it constitutes the most important source of spectral interferences in ICP–AES analysis of biological material [52].

Table 5.1. Composition of Human Body Fluids: Range, or Mean Value, of the Content of some Major Elements and Iron [93]

	Element (mg/L)					
Fluid	Na	K	Ca	Mg	P	Fe
Amniotic fluid	3100	160	70	17	20–40	
Breast milk (ripened)	60–450	350–750	170–650	20–60	60–270	0.2–0.8
Cerebrospinal fluid	3000–3600	80–180	40–50	5–50	10–25	0.2–0.6
Gall (bladder)	3300–8300	300–700	70–700	15–30	1400	0.6–3.8
Gall (liver)	3000–3800	100–500	60–80	15–30	60–250	0.4–3.1
Gastric juice	400–1600	250–650	40–100	3–40	5–180	
Pancreas juice	3200–3300	160–350	40–100	12	0–50	
Saliva	100–800	400–1600	40–110	2–15	120–290	
Serum (adults)	3000–3500	120–200	80–120	15–30	90–150	0.75–1.75
Sweat	0–2400	70–700	5–120	0.5–50	3–6	0.6–2.3
Urine (adults, 24 h)	2700–5100	1300–3200	130–330	50–200	800–2000	$\ll 1$
Seawater[a]	10^4	400	400	1300		

[a]Seawater has been added for comparison (see Section 5.3.5).

Table 5.2. Composition of Human Tissues: Range, or Mean Value, of the Content of some Major Elements and Iron [93]

	Element (mg/kg wet tissue)					
Tissue	Na	K	Ca	Mg	P	Fe
Bone (adults)	1800–6000	500–3000	264×10^3	3900	113×10^3	110–170
Brain (whole)	1200–1900	2200–3300	80–100	100–140	1600–3400	
Hair			1800–5000	10–100		0.8–170
Heart (whole)	1300–1500	1900–2600	50–170	130–160	1400–1500	
Kidney	1700–1900	2150–2250	140–160	100–110	1700–1900	3.3–32
Liver	900–1400	2300–3000	50–90	120–190	1700–2700	28–300
Lung	1700	1900–2150	250	60	1300–1600	150–200
Muscle (skeleton)	800–1200	2250–3600	50–90	180–250	1400–2000	
Nails				20–120		
Placenta	2250	1600	250	80	930	
Skin	1500–2000	900–1800	190–230	40–90	400–1100	
Spleen		3200	85	160	2200	85–170

5.3.2.2. Dry-Ashing

Dry-ashing procedures include ashing at high temperature (400–800°C) and normal (air) pressure, ashing at lower temperature (100–200°C) and reduced pressure of the gaseous oxidant, and several combustion-like procedures. In analytical practice, ashing at high temperature is most frequently applied.

5.3.2.2.1. High Temperature Ashing

High-temperature dry-ashing is usually carried out in a muffle furnace. The operation generally proceeds in three or four steps [53], namely, drying at 105–150°C, preashing from 200 to 400°C, main ashing between 450 and 550°C, and, if after treatment is necessary and involatile analytes are involved, additional ashing in the 700–800°C temperature range.

Classical dry-ashing at higher temperature is mainly combustion by air, combined with pyrolysis [53]. Muffle furnace ashing is used with and without extra air supply [54, 55]. In the former case the sample is glowing and reaches a higher temperature than the air temperature in the furnace. The temperature of the sample also depends on the thickness of the sample layer and the extent of the air supply. A notable advantage of the application of extra air supply is the substantial reduction of the ashing times that can be achieved. However, as a result of the poorly controllable actual ashing temperatures, the procedure becomes more critical, especially when volatile elements like cadmium and lead have to be determined. In the case that ashing is performed without extra air supply, pyrolysis is an important factor in the ashing mechanism. Consequently, the temperature of the sample is then closer to the adjusted furnace temperature. Nevertheless, also under these conditions considerable temperature gradients can still occur in the furnace [56]. In addition to the parameters of ashing temperature, exposure time, and air supply, some other experimental variables have to be considered, namely, the type of the crucible material, the use of ashing aids, and the dissolution procedure of the ash. Porcelain crucibles are frequently applied but their usage is not recommendable because of possible sample contamination that can be caused by the glaze [57, 58]. Further, element- and matrix-dependent retention losses can occur. Rentention losses can also occur for other types of crucible material, for example, fused silica and Pt. This aspect becomes even more complex because the retention behavior also depends on the condition of the surface of the crucible material. Thus Gorsuch [59] examined the retention behavior of new and etched vitreous silica and Pt crucibles expecially for the elements Cu and Pb. Koirtyohann and Hopkins [57] performed similar experiments for Cd, Cr, Fe, and Zn in new and etched porcelain, Vycor (96% silica), and Pt crucibles. Platinum came out rather favorably, but Hamilton et al. [56] found deposits of Pb on this material.

Ashing aids are used to prevent volatilization losses and to obtain carbon-free and or bulky ashes. Bulky ashes can be handled more easily and are less susceptible to retention losses. Frequently employed ashing aids are sulfuric acid and nitric acid. Sulfuric acid is used to prevent losses of, for instance, cadmium and lead [59, 60] and nitric acid for obtaining carbon-free ashes. Examples of ashing aids that create bulky ashes are magnesium nitrate and sodium carbonate [59, 61]. Various acids and mixtures of acids are used to dissolve the ash resulting from dry-ashing procedures. Commonly used solvents are hydrochloric acid [59, 62], nitric acid [62], sulfuric acid [63], and mixtures of hydrochloric and nitric acids [60]. Aqua regia is recommended for dissolving ashes with a high content of phosphates, for example, ashed bone tissue [62].

According to Sansoni and Panday [53], the main advantage of muffle furnace ashing is its simplicity. The method also requires little supervision and is capable of ashing large numbers of samples in routine operation. Notable disadvantages are volatilization and retention losses, long ashing times, and contamination caused by impurities introduced through air and the ashing container.

5.3.2.2.2. Ashing at Lower Temperatures and Reduced Pressure

In dry-ashing methods operated at reduced pressure (70–1000 Pa), the gaseous oxidant, usually oxygen, is activated by a high-frequency electromagnetic field. Two frequencies are used to activate the oxidant, 27.12 and 2450 MHz. The 27.12 MHz frequency is used in equipment called low-temperature asher (LTA)

[63–71]. The ashing temperatures observed for LTA-conditions range from 100 to 200°C. The activated oxidant flows through ashing chambers in which fused silica sample crucibles are placed. When oxygen is used as an oxidant, ashing proceeds rather slowly. Ashing of 1–4 g of freeze-dried soft animal tissue requires 30–40 h [64, 65]. The speed of ashing is greatly increased when a mixture of oxygen and tetrafluoromethane is used [66]. A similar accelaration effect was found by Hamilton et al. [67] when Teflon (PTFE) crucibles instead of fused silica or Pt crucibles were used. The acceleration of the ashing process is most probably due to the action of gaseous fluorine compounds that are released from the PTFE crucibles under LTA conditions. Walsh et al. [68] observed losses of As from air particulates and several synthetic sea salt matrices. Some loss of As was also observed from human lung [69] and the complete recovery of Se is questionable. However, many elements, including Cd and Pb, can be successfully determined in various types of biological material if LTA is employed [72–75].

In the equipment in which the 2450-MHz microwave radiation is used, an oxidant-flushed reaction tube of fused silica contains the dry sample (up to 1 g). The reaction tube is placed in a microwave cavity [70, 71, 76]. Moderate temperatures are supposed to prevail under these ashing conditions, but Ross and Umland [77] found local temperatures of 730°C. On the other hand, the system is well suited for the use of a cold trap for the condensation of volatile products.

5.3.2.2.3. Procedures Involving Combustion

This group of procedures includes combustion of the sample in the so called "Trace-O-Matt," the Schöninger oxidation, and the oxidation according to Berthelot.

"The Trace-O-Matt" [78] consists of a small volume (75 mL) combustion chamber and a liquid nitrogen-cooled condenser. The pelleted sample is burned in a stream of oxygen at atmospheric pressure. Heating is performed by IR lamps. The combustion products are condensed in the cooling unit. The method is rather rapid (reaction time is about 40 min) and enables the digestion of 1-g samples [78].

In the Schöninger procedure [79–82], the dry sample is subsequently wrapped in filter paper, placed in a Pt holder, ignited, and inserted quickly into a flask that is filled with oxygen and contains a suitable liquid for the absorption of the combustion products. Only small samples (25–50 mg) can be digested in this way. Larger samples, up to 5-g dry weight, can be digested when the combustion is carried out at higher oxygen pressure (2.5–4 MPa). In a vessel originally developed by Berthelot [83] and which is provided with an external ignition mechanism, the sample is burned within a few seconds [84, 85]. Obviously, contamination is to be expected from the metal the vessel is made of.

5.3.2.3. Wet-Ashing

Wet decomposition of biological materials can be performed at atmospheric pressure in open systems or at higher pressure in closed systems. Closed systems are used to enable operation at temperatures above the boiling point of the reagent used. The reagents commonly employed in wet-ashing procedures include mainly strong oxidizing mineral acids and hydrogen peroxide. Very often two or more of these reagents are applied in combination because of the inherent advantages of each reagent. The main advantage of wet-ashing over dry-ashing is its speed. On the other hand, wet-ashing methods require the use of comparatively large quantities of reagents that may lead to high blank values or to the necessity of employing excessively expensive reagents.

5.3.2.3.1. Wet Digestion in Open Systems

Open reaction systems comprise simple vessels like Erlenmeyer and Kjeldahl flasks whether or not provided with reflux condensers. Middleton and Stuckey [86] described the successful digestion of a great many substances of biological origin employing a mixture of nitric and sulfuric acid. According to their procedure, the acid mixture is added to the samples and then the sludge is heated to dryness. If charring occurs, fresh portions of concentrated nitric acid are added to the dark-colored ash and heating to dryness is repeated until a white ash results. The combination of nitric and sulfuric acid is also used in the so-called Bethge apparatus [87]. This equipment is provided with a reflux condenser for making optimum use of the nitric acid [59, 61]. The presence of much calcium in the sample precludes the use of sulfuric acid because of precipitation of calcium sulfate and the risk of coprecipitation of trace elements. Compared with HNO_3–H_2SO_4 mixtures, a more powerful combination of reagents is HNO_3–$HClO_4$ [59, 61, 88–91]. The danger associated with the use of perchloric acid stems from the possible formation of unstable perchlorates. Recommendations for safe handling of perchloric acid can be found in [92, 93].

Hydrogen peroxide (50% w/w) is mainly used in combination with sulfuric acid. This combination has found wide application for the wet-ashing of biological material [94–99]. The advantage of the hydrogen peroxide digestion is its strong oxidizing power. Provided the procedure is carefully executed, rapid and complete ashing can be achieved. Excessive charring should be avoided because the resulting products may be very resistant to further oxidation [93].

5.3.2.3.2. Wet Digestion in Closed Systems

When wet digestions are carried out in closed systems (autoclaves), the acids used can be heated to a higher temperature, resulting in an increase of oxidizing power. With high-pressure methods, however, the maximum amount of sample that can be processed is strongly limited [100–102]. A notable advantage of digestions carried out in closed systems is that volatilization losses can be minimized. Pressure digestions are usually carried out in so-called "Teflon bombs." The autoclave-type reaction vessels consist of a PTFE crucible and lid which are placed in a stainless steel vessel that can be tightly closed. Normally, pressures of about 5 MPa are reached. Teflon bombs were originally used for decomposition of silicates [103] and later also for the digestion of biological material of widely different nature [100, 102, 104–108]. For the determination of volatile elements, for example, mercury, instead of PTFE glassy carbon is used [109] because this material is less porous. Inner vessels made of quartz and titanium have also been employed [110]. Nitric acid is the most frequently applied oxidizing agent in wet-ashing by pressure digestion. Sometimes mixtures are used, including sulfuric acid [106, 111], perchloric acid [107, 108], and hydrochloric acid [111].

Sansoni and Panday [53] extensively reviewed ashing in trace element analysis of biological material. Their study, which comprises over 400 literature citations, is warmly recommended to analysts who consider the application of dry- or wet-ashing for trace constituent determinations by ICP–AES.

5.3.3. Solubilization and Leaching Methods

5.3.3.1. General

To be suitable for ICP–AES analysis, solubilizers must satisfy two prerequisites, namely, efficient dissolution of the sample and efficient nebulization of the sample solution. For biological material of animal origin, it is obvious that with regard to the first demand the applicability of water as a sample solubilizer is restricted to body fluids, whereas the nebulization properties of the aqueous sample solutions, or better sample dilutions, depend on the dilution factor applied. In the case water is used as an analyte leachant rather than as a sample solubilizer, its applicability covers a wide range of biological tissues. Solubilizers in the real sense of the word are quarternary ammonium hydroxides. The use of tetramethylammonium hydroxide (TMAH) is typical, and in aqueous or organic solvents TMAH can solubilize a wide variety of soft tissues.

5.3.3.2. Dilution with Water

For the determination of Al in body fluids by ICP–AES, Allain and Mauras [112] applied dilution factors of 4 and 10 for urine and blood, respectively. The rather low dilution indicated, however, the use of the standard addition technique for calibration. Schramel and Klose [113] employed two dilution factors for ICP–AES analysis of six elements in human and animal serum. A dilution factor of 10 was used for the determination of Cu, Fe, and Zn, and a factor of 100 for Ca, Mg, and Na. Calibration on the basis of acidified aqueous standard solutions appeared to be satisfactory for serum analysis. For the determination of Ca in reconstituted reference serum, van Deijck et al. [114] employed high-pressure digestion, low-temperature ashing, and dilution with water as sample pretreatment methods for ICP–AES analysis; 100- and 1000-fold diluted samples were analyzed. Statistical treatment of the data showed that each procedure yielded a valid estimate of the target concentration value. Hence, for practical reasons, the dilution procedure is preferable.

5.3.3.3. Extraction with Aqueous Leachants

Employing strongly diluted mineral acids, Hinners [115] studied the extraction of nine metals from National Bureau of Standards (NBS) bovine liver: Standard Reference Material (SRM) 1577. According to the author, acid extraction is expected to prove effective for many tissues since biological systems share common metal-binding mechanisms. This expectation was more or less confirmed by leaching experiments performed in our laboratory [93]. From the results it could be concluded that extraction is certainly not to be considered as a multielement procedure but can be very useful for the determination of specific elements in a large series of samples of the same type. Where bovine liver is concerned, incomplete extraction was found for Cu and Fe whereas satisfactory results were obtained for Cd, Mn, Pb, and Zn. For the determination of Pb in whole blood, Stoeppler et al. [116] treated aliquots of fingerprick and venous samples with nitric acid for deproteinization and matrix modification. After centrifuging, the supernatant was taken for analysis. Various analytical techniques were employed comprising atomic spectrometry and electrochemistry. The method proved very reliable in routine applications and the approach is in principle also useful for multielement determination by ICP–AES.

5.3.3.4. Solubilization

Through their intensive hydrolytic action, quarternary ammonium hydroxides (QAH) are capable of the transformation of animal tissue into a solution-like form. In the past decade various tissue solubilizers, based on QAHs, were put on the market under the trade names Lumaton (Kurner, Neuberg, FRG), Soluene-100, and Soluene-350 (Packard, Warrenville, IL). Soluene-100 was developed to obtain complete dissolution of biological material in hydrocarbon solvents whereas Soluene-350 incorporates a high water solubilization capacity. Lumaton is strongly alkaline and contains organic solvent with a flash point of 6°C. In addition to animal tissue, organic material of other origin is solubilizable with Lumaton. Tetramethyl ammonium hydroxide is very effective for animal tissue and can be used in organic as well as in aqueous solution. So far, tissue solubilization is virtually exclusively employed as a sample digestion method for atom absorption spectrometric analyses. Thus, Alt and Massmann [117] employed Lumaton for the determination of Pb in blood. Soluene-100 was used by Jackson et al. [118] for the determination of Cu, Fe, and Zn in rat liver, brain, and plasma and rabbit plasma. Khera and Wibberley [119] used Soluene-350 in toluene for Pb determination in human placenta. Employing alcoholic solutions of TMAH, Gross and Parkinson [120] determined five metals (Cd, Cu, Mn, Pb, Zn) in 20 different human tissues. For the determination of the endogeneous tissue levels of Cd, Cu, Pb, and Zn of rat liver, kidney, and hair, Murthy et al. [121] made use of aqueous TMAH solution. As far as sample digestion is concerned, the investigators [117–121] positively assessed the tissue solubilization method, especially with respect to the ease of operation and the absence of the hazard of analyte loss during sample digestion.

In the author's laboratory, ICP–AES trace analysis of NBS (SRM 1577) bovine liver was performed using a 10% (w/w) aqueous TMAH solution for sample digestion [93]. For four out of the six test elements considered, Cu, Fe, Mn, and Zn, acceptable and even good results were obtained. However, the strong sample dilution (100-fold) which was required for efficient nebulization, precluded the determination of the elements present at a lower concentration level—Cd and Pb. Another important disadvantage of the method is that standard addition technique for calibration must be used.

5.3.4. Comparison of Digestion Methods

Comparative studies of digestion methods must be confined to a very limited number of procedures, types of tissues, and/or analytical criteria. This is inherent in the subject: There exists a multitude of digestion methods, each prone to numerous possible modifications, while the variety of biological tissues, too, is bewildering; finally rather different criteria may be applied in practical analysis. Thus, Stoeppler [122] compiled for the determination of Ni in biological samples no less than 70 procedures, the majority of which are applicable to ICP analysis.

So far, the results of only a few more or less extensive comparative studies of sample preparation procedures for ICP–AES analysis of biological materials have become available [123–127]. Dahlquist and Knoll [123] compared wet digestion (HNO_3–$HClO_4$) and dry ashing (500°C) for the determination of 19 elements in animal and botanical materials. The only serious limitation of the wet digestion procedures was the poor solubility of the perchlorates of ammonium and the alkali metals, which required about 50-fold dilution of the sample. However, dry ashing of tissue and subsequent dissolution in hydrochloric acid allowed a dilution factor of between 5 and 10. McQuaker et al. [124] described procedures using HNO_3–$HClO_4$ and HNO_3–HF–$HClO_4$ for analysis of 13 elements in plant and animal tissue. These procedures are characterized by predigestion with HNO_3. In this way the amount of $HClO_4$ consumed in the oxidation phase is minimized and also kept constant, thus avoiding acid effects in ICP analysis [128]. Munter et al. [125] compared different ashing methods for the determination of 14 elements in a variety of animal and food materials. Their results demonstrated that dry ashing between 475 and 500°C without ashing aids is the best method for determining B, Ca, Cd, Cu, K, Mg, Na, Ni, P, Pb, and Zn in easily destructable animal tissue, whereas wet-ashing (HNO_3–$HClO_4$) is necessary for satisfactory recovery of Al, Cr, and Fe in difficult materials, such as bovine liver and rat muscle tissue. Ward et al. [126] used HNO_3–H_2O_2, HNO_3–$HClO_4$, and dry-ashing for the determination of 19 elements in bovine liver and various agricultural products. The authors suggest that the ashing temperature used (550°C) might have been too high so that losses of volatile elements, such as As, Se, and Pb, could have occurred.

Another disadvantage frequently observed with dry-ashing was low recovery for Al, Cr, and Fe. Other important observations made in this comparative study were that the HNO_3–H_2O_2 wet ash preparation needed more operator intervention compared to dry-ashing and that a high dilution of the samples is required to give solutions that can be nebulized for several hours of continuous operation without building up of fats in the nebulizing system. Also in the author's laboratory, the performance of digestion procedures was established; these included high- and lower-temperature dry-ashing, wet-ashing in open and closed systems, extraction, and solubilization [127]. Apart from accuracy and precision, several criteria of special interest for practical trace analysis were applied. NBS (SRM 1577) bovine liver served as a test sample. Six elements (Cd, Cu, Fe, Mn, Pb, Zn) were determined simultaneously. The following features became apparent.

1. As to accuracy and precision, all the procedures examined had significance as a digestion method for ICP–AES analysis of bovine liver.
2. None of the procedures could be marked as the "best one" in every respect.
3. Copper and Fe were partly extracted by diluted nitric acid. The extraordinary good reproducibility of the recoveries found for these elements justifies further investigations in view of chemical speciation.
4. The precision of the results of the analyses of the digests resulting from the various procedures was dominated by the ratio of the analyte concentration of the digest and the ICP detection limit of the analyte; the effect exerted by the nature of the procedures on the precision was ambivalent.
5. The main source of sample contamination was constituted by reagents impurities.
6. The sample dilution associated with the digestion procedures examined may preclude direct ICP–AES analysis of trace elements.

Consequently, an important criterion for selecting a sample treatment procedure is its suitability for preconcentration of the digests. In addition to analyte preconcentration, also analyte–matrix separation is often required. This is due to the relatively high contents of alkali and alkaline earth metals of biological materials (cf. Tables 5.1 and 5.2). If highly concentrated sample solutions are analyzed, the salts easily cause nebulizer instability as well as spectral interferences. Accordingly, preconcentration procedures that at the same time effect a separation of the traces from the bulk of the matrix, that is, "enrichment procedures" [129], should be preferably used in ICP analysis of biological materials.

Table 5.3 gives an impression of concentration levels of trace elements of clinical interest in some selected human body fluids and tissues. The concentration values pertain to normal adult subjects. With the exception of Ni, all data were taken from a recent extensive pilot study, aimed at the establishment

Table 5.3. Range of Frequent Values or Mean Value[a] of Trace Element Concentrations[b] in some Adult Human Body Fluids and Tissues of Clinical Interest. Compiled from Iyengar [130] and Stoeppler [122].

Sample type	As	Cd	Co	Cr	Cu
Whole blood	2–7	0.3–1.2	40	20	800–1400
Blood serum	1.7		0.1	0.1–0.2	800–1400
Milk	0.25–30	0.7	0.2	1.0–1.5	250–400
Urine	10	0.45	1	0.2	30–60
Liver	5–15	300	6	8	500–800
Hair	150–300	400–1000	30	300–800	15.10^3–25.10^3
	Fe	Hg	Mn	Mo	Ni
Whole blood	4.10^5–5.10^5	1	8–12	0.8	3–15
Blood serum	800–1200		0.5–1.0	0.5	0.3–10
Milk	350–600	1–3	3–6	1	
Urine	100	0.1	0.65	30	0.3–10
Liver	15.10^4–25.10^4	30–150	10^3–2.10^3	360	5–13
Hair	3.10^4–6.10^4	500–2000	500–1500	50	10^2–10^4
	Pb	Se	Zn		
Whole blood	90–150	90–130	6–7		
Blood serum		75–120	800–1100		
Milk	3	15–25	1500–2000		
Urine	6	25–50	400–600		
Liver	350–550	250–400	4.10^4–6.10^4		
Hair	300	500–1000	15.10^4–25.10^4		

[a]Mean of lowest regional values.

[b]Concentration is expressed in $\mu g \cdot L^{-1}$ (fluids) or $\mu g \cdot kg^{-1}$ (tissues).

of reference values, which was compiled and discussed by Iyengar [130]. The concentrations presented in Table 5.3 should not be considered as the ambiguous but frequently used so-called "normal values." The significance of the concepts "normal value," "tolerance limit," and "reference limit" of trace elements in biological systems is thouroughly discussed by Iyengar [130] and Hemel et al. [131]. In the present context Table 5.3 merely serves the purpose of providing concentration levels that one might have to deal with in the analytical practice of elemental analysis of biological–clinical material. It should be stated, however, that the analyte concentration in the eventual sample digest solution rather than in the original sample is of importance for the possible need of the application of enrichment procedures. Consequently, application of such procedures becomes relevant when the limits of determination of the analytes valid for the specific sample digest solution [132] exceed the anticipated analyte concentration in that solution.

5.3.5. Digestion Followed by Analyte Preconcentration and Matrix Separation

5.3.5.1. Introduction

Analogous to sample digestion methods, analyte preconcentration procedures initially developed for atomic absorption spectrometry (AAS) can often be used advantageously in ICP–AES. In atomic spectrometry, chelating ion-exchange and solvent extraction are the most widely spread methods for preconcentration. The theoretical principles underlying these methods and compilations of numerous illustrative applications, including simultaneous multielement preconcentration and matrix separation, can be found in [133–135]. But so far, only a few studies are available dealing with the suitability for preconcentration of digests of biological materials obtained with specific sample digestion procedures. However, since the major metallic elements that constitute the matrix of both animal material and seawater are of the same nature (see Tables 5.1 and 5.2), procedures that are suitable for trace element preconcentration and matrix separation of seawater may also be applicable to solutions of mineralized biological fluids and tissues. In the past decade ample experience has been gained with preconcentration of trace elements from seawater.

Taking full advantage of the simultaneous multielement capability of ICP–AES, group-wise preconcentration of trace elements is highly preferred. In the following, a few typical examples are detailed. All of them refer to seawater.

5.3.5.2. Chelating Ion-Exchange Procedures

Kerfoot and Crawford [136] have described the use of a high-surface-area chelating ion-exchange material sandwiched between membrane filters for preconcentration of transition metals from seawater. With an extraction of 200-mL samples, ICP detection limits for Be, Cd, Co, Cr, Cu, Fe, Mn, Ni, Ti, and Zn were below 1 $\mu g \ L^{-1}$ with a column preconcentration and matrix separation procedure employing Chelex-100, Kingston et al. [137] achieved a concentration factor of 100 for Cd, Co, Cu, Fe, Mn, Ni, Pb, and Zn. Complete separation of the matrix metals Ca, Mg, Na, and K from the analyte elements was obtained by elution of the column with ammonium acetate in the pH range 5.0–5.5. The preconcentration and separation thus achieved permitted the determination of the analytes at the subnanogram level in various types of seawater using graphite furnance AAS.

Guedes da Mota et al. [138] obtained a concentration factor of 300–500 for Cu, Pb, and Zn by percolating buffered sea-water through a column filled with immobilized diamine triacetic acid (ED3A) on controlled pore glass (CPG). Using flame AAS, the standard deviations (four analyses) for the elements in the normal concentration range of 2–6 $\mu g \cdot L^{-1}$ were 2–5% for Cu, 5% for Pb, and 1–10% for Zn. For the simultaneous determination of trace metals in seawater by ICP–AES, Berman et al. [139] used the same chelating ion-exchange preconcentration method as described by Kingston et al. [137] for graphite furnace AAS. Good results were obtained for Cu, Fe, Mn, Ni, and Zn in relatively unpolluted coastal seawaters. However, the concentration levels of Cd, Co, Cr, and Pb in the concentrates remained below values at which reliable ICP–AES analysis can be made.

Considering that the iminodiacetate containing resin Chelex-100 suffers from the fact that removal of sequestered Ca and Mg requires careful washing procedures and further that Chelex-100 permits no greater sample flow rates than 1–2 $mL \cdot min^{-1}$ [137, 140], Sturgeon et al. [141] examined on silica gel immobilized 8-hydroxyquinoline (8-HOQ) for preconcentration of Cd, Co, Cu, Fe, Mn, Ni, and Zn from seawater prior to determination by graphite furnace AAS. Employing a sample flow rate of 15 $mL \cdot min^{-1}$, a matrix-free analyte concentrate with a concentration factor of 50 for near-shore samples and of 90 for open ocean waters could readily be realized. The obtained analytical results compared well with accepted values.

5.3.5.3. Extraction Procedures

Watanabe et al. [142] have described a preconcentration method for Cd, Cu, Fe, Mn, Ni, Pb, and Zn from seawater prior to their determination by ICP–AES. The method involves complexation in batch of the metal ions with 8-HOQ followed by (column) adsorption of C_{18}-bonded silica gel. The samples are passed through the column with a rate of 10–14 $mL \cdot min^{-1}$. The metal chelates of 8-HOQ are then eluted from the column with 5 mL of methanol. Finally, the organic concentrate is converted into an aqueous concentrate of the same volume which serves as analysis solution. The results for Cu, Fe, Mn, and Ni were in close agreement with those obtained by furnace AAS and isotope dilution, spark source mass spectrometry. But owing to high blanks and a still too low concentration factor (200), Cd, Pb, and Zn could not be determined. Sturgeon et al. [143] used the same preconcentration method for the determination of the same elements (plus cobalt) by graphite furnace AAS. The authors reported excellent rejection of Ca and Mg and satisfactory analytical results. However, owing to the strongly varying recovery of Pb from sea-water, this element could not be determined.

A method based on dithiocarbamate preconcentration for the simultaneous determination of Cd, Cu, Fe, Mo, Ni, V, and Zn in seawater by ICP–AES is described by McLeod et al. [144]. According to this method, the trace metals are extracted from the sample with a mixture of ammonium tetramethylene dithiocarbamate (APDC) and diethylammonium diethyldithiocarbamate (DDDC) in chloroform and backextracted into nitric acid. The concentrations of Cd, Cu, Ni, and Zn, found at the $ng \cdot L^{-1}$ level, were consistent with the results of the independent AAS analysis, whereas the detection limits obtained for all of the elements considered proved to be sufficiently low for the determination of their normal concentration levels in ocean waters.

Seeverens et al. [145] examined the applicability of *N*, *N'*-diphenylthiourea (DPTU) and triphenylphosphine (TPP) for preconcentration by solvent extraction of precious metals prior to their determination by ICP–AES. With both complexing agents, the elements of special interest in the biological–clinical field, namely Ag, Au, and Pt (see Section 5.1), were selectively extracted as a group with high efficiency and excellent matrix separation from aqueous test solutions having compositions similar to seawater. For extraction of the DPTU and TPP metal complexes, 1,2-dichloroethane (DCE) and chloroform proved to be suitable, but for direct ICP analysis of the organic concentrates preference was given to DCE. However, after establishing proper experimental conditions for ICP analysis of volatile organic solutions, direct analysis of chloroform concentrates can be performed routinely [146].

5.3.5.4. *Digests of Biological Standard Reference Materials*

Jones et al. [147] applied the Chelex-100 ion-exchange procedure for seawater as described by Kingston et al. [137] for preconcentration of trace transition elements from digest solutions obtained with binary acid (HNO_3–$HClO_4$) and ternary acid (HNO_3–$HClO_4$–H_2SO_4) wet-ashing methods. As test samples of animal nature served the NBS reference materials oyster tissue (SRM 1566), bovine liver (SRM 1577), and albacore tuna (RM 50). From both the binary and ternary acid digests, Ca, Mg, and K were removed quantitatively from the resin before the final nitric acid eluate containing the analytes. Sodium was not measured by ICP–AES because no direct-reading channel was available, but no visible 589-nm emission was observed. Phosphorus was removed with at least 90% efficiency from the analyte solutions. By comparison with NBS-certified values, recoveries of the following trace elements could be established: Cd, Cr, Cu, Fe, Mn, Mo, Ni, Pb, and Zn. The majority of the analyses showed good recoveries of the trace elements but poor, and occasionally erratic, recoveries were found for Cr in oyster and liver samples and for Mn in all of the three animal sample types considered. Further, the recovery of Fe from bovine liver was too low. However, neither the satisfactory results nor the poor or erratic ones pointed to a preference of the wet-ashing sample digestion procedures applied. In a subsequent study [148], Jones and O'Haver expanded the original investigation [147] to examine in more detail the effect of the pH of the eventual sample solution and the acid digestion conditions on the Chelex-100 reaction. In the pH range of 3.3–4.7, substantial Chelex-100 retention differences were observed for several elements between the binary and ternary acid digestion method. Between pH 4.7 and 6.3, either digestion was satisfactory for most elements examined. Manganese retention by the resin was strongly affected by the pH, whereas it was recovered quantitatively only at a pH of about 6. It was concluded that either the binary or the ternary acid digestion procedure may be used provided complete mineralization of the sample is realized. However, the ternary acid procedure appeared to be more consistently reliable.

By identifying various organic compounds in the residue after wet-ashing with perchloric acid, Martinie and Schilt [149] showed that biological samples may not be completely mineralized by wet ashing. This suggests that elements that are strongly complexed in biological tissue, such as iron present in the hemal part of liver, remain partially complexed. The work of Pella et al. [150] points in the same direction. These authors examined the recovery of Cu and Fe from bovine liver by using perchloric acid digestion at various temperatures. Their results indicate that temperatures of 300°C or higher are required to break up complexes of metals in biological samples when wet-ashing procedures involving perchloric acid are applied. In this connection, it is striking that the recovery of Fe from bovine liver obtained by leaching with strongly diluted nitric acid (see Section 5.3.3) or by digestion with the binary mixture of concentrated acids at boiling temperature [147] is much the same.

Van Deijck et al. [151] examined the extent to which the specific advantages of ICP–AES [152,153] compensate for the disadvantages associated with sample treatment and, in particular, preconcentration. They tested sample digestion by LTA and muffle furnace ashing (see Section 5.3.2) in the framework of ICP–AES determination of Cd and Pb in a whole blood candidate SRM. These methods were applied in combination with an extraction procedure for Fe and a column preconcentration procedure using CPG-immobilized 8-HOQ. The concentration factor of 5 obtained with the column procedure is not sufficient for the determination of normal concentrations of Cd and Pb in blood. However, as the analyte concentrates obtained by the column procedure were essentially free of matrix constituents, volume reduction of the concentrates by evaporation enabled further analyte concentration up to a factor of 50. The analysis routes, including LTA, yielded inaccurate results. This is most probably due to the limited mineralization capability of LTA; after 48 h of ashing, the fraction of the organic material still present in the residue ranged from 5 to 20%. The results obtained with thermal ashing (520°C, 12 h) compared well with the assigned values of two different reference whole blood samples.

From the summarized results of analyte preconcentration in biological digests of SRMs, we can conclude that enrichment procedures suitable for seawater can be applied to biological digests provided complete mineralization is guaranteed. It should be borne in mind, however, that biological materials digested by methods other than thermal ashing show a markedly high resistance to exhaustive oxidation. This also holds for wet-ashing in closed systems. Thus, Stoeppler et al. [101] reported for the pressurized decomposition of whole blood (HNO_3, 140 min, max temp. 160°C) only 65% oxidation of the initial amount of carbon. Pressure digestion of bovine liver (HNO_3, 2 h, 140°C), performed in the author's laboratory [127], showed that after 20 min the amount of carbon was reduced to about half of the initial content, whereas after 2 h, <10% carbon was left.

Strictly speaking, complete mineralization is a prerequisite only when preconcentration is applied by methods such as chelating ion-exchange or solvent extraction. However, to avoid irreversible coagulation upon contact with water of the digest solution, a high degree of sample oxidation is required anyhow [127].

5.3.5.5. *Digests of Real Biological Materials*

For enrichment of trace elements, the group of Barnes [154] successfully applied two novel chelating resins, namely, poly(dithio-carbamate) (PDTC) and poly(acrylamid-oxine) (PAAO), prior to the ICP–AES analysis of urine [155–158], dialysis fluid [158], serum [159], and bone [160]. For metals strongly chelated by the resins, dilution of liquid samples may be sufficient, but for weakly chelated metals or metals not free to complex owing to their binding to organic compounds in the sample, the samples must be digested completely [154]. Both resins, which can be used in the batch mode, complex heavy metals without reaction with alkali and alkaline earth elements. Hence, the critical and time-consuming washing procedures that are required for the removal of Ca and Mg from materials like Chelex-100 or 8-HOQ can be omitted.

When using PDTC, total digestion of the resin in nitric acid and hydrogen peroxide is preferred [157]. But the sequestered trace elements on the PAAO resin can be recovered by leaching with acid or other suitable reagents [158]. Fifty-two elements are known to complex with PDTC resin and at least 31 elements are known to chelate with PAAO resin [154]. The PDTC resin is also effective for column use. With a solution flow rate of about 2 mL · min^{-1}, volumes between 25 and 5000 mL have been employed. Since the uptake for a number of elements strongly depends on their oxidation state, for example, Fe(III) and Cr(VI) complex with PDTC whereas Fe(II) and Cr(III) do not [161,162], the resins also enable speciation, in addition to analyte concentration and matrix separation. Thus, Mianzhi and Barnes [158] developed a technique for the ICP–AES determination of Cr(VI) and Cr(III) in urine using both resins. The differential determination of trace amounts of Cr(VI) and Cr(III) is important, especially in clinical chemistry, because of the toxicity of Cr(VI) species.

It is the experience of the author that the properties of commercially available ion-exchange chelating materials often differ considerably for different production batches. Thus, capacity differences of the order of a factor of 2 are no exception. However, since the PDTC and PAAO resins can easily be synthesized by the user [158], it may be anticipated that the dependence of the resin properties on the production process can be reduced or even eliminated.

5.4. ANALYTICAL PRACTICE

5.4.1. Sample Introduction

5.4.1.1. General

For the analysis of biological materials, the analytes are introduced into the plasma either in the form of partly or completely desolvated aerosols or as vapors. The aerosols are commonly generated by pneumatic nebulization of the solution into which the sample is converted. The main shortcomings of pneumatic nebulization are the high consumption of the sample solution and the low sample transport efficiency [163]. Because of its ease of operation, however, this mode of sample introduction is by far the most frequently used technique. Although ultrasonic nebulization, too, can be used, this technique has not received widespread acceptance for ICP–AES analysis of biological samples. The reasons for the low interest in ultrasonic nebulization for routine analysis have been recently summarized by Mermet and Hubert [19].

With conventional pneumatic nebulizers, the consumption rate of the sample solution is of the order of 1 mL · min^{-1}. This means that the ICP–AES analysis of considerably smaller volumes, which is of special importance in clinical chemistry, encounters a serious difficulty. For direct analysis of microliter volumes special microsample techniques have been developed [164–167]. As early as 1973, Kniseley et al. [165] developed a microliter sample introduction system for ICP–AES trace analysis of whole blood and serum. The reported detection limits, for example, 2 ng · mL^{-1} for Pb in whole blood, were too optimistic, however.

Main problems met with the application of such techniques are the following [168].

1. The difficulty of manipulating discrete microsamples by means of capillary tubes or syringes, which leads to deterioration of both precision and accuracy.
2. The relatively large dead volume existing in many injection systems, which results in considerable waste because the total sample volume necessary for analysis often exceeds the nominal injection volume by a substantial amount.

To overcome these practical limitations, sample introduction methods, which are essentially flow injection analysis (FIA) systems [168–170], are used for the analysis of biological materials. In such arrangements microliter samples are injected into a flowing solvent stream that is continuously nebulized. An additional advantage of sample introduction by flow injection is the potential for a high sample throughput. Thus Alexander et al. [169] applied this technique to electrolyte analysis of serum and achieved for 10-μL samples an injection rate of 240 h^{-1}.

As mentioned previously, apart from the relatively large sample consumption, the low sample transport efficiency constitutes a second main shortcoming of conventional liquid sample introduction. For the large majority of pneumatic nebulizers in current use for ICP–AES, the transport efficiency ranges between 0.5 and 4% [163] (cf. Chapter 8). However, when the analytes are introduced in the form of vapor, significantly higher transport efficiencies can be attained. The following vapor introduction techniques have been employed for the analysis of biological–clinical materials.

Conversion of the analytes into their hydrides [126,156,171,172].

Formation of volatile analyte–chelates [173,174].

Electrothermal vaporization of the sample [175–181].

5.4.1.2. Hydride Technique

The determination of trace concentrations of the hydride forming elements is of great importance in clinical chemistry because these elements occur in biomedical materials as essential trace elements (Se, Sn), nonessential trace elements used therapeutically (Bi), and toxic trace elements (As, Sb, Te), (cf. section 5.1.). However, for conventional sample solution nebulization the detection limits of the majority of these elements are as high as 0.1 mg $\cdot$ L^{-1}. When the analytes are introduced into the ICP in the form of their hydrides, the detection limits of As, Sb, Se, and Sn are three orders of magnitude and those of Bi and Te two orders of magnitude better compared with solution nebulization. For a detailed comparison of typical hydride detection limits and values for normal liquid sample introduction, the reader is referred to the survey on sample introduction techniques for atomic spectrometry by Browner and Boorn [182].

Nakahara [171] recently reviewed hydride generation methods for atomic spectrometric analysis, including ICP–AES determination of hydride-forming elements in biological materials. Several reducing agents and sources of nascent hydrogen have been investigated for the reduction of the elements of interest to their hydrides. The sodium tetrahydroborate ($NaBH_4$)–acid system was found to be superior to the metal–acid reduction systems with respect to, inter alia, reduction yield and reaction time [171]. Of special importance for ICP–AES is the greater potential for simultaneous multielement hydride generation of the $NaBH_4$–acid system than the metal–acid reactions.

Two approaches are used to introduce the hydrides of the analyte elements. In the batch approach, the hydride is collected above the reaction liquid and then rapidly swept into the ICP [156]. In the continuous approach, the hydrides are formed in a continuously flowing stream of reagents [172].

Interferences may result from interelement effects. The presence of Cu or Ni, for example, causes low recovery of some hydride-forming elements, especially Se and Te. Mianzhi and Barnes [159], who thoroughly investigated interference from Cu on hydride formation, found that the effect of Cu on Se-hydride evolution depends not only on the Cu concentration level but also on the Se concentration. An additional problem inherent in hydride generation is the adverse effect on the reaction by-products (hydrogen, water) on the stability of medium- and low-power ICPs. These problems can be overcome by specific chemical treatment of the sample solution prior to analysis (see Section 5.4.2) and by employing separation and low-temperature condensation techniques to exclude the unwanted reaction by-products [171].

In addition to the notable improvement of detection power, the utility of hydride generation is supported by its potential to contribute to element speciation. Variation of the reduction conditions can afford selective determination of the oxidation state of hydride-forming elements, for example, As(III) and As(V). The greatest difficulty in speciation is posed by the method of sample digestion because such methods should not change the original valency of the analytes in the sample [183].

5.4.1.3. Volatile Metal Chelates

Solvent extraction of volatile metal diketonate complexes has been used extensively in the gas chromatographic analysis of metals. Meanwhile, a few studies have illustrated the feasibility of direct chelation of some metals from various biological–clinical materials [173,174]. Selective removal of metal species from the sample matrix through a direct chelation reaction and the use of the ICP as a spectroscopic detector for the resulting volatile metal chelate vapors offers, in principle, significant advantages, including avoidance of extensive sample preparation and the possibility of using small samples. Black and Browner [173] employed vapor-phase introduction into the ICP of volatile chelate complexes of Cr, Fe, and Zn from human blood serum and NBS bovine liver (SRM 1577). The complexes were generated by direct chelation reaction of the analyte metals of the sample with the chelate ligands trifluoro-acetylacetone or hexafluoro-acetylacetone. These ligands were chosen, among other things, because of the excellent thermal stability and high volatility of their known metal chelate complexes.

Two different approaches were used for the chelation reactions. According to one procedure the sample is placed in a vessel, chelate ligand is added, the system is heated, and the resulting vapor is flushed into the plasma with the Ar carrier gas. The second procedure involved the use of sealed ampules into which sample and chelate ligand are heated. The authors obtained with their metal chelate technique for Fe a gain in detection power of a factor of 50 compared to conventional pneumatic nebulization, whereas such a gain for Cr and Zn was considerably smaller. However, the range of elements that will react and volatilize quantitatively is not yet fully explored. Therefore, the method is hard to assess at its present state of development. Nevertheless, it may be anticipated that the volatile chelate–ICP method can be a valuable supplement to conventional sample introduction; the more so because formation and volatilization of metal chelates from the sample matrix permits increasing the selectivity of the analyte determination with respect to potentially interfering elements present in biological samples.

5.4.1.4. Electrothermal Vaporization (ETV)

The electrothermal sample vaporizers currently in use for the analysis of biological–clinical materials are largely, whether or not modified, graphite furnaces originally designed for electrothermal atomization–atomic absorption spectrometry (ETA–AAS). Just as in ETA–AAS, microliter (10–50 μL) aliquots of the sample solution or sub-milligram amounts of solid samples are used in ETV–ICP–AES. The essential difference between small volume sample introduction by conventional nebulization and correspondingly small amount sample introduction through ETV is that the loss of sensitivity inherent in conventional nebulization is compensated or even outweighed. This improvement, which results from the high sample transport efficiency associated with ETV, permits trace analysis of microsamples with detection limits comparable to or even better than those attainable with continuous nebulization.

The main problems with ETV sample introduction arise from effects exerted by matrix constituents on the thermal evaporation of the analytes and the processes that can hamper the analyte transport from the vaporizer to the ICP—the formation of refractive carbides on the graphite walls of the vaporizer or the loss of volatile analytes on the tubing between vaporizer and plasma. Considering the nature of these problems, one should realize, however, that the matrix effects in ETA–AAS, which are dominated by atomization efficiency interferences [184], can influence the vaporization behavior in ETV in a totally different way. This is clearly understandable on the basis of the fundamentally different optimization aims that have to be pursued in the two techniques. Regarding this, Browner and Boorn [185] remarked: "With ETV the criterion for optimum sample transport is the production of highly dispersed particulate matter that may be transported in the gas stream between the vaporizer and the plasma. This is in marked contrast to the situation in ETA–AAS where efficient atomization in the furnace is essential for accurate analysis. With the ICP the desired species are not atomic but molecular."

Thus, Aziz et al. [177] showed in their study on the ETV–ICP–AES analysis of biological–clinical materials that analyte volatilization strongly depends on widely varying and hardly predictable matrix effects. As a consequence of the severe matrix effects observed, trace analysis of bovine liver and blood serum required calibration by standard addition.

5.4.2. Illustrative Examples of Analytical Techniques

In the present section, we consider illustrative examples of techniques typical for the more or less frequently used modes of sample introduction employed for ICP–AES analysis of biological–clinical materials. The studies, selected from the current literature, elucidate the various techniques together with limitations and associated problems. The examples successively feature the following items.

1. Continuous pneumatic nebulization, direct simultaneous multielement analysis (human kidney and liver).
2. Continuous pneumatic nebulization, spectral interferences, matrix effects, simultaneous multielement analysis in fractions of the sample solution, simultaneous analyte enrichment (animal bone).

3. Continuous pneumatic nebulization, simultaneous multielement trace analysis, group-wise analyte enrichment (human urine).
4. Discrete pneumatic nebulization, microliter sample solution introduction, sequential trace element analysis, matrix effects (human blood serum).
5. Continuous pneumatic nebulization, microliter sample solution introduction, flow injection, sequential trace element analysis (bovine liver).
6. Hydride generation, trace element analysis, acids and interelement interference effects (bovine liver).
7. Electrothermal sample vaporization, trace element analysis (human milk).

1. The simultaneous determination of 20 elements in large numbers of autopsy samples of human kidney (cortex and medulla) and liver, which has been reported by Subramanian and Meranger [186], constitutes a typical example of simultaneous multielement analysis of clinical–biological materials using conventional nebulization. The autopsies were done within 48 h after death using stainless-steel scalpels. The scalpels were cleaned until the concentrations of the analyte elements were below the ICP detection limits. The NBS bovine liver (SRM 1577) was used as the reference sample. Autopsy samples (1–2 g) were taken, frozen, and shipped to the laboratory in polyethylene bags and stored at −20°C until analyzed. Prior to analysis, the samples were thawed and then washed three times with high-purity water to remove residual blood. Since the entire sample was used for analysis, homogenization could be omitted. The samples were digested by heating with 10 mL of a mixture of HNO_3–$HClO_4$ (4 : 1, v/v) on a sand bath. The resulting damp residue was dissolved in concentrated HCl and subsequently made up to 10 mL with 0.5 *M* HCl. The sample solution was then peristaltically pumped at a rate of 2–2.5 $mL \cdot min^{-1}$ to a cross-flow nebulizer. The resulting emission signals were detected by a polychromator. The wavelengths of the analysis lines were chosen to provide a reasonable compromise between optimum sensitivity and minimum spectral interference. The concentration values found for the major elements Ca, K, Mg, Na, and P and for the trace elements Cd, Cu, Fe, Mn, and Zn, generally agreed with those given in the compilation of Iyengar et al. [187] for normal adults. However, in all the kidney and liver samples, the concentrations of Ag, Ba, Be, Co, Cr, Mo, Ni, Sr, Th, and Ti were found to be below the detection limit for these elements of the method applied.

2. Lee [188] has amply reported on matrix effects and spectral interferences encountered in multielement analysis of animal bone. The unavoidable presence of large quantities of Ca and P in the bone solution resulted in suppression of the signal intensity of many ionic and atomic lines. The author also observed severe spectral interferences, for example, spectral overlap of the Ca(II) 180.739-nm line on the S(I) 180.735-nm analysis line and apparent emission recorded at the Cu(I) 324.75-nm analysis wavelength caused by scattering and reflection of Ca(II) 317.93-nm radiation. Direct multielement determination of major (Ca, P) and minor (K, Mg, Na, S) elements could be performed by appropriate dilutions of the nitric acid-digested bone material. For the determination of S a correction factor of 0.013 kg $\cdot$ mL^{-1} "apparent S" for every $\mu g \cdot mL^{-1}$ Ca in solution was used. However, to minimize matrix effects by dilution of the sample solution often results in lowering the original trace analyte concentrations below their limits of determination.

Because of the high Ca concentration and the low levels of some trace elements of interest, the author employed an enrichment procedure. Thus, Cu and Mo were determined after preconcentration and separation from matrix elements on the strong cation-exchange resin Biorad Ag-MP-50. A column was packed with 10 g of the resin and 25 mL of nitric acid-digested bone solution (corresponding to about 2 g of freeze-dried bone material) was washed onto the resin with 0.1 *M* HCl. Al, Cu, Mo, Ni, Pb, and Zn could be consistently recovered quantitatively as a group when eluted with 3 *M* HCl–methanol (1 : 1; v/v). Ca was strongly retained whereas phosphate passed freely through the column. The fraction containing the trace elements was evaporated to near dryness and the residue dissolved in 5 mL of 2 *M* HCl. This sample solution volume, which corresponds to 2 g of bone material, contained less than 0.1 mg of each of Ca and phosphate. The International Atomic Energy Agency (IAEA) animal bone candidate reference material (H-5) was used to assess the overall performance of the technique and digestion procedure in terms of recovery and precision of the simultaneous determination of major, minor, and trace elements. The obtained analytical results proved to be in accordance with the information values given in the first progress report.

3. Barnes and Genna [155] demonstrated the separation and concentration capabilities of a poly(dithio-carbamate) resin for the determination of trace elements in urine. Their approach takes advantage of the selective concentration of specific trace metals provided by this resin to separate desired trace metals from interfering concomitants—alkali and alkaline earth elements and the organic matrix of the urine. The resin was synthesized in the laboratory according to a simple procedure. Silanized disposable pipet columns were loaded with 0.1 g of the resin. Urine samples from local volunteers were collected over concentrated hydrochloric acid (50 mL HCl/L of urine), filtrated through membrane filters, and refridgerated. For analysis, 250-mL aliquots of urine from 2-L samples were placed in silanized glass sample reservoirs attached to the prepared column and allowed to flow through the resin at approximately 2.5 mL $\cdot$ min^{-1}. After sample passage, the sequestered trace metals containing resin was transferred from the column into a 10-mL beaker by forcing air from a pipet bulb placed at the constricted end of the column. One milliliter of 1:1 (v/v) HNO_3–H_2SO_4 was added to the resin and the contents of the beaker were heated gently on a hot plate, usually 2–3 min, until a clear solution was obtained. The clear solution was transferred to a 2-mL volumetric flask and diluted to volume with water. According to this approach, a concentration factor of 125 is attained for the trace elements, provided a recovery of 100% is achieved. Quantitative recoveries were obtained for Cd, Cu, Hg, Ni, Pb, and U when sequestered at pH 6 and for Bi, Se, Sn, and Te when sequestered at pH 1. The authors demonstrated further that a simple acid-matched aqueous solution is an adequate solvent for the reference solutions.

4. Aziz et al. [167] developed a sample introduction technique for the analysis of small volume samples (50–200 μL) with a 3-kW Ar/N_2 ICP (cf. Part 1, Section 6.2.7). The authors applied the method to the sequential determination of trace elements (Cu, Fe, Mg) in human blood serum and illustrated the accuracy of the method by the analysis results for Ca, Cu, Fe, Mg, and Zn in a series of certified test samples comprising Cation-cal TM, Monitrol I, and Monitrol II from Dade Division, American Hospital Supply Corporation (Miami, FL) and Presilip from Boehringer Mannheim GmbH, Diagnostica (F.R.G). The serum samples were diluted with Hermann solution (Merck No. 9975) containing 0.1 *M* HNO_3 and all reference solutions were made accordingly. Sample aliquots were injected manually with the aid of Eppendorf micropipettes, or by an automatic injection device, into a PTFE sample funnel with a depth and diameter of 10 mm. The funnel was connected through polyethylene tubing to a conventional pneumatic nebulizer. The analytical signals were detected by a monochromator. The detection limits for the test elements were determined after optimizing the injection volume and the aerosol carrier gas flow rate with respect to the line-to-background intensity ratio and the analytical precision. Maximum line-to-background ratios were already obtained at 50- to 100-μL sample volume. At a 50-μL injection volume the analytical precision was optimum. When the nebulizer was operated at an aerosol carrier gas flow rate of 2.3 L min^{-1}, the detection limits of the five test elements were a factor of 2–10 poorer compared with those obtained with ICP–AES methods using continuous nebulization combined with integration of the steady-state analytical signals. Subsequent injections could be made at intervals of 30 s without risk of memory effects. The matrix effects caused by the Na present in human serum (about 3.2 g $\cdot$ L^{-1}) required sample dilution of at least 1 + 4 for direct trace determination. The results of analysis agreed reasonably well with the certified values of the test samples. From the analytical performance of the injection technique described it may be concluded that this technique is of value for trace analyses in small volume samples as they are often encountered in the biological–clinical field.

5. Faske et al. [168] showed the suitability of microliter sample introduction by a simple flow injection system to the trace analysis (Cu, Fe, Mn, Zn) of NBS bovine liver (SRM 1577). The arrangement employed was a medium dispersion flow injection analysis (FIA) system by which microliter samples were injected into a flowing solvent stream. As a result of the continuous nebulization of the sample carrier stream, the physical state of the plasma is not disturbed when the sample "plug" passes the ICP. Samples were injected into a commercially available syringe loading sample injection device fitted with interchangeable sample loops with volumes of 10, 20, 50, and 100 μL. For sample introduction into the ICP, a conventional cross-flow pneumatic nebulizer was used and a monochromator provided with a pico ammeter for recording the transient analytical signals. The bovine liver was digested in a 2 : 1 (v/v) mixture of nitric to sulfuric acid. A portion of the sample (about 0.3 g) was placed in a 50-mL round-bottom boiling flask and refluxed with the acid mixture (10 mL) for 1–2 h, after which the contents of the flask were transferred to a 100-mL volumetric flask and diluted to volume with water. Based on 20-μL sample solution injections and peak height measurement of the transient signals, the following analytical results were obtained. The recoveries found for Fe, Mn, and Zn did not significantly deviate from 100%, but for Cu the recovery was only 80%. The authors assumed that the low recovery of Cu is probably due to volatilization losses during the sample digestion. The detection limits of Cu, Fe, and Mn were only about five times poorer than those for continuous sample introduction. The detection limit of Zn, however, was deteriorated by a factor of 15.

6. Nakahara and Kikui [172] described a method for the determination of traces of Se, based on sodium tetrahydroborate reduction of Se and continuous introduction of the formed hydrogen selenide into the ICP by a commercially available flowthrough hydride generator. Previously acidified reference or sample solutions and alkaline sodium tetrahydroborate solution were continuously introduced into the hydride generator. The hydrogen selenide and excess hydrogen formed were introduced into the ICP by a flow of Ar carrier gas. The drain outlet of the employed conventional spray chamber served as the inlet of the carrier gas. The authors investigated the effects of a variety of acids and concomitants on the analytical performance of the method; 2 *M* hydrochloric acid turned out to be the most suitable reaction medium and standard addition was recommended for accurate Se determination. Added to a 1000-fold weight ratio with respect to Se, the following elements did not significantly depress the Se emission intensity: Al, B, Ba, Be, Ca, Ce, Cr(III), Cs, Ga, K, La, Li, Mg, Mn, Na, P, Rb, Si, Sr, Th, Ti, T1, W, and Y. For optimum operating conditions, which included a sample solution flow rate of 17 $mL \cdot min^{-1}$, the detection limit of Se attained with the most sensitive line, Se(I) 196.026 nm, was 0.6 $ng \cdot mL^{-1}$. This is an improvement of about two orders of magnitude compared with the detection limit for conventional sample nebulization. The method was applied to the determination of Se in NBS bovine liver (SRM 1577). Three-gram liver samples were digested in 45 mL of a mixture of nitric, perchloric, and sulfuric acid (3:2:1, v/v) containing 0.67 L^{-1} of NH_4VO_4 as an oxidative catalyst. After cooling and dilution to 15 mL with hydrochloric acid the solution was heated to 90°C for 10 min to reduce Se(VI) to Se(IV). Finally, the sample solution was transferred to a 100-mL volumetric flask and diluted to volume with hydrochloric acid and water to give an acidity of 2 *M* HCl. The recovery of Se added was found to be 71%. The authors attributed the low recovery to losses during the digestion procedure and suggested therefore the use of PTFE-sealed bombs for digestion of samples with Se as an analyte element. However, the reproducibility of the losses was sufficient (71 ± 4%) to allow accurate analysis. From the results obtained one concludes that obviously the same holds for the reproducibility of the effect of the 200-fold weight ratio of Cu to Se occurring with the bovine liver test sample. Thus the Se concentration found with the proposed method (1.13 ± 0.05 $\mu g \cdot g^{-1}$) was not significantly different from the certified value (1.1 ± 0.1 $\mu g \cdot g^{-1}$).

7. For the determination of Ni in nearly 200 human milk samples Camara Rica and Kirkbright [176] used the electrothermal vaporization technique (ETV) for analyte introduction into the ICP. A carbon rod atomizer provided with a depression in the graphite rod served as a vaporizer. The samples originated from different countries and different sociological groups in widely varying areas of the world. The authors tested various sample pretreatment and digestion procedures. Their eventual method of choice involved solubilization using tetramethylammonium hydroxide (TMAH). Approximately 0.1-g samples of the freeze-dried solid human milk, collected under a well-defined protocol specified by the International Atomic Energy Agency–WHO program, was treated with 0.1 mL of TMAH solution (25%, w/w) and 2 mL of water. The mixture was very gently heated to dryness and the residue dissolved in water and diluted to 2 mL. Of the resulting sample solution, 10 μL was transferred to the depression in the graphite rod. After desolvation (100°C, 30 s), ashing (900°C, 40 s), and vaporization (2800°C, 1 s), the analyte together with the matrix elements was vaporized into the Ar carrier gas stream and swept into the plasma. An examination of the effect of the presence of Ca, Cu, Fe, K, Mg, Mn, Na, P, and Zn (up to 1 $mg \cdot mL^{-1}$) showed that the only matrix effects observed were those from Cu, Mn, and Mg. These elements increased the atomic emission intensity of Ni at the analysis wavelength of 341.476 nm. The anions chloride, nitrate, acetate, and phosphate exhibited no effect on the emission intensity of Ni. For analysis the method of standard addition was employed using the peak intensity as the analytical signal. The Ni concentrations found ranged from 30 to 420 $\mu g \cdot kg^{-1}$ of human milk. The attained precision was concentration dependent. The relative standard deviations (RSDs) for Ni contents <100, 100–200, and >200 $\mu g \cdot kg^{-1}$ were 17, 10, and 7%, respectively.

5.4.3. Materials Analyzed and Elements Determined

5.4.3.1. Chronological Review

Table 5.4 lists, in a chronological order, studies dealing with ICP–AES element determination in biological-clinical materials. The selection is primarily based on literature readily available to analytical chemists. When the employed sample introduction technique is not mentioned explicitly, conventional pneumatic nebulization has been used, which means that the sample solution is continuously fed to the ICP and steady-state signals are measured. "Microliter-sample introduction" points to nebulization of small-volume sample solutions, whereas the procedure that indeed also involves microliter sample solution consumption, but where thermal vaporization is used, has been indicated by "ETV" (cf. Section 5.4.1). The additional remarks do not always pertain to the main objective of the cited study but may refer to specific analytical aspects in the case such aspects are considered in the paper in question and elucidated in the context of analytical practice.

A glance at Table 5.4 reveals the markedly long period of time that elapsed after the first studies on elemental analysis of biological materials had been published and the annual appearance of a more or less constant number of papers on this subject.

5.4.3.2. Determination of Molecular Species

The performance of the ICP as an element specific detector for high-performance liquid chromatography (HPLC) has already been demonstrated in the late 1970s [217]. Since the sample uptake of the ICP for conventional nebulization compares well with the flow rate of effluent from HPLC, interfacing HPLC with ICP is easily realized. Thus, Morita and Uehiro [218] showed the separation and quantitative analysis of phosphate compounds by using a HPLC–ICP–AES system, whereas Yoshida et al. [219] employed that system for the determination of 12 common 5′-ribonucleotides by observing the P(I) 213.547-nm integrated emission intensity. The transient intensity signal proved to be proportional to the number of the P atoms in all of the nucleotides.

5.4.3.3. Speciation

The determination of nucleotides can, of course, also be considered as determination and speciation of the analyte P. For reasons indicated in Section 5.1, element speciation is of great importance in clinical chemistry. However, thus far only little could have been achieved in this respect because the large majority of the required sample digestion procedures inevitably leads to unknown changes of the analyte distribution over the various physiochemical species in the orig-

Table 5.4. Chronological Review of ICP–AES Application to the Analysis of Biological-Clinical Materials

Matrix	Analyte	Author	Remarks and Reference
Blood	Ag, Al, Cu, Fe, Mg, P, Pb, Si	Greenfield and Smith (1972)	Microliter-sample introduction; exploratory study [164]
Blood, serum	Various (11)	Kniseley et al. (1973)	Microliter-sample introduction [165]
Blood, urine	Al	Allain and Mauras (1979)	Comparison with furnace AAS [112]
Hair, liver, muscle tissue	Various (11)	Munter et al. (1979)	Various ashing procedures; comparison with AAS methods [125]
Urine	Various (10)	Barnes and Genna (1979)	Simultaneous analyte enrichment by poly(dithiocarbamate) resin [155]
Urine	Various (13)	Haas Jr. et al. (1979)	Application of Ga and Y as internal standard elements [189]
Breast milk	B	Schramel (1979)	Ultrasonic and pneumatic nebulization [190]
Urine, blood	Ba	Mauras and Allain (1979)	Comparison with flame AAS and AES and furnace AAS and AES [191]
Liver	Various (29)	Ward et al. (1980)	Comparison of sample digestion methods; hydride generation cell [126]
Blood	Al, Si	Mauras et al. (1980)	Sampling by plastic syringes or by free flow [192]
Serum	Al	Schramel et al. (1980)	Sample dilution with Herrmann solution [193]
Blood, serum, plasma	Various (16)	Pehlivanian et al. (1980)	Selection of analysis lines [194]
Blood, urine, faeces	Al, Si	Lichte et al. (1980)	Limiting factors for precision [195]
Plasma	Various (16)	Pehlivanian (1980)	Detection limits [196]

Table 5.4. (*Continued*)

Matrix	Analyte	Author	Remarks and Reference
Serum	Ca, Cu, Fe, Mg, Na, Zn	Schramel and Klose (1981)	Sequential determination. Sample consumption [113]
Blood, serum	Ca, Cu, Fe, K, Mg, Na, P, Zn	Uchida et al. (1981)	Microliter-sample introduction; application of Y as an internal standard element [166]
Serum	Cu, Fe, Mg	Aziz et al. (1981)	Microliter-sample introduction; matrix effects caused by Na [167]
Liver, serum	Cr, Fe, Zn	Black and Browner (1981)	Volatile metal–chelate introduction [173]
Skin, serum, liver	Al, Cr, Cu, Fe, Mn, Zn	Black et al. (1981)	Introduction of xylene solvent solutions of metal–chelate complexes [174]
Muscle tissue, blood	Mn, Ni	Camara Rica et al. (1981)	ETV-sample introduction [175]
Teeth	Various (20)	Kluckner and Brown (1981)	Comparison with flame–AAS [197]
Serum	Various (21)	Mermet et al. (1981)	Comparison of ICP–AES detection limits with average concentrations [198]
Hair	Various (19)	Quan et al. (1981)	Useful dilution levels of the sample digest solution for groups of analytes [199]
Liver, oyster tissue, albacore tuna	Various (15)	Jones et al. (1982)	Chelex-100 separation scheme; comparison of binary and ternary acid sample digestion [147]
Serum	Ca, Mg, Fe, K, Na	Alexander et al. (1982)	FI-sample introduction [169]
Breast milk	Ni	Camara Rica et al. (1982)	Sample solubilization by TMAH [176]
Liver, serum	Cd, Mn, Pb, Zn	Aziz et al. (1982)	ETV-sample introduction; matrix effects [177]

Table 5.4. (*Continued*)

Matrix	Analyte	Author	Remarks and Reference
Kidney, liver	Various (20)	Subramanian and Méranger (1982)	Accuracy and precision [186]
Liver	Be	Schramel and Xu-Li-Qiang (1982)	Spectral interferences [200]
Faeces	Various (11)	Fisher and Lee (1982)	Analysis of faeces containing Cr-marker [201]
Serum	Cu, Zn	Herber et al. (1982)	Comparison with furnace–AAS [202]
Liver	Al, Cd, Cu, Mn, Zn	Schramel and Xu-Li-Qiang (1982)	Matrix effects and spectral interferences [203]
Urine	As, Bi, Sb, Se, Sn, Te	Fodor and Barnes (1983)	Hydride introduction. Separation and concentration properties of poly(dithiocarbamate) resin [156]
Urine	Various (11)	Barnes et al. (1983)	Enrichment of poly(dithiocarbamate) resin; speciation [157]
Urine	Cr (VI), Cr(III)	Mianzhi and Barnes (1983)	Speciation; selective enrichment of Cr [158]
Bone	Various (20)	Mahanti and Barnes (1983)	Hydride (As, Se) and elemental (Hg) introduction; ETV of the other elements, enrichment [160]
Urine	Various (13)	Barnes et al. (1983)	ETV-sample introduction; modified commercial graphite rod electrothermal vaporizer [178]
Bone	Various (11)	Lee (1983)	Cu and Mo enrichment; spectral interference; matrix effects caused by Ca [188]
Blood, urine	Si	Allain et al. (1983)	Study of digestive Si-absorption in humans [204]
Serum brain, liver	Sr	Sanz-Medel et al. (1983)	Comparison with flame AAS and AES [205]
Plasma, urine	Al, Si	Mauras et al. (1983)	Study of digestive Si-absorption in man [206]

Table 5.4. *(Continued)*

Matrix	Analyte	Author	Remarks and Reference
Blood, urine	Ba	Mauras et al. (1983)	Study of digestive Ba-absorption in man [207]
Breast milk	Ni	Barnett et al. (1983)	ETV-sample introduction; extraction of Ni with dithizone–chloroform solution [208]
Bone	Ba, Ca, Fe, K, Na, Sr, Zn	Brätter et al. (1983)	Comparison with INAA; matrix matching [209]
Serum	Ca	van Deijck et al. (1984)	Accuracy; comparison of sample digestion methods [114]
Serum, blood liver	As, Cd, Co, Cu, Fe, Mo, Se, V, Zn	Mianzhi and Barnes (1984)	Hydride generation, batch wise introduction; high-pressure sample digestion [159]
Serum	Ca, Cu, Fe, K, Li, Mg, Na, Zn	McLeod et al. (1984)	FI-sample introduction; matrix effects [170]
Hemodialysis solution, serum, urine	Al, Si	Matusiewicz and Barnes (1984)	ETV-sample introduction; comparison of calibration functions [179]
Plasma, liver	Various (10)	Blakemore et al. (1984)	ETV-sample introduction, liquid (5–10 μL) and solid (0.5 mg) sample vaporization [180]
Bone, muscle, blood, liver	B, P, S	Pritchard and Lee (1984)	Corrections for spectral interferences [210]
Faeces	Cr	Roofayel and Lyons (1984)	Determination of marker chromium [211]
Human stones	Ca, Mg, P	Wandt et al. (1984)	Routine determination [212]

Table 5.4. (*Continued*)

Matrix	Analyte	Author	Remarks and Reference
Serum	Ni	Hayman et al. (1984)	ETV-sample introduction; comparison with liquid scintillation counting detection [213]
Liver	Cu, Fe, Mn, Zn	Faske et al. (1985)	FI-sample introduction [168]
Liver	Se	Nakahara and Kikui (1985)	Hydride generation, continuous introduction mode [172]
Urine	Cr, Cu, Ni	Matusiewicz and Barnes (1985)	ETV-sample introduction; sample deposition on vaporizer in the form of an aerosol [181]
Cerebrospinal fluid, plasma	Various (13)	Bourrier-Guerin et al. (1985)	Large numbers of samples [214]
Blood, dialysis fluid	Al	Mauras and Allain (1985)	Automatic method, application of Cs as matrix modifier and Ga as internal standard element [215]
Various hard and soft animal tissues	Ca, Cu, Fe, K, Mg, Mn, Na, P, Zn	Wolnik et al. (1985)	Preparation of laboratory control samples; comparison of sample digestion procedures [216]

inal sample and the ultimate sample solution to be analyzed. Thus, Robberecht and Deelstra [220] questioned whether the addition of concentrated hydrochloric acid to urine, as was done by Fodor and Barnes [156] in their oxidation state speciation study of Se in urine, could have resulted in a definite redistribution of the original species equilibrium. Weigert and Sappl [183] conducted experiments with rats to determine the behavior of As at low dosage. Because adequate reference materials with a certified As(III) content are not available, the authors used organic model-material with admixtures of As(III) and As(V) for studying the influence of sample digestion methods on the valency of As. All mineralization methods used were found to change the original valency of As to a degree that was not reproducible.

Methods that do not require any chemical treatment of the sample prior to analysis are based on selective thermal vaporization of the sample constituents. Analyte species in the sample vaporize at different temperatures, and each species, when introduced into a plasma, produces an emission peak at a vaporization temperature characteristic of the analyte species and the sample type [221]. Thus far such methods have not been applied in ICP–AES analysis of biological-clinical materials. Nevertheless, because of their potential for element speciation and possible use in the field of present interest, a brief discussion of some aspects is given here. Prack and Bastiaans [222] described an evolved gas analysis (EGA)–ICP technique capable of speciating inorganic compounds in solid samples. Samples of 10^{-3}–10^{-5} g were gradually heated to 2300°C in a graphite sample cup that was moved in a controlled manner into the ICP. Identification of a given compound was based on the position of the sample at the time of evolution of the element. The decomposition behavior of various compounds of each of the elements Cd, Hg, Pb, and V in test samples was characterized. According to the authors, chemical reactions between sample components and the walls of the graphite cup limit the separation of the sample, but the use of pyrolytic graphite may alleviate this problem. By means of thermal vaporization using a furnace provided with a quartz sample holder, Hanamura et al. [221] separated inorganic, organic, and metalorganic species and detected elemental emission in a microwave plasma. The precision of the characteristic appearance temperature was ± 2°C. The system was used to measure C, H, N, O, and Hg in solid biological materials. Considering the results, this EGA–sample introduction method certainly merits testing in combination with ICP, especially when the evolved gas furnace allows higher maximum sample temperatures.

5.4.3.4. Merits of ICP–AES

Currently used analytical methods generally are supplemental rather than competitive. Hence, assessing the significance of the analytical performance of a specific method is a ticklish matter unless the method is considered in connection with a well-defined analytical problem targeted to be solved with that method. Therefore, the merits of ICP–AES for the analysis of biological-clinical materials are discussed here on the basis of real practical analytical tasks, whereby ICP–AES has eventually been selected as the method of choice.

Using graphite furnace–AAS for the determination of Al in blood and urine Allain and Mauras [112] experienced the appearance of frequent erratic peaks. The poor consistency of the method sometimes obliged the authors to inject each sample three or even five times over. Since no difficulties of this kind were encountered during hundreds of blood and urine analyses with ICP–AES, it was finally decided to drop graphite furance–AAS for determination of Al.

Munter et al. [125] reported that in their service laboratory ICP–AES had become an ideal replacement for both rotating-disc electrode spark emission and flame absorption spectrometry. More than 15,000 samples were submitted annually for flame–AAS. About half a year after installation of an ICP spectrometer system, all samples formerly received for spark emission and most of those for flame–AAS were performed by ICP–AES. According to the authors a main advantage of ICP–AES for trace element analysis of biological samples was the relief from time-consuming dilutions, additions of releasing agents, and ionization buffers, coupled with preservation of the multielement capability of spark emission.

For the simultaneous determination of major, minor, and trace elements in biological materials Ward et al. [126] excluded more traditional methods in favor of ICP–AES on the basis of the following considerations. Spark emission, although a very rapid technique is not sufficiently sensitive. For AAS, and especially furnace–AAS, the reverse is true. The technique is very sensitive but the sample throughput is, however, extremely slow. Spectrophotometric procedures take as much time as atomic absorption but are less specific. Wet chemical procedures are also very time-consuming and provide accurate analyses only for the major elements. Neutron activation analysis and spark source mass spectrometry meet the analytical demands but require expensive instrumentation.

For the determination of 20 elements in teeth, Kluckner and Brown [197] considered a variety of analytical techniques including X-ray fluorescence, anodic stripping voltammetry, flame–AAS, and arc emission spectrometry. The authors chose ICP–AES because that method offered the possibility of simultaneous determining many elements in teeth without the extensive sample preparation required by other procedures. Flame–AAS was only used to provide comparison data for some of the analyte elements.

Herber et al. [202] compared the analytical performance of ICP–AES and furnace-AAS for the determination of Cu and Zn in human blood serum. The authors arrived at the conclusion that furnace–AAS is very suitable for monitoring and screening in toxicology and clinical-chemistry, whereas ICP–AES is to be used when more precise determinations are required.

Sanz-Medel et al. [205] performed a comparison study on the analytical performance of flame–AAS, flame–AES, and ICP–AES for the determination of Sr in animal serum and soft tissues. ICP–AES appeared to be the best method because it provided a sensitivity of about 100 times better than the next most sensitive method, a linear calibration graph over five orders of magnitude, good precision in real sample analysis, and virtual absence of spectral or chemical interferences.

For the determination of Ca, Mg, and P in human kidney stones, Wandt et al. [212] summarized element-sensitive methods currently in use in stone research, which, in addition to wet chemical methods, include X-ray fluorescence, flame–AAS and flame–AES, electron microprobe analysis, and activation analysis. Although ICP–AES had not yet been used routinely for stone analysis, the authors preferred that method because of the markedly reduced or nonexistent classical Ca and PO_4 solute vaporization interference and the sufficient excitation of phosphorus in the plasma.

5.4.3.5. Research

To satisfy present and near-future needs, research should include the following aspects.

Sampling; homogeneity control of microsamples of various nature.

Optimization and automation of sample digestion procedures.

Groups-wise trace element enrichment in bulk samples.

Improvement of methods for trace analysis in microsamples.

Application of hybrid techniques; especially HPLC-ICP coupling for trace element determination and speciation in protein fractions.

REFERENCES

1. G. Pethes, "The need for trace element analyses in animal sciences," in *Elemental Analysis of Biological Materials*, International Atomic Energy Agency, Technical Report Series No. 197, Vienna (1980) p. 3.
2. K. M. Hambidge, "The need for trace element analyses in the medical sciences," in *Elemental Analysis of Biological Materials*, International Atomic Energy Agency, Technical Report Series No. 197, Vienna (1980) p. 19.
3. N. Rao Maturo, "The need for more information on the trace element content of foods for improving human nutrition," in *Elemental Analysis of Biological Materials*, International Atomic Energy Agency, Technical Report Series No. 197, Vienna (1980) p. 29.
4. E. I. Hamilton, "The need for trace element analyses of biological materials in the environmental sciences," in *Elemental Analysis of Biological Materials*, International Atomic Energy Agency, Technical Report Series No. 197, Vienna (1980) p. 39.
5. E. I. Hamilton, *Sci. Total Environ.* **5,** 1 (1976).
6. E. G. Underwood, *Trace Elements in Human and Animal Nutrition*, 4th Edition, Academic, New York (1977).
7. K. S. Schwartz, "Essentiality versus toxicity of metals," in S. S. Brown, ed., *Clinical Chemistry and Chemical Toxicology of Metals*, Elsevier/North Holland Publ. Comp., Amsterdam-New York-Oxford (1977) vol. 1, p. 3.

8. G. H. Morrison, *CRC Crit. Rev. Anal. Chem.* **8,** 287 (1979).
9. H. T. Delves, *Prog. Analyt. Atom. Spectrosc.* ***4,*** 1 (1981).
10. P. W. J. M. Boumans and F. J. De Boer, *Spectrochim. Acta* **30B,** 309 (1975)
11. P. W. J. M. Boumans and F. J. De Boer, *Spectrochim. Acta* **31B,** 355 (1976).
12. P. W. J. M. Boumans, invited lecture presented at the 25th Candian Spectroscopy Symposium, Mont Gabriel, Quebec, 28–29 September 1978.
13. P. W. J. M. Boumans, R. J. McKenna and M. Bosveld, *Spectrochim. Acta* **36B,** 1031 (1981).
14. P. W. J. M. Boumans and M. Ch. Lux-Steiner, *Spectrochim. Acta* **37B,** 97 (1982).
15. M. Thompson, B. Pahlavanpour, S. J. Walton, and G. F. Kirkbright, *Analyst* **103,** 568 (1978).
16. W. B. Robbins and J. A. Caruso, *Anal. Chem.* **51,** 889A (1979).
17. K. A. Wolnik, F. L. Fricke, M. H. Hahn, and J. A. Caruso, *Anal. Chem.* **53,** 1031 (1981).
18. M. H. Hahn, K. A. Wolnik, F. L. Fricke, and J. A. Caruso, *Anal. Chem.* **54,** 1048 (1982).
19. J. M. Mermet and J. Hubert, *Prog. Anal. Atom. Spectrosc.* **5,** 1 (1982).
20. E. J. Maienthal and D. A. Becker, A Survey of Current Literature on Sampling, Sample Handling, and Long Term Storage for Environmental Materials. NBS Technical Note No. 929, Washington, DC (1976).
21. B. Sansoni and G. V. Iyengar, Sampling and Sample Preparation Methods for the Analysis of Trace Elements in Biological Material, Invited Review Paper, International Symposium on Nuclear Activation Techniques in the Life Sciences, International Atomic Energy Agency, Vienna, 22–26 May 1978.
22. A. Speecke, J. Hoste, and J. Versieck, ''Sampling of biological material,'' in P. D. LaFleur, ed., *Accuracy in Trace Analysis: Sampling, Sample Handling, Analysis* NBS Special Publication No. 422, Washington DC (1976) vol. 1, p. 299.
23. J. Versieck, A. Speecke, J. Hoste, and F. Barbier, *Clin. Chem.* **19,** 473 (1973).
24. D. Mikac-Devic, S. Nomoto, and F. W. Sunderman Jr., *Clin. Chem.* **23,** 948 (1977).
25. J. Webb, W. Niedermeier, J. H. Griggs, and T. N. James, *Appl. Spectrosc.* **27,** 343 (1973).
26. T. E. Gills and L. T. McClendon, *J. Radioanal. Chem.* **39,** 285 (1977).
27. J. S. Hislop and A. Parker, *Analyst* **98,** 694 (1973).
28. E. C. Kuehner, R. Alvarey, P. J. Paulsen, and T. J. Murphy, *Anal. Chem.* **44,** 2050 (1972).
29. J. W. Mitchell, *Anal. Chem.* **45,** 492A (1973).
30. R. W. Karin, J. A. Buono, and J. L. Faschino, *Anal. Chem.* **47,** 2296 (1975).
31. J. R. Moody and R. M. Lindstrom, *Anal. Chem.* **49,** 2264 (1977).
32. P. Valenta, H. Rützel, H. W. Nürnberg, and M. Stoeppler, *Fresenius Z. Anal. Chem.* **285,** 25 (1977).

33. T. E. Gills, H. L. Rook, P. D. LaFleur, and G. M. Goldstein, Evaluation and Research of Methodology for the National Environmental Specimen Bank, EPA-600/1-78-015, Environmental Health Effects Research Series (1978).
34. L. Mart, Ph.D. Thesis, University of Aachen (1978).
35. R. Massee and F. J. M. J. Maessen, *Anal. Chim. Acta* 127, **181** (1981).
36. J. Golimowski, P. Valenta, M. Stoeppler, and H. W. Nürnberg, *Fresenius Z. Anal. Chem.* **290,** 107 (1978).
37. M. Stoeppler, K. May, and C. Mohl, Kristiansand Conference on Nickel Toxicology, Studies of the Direct Extraction of Nickel from Urine, Abstracts of Scientific Papers, p. 4. (1978).
38. G. L. Fisher, L. G. Davies, and L. S. Rosenblatt, "The effects of container composition, storage duration and temperature on serum mineral levels, in P. D. LaFleur, ed., *Accuracy in Trace Analysis: Sampling, Sample Handling, Analysis,* vol. 1, NBS Special Publication No. 422, Washington DC (1976) p. 575.
39. F. A. Ibbott, "Sampling for clinical chemistry," in P. D. LaFleur, ed., *Trace Analysis: Sampling, Sample Handling, Analysis,* vol. 1, NBS Special Publication No. 422, Washington, DC (1976) p. 353.
40. S. H. Omang and O. D. Vellar, *Z. Anal. Chem.* **269,** 177 (1974).
41. P. S. Tjioe, J. J. M. De Goey, and J. P. W. Houtman, *J. Radioanal. Chem.* **37,** 511 (1977).
42. H. J. M. Bowen and D. Gibbons, *Radioactivation Analysis*, Oxford University Press, Oxford (1963) pp. 134–135.
43. I. Y. Donev, *J. Radioanal. Chem.* **39,** 317 (1977).
44. T. Kneip, M. Kleinman, R. Riddick, and D. Bernstein, Trace Elements in Human Tissues, International Workshop on Biological Specimen Collection, Commission of the European Communities, World Health Organisation, United States Environmental Protection Agency, Luxembourg (1977).
45. G. V. Iyengar, *Radiochem. Radioanal. Lett.* **24,** 35 (1976).
46. G. V. Iyengar and K. Kasperek, *J. Radioanal. Chem.* **39,** 301 (1977).
47. J. L. M. De Boer and F. J. M. J. Maessen, *Anal. Chim. Acta* **117,** 371 (1980).
48. S. Salmela, E. Vuori, and J. O. Kilpiö, *Anal. Chim. Acta* **125,** 131 (1981).
49. V. D. Mattera, Jr., V. A. Arbige, Jr., S. A. Tomellini, D. A. Erbe, M. M. Doxtader, and R. K. Forcé, *Anal. Chim. Acta* **124,** 409 (1981).
50. G. S. Assarian and D. Oberleas, *Clin. Chem.*, **23,** 1771 (1977).
51. W. M. Mertz, in R. M. Barnes, ed., *Developments in Atomic Plasma Spectrochemical Analysis*, Heyden, Philadelphia (1981) p. 635.
52. E. Michaud and J. M. Mermet, *Spectrochim. Acta* **37B,** 145 (1982).
53. B. Sansoni and V. K. Panday, in J. Fachetti, ed., *Analytical Techniques for Heavy Metals in Biological Fluids*, lectures of a course held at the Joint Research Centre, Ispra, Italy, 22–26 June 1981, Elsevier, Amsterdam (1982) p. 91.
54. B. Boppel, *Z. Anal. Chem.* **266,** 257 (1973).
55. B. Boppel, *Z. Anal. Chem.* **268,** 114 (1974).

56. E. I. Hamilton, M. J. Minski, and J. J. Cleary, *Analyst* **92,** 257 (1967).
57. S. R. Koirtyohann and C. A. Hopkins, *Analyst* **101,** 807 (1976).
58. J. L. St. John, *J. Assoc. Analyt. Chem.* **24,** 848 (1941).
59. T. T. Gorsuch, *Analyst* **84,** 135 (1959).
60. M. Feinberg and C. Ducauze, *Anal. Chem.* **52,** 207 (1980).
61. Analytical Methods Committee, *Analyst* **85,** 643 (1960).
62. E. E. Menden, D. Brockman, H. Choundhury, and H. G. Petering, *Anal. Chem.* **49,** 1644 (1979).
63. J. Locke, *Anal. Chim. Acta* **104,** 225 (1979).
64. G. J. Lutz, J. S. Stemple, and H. L. Rook, *J. Radioanal. Chem.* **39,** 277 (1977).
65. T. H. Lockwood and L. P. Limtiaco, *Am. Ind. Hyg. Ass. J.* **36,** 57 (1975).
66. G. F. Carter and W. B. Yeoman, *Analyst* **105,** 295 (1980).
67. E. I. Hamilton, M. J. Minski, and J. J. Cleary, *Sci. Total Environ.* **1,** 1 (1972).
68. P. R. Walsh, J. L. Fashing, and R. A. Duce, *Anal. Chem.* **48,** 1012 (1976).
69. T. Tarumoto and H. Freiser, *Anal. Chem.* **47,** 180 (1975).
70. G. Kaiser, P. Tschöpel, and G. Tölg, *Z. Anal. Chem.* **253,** 177 (1971).
71. G. Kaiser, E. Gallath, P. Tschöpel, and G. Tölg, *Z. Anal. Chem.* **259,** 257 (1972).
72. J. W. Mair Jr. and H. G. Day, *Anal. Chem.* **44,** 2015 (1972).
73. M. E. Tatro, W. L. Raynolds, and F. M. Costa, *At. Absorption Newslett.* **16,** 143 (1977).
74. R. Djudzman, E. E. Vanden, and P. de Moerloose, *Analyst* **102,** 688 (1977).
75. G. J. Lutz, J. S. Stemple, and H. L. Rook, NBS Spec. Publ. No. 501, 33 (1978).
76. G. Kaiser, D. Götz, G. Tölg, G. Knapp, B. Marchin, and H. Stitzy, *Fresenius Z. Anal. Chem.* **291,** 278 (1978).
77. W. Ross and F. Umland, *Talanta* **26,** 727 (1979).
78. G. Knapp, S. E. Raptis, G. Kaiser, G. Tölg, P. Schramel, and B. Schreiber, *Fresenius Z. Anal. Chem.* **308,** 97 (1981).
79. W. Schöniger, *Mikrochim. Acta* 123 (1955).
80. W. Schöniger, *Z. Anal. Chem.* **181,** 28 (1961).
81. P. Gouverneur and C. D. F. Eerbeek, *Anal. Chim. Acta* **27,** 303 (1962).
82. A. M. G. Macdonald, *Analyst* **86,** 3 (1961).
83. M. Berthelot, *Comp. Rend.* **129,** 1002 (1899).
84. M. Fujita, Y. Takida, T. Terao, O. Hiskins, and T. Ukita, *Anal. Chem.* **40,** 2024 (1968).
85. E. Scheubeck, A. Nielsen, and G. Iwantscheff, *Fresenius Z. Anal. Chem.* **294,** 398 (1979).
86. G. Middleton and R. E. Stuckey, *Analyst* **79,** 39 (1954).
87. P. O. Bethge, *Anal. Chim. Acta* **10,** 317 (1954).
88. C. Feldman, *Anal. Chem.* **46,** 1606 (1974).

89. G. F. Smith, *Anal. Chim. Acta* **8,** 397 (1953).
90. G. F. Smith, *Analyst* **80,** 16 (1955).
91. G. F. Smith, *Anal. Chim. Acta* **17,** 175 (1957).
92. Analytical Methods Committee, *Analyst* **84,** 214 (1959).
93. J. L. M. de Boer, Ph.D. Thesis, University of Amsterdam (1984).
94. P. Schramel, *Anal. Chim. Acta* **67,** 69 (1973).
95. D. Polley and V. L. Miller, *Anal. Chem.* **27,** 1162 (1955).
96. R. P. Taubinger and J. R. Wilson, *Analyst* **90,** 429 (1965).
97. J. L. Down and T. T. Gorsuch, *Analyst* **92,** 398 (1967).
98. Analytical Methods Committee, *Analyst* **92,** 403 (1967).
99. K. W. Budna and G. Knapp, *Fresenius Z. Anal. Chem.* **291,** 116 (1978).
100. M. Stoeppler and F. Backhaus, *Fresenius Z. Anal. Chem.* **291,** 116 (1978).
101. M. Stoeppler and K. P. Müller, and F. Backhaus, *Fresenius Z. Anal. Chem.* **297,** 107 (1979).
102. K. Eustermann and D. Seifert, *Fresenius Z. Anal. Chem.* **285,** 253 (1977).
103. B. Bernas, *Anal. Chem.* **40,** 1682 (1968).
104. L. Kötz, G. Kaiser, P. T. Tschöpel, and G. Tölg, *Z. Anal. Chem.* **260,** 207 (1972).
105. B. Krinitz and W. Holak, *J. Assoc. Analyt. Chem.* **57,** 568 (1974).
106. A. Faanhof and H. A. Das, *Radiochem. Radioanal. Lett.* **30,** 405 (1977).
107. A. van Eenbergen and E. Brunninx, *Anal. Chim. Acta* **98,** 405 (1978).
108. C. Iida, T. Uchida, and I. Kojima, *Anal. Chim. Acta* **113,** 365 (1980).
109. L. Kötz, G. Henze, G. Kaiser, S. Pahlke, M. Veber, and G. Tölg, *Talanta* **26,** 681 (1979).
110. R. Uhrberg, *Anal. Chem.* **54,** 1906 (1982).
111. P. E. Paus, *At. Absorption Newslett.* **11,** 129 (1972).
112. P. Allain and Y. Mauras, *Anal. Chem.* **51,** 2089 (1979).
113. P. Schramel and B. J. Klose, *Fresenius Z. Anal. Chem.* **307,** 26 (1981).
114. W. van Deijck, J. Balke, and F. J. M. J. Maessen, *Fresenius Z. Anal. Chem.* **317,** 121 (1984).
115. T. A. Hinners, *Fresenius Z. Anal. Chem.* **277,** 377 (1975).
116. M. Stoeppler, K. Brandt, and T. C. Rains, *Analyst* **103,** 714 (1978).
117. F. Alt and H. Massmann, *Spectrochim. Acta* **33B,** 337 (1978).
118. A. J. Jackson, L. M. Michael, and H. J. Schumacher, *Anal. Chem.* **44,** 1064 (1972).
119. A. K. Khera and D. G. Wibberley, *Proc. Analyt. Div. Chem. Soc.* **13,** 340 (1976).
120. S. B. Gross and E. S. Parkinson, *At. Absorption Newslett.* **13,** 107 (1974).
121. L. Murthy, E. E. Menden, P. M. Eller, and H. G. Petering, *Anal. Biochemistry* **53,** 365 (1973).

122. M. Stoeppler, in J. O. Nriagu, ed., *Analysis of Nickel in Biological Materials and Natural Waters, Nickel in the Environment* ch. 29, Wiley, New York, (1980) p. 675.

123. R. L. Dahlquist and J. W. Knoll, *Appl. Spectrosc.* **32,** 1 (1978).

124. N. R. McQuaker, D. F. Browner, and P. D. Kluckner, *Anal. Chem.* **51,** 1082 (1979).

125. R. C. Munter, R. A. Grande, and P. C. Ahn, *ICP Information Newslett.* **5,** 368 (1979).

126. A. F. Ward, L. F. Marciello, L. Carrara, and V. J. Luciano, *Spectrosc. Lett.* **13,** 803 (1980).

127. J. L. M. de Boer and F. J. M. J. Maessen, *Spectrochim. Acta* **38B,** 739 (1983).

128. F. J. M. J. Maessen, J. Balke, and J. L. M. de Boer, *Spectrochim. Acta* **37B,** 517 (1982).

129. A. Mizuike, *Fresenius Z. Anal. Chem.* **319,** 415 (1984).

130. G. V. Iyengar, Concentrations of 15 trace elements in some selected adult human tissues and body fluids of clinical interest from several countries: results from a pilot study for the establishment of reference values. Report from the Institute of Medicine, Juelich Nuclear Research Center, Federal Republic of Germany. Berichte der Kernforschungsanlage Jülich—Nr. 1974. February 1985.

131. J. B. Hemel, R. F. Hindriks, and W. van der Slik, *J. Autom. Chem.* **7,** 20 (1985).

132. P. W. J. M. Boumans and J. J. A. M. Vrakking, *Spectrochim. Acta* **40B,** 1085 (1985).

133. K. Bächmann, *C.R.C. Crit. Rev. Anal. Chem.* **12,** 1 (1981).

134. J. Minczewski, J. Chwastowska, and R. Dybczynski, *Separation and Preconcentration Methods in Inorganic Trace Analysis*, Wiley, New York, published by Ellis Horwood, Chichester, England (1982).

135. M. S. Cresser, *Solvent Extraction in Flame Spectroscopic Analysis.* Butterworth, London (1978).

136. W. B. Kerfoot and R. L. Crawford, *ICP Information Newslett.* **2,** 289 (1977).

137. H. M. Kingston, I. L. Barnes, T. J. Brady, T. C. Rains, and M. A. Champ, *Anal. Chem.* **50,** 2064 (1978).

138. M. M. Guedes da Mota, M. A. Jonker, and B. Griepink, *Fresenius Z. Anal. Chem.* **296,** 345 (1979).

139. S. S. Berman, J. W. McLaren, and S. N. Willie, *Anal. Chem.* **52,** 488 (1980).

140. R. E. Sturgeon, S. S. Berman, J. A. H. Desaulniers, and D. S. Russell, *Talanta* **27,** 85 (1980).

141. R. E. Sturgeon, S. S. Berman, S. N. Willie, and J. A. H. Desaulniers, *Anal. Chem.* **53,** 2337 (1981).

142. H. Watanabe, K. Goto, S. Taguchi, J. W. McLaren, S. S. Berman, and D. S. Russell, *Anal. Chem.* **53,** 738 (1981).

143. R. E. Sturgeon, S. S. Berman, and S. N. Willie, *Talanta* **29,** 167 (1982).

144. C. M. McLeod, A. Otsuki, K. Okamoto, H. Haraguchi, and K. Fuwa, *Analyst* **106,** 419 (1981).
145. P. J. H. Seeverens, E. J. M. Klaassen, and F. J. M. J. Maessen, *Spectrochim. Acta* **38B,** 727 (1983).
146. F. J. M. J. Maessen, G. Kreuning, and J. Balke, *Spectrochim. Acta* **41B,** 3 (1986).
147. J. W. Jones, S. G. Capar, and T. C. O'Haver, *Analyst* **107,** 353 (1982).
148. J. W. Jones, T. C. O'Haver, *Spectrochim. Acta* **40B,** 263 (1985).
149. G. D. Martinie and A. A. Schilt, *Anal. Chem.* **48,** 70 (1976).
150. P. A. Pella, H. M. Kingston, J. R. Sieber, and L. Feng, *Anal. Chem.* **55,** 1193 (1983).
151. W. van Deijck, M. J. Lips, and F. M. J. Maessen, *Fresenius Z. Anal. Chem.* **317,** 858 (1984).
152. R. M. Barnes, *CRC Crit. Rev. Anal. Chem.* **7,** 203 (1978).
153. S. Greenfield, *Pure Appl. Chem.* **52,** 2509 (1980).
154. R. M. Barnes, *Biological Trace Element Research* **6,** 93 (1984).
155. R. M. Barnes and J. S. Genna, *Anal. Chem.* **51,** 1065 (1979).
156. P. Fodor and R. M. Barnes, *Spectrochim. Acta* **38B,** 229 (1983).
157. R. M. Barnes, P. Fodor, K. Inagaki, and M. Fodor, *Spectrochim. Acta* **38B,** 245 (1983).
158. Z. Mianzhi and R. M. Barnes, *Spectrochim. Acta* **38B,** 259 (1983).
159. Z. Mianzhi and R. M. Barnes, *Appl. Spectrosc.* **38,** 635 (1984).
160. H. S. Mahanti and R. M. Barnes, *Anal. Chim. Acta* **151,** 409 (1983).
161. A. Miyazaki and R. M. Barnes, *Anal. Chem.* **53,** 299 (1981).
162. A. Miyazaki and R. M. Barnes, *Anal. Chem.* **53,** 364 (1981).
163. F. J. M. J. Maessen, P. C. Coevert, and J. Balke *Anal. Chem.* **56,** 899 (1984).
164. S. Greenfield and P. B. Smith, *Anal. Chim. Acta* **59,** 341 (1972).
165. R. N. Kniseley, V. A. Fassel, and C. C. Butler, *Clin. Chem.* **19,** 807 (1973).
166. H. Uchida, Y. Nojiri, H. Haraguchi, and K. Fuwa, *Anal. Chim. Acta* **123,** 57 (1981).
167. A. Aziz, J. A. C. Broekaert, and F. Leis, *Spectrochim. Acta* **36B,** 251 (1981).
168. A. J. Faske, K. R. Snable, A. W. Boorn, and R. F. Browner, *Appl. Spectrosc.* **39,** 542 (1985).
169. P. W. Alexander, R. J. Finlayson, L. E. Smythe, and A. Thalib, *Analyst* **107,** 1335 (1982).
170. C. W. McLeod, P. J. Worsfold, and A. G. Cox, *Analyst* **109,** 327 (1984).
171. T. Nakahara, *Prog. Anal. Atom. Spectrosc.* **6,** 163 (1983).
172. T. Nakahara and N. Kikui, *Spectrochim. Acta* **40B,** 21 (1985).
173. M. S. Black and R. F. Browner, *Anal. Chem.* **53,** 249 (1981).
174. M. S. Black, M. B. Thomas, and R. F. Browner, *Anal. Chem.* **53,** 2224 (1981).

175. C. Camara Rica, G. F. Kirkbright, and R. D. Snook, *At. Spectrosc.* **2,** 172 (1981).

176. C. Camara Rica and G. F. Kirkbright, *Sci. Total Environ.* **22,** 193 (1982).

177. A. Aziz, J. A. C. Broekaert, and F. Leis, *Spectrochim. Acta* **37B,** 369 (1982).

178. R. M. Barnes and P. Fodor, *Spectrochim. Acta* **38B,** 1191 (1983).

179. H. Matusiewicz and R. M. Barnes, *Spectrochim. Acta* **39B,** 891 (1984).

180. W. M. Blakemore, P. H. Casey, and W. R. Collie, *Anal. Chem.* **56,** 1376 (1984).

181. H. Matusiewicz and R. M. Barnes, *Spectrochim. Acta* **40B,** 41 (1985).

182. R. F. Browner and A. W. Boorn, *Anal. Chem.* **56,** 875A (1984).

183. P. Weigert and A. Sappl, *Fresenius Z. Anal. Chem.* **316,** 306 (1983).

184. F. J. M. J. Maessen, J. Balke, and R. Massee, *Spectrochim. Acta* **33B,** 311 (1978).

185. R. F. Browner and A. W. Boorn, *Anal. Chem.* **56,** 787A (1984).

186. K. S. Subramanian and J. C. Méranger, *Sci. Total Environ.* **24,** 147 (1982).

187. G. V. Iyengar, W. E. Kollmer, and H. J. M. Bowen, *The Elemental Composition of Human Tissues and Body Fluids*, Verlag Chemie, Weinheim, New York (1978).

188. J. Lee, *Anal. Chim. Acta* **152,** 141 (1983).

189. W. J. Haas, Jr., V. A. Fassel, F. Grabau IV, R. N. Kniseley, and W. L. Sutherland, "Simultaneous determination of trace elements in urine by inductively coupled plasma-atomic emission spectrometry," in *Ultratrace Metal Analysis in Science and Environment*. ch. 8, American Chemical Society, Washington, DC (1979) p. 91.

190. P. Schramel, *Z. Lebensm. Unters. Forsch.* **169,** 255 (1979).

191. Y. Mauras and P. Allain, *Anal. Chim. Acta* **110,** 271 (1979).

192. Y. Mauras, P. Riberi, F. Cartier, and P. Allain, *Biomedicine* **33,** 228 (1980).

193. P. Schramel, A. Wolf, and B. J. Klose, *J. Clin. Chem. Clin. Biochem.* **18,** 591 (1980).

194. E. Pehlivanian, J. M. Mermet, and J. Robin, *J. Biophys. Med. Nucl.* **4,** 247 (1980).

195. F. E. Lichte, S. Hopper, and T. W. Osborn, *Anal. Chem.* **52,** 120 (1980).

196. E. Pehlivanian, Labo-Pharma, *Problèmes et Techniques*, No. 295, 129 (1980).

197. P. D. Kluckner and D. F. Brown, in R. M. Barnes, ed., *Developments in Atomic Plasma Spectrochemical Analysis*, Heyden, Philadelphia (1981) p. 713.

198. J. M. Mermet, E. Pehlivanian, and J. Robin, in R. M. Barnes, ed., *Developments in Atomic Plasma Spectrochemical Analysis*, Heyden, Philadelphia (1981) p. 718.

199. H. Quan, M. Morita, T. Uehiro, K. Okamoto, and K. Fuwa, *9th Intern. Conf. Atom. Spectrosc. and XXII Coll. Spectrosc. Intern., Tokyo 1981*, Abstracts, p. 373.

200. P. Schramel and Xu-Li-Qiang, *Anal. Chem.* **54,** 1333 (1982).

201. M. T. Fisher and J. Lee, *Anal. Chim. Acta.* **139,** 333 (1982).

202. R. F. M. Herber, H. J. Pieters, and J. W. Elgersma, *Fresenius Z. Anal. Chem.* **313,** 103 (1982).
203. P. Schramel and Xu-Li-Qiang, *ICP Information Newslett.* **7,** 429 (1982).
204. P. Allain, A. Cailleux, Y. Mauras, and J. C. Renier, *Thérapie* **38,** 171 (1983).
205. A. Sanz-Medel, R. Rodriguez-Roza, and C. Perez-Conde, *Analyst* **108,** 204 (1983).
206. Y. Mauras, J. C. Renier, A. Tricard, and P. Allain, *Thérapie* **38,** 175 (1983).
207. Y. Mauras, P. Allain, M. A. Roques, and C. Caron, *Thérapie* **38,** 107 (1983).
208. N. W. Barnett, L. S. Chen, and G. F. Kirkbright, *Anal. Chim. Acta* **149,** 115 (1983).
209. P. Brätter, K. P. Berthold, and P. E. Gardiner, *Spectrochim. Acta* **38B,** 221 (1983).
210. M. W. Pritchard and J. Lee, *Anal. Chim. Acta* **157,** 313 (1984).
211. R. L. Roofayel and D. J. Lyons, *Analyst* **109,** 523 (1984).
212. M. A. E. Wandt, M. A. B. Pougnet, and A. L. Rodgers, *Analyst* **109,** 1071 (1984).
213. B. P. Hayman, D. M. L. Goodgame, and R. D. Snook, *Analyst* **109,** 1593 (1984).
214. L. Bourrier-Guerin, Y. Mauras, J. L. Truelle, and P. Allain, *Trace Elements in Médicine* **2,** 88 (1985).
215. Y. Mauras and P. Allain, *Anal. Chem.* **57,** 1706 (1985).
216. K. A. Wolnik, J. I. Rader, C. M. Gaston, and L. F. Fricke, *Spectrochim. Acta* **40B,** 245 (1985).
217. C. H. Gast, J. C. Kraak, H. Poppe, and F. J. M. J. Maessen, *J. Chromatogr.* **185,** 549 (1979).
218. M. Morita and T. Uehiro, *Anal. Chem.* **53,** 1997 (1981).
219. K. Yoshida, H. Haraguchi, and K. Fuwa, *Anal. Chem.* **55,** 1009 (1983).
220. H. J. Robberecht and H. A. Deelstra, *Talanta* **31,** 497 (1984).
221. S. Hanamura, B. W. Smith, and J. D. Winefordner, *Anal. Chem.* **55,** 2026 (1983).
222. E. R. Prack and G. J. Bastiaans, *Anal. Chem.* **55,** 1654 (1983).

CHAPTER

6

APPLICATIONS: ORGANICS

ANDREW W. BOORN,

SCIEX
Thornhill, Canada

and

RICHARD F. BROWNER,

School of Chemistry, Georgia Institute of Technology
Atlanta, Georgia

6.1. Introduction
- 6.1.1. Solvent Extraction
- 6.1.2. Direct Dissolution of Organic Matrices
- 6.1.3. HPLC–ICP Interfacing
- 6.1.4. Analysis of Organic Vapors

6.2. Nebulization of Organic Liquids
- 6.2.1. Primary Drop Size Distribution
- 6.2.2. Calculated Primary Median Drop Diameter
- 6.2.3. Tertiary Drop Size Distribution
- 6.2.4. Evaporation
- 6.2.5. Transport Efficiency

6.3. ICP Operating Conditions for Organic Samples

6.4. Tolerance of Plasma to Organic Solvents
- 6.4.1. High-Powered Plasmas
- 6.4.2. Intermediate-Powered Plasmas
- 6.4.3. Low-Powered Plasmas

6.5. Instrumental Considerations
- 6.5.1. Torch Design
- 6.5.2. Solvent Uptake
- 6.5.3. Nebulizer Design
- 6.5.4. Aerosol Vapor Thermostatting

6.6. Organic Compound Spectra in Plasma
- 6.6.1. Spectral Height Profiles

The preparation of this chapter, and parts of the work described therein, was supported by the National Science Foundation under Grant No. CHE85-03090.

6.7. Signal Enhancements
6.7.1. Basic Concepts
6.7.2. High-Powered Plasmas
6.7.3. Intermediate-Powered Plasmas
6.8. Applications
6.8.1. Standards
6.8.2. Detection Limits
6.8.3. Organic Vapor Introduction
6.8.4. Gas Chromatography Interfacing
6.8.5. Liquid Chromatography Interfacing
6.9. Conclusions
References

6.1. INTRODUCTION

The great majority of samples analyzed by inductively coupled plasma–atomic emission spectrometry (ICP–AES) are inorganic—for example, metals, rocks, and so on. Even when the elements to be determined are contained in an organic matrix, the sample is often reduced to an inorganic form by ashing or digestion prior to analysis. Typical examples are agricultural materials and foods. However, there are certain samples for which direct introduction of organic sample to the ICP, without prior ashing or digestion, is necessary. In addition, there are certain inorganic samples for which extraction from an aqueous solution into an organic solvent may be advantageous. Consequently, organic solvents have been of importance in ICP-AES from its inception, and their use is described in several early studies [1–4].

The four basic types of organic samples analyzed by ICP spectrometry are 1 solvent extracts, 2 samples directly dissolved in organic solvents, 3 normal-phase high-performance liquid chromatography (HPLC) effluents, and 4 organic vapors. This chapter will describe four aspects of organic sample introduction: 1 the situations where organic sample analysis is appropriate, 2 the instrumental and theoretical implications of organic sample introduction, 3 analytical figures of merit, and 4 alternative sample preparation procedures.

6.1.1 Solvent Extraction

Solvent extraction is a well-known and widely used technique in analytical chemistry. Its usual aim is to remove analyte from a sample matrix that would otherwise cause an interference in the subsequent determination. Extractions are usually accomplished by complexing the analyte with an organic chelating agent, such as diethyldithiocarbamate (DDC), ammoniumpyrrolidinedithiocarbamate (APDC), 8-hydroxyquinoline (oxine), or a β-diketone, such as acetylacetone. The metal chelate formed is sparingly soluble in aqueous media, and when an organic solvent is present, it is readily extracted into the organic layer. Careful selection of the chelating agent, its concentration, and the pH of the aqueous sample solution are all necessary for efficient and selective analyte extraction. Organic solvents commonly used for solvent extraction include aromatic hydrocarbons, alcohols, ketones [for example, 4-methyl-2-pentanone (MIBK)], and chlorinated hydrocarbons (for example, chloroform and carbon tetrachloride).

Two of the most frequently stated advantages of the ICP as a source for emission spectrometry are the generally good freedom from interference and the low detection limits. However, in certain instances detection limits possible with the ICP are still inadequate, especially when the sample has a complex matrix. In these circumstances, solvent extraction can provide valuable analyte preconcentration. There are also many sample matrices that present difficulties in their own right. Problems may arise due to the high dissolved solid content of the sample solution, which can cause erratic nebulization and salt build-up on the torch aerosol injector tip. Other matrix-induced problems include spectral interferences, which result in measurement errors. Also, in some instances analyte emission enhancement, and occasionally emission suppression, may result from complex sample matrices. A typical example of a problem where solvent extraction from the sample matrix is necessary to ensure reasonable accuracy is trace element analysis of geological samples.

Although there have been relatively few literature reports of solvent extraction combined with ICP spectrometry, large numbers of industrial and government laboratories use solvent extraction on a routine basis, particularly in situations where methods have been carried over directly from earlier AAS procedures. The higher sample throughput possible with ICP–AES resulting from its simultaneous multielement capability, can provide a substantial advantage over flame AAS.

6.1.2. Direct Dissolution of Organic Matrices

Petroleum products are the largest single group of organic samples analyzed using ICP spectrometry. This category encompasses various crude oils, refined products, and used lubricating oils. Generally, these samples may be nebulized directly after dilution with an appropriate organic solvent. A 1:10 dilution is commonly used, although the amount of solvent necessary is generally determined by the viscosity of the sample. A workable viscosity range for ICP pneumatic nebulizers with natural solution aspiration is 1–3 cp. When the nebulizer is fed with a pump, which is a preferred situation, the viscosity of the samples is less critical. However, there remains a need for some viscosity control by dilution or warming to ensure efficient nebulization. This is because solution viscosity influences the aerosol drop size distribution produced by the nebulizer. Solvent viscosity, together with other sample physical properties and their influence on nebulizer operation and aerosol drop size, will be discussed in general terms in Chapter 8. They will be also discussed in the specific context of organic solvents later in this chapter.

Monitoring of wear metal concentrations in used aircraft engine lubricating oils is widely used as an indication of bearing wear in various engine components. Regular monitoring of the presence these metals in the oil can reveal excessive wear of engine components and warn of potential engine failure. The importance of early detection in such situations is obvious. To this end, the U.S. Air Force alone analyzes over one million oil samples annually [4]. A variety of instrumental techniques are used, including ICP–AES.

6.1.3. HPLC–ICP Interfacing

The ICP has been used to good advantage as an element-selective detector for HPLC. Most reports have described aqueous solvent-based separations, using either reversed-phase, size exclusion (gel permeation), or ion-exchange chromatography. However, several groups have also shown the viability of combining normal-phase (organic solvent) separations with ICP elemental detection. Gradient elution has also been attempted. Clearly, the potential for HPLC–ICP spectrometry is quite substantial. The ability to obtain simultaneous, multielement analysis of chromatographically separated components has great potential for clinical, environmental, and energy-related applications. In all these situations, the ability to provide speciation information gives an added dimension to the usual quantitative data.

6.1.4. Analysis of Organic Vapors

Organic species, introduced to the ICP in the vapor phase, may originate from gas chromatography effluents or as volatile organic species generated by direct chemical reaction with the sample. In gas chromatographic analysis, the primary interest in GC–ICP interfacing has been in the area of empirical and molecular formula determination. The ICP has proved itself in this capacity to be a rather more robust source than the microwave induced plasma (MIP) (see Part 1, Chapter 2). Applications in this area have been extended to elements with strong near-infrared lines, such as oxygen and the halogens.

6.2. NEBULIZATION OF ORGANIC LIQUIDS

The most common method of sample introduction in ICP–AES is pneumatic nebulization (see Chapter 8 and Part 1, Chapter 6). The major implications of the operational parameters of pneumatic nebulizers, as they affect the analytical performance with organic samples, are considered in this section.

6.2.1. Primary Drop Size Distribution

The best available estimate of the median drop diameter produced by a pneumatic nebulizer, is obtained from the equation due to Nukiyama and Tanasawa [5]:

$$d_S = \frac{585}{V}\left(\frac{\sigma}{\rho}\right)^{0.5} + 597\left[\frac{\eta}{(\sigma\rho)^{0.5}}\right]^{0.45}\left[10^3\,\frac{Q_l}{Q_g}\right]^{1.5} \tag{6.1}$$

where V is the velocity difference between gas and liquid flows to the nebulizer ($m \cdot s^{-1}$) σ, the surface tension of the liquid ($dyn \cdot cm^{-1}$) ρ, the liquid density ($g \cdot cm^3$) η, the liquid viscosity (poise) Q_l and Q_g the volume flow rates of liquid and gas respectively ($cm \cdot s^{-1}$), and d_S, the Sauter median (volume-to-surface area ratio) diameter (μm) of the aerosol produced. In terms of transport of sample to the plasma, it is the mass median diameter (d_m) of the aerosol that is of interest because this relates directly to the transport efficiency of analyte to the plasma. For spherical drops of unit density, d_S and d_m are identical. The densities of organic solvents commonly used in atomic spectroscopy fall in the range of 0.8–1.6 $g \cdot cm^{-3}$. Hence, with organic solvents, the assumption that the d_S value calculated from the Nukiyama and Tanasawa equation is identical with the d_m value needed for correlation with signal magnitudes is not precisely true. In spite of this limitation, predictions made using the equation can still be valuable in helping to explain trends in nebulizer performance.

Solvent density appears as a term in the Nukiyama and Tanasawa equation. Consequently, the primary mean drop size (d_S) produced by a nebulizer will itself vary with solvent density. Similarly, surface tension and viscosity variations between different organic solvents will cause shifts in the primary drop size produced by the nebulizer.

The values of surface tension, viscosity, and density for all the solvents whose use has either been reported in, or seems possible in, ICP–AES are given in Table 6.1. Densities cover the range 0.8–1.6 $g \cdot cm^{-3}$, surface tensions cover the range 16–40 $dyn \cdot cm^{-1}$, and viscosities cover the range 0.6–4.5 cP. Comparing these to the values for water ($\rho = 1.0\ g \cdot cm^{-3}$; $\sigma = 73\ dyn \cdot cm^{-1}$; and $\eta = 1.0$ cP) leads to the expectation of appreciably different nebulizer performance for organic solvents.

6.2.2. Calculated Primary Median Drop Diameter

The Nukiyama and Tanasawa equation [Eq. (6.1)] was used, together with the data from Table 6.1, to calculate the primary median drop diameters, d_S, of nine common organic solvents at a variety of aspiration rates [6]. Results for

Table 6.1. Physical Properties for Selected Organic Solvents

Solvent	Density, ρ ($g \cdot cm^{-3}$)	Surface Tension, σ ($dyn \cdot cm^{-1}$)	Viscosity, η (cP)
Methanol	0.7915	22.6	0.578
Ethanol	0.7893	22.8	1.20
Propanol	0.8035	23.8	2.26
Butanol	0.8098	24.6	2.95
Pentanol	0.8144	25.6	4.50
Pentane	0.6263	16.0	0.240
Hexane	0.6603	18.4	0.326
Heptane	0.6838	20.9	0.409
150 Octane	0.6980	21.3	0.571
Decane	0.7299	23.9	0.907
Cyclohexane	0.7786	25.5	1.02
Benzene	0.8787	28.9	0.652
Toluene	0.8669	28.5	0.590
Xylene	0.8685	29.1	0.690
Nitrobenzene	1.203	43.9	2.01
Chlorobenzene	1.106	33.5	0.800
Acetone	0.7899	23.7	0.316
4-Methyplentan-2-one (MIBK)	0.7978	23.6	0.542
Diethyl ether	0.7137	17.0	0.233
Tetrahydrofuran	0.8892	26.4	0.412
Acetonitrile	0.7857	29.3	0.360
Pridine	0.9819	38.0	0.974
Dimethylformamide	0.9487	36.8	0.891
Methylene chloride	1.327	26.5	0.431
Chloroform	1.483	27.1	0.569
Carbon tetrachloride	1.594	27.0	0.955
Carbon disulfide	1.263	32.3	0.363
Dimethylsulphoxide	1.01	43.5	0.982
Ethyl acetate	0.9005	24.3	0.477
Propyl acetate	0.8878	24.3	0.585
Water	0.9982	72.8	1.002

water are included for reference purposes. Nebulizer dimensions vary from type to type, which in turn leads to variations in gas velocity and gas and liquid flow rates. The results shown in Fig. 6.1 were calculated for a Meinhard concentric all-glass nebulizer. Curves calculated for an adjustable cross-flow nebulizer (manufactured by PlasmaTherm, Inc.) were quite similar, and so are not shown here. A typical ICP nebulizer gas flow rate of 1.0 $L \cdot min^{-1}$ was assumed for both nebulizers. The calculation for the cross-flow nebulizer makes one as-

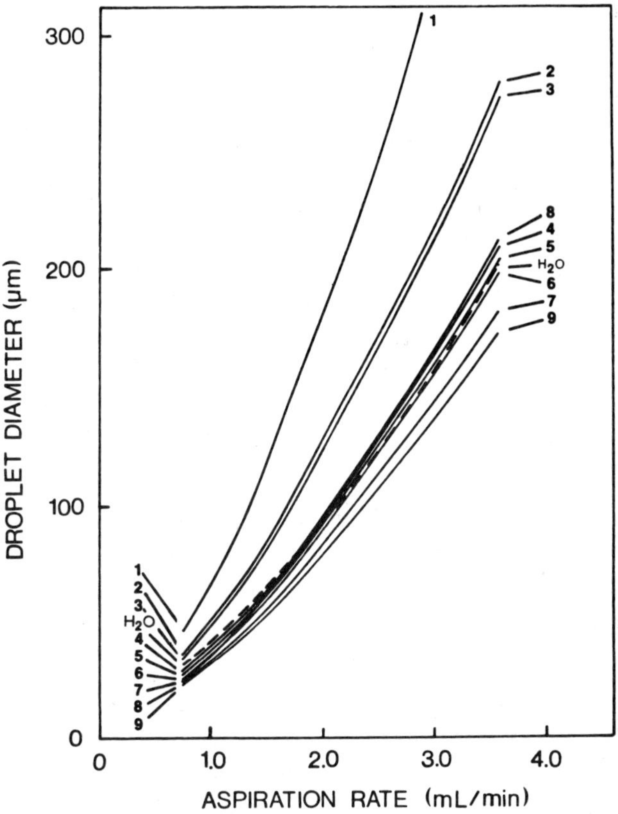

Figure 6.1. Calculated primary median drop diameters (d_s) for various organic solvents. Concentric glass nebulizer operated at 1.00-L · min^{-1} gas flow. 1 = butanol; 2 = nitrobenzene; 3 = ethanol; 4 = xylenes; 5 = benzene; 6 = MIBK; 7 = ethyl acetate; 8 = carbon tetrachloride; 9 = chloroform. [6]. [Reprinted by permission from A. W. Boorn and R. F. Browner, *Anal. Chem.* **54,** 1402 (1982). Copyright (1982), American Chemical Society.]

sumption additional to the normal approximations that are commonly accepted in applying the Nukiyama and Tanasawa equation to atomic spectrometry nebulizers, namely, that the equation, which was developed for a particular pattern of concentric nebulizer, is applicable also to cross-flow nebulizers. There is no a priori justification for this conclusion. The validity of the assumption is based solely on the empirical observation of a reasonably good fit between predicted and observed trends in the behavior of concentric and cross-flow nebulizers.

Calculations of this type bring out several crucial points regarding Eq. (6.1) which are worthy of note. In contrast to nebulizers used in flame atomic spectrometry (absorption, emission, and fluorescence), ICP nebulizers use significantly (5 to 10 times) lower gas flows, with correspondingly lower liquid flows. The first term in the equation dominates at high gas flows, typical for atomic absorption spectrometry (AAS) nebulizers, while the second term has minimal effect. With ICP nebulizers, which run at lower gas flows, the second term dominates. Because the second term shows the greatest dependence on solvent properties such as density and viscosity, ICP nebulizers have more predicted sensitivity to changes in solvent properties than AAS nebulizers.

The nebulizer gas flow is, of course, the overwhelmingly dominant factor in Eq. (6.1) and exerts the greatest influence on drop diameter. Consideration of Fig. 6.1 and the density, surface tension, and viscosity values in Table 6.1 allows some generalizations to be made on the relative importance of the organic solvent's physical properties in ICP–AES applications.

An increased aspiration rate leads to production of larger drops because the same disruptive energy of the gas jet is spread among a larger volume of material [that is, in Eq. (6.1), V decreases and Q_l/Q_g increases]. This variation of uptake rate independent of the nebulizer gas flow is important in ICP–AES because of the widespread and highly recommended use of pumps to control sample flow rate. Use of a pump eliminates any viscosity dependence of V and Q_l. However, viscosity still has the greatest influence on primary drop diameter, since it is present in the dominant second term of Eq. (6.1). Ethanol, nitrobenzene, and butanol have appreciably greater viscosities than water, whereas chloroform has a much lower viscosity. This results in the widely different calculated drop diameters shown in Fig. 6.1.

It is important to bear in mind that only the primary drop diameter has been considered so far in this discussion. Although the value of the primary drop diameter is necessary in the discussion of transport and evaporation processes, it is important to recognize that it does not represent the diameter of the drops actually entering the plasma. The tertiary or preatomization drop size distribution describes the aerosol reaching the plasma.

6.2.3. Tertiary Drop Size Distribution

For the spray chamber designs most commonly used with the ICP, few drops with diameters greater than 10 μm reach the plasma [7]. The various processes that modify the primary drop size distribution, namely impaction, gravitational settling, turbulence, and evaporation are dominated by nebulizer gas flow rate and spray chamber design. These will be discussed in detail in Chapter 8. Only aspects particularly relevant to organic solvents will be discussed in this section.

All the aerosol-modifying effects, except evaporation, involve direct loss of drops from the aerosol produced at the nebulizer. A specific nebulizer–spray chamber–torch system at the same nebulizer gas flow rate and liquid aspiration rate will exhibit definite cut-off diameters in terms of the size of aerosol drops that will pass through it. Solvent (or sample) density will be of prime importance in determining preatomization drop size when comparing different solvents. Most organic solvents have densities less than that of water (0.9982 g · cm^{-3} at 25°C), although some chlorinated hydrocarbons, nitrobenzene, and chlorobenzene have densities greater than that of water. With decreasing solvent density, the impaction and gravitational cut-off diameters will increase, allowing transport of larger drops to the plasma. Aerosols of denser solvents will have lower transport efficiency to the plasma, unless evaporation of solvent from the drop surface acts to reduce significantly the median drop diameter.

The values of the cut-off diameters for various ICP spray chambers have been estimated for the various drop loss mechanisms [7] and the preatomization drop diameter for aqueous aerosols reported [7, 8]. Direct measurements of organic aerosol preatomization drop diameters are not usually possible by the same technique, owing to the significant role of evaporation with these solvents. Drop size measurements with highly involatile organic materials, such as dioctylphthalate (DOP), yielded preatomization mass median drop diameters of around 2–3 μm for cross-flow and concentric glass ICP nebulizers, with few drops larger than 10 μm reaching the plasma, as predicted by the modeling of the aerosol transport system.

6.2.4. Evaporation

Evaporation of solvent from the surface of aerosols has been proposed as a major effect in determining the preatomization drop diameter in ICP–AES [9]. In this work, evaporation curves were calculated using Eq. (6.2):

$$d_t = (d_o^3 - E)^{1/3} \qquad (6.2)$$

where d_o is the initial aerosol drop diameter (μm), and d_t the diameter at time t(s) after formation. The evaporation factor E is given by Eq. (6.3):

$$E = 48D_V\sigma P_S M^2(\rho RT)^{-2} \qquad (6.3)$$

where, D_V is the diffusion coefficient for solvent vapor, P_S the saturated pressure, M the molecular weight of the solvent, R the gas constant, T the absolute temperature, σ the surface tension, and ρ the density.

This equation involves many simplifying assumptions, such as isothermal laminar flow conditions, which cannot be present in the spray chamber used in ICP–AES. However, the trends indicated by the magnitude of E should be valid for diffusion limited evaporation processes.

Values of E calculated in reference [9] are included in the second column of Table 6.2. Examples of the evaporation curves are shown in Fig. 6.2, for carbon tetrachloride, 4-methyl-2-pentanone (MIBK), methanol, and xylenes. These curves are based on varying initial drop diameters, in the range that is likely to

Table 6.2. Limiting Aspiration Rates for Selected Organic Solvents[a] [6]

Solvent	Evaporation Factor ((E), $\mu m^3/s$)	Limiting Aspiration Rate ($mL \cdot min^{-1}$)
Methanol	47.2	0.1
Ethanol	45.6	2.5
Propanol	22.5	3.0
Butanol	9.30	>5.0
Pentanol	5.27	>5.0
Pentane		<0.1
Hexane	298	<0.1
Heptane	165	0.2
Isooctane		0.5
Decane	8.90	2.0
Cyclohexane	179	<0.1
Benzene	164	<0.1
Toluene	58.4	1.0
Xylene	18.5	4.0
Nitrobenzene	4.09	>5.0
Chlorobenzene	23.8	3.0
Acetone	264	0.1
4-Methylplentan-2-one (MIBK)	77.3	3.0
Diethyl ether	771	<0.1
Tetrahydrofuran		<0.1
Acetonitrile		0.2
Pyridine		1.0
Dimethylformamide		0.5
Methylene chloride	557	2.0
Chloroform	321	3.0
Carbon tetrachloride	164	>5.0
Carbon disulfide	399	0.5
Dimethyl sulfoxide		2.0
Ethyl acetate	157	1.5
Propyl acetate		2.5
Water	13.1	

[a]ICP operating conditions are 1.75-kW rf power, 12 $L \cdot min^{-1}$ outer Ar, 2.0 $L \cdot min^{-1}$ intermediate Ar, and a PlasmaTherm cross-flow nebulizer operated at 1.00 $L \cdot min^{-1}$ Ar.

exit the spray chamber, based on its cut-off diameter characteristics; a nebulizer gas flow rate of 1.0 $L \cdot min^{-1}$ was assumed, which gives a typical drop transit time, from nebulizer tip to approximately 15 mm above the load coil, of about 700 ms.

As can be clearly seen in Fig. 6.2, evaporation from drops of 10 μm or less in diameter in aerosols of carbon tetrachloride, MIBK, and methanol is quite

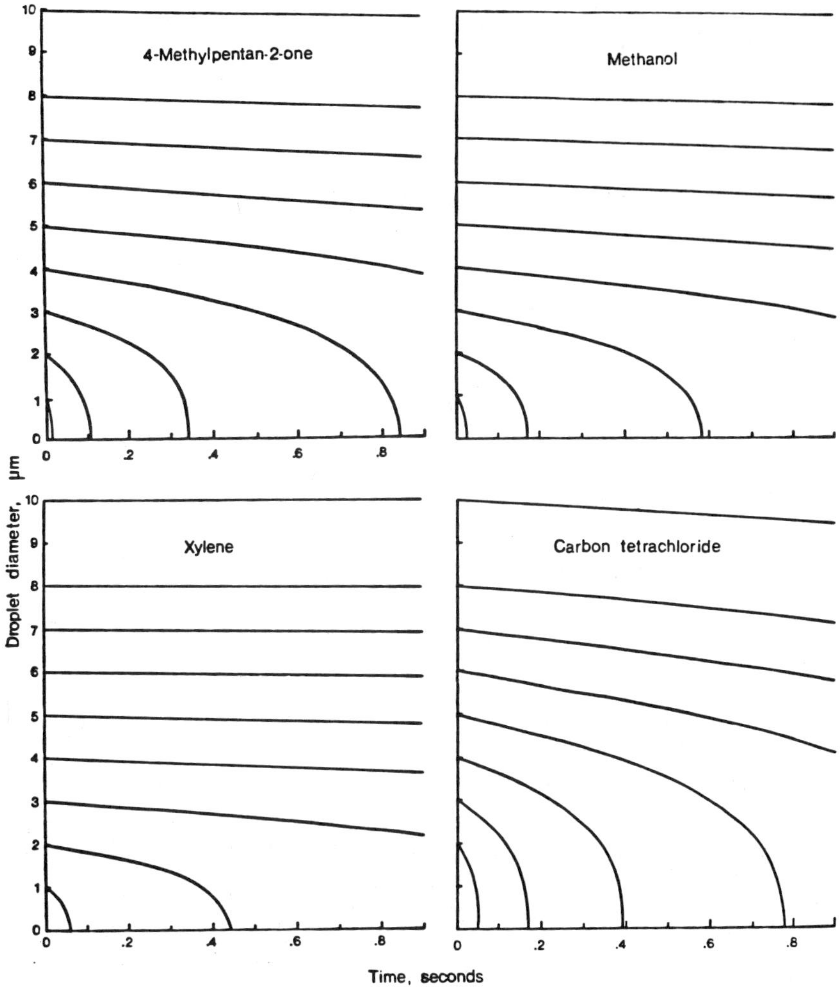

Figure 6.2. Calculated evaporation curves for various organic solvents. Data calculated from Eq. 2. [6]. [Reprinted by permission from A. W. Boorn and R. F. Browner, *Anal. Chem.* **54,** 1402 (1982). Copyright (1982), American Chemical Society.]

significant during the time they are in transit to the plasma. The curves for xylenes indicate that only the smallest drops will be reduced further in size by evaporation. Xylene evaporates at nearly the same rate as water (see Table 6.2). Hence, in comparing the two organic solvents most widely used in ICP work (that is, MIBK and xylenes) with water, evaporation may account for differences in transport efficiency in the case of MIBK but not for xylenes.

The major potential value of these curves, taken with the evaporation characteristics (E values) of organic solvents, comes in estimating transport efficiencies, predicting signal enhancements compared to an aqueous solution, and explaining ICP stability with organic solvents. These three factors are the most important in considering organic solution nebulization into the ICP; they are considered individually in subsequent sections.

6.2.5. Transport Efficiency

The crucial factor in assessing ICP sample introduction systems is, of course, the amount of sample transported to the plasma. Traditionally, this has been expressed in terms of the transport efficiency, ϵ_n (%), which is simply the fraction of the aspired sample which enters the plasma. Recently, the use of the W parameter ($\mu g \cdot s^{-1}$) to represent the mass of analyte reaching the plasma per second has been proposed [7]. The methods of obtaining ϵ_n and W data, and their various merits, are discussed in Chapter 8. For the purposes of this discussion it is sufficient to note that, historically, organic solvents have been assumed to increase transport efficiency compared with aqueous solutions, by reducing the primary median drop diameter [3]. This reduction arises either by a significantly reduced aspiration rate due to increased viscosity or differences in density and surface tension; in many cases, it is due to a combination of both effects. Certainly in flame spectrometry with typical pneumatic nebulizers, the aspiration rate effect is dominant.

Pneumatic Nebulizers. In the case of ICP pneumatic nebulizers, the effect of viscosity will be dominant in dictating primary drop size. In many instances, this will significantly affect the quantity of material contained in the preatomization drop size distribution. However, it is now almost universally accepted that to obtain reliable nebulizer operation and reproducible analytical results, the aspiration rate should be controlled, most conveniently by a peristaltic pump.

Owing to the multiplicative nature of their action, it is difficult to quantify the relative importance of the three major factors which influence organic sample transport efficiency: 1 The primary drop size distribution produced by the nebulizer, 2 the effect of solvent density on aerosol loss in the spray chamber, and 3 the change in drop size distribution resulting from solvent evaporation.

However, in comparing water and organic solvents, the main influence on drop size distribution is likely to be the combination of differences in surface tension and aspiration rate. Most organic solvents have significantly lower surface tension than water (Table 6.1). For example, with nitrobenzene, for which evaporation is negligible (Table 6.2), any changes in ϵ_n compared with water should reflect, primarily, changes in the primary drop distribution.

In contrast to the differences between aqeuous and organic solvents, there is little difference between the surface tensions of the common organic solvents (Table 6.1). Here, dominant effects in determining relative drop size distribu-

Table 6.3. Transport Efficiencies for Various Solvents Using an ICP Crossflow Nebulizer [9]

Solvent	Surface Tension at 20°C ($dyn \cdot cm^{-1}$)	Nebulization Efficiency (%) Controlled Aspiration[a]	Free Aspiration	(Rate)[b]
Benzene	28.8	22	27	(1.1)
Xylenes	29.1	8	14	(1.0)
Nitrobenzene	43.9	6.4	15	(0.2)
Water	72.7	1.5		

[a]Aspiration rate controlled at 1.8 $mL \cdot min^{-1}$ by syringe pump.
[b]Aspiration rate in milliliters per minute ($mL \cdot min^{-1}$).

tions, and hence transport efficiencies, must be a direct consequence of nebulizer aspiration rate and solvent evaporation rate. The relative magnitude of these effects will depend on the differences in evaporation factor E and viscosity of the two solvents (Table 6.2).

Table 6.3 shows transport efficiencies for three organic solvents determined by collection of the drain liquid. This technique, which generally is not recommended for measuring ϵ_n, was used in this case to enable volatile solvents to be studied [9]. Volatile solvents are not readily handled by the preferred cascade impactor or total filter technique [10].

The reproducibility of the technique is relatively poor, but certain trends are nonetheless evident. If a fixed aspiration rate is considered (for example, 1.8 $mL \cdot min^{-1}$), the order of transport efficiencies is benzene > xylene > nitrobenzene, which is the same as their respective evaporation factors (Table 6.2), indicating that evaporation is the dominant effect. As the aspiration rate decreases, transport efficiency increases, the increase for benzene being the largest due to the increased evaporation. Comparing the transport efficiencies of the three organic solvents at a flow rate of 1.8 $mL \cdot min^{-1}$ to that of water, the following general conclusions may be drawn. First, the differences between water, benzene, and xylenes are largely due to their different evaporation rates (Table 6.2), their primary median drop sizes being predicted to be very similar (Fig. 6.1). Second, comparing nitrobenzene with water, the higher viscosity and density of the former (Table 6.2) would indicate a larger primary drop size, and greater gravitational and impaction aerosol losses, respectively. The transport efficiency of nitrobenzene would therefore be predicted to be less than that of water. Experimentally, nitrobenzene is observed to give a higher transport efficiency than water. This may be attributed to the large disparity in surface tension between the two solvents (Table 6.2), which acts to reduce the difference in primary drop diameter between them.

6.3. ICP OPERATING CONDITIONS FOR ORGANIC SAMPLES

A survey of the literature describing analysis of organic samples and solutions by ICP–AES reveals a variety of different operating conditions for the various types of ICP equipment used. Operating conditions are listed in Table 6.4 when available, in approximate chronological order of appearance of the reports. This table includes the following information:

- radio frequency (rf) generator operating frequency and design (C indicates crystal-controlled, F indicates free-running)
- Radio frequency (rf) power; in most uses this is the "incident power" measured by an in-line meter and does not represent the actual power coupled to the plasma
- The outer, intermediate, and neubilizer or carrier gas flows ($L \cdot min^{-1}$)
- The nebulizer design, if given
- The sample aspiration (uptake) rate
- The viewing height in millimeters above the top of the load coil
- The literature reference

The major logical division in the reported operating conditions is due to the type of plasma used, that is, high-power (≥3 kW), typically nitrogen–argon, plasmas, intermediate-power (0.75–3 kW) all-argon plasmas, and low-power (≤0.75 kW) all-argon plasmas. These types of plasmas are considered separately in the tables, and differences between their modes of operation and tolerance to organic solvent introduction are also discussed in the following sections.

6.4. TOLERANCE OF PLASMA TO ORGANIC SOLVENTS

The ability of the ICP to tolerate organic solvents, while still maintaining a stable plasma with high excitation temperature, is a function of the following variables: 1 plasma type, for example, high, intermediate, or low power plasma, 2 plasma gas composition, 3 organic solvent type, and 4 solvent loading in plasma.

6.4.1. High-Powered Plasmas

The first practical applications of the ICP to organic samples were carried out by Greenfield et al. [2, 3], using a high-powered, nitrogen–argon plasma, operating at a power of 5.5 kW in the plasma. In this and subsequent work by

Table 6.4. ICP Operating Conditions Employed with Organic Samples

Generator frequency / (MHz) and design[a]	rf Power (kW)	Outer[b] Gas Flow (L · min^{-1})	Intermediate[b] Gas Flow (L · min^{-1})	Nebulizer[b] Gas Flow (L · min^{-1})	Nebulizer Design[c]	Sample Aspiration Rate (mL · min^{-1})	Viewing Height (mm above load coil)	References
7 *F*	5.5	64(N_2)	15	2–35	C(AA type)	1–25 μL		2
7 *F*	6.0	64(N_2)	15	2–35	heated CF	discrete		3
~30		17	1	0.9	C(AA type)	1.0		4
27.12 C	2.1							
27.12 C					CF			11
27.12 C	2.1	15	1.4	1	CG		14	12
27.12 C	1.80	18	0.8	>2.0	Heated block modified AA type	2.5	18	13
27.12 C	1.8	20	1	0.6	CF	0.95	18	14–17
27.12 C	1.9	24	0	0.8	CF	1.2	17	18
27.12 C	1.7	10.5	0(?)	0.7	CG	1.8	18.5	19
27.12 C	1.9	Reported as pressure			CG	0.65	18.5	20
27.12 C	1.3–1.8	10;15	0(?)	0.8	(?)		10	21
27.12 C	1.7	12	0–2	1.0	CF	Variable		9
27.12 F	1.5	12(N_2)	10.8	1.0	CF	3.4	8	22
27.12 C	1.7	20	0.8	1.9	C(AAS type) stream-split	1.25	18	23
27.12 C	1.4				CG	1.0;0.5		24,25
27.12 C	1.85	14.5	1.5	1.2	CF	0.75	16	26

Table 6.4. (*Continued*)

Generator frequency / (MHz) and design[a]	rf Power (kW)	Outer[b] Gas Flow (L · min^{-1})	Intermediate[b] Gas Flow (L · min^{-1})	Nebulizer[b] Gas Flow (L · min^{-1})	Nebulizer Design[c]	Sample Aspiration Rate (mL · min^{-1})	Viewing Height (mm above load coil)	References
27.12 C	1–1.75				CG			27
27.12 C	4	25(N_2)	6	0.4	CG	1	4–8	28
50 F	1.7	18	0.9	0.8	CF	0.8–1.4	15	29
27.12 C	1.75	12	2.0	1.00	CF and CG	Various	Various	6
27.12 C	1.2–1.75	15		Reported as psig		Ca 1	17	30
27.12 C	1.2	18	1.0	0.5		0.5		31
27.12 C	1.0	12	0.2		CG	Not given	13	32
27.12	2.0	19	2	1	Heated, modified Babington	5	30	33
32.9 F	2.5–3.7			0.48		0.1–1.1	15	34
27.12 C	1.1–1.75	20	Not separately controlled	0.8	CF (sapphire)	0.1–2.0	Various	35
27.12 C	1.6	12	2.2	0.7	CG	1.0	14	36
40.68 C	0.8–1.0	17	2.3	0.5–0.7	CF	0.5	20	37
	1.2–1.6	12.5–16.0	0.75–1.25	0.65–0.75		1.8	14–18	38–40
27.12 C	1.4	15	1.4	1.0	CG	4.7	21	41
50 F	1.45–1.6	17–19.5	1.1	0.4–0.8	CG and CF		20–25	42
50 F	1.3	33 (N_2 incl. 1% O_2)	6	1.5–2.9				43
27.12 C	1.9	25	1.5	0(?)	CF	1.0	20	44

27.12 C								45
27.12 C		8	2	0.8	CG		10–20	46
27.12 C	1.0–1.8	17–20		0.2	Fritted disk	1.0		47
27.12 C	0.75–2.25	7–20	0.5–2.5	0.5	MAK	0.5		48
27.12 C	1.75	16	1.6	0.6	CF	1.0	15	49
27.12 C	1.75	24	1.4	0.5			5.5	50
27.12 C	1.75	20	1.5	0.75	US	1	15	51
27.12 C	1.4	16	0.8	0.5	GC	1.6	15	52
27.12 C	1.0	13		0.4	ET		15	53
27	0.5	12	0.3	1.9	GC		9	54
27.12	1.6	12	1.2	1.2	Vapor Intro.		15	55
27.12	0.80	12	0.5	0.9	GC		9	56
27	1.75	24	1.4	0.5	GC		5.5	57
27.12	1.0				CF		12	5

[a]F is free running; C is crystal controlled.
[b]Argon unless otherwise indicated.
[c]Nebulizer Types: C is Concentric; CG is concentric glass monolithic; CF is cross-flow; GC is GC effluent; US is ultrasonic; and ET is electrothermal.

other groups, the dominant feature noticeable with the high-powered plasma is the far greater tolerance of such plasmas to organic solvents. In general, studies find that larger quantities of a wide range of organic solvents may be tolerated very readily in the plasma without further modifications [2, 3, 22, 28, 58]. In the initial study by Greenfield et al., a conventional atomic absorption nebulizer was used, together with a heated spray chamber, which generates an even higher vapor loading of organic solvent than would normally be found with an unheated or a cooled spray chamber. These authors used two different free-running generators in their studies [58], one running at 36 MHz and one at 7 MHz. It was found that the frequency stability of the generators was far less with the introduction of organic solvents than with the introduction of aqeuous solvents. However, frequency changes were lower when the high-power plasma was used. A study of background emission spectra showed good correlation between the reduction of molecular band emission from C_2 species and increased power in the plasma. A more detailed discussion of spectral features is presented in Section 6.5. With the high-powered plasma, there appears to be very little difference in behavior for a wide range of solvent types, and it appears that solvents such as methanol may be as readily introduced as solvents such as xylenes.

The study of Ohls and Sommer [22] followed closely the work of Greenfield et al., but in addition included a comparison of detection limits between nitrogen–argon plasmas and argon–argon plasmas (cf. Chapter 1). In this study, it was found that the detection limits obtainable with intermediate-powered (1.5-kW level) plasmas of both types were very similar. Furthermore, with the exception of a few highly refractory elements, such as Si and Ti, detection limits were very similar in organic solvents for intermediate-powered and high-powered plasmas. It was further found that the replacement of nitrogen with air in the nitrogen–argon plasma could also improve the tolerance of the plasma to organic solvents. In this situation, it may be assumed that the oxygen present in the air serves to oxidize carbon-containing molecular species to CO, and thus increase the tolerance of the plasma to the organic solvent even further.

In an additional study of high-powered plasmas, by Broekaert et al. [28], it was found that the use of the high-powered nitrogen–argon plasma (4 kW) produced comparable precision to normal aqueous sample introduction, when using xylene as a solvent.

6.4.2. Intermediate-Powered Plasmas

The intermediate-powered argon ICP has been studied by the majority of workers using organic solvents in ICP–AES. With this type of plasma, the tolerance of the plasma to organic sample introduction is greatly more circumscribed than with the higher-powered nitrogen- or air-containing plasmas.

A detailed study by Boorn and Browner [6] showed that there was a wide range of tolerance to organic sample introduction from different classes of solvent. Table 6.2 shows the limiting aspiration rates for sample introduction of a wide range of organic solvents to a 27.12-MHz argon plasma, operated at 1.75 kW forward power. The plasma was found to be very tolerant to certain solvents. For example, xylenes, nitrobenzene, and chlorobenzene could be tolerated at solvent uptake rates in excess of 4 mL · min^{-1}. Other solvents, such as methanol, diethyl ether, tetrahydrofuran, and benzene could be tolerated only at flow rates ≤ 0.1 mL · min^{-1}. In general it was found that a slightly higher forward rf power of 1.75 kW, in comparison with the normal operating power of 1.25 kW was advantageous for improving plasma stability. A further increase in power to the generator limit of 2.5 kW produced only a marginal improvement in plasma tolerance to organic solvent introduction for those solvents for which the plasma was particularly sensitive, such as benzene and methanol.

The ability of the plasma to tolerate organic solvents [6] was considered in terms of the evaporation of the aerosol drops produced by the nebulizer [see also Boorn et al., (9)], as described by Eqs. (6.2) and (6.3) in Section 6.2.4. The evaporation factors, *E*, shown in Table 6.2, show reasonably good correlation with the tolerance of the plasma for the particular solvent. Exceptions to these general rules were found for the chlorinated hydrocarbons and the alcohols. While solvents such as methylene chloride, chloroform, and carbon tetrachloride have high evaporation factors, the plasma is also able to tolerate high aspiration rates of these solvents. It is possible that this is due to particle and gas dynamic factors in the sample injection region of the plasma. The lack of tolerance of the plasma to methanol and ethanol is more difficult to explain. Recent work of Maessen et al. [50] appear to give better insight into this matter (see also Section 4.7.8 in Part 1).

The effects of organic solvents introduced to the plasma on the plasma excitation processes can be qualitatively correlated with the visual appearance of the plasma. In general terms, intermediate-power plasmas show substantial green C_2 (Swan) band emission around the base of the plasma, as well as the central sample injection region. The C_2 emission also extends into the outer energy addition region of the plasma. The extent of the C_2 emission in the outer region of the plasma generally shows a positive correlation with the volatility of the solvent. The influence of alcohols on the plasma is difficult to explain, but the instability results from a gradual quenching of the luminous plasma core from beneath the load coil, which finally reaches the core region at which point the plasma extinguishes. During this process, the plasma resistance, as indicated by the reflected power, increases progressively. The extent of the plasma instability, which may result in complete plasma quenching, is partly controlled by the quality of the matching network of the plasma coupling box to match accurately changes in the plasma impedance induced by the presence of organic compounds. In the case of alcohols, reflected power gradually rises until it reaches approximately 200 W, at which point the automatic matching network may no longer be able to adjust the reflected power to within acceptable limits. It is likely that much of this specific difficulty can be solved by improvements in the quality of impedance matching networks.

Nisamaneepong et al. [47], found that the tolerance of the plasma to organic solvents, particularly methanol, is increased with the use of a fritted disk nebulizer (see also Section 6.5.3). It was found to be possible to tolerate solvent flow rates of up to 1.2 mL · min^{-1} of 100% methanol with this system. It is not entirely clear why this should be so.

Gas Flows. In general, operating conditions with intermediate power argon plasmas require, in addition to the use of somewhat higher power, the use of intermediate gas flow. Typical gas flow conditions are therefore: 12 L · min^{-1} outer argon; 2 L · min^{-1} intermediate argon; 1 L · min^{-1} nebulizer argon. Increasing the nebulizer gas flow increases the velocity of the gas entering the plasma, and tends to increase the tolerance of the plasma to solvent. The negative influence of the increase in nebulizer gas flow is that the residence time of the analyte in the plasma is decreased, which results in both a decrease in analytical signal and a shift in optimum viewing height in the plasma to a position higher in the tail flame. This causes a reduction in measurement precision. However, in marginal cases, where vapors are to be introduced directly into the plasma, such as GC effluents or volatilized metal chelate vapors, the use of a high gas velocity, which can also be accomplished through the use of a smaller plasma torch injection orifice (for example, 1 mm i.d. vs. 2 mm i.d. traditionally used), can be beneficial.

Solvent Mixtures. The tolerance of the ICP to solvent mixtures is greater for organic solvents diluted with water than for pure solvents alone. This is of particular importance in liquid chromatography (LC)–ICP interfacing, where the mixed solvents that are typically used in reversed-phase separations and gradient elution work typically have an aqueous component. For instance, it has been found [6] that a 75% methanol/25% water mixture may be aspirated at aspiration rates up to 3 mL · min^{-1} routinely, in comparison with pure methanol, which can only be tolerated at a maximum flow rate of approximately 0.1 mL^{-1}. This has been confirmed by other workers, including Nisamaneepong et al. [47], working with a fritted disk nebulizer (see [59] and Part 1, Section 4.7.8, however).

Use of Oxygen in Argon Plasma Gas. It has been demonstrated that oxygen can be beneficial in reducing background emission spectra from plasmas containing organic solvents (see Section 6.5.1.) An additional benefit arising from the addition of oxygen to the outer gas flow can be an improved tolerance of the plasma for troublesome solvents [58]. Magyar et al. [43], found that the addition of 1% v/v oxygen to the outer gas flow reduced background emission by 10%, and increased sensitivity by a factor of between 10 and 20 times. Furthermore, with chloroform solvent, the oxygen reduced significantly the high level of graphitic carbon otherwise formed in the plasma.

Excitation Conditions. The excitation conditions in the intermediate-power ICP have been examined extensively by Blades and Caughlin [46]. By careful height and radial profile measurements of excitation temperatures and electron densities in an argon plasma containing xylenes, they established that excitation conditions were comparable at 1.75 kW of forward power to a water containing plasma at 0.5 kW. This 0.5 kW difference agrees precisely with the conclusions reached earlier by Boumans and Lux-Steiner [47], based on emission intensity data. For the organic ICP, temperature height profiles did not exhibit the typical sharp peak at 15 mm found for "hard" lines with aqeuous sample introduction (cf. Part 1, Section 4.7.2–4.7.7), but increased steadily up to approximately 20 mm above the load coil. It was suggested that this difference in spatial behavior was probably due to the transfer of discharge energy to the C_2 dissociation equilibrium in the plasma. With an organic ICP, C_2 dissociation is essentially complete at approximately 15 mm above the load coil. Above this height, the excitation temperature of the discharge tends to move toward the values characteristic of an aqueous ICP.

6.4.3. Low-Powered Plasmas

The studies of Ng et al. [48] with a low-power, low-flow, all-argon ICP indicate that this novel design of torch is also able to tolerate organic sample introduction somewhat better than a typical intermediate-powered plasma system. This torch was operated at 27.12 MHz, with forward powers ranging from 0.75 to 1.25 kW. It is noticeable that the reflected powers did not exceed 10 W for all solvents used. This indicates that a good coupling efficiency between the rf power and the plasma gas existed. The external torch dimensions were essentially those of a conventional intermediate-power torch (cf. Part 1, Section 5.4). However, the annular space between the outer and the intermediate tubes was substantially reduced. In the low-power torch, the annular space was reduced from the normal 1.0–1.6 mm to 0.35 mm. This torch was used in conjunction with a high-precision "MAK" nebulizer [60] (see Part 1, Section 6.2.2).

The operating conditions determined to be optimum for organic samples with this torch and nebulizer varied substantially with solvent type. Taking water as a reference point, with a forward rf power of 0.75 kW, an outer gas flow of 7 $L \cdot min^{-1}$, and an intermediate gas flow of 0.5 $L \cdot min^{-1}$, optimum conditions covered quite a wide range. For example, values for benzene were: forward power 0.75 kW, outer gas flow 12 L · min^{-1}, and intermediate gas flow 2.5 L · min^{-1}. For acetone, the corresponding values were: forward power, 2.25 kW; outer gas flow, 16 L · min^{-1}; intermediate gas flow, 2.5 L · min^{-1}. All the common solvents, such as xylenes and MIBK, were operated with a forward power of 1 kW, with reflected power < 10 W, outer gas flows of 7–8 L · min^{-1}, and intermediate gas flows of 1.25 L · min^{-1}. Precision data for a number of elements, including both hard and soft lines, ranged from 0.35–0.79% RSD for an integration time of 10 s. It was also noticed that the change in detection limits found on switching between aqueous and organic solvents was less than with typical intermediate-power plasma experience [47], when plasma conditions were optimized for each individual solvent. While the majority of the work was carried out using MIBK and xylenes as solvents, there was apparently quite acceptable performance when using a wide range of other solvents.

6.5. INSTRUMENTAL CONSIDERATIONS

Typical operating conditions for various sample types are shown in Table 6.4. A discussion of design criteria for the components of these systems is presented in the following section.

6.5.1. Torch Design

The requirements placed on the torch by the use of organic solvents are somewhat different, and slightly more stringent than those applicable with aqueous sample introduction. The tendency for organic solvents to form sooty deposits on torch surfaces can lead to practical operating difficulties. This effect is particularly prevalent for aromatic solvents. The majority of work with organic solvent introduction has been carried out using conventional torches. However, Boumans and Lux-Steiner [29] have described a novel torch, which is designed to improve the aerodynamics of sample injection into the argon streams of the torch. This is accomplished by the use of carefully designed components to the torch that are based on the principles required for smooth aerodynamic flow.

Earlier studies by Greenfield et al. [3] had also shown that the design of the sample injector tube geometry was important for reducing carbon buildup at the torch tip. They found that replacing the initial sample tube design, which had a large flat area on the top, with a more conical injector tip was beneficial in maintaining stable injection of sample into the central zone of the toroid, and minimizing carbon buildup. Boumans and Lux-Steiner [29] considered a number of injection tube geometries including those described by Greenfield et al.

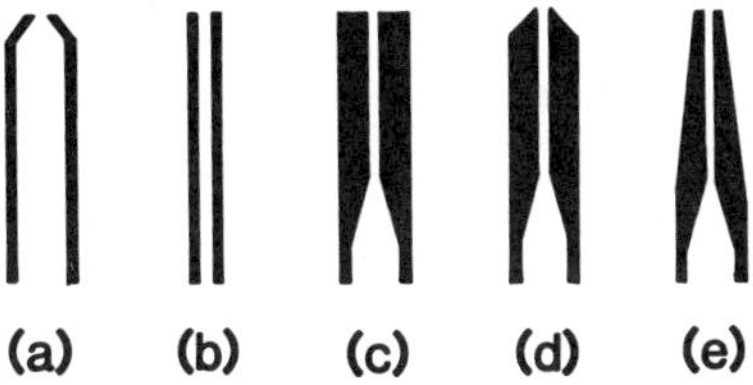

Figure 6.3. Sample injection tube geometries for organic sample introduction. (e) = optimum design; (a)–(d) = designs leading to carbon buildup, and/or poor plasma penetration. [29]. [Reprinted by permission from P. W. J. M. Boumans and M. C. Lux-Steiner, *Spectrochim. Acta* **37B,** 97 (1982). Copyright (1982), Pergamon Journals.]

The final optimum geometry was determined to be with a highly tapered design (Fig. 6.3), which has minimal surface area for carbon buildup and maintains the smoothest aerodynamic flow for the mixing both of carrier gas, outer gas, and intermediate gas flow. The plasma torch described by Boumans and Lux-Steiner (Fig. 6.4) is also described as being developed for general analytical use for aqueous solutions. The system described is in fact part of a demountable torch system, similar in many respects to the design of Mermet and Trassy [61].

Hauser and Blades [62] described an ICP torch designed specifically for the determination of oxygen in organic solvents. Their torch, which was of the demountable type, had an extended outer tube which extended the region with minimal oxygen diffusion from the external atmosphere to a position higher in the plasma. This allowed oxygen emission spectra to be observed from oxygen-containing compounds introduced into the plasma in organic solvents. With this torch design, they determined that carbon buildup was less troublesome than that found with a conventional torch.

Barrett and Pruszkowska [49] investigated the influence of injector tube diameter on the stability of the plasma, while nebulizing several organic solvents. While investigating the 213.86-nm Zn(I) line, using xylenes as solvent, it was found that the net relative intensity reached a peak value at lower gas flow for the smaller i.d. injection tubes. In general, it was found that the smaller i.d. injector tubes resulted in improved plasma stability, and hence in lower background noise. However, this must be balanced against the signal reduction which results from the increased injection velocity, because of the reduced analyte plasma residence time. Overall, a net slight improvement in detection limit for the 0.7- to 0.8-mm i.d. tube was found, compared to the 2.0 mm i.d. tube.

A detailed study of the effect on plasma impedance of increasing the percentage of methanol in a methanol–water mixture showed that the plasma resistance increased with increasing methanol concentration, and also with increasing rf power [63]. These precise measurements tie in well with reflected power and other more empirical data reported by many workers.

6.5.2. Solvent Uptake

To cope with solvents of varying viscosity, and hence varying natural uptake rate, it is generally found to be convenient in ICP–AES to use peristaltic pump-

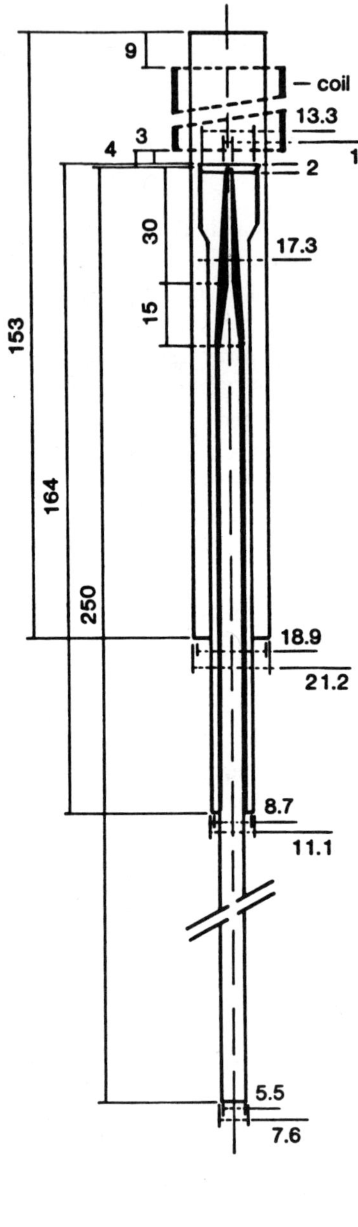

Figure 6.4. Plasma torch design for organic samples [Boumans and Lux-Steiner (29)]. [Reprinted by permission from P. W. J. M. Boumans and M. C. Lux-Steiner, *Spectrochim. Acta* **37B,** 97 (1982). Copyright (1982), Pergamon Journals.]

ing. However, it is unfortunately true that many organic solvents cause the properties of flexible tubing to change rapidly, either by removing the plasticizers, and making it inflexible, or by actually dissolving the tubing itself. Even supposedly solvent-resistant varieties change their properties after a short period of time, negating the prime purpose of pumping, which is to maintain a stable liquid flow to the nebulizer. Certain solvents, such as chloroform and methylene chloride, are essentially impossible to pump even for a limited time with any degree of reliability. As a consequence, this leaves open three possible modes of sample introduction with organic solvent that are useful for practical routine operation. These are 1 normal (nonpumped) aspiration, 2 syringe pumping, and 3 positive displacement pumping.

In the simplest case, the normal sample introduction tubing, generally made of rigid polyethylene, can be used for direct sample injection, using the pressure drop of the nebulizer to control the liquid flow rate. Further control is provided by adjusting the internal diameter and length of the uptake tubing. This method is perfectly satisfactory, provided that solvent viscosity does not vary significantly from sample to sample, or between samples and standards. However, other effects in the plasma caused by changes in solvent composition can exert a far more dramatic effect on analytical signals than minor changes in uptake rate. The importance of maintaining a constant uptake rate with organic solvents should not, therefore, be overrated.

If pumped operation is required for "aggressive" solvents, the only feasible approach is to use either a direct syringe injection device, or to use a positive displacement approach. Direct syringe injection suffers from the difficulty of reloading the syringe between samples, and problems associated with the limited pumping capacity of the syringe. Direct syringe injection however, has a practical application in a flow injection mode, in which a large volume syringe is pumped, and small samples are injected into the flowing stream. Another alternative is the positive displacement approach, which can involve either liquid or gas displacement. With liquid displacement systems, the solvent (for example, chloroform) does not come directly into contact with flexible tubing, rather it is displaced by a flow of an immiscible solvent, such as water. The water is pumped continuously and directly by the peristaltic pump. This approach, described only for AAS as yet, is clearly quite feasible for the ICP. Boumans and Lux-Steiner [29] described the use of a different liquid positive displacement system for feeding organic sample into the plasma torch. This involved sucking solvent into a reservoir, and then reversing the liquid flow to displace the solvent into the plasma. It was claimed that no noticeable pulsation was observed. Flow injection systems using gas positive displacement technology are currently available commercially and seem quite suitable for organic sample introduction, although their use has not yet been described.

6.5.3. Nebulizer Design

Pneumatic. Normal pneumatic nebulizers (see Part 1, Sections 6.2.1 and 6.2.2) may be used for organic sample introduction, without further modification. The use of all conventional types has been described. Current experience does not favor one design over another for any reasons associated specifically with the introduction of organic samples.

Fritted Disk. The fritted disk nebulizer (see Part 1, Section 6.2.4) operates in such a manner that a very fine aerosol is produced by pumping liquid from an injector tube onto the surface of a fine mesh-size glass fritted disk, while passing argon through from the other side of the disk. The use of this nebulizer for organic sample introduction has been described by Nisamaneepong et al. [47]. These authors found a substantial improvement in the tolerance of the argon ICP to organic sample introduction when using this nebulizer in comparison to a conventional pneumatic nebulizer. As has been described earlier, a primary consideration in this situation appears to be that solvent evaporation occurs more rapidly and more completely than with a conventional pneumatic nebulizer. Presumably, this is a consequence of the initial drop size of the aerosol being less than with a conventional pneumatic nebulizer [47]. A negative factor with frittered disk nebulizers is their need for a substantial washout time between samples, which is an inherent function of the design. A typical wash cycle was found to be about 80 s, to reduce signal to background level, at a solution pumping rate of 1 mL $\cdot$ min^{-1}, compared to $\leq$40 s with conventional nebulizers and spray chambers. The nebulizer operates rather differently from conventional pneumatic nebulizer–spray chamber combinations, in that it shows a nearly linear relationship between analyte pump rate and analyte mass transport. Pneumatic nebulizer–spray chambers generally show less sensitivity to changes in solution flow rates.

It was also observed for a range of water–methanol mixtures that the emission intensity was generally quite sensitive to nebulizer flow rate. Doubling the nebulizer flow rate from 0.25 to 0.5 L $\cdot$ min^{-1} produced an increase in net emission intensity of approximately an order of magnitude. Linear dynamic range for the nebulizer was found to be comparable for both organic and aqueous solutions, extending over more than three orders of magnitude. Detection limits were found to be approximately an order of magnitude poorer in 100% methanol than in 100% water. The difference between detection limits in 60% methanol and in water was approximately a factor of three.

Ultrasonic. In general, with aqueous sample introduction using ultrasonic nebulization (see Part 1, Section 6.3), a substantial improvement in detection limits is only achieved after effective desolvation. It would be anticipated that as the influence of organic solvents on plasma excitation conditions has been found to be even greater than for aqueous solvents, desolvation would be a necessity for organic sample introduction. There has been one study of such a system [51], in which 50 organic solvents were tested. In this study, the expected correlation between solvent volatility (hence vapor pressure) and ease of introduction to the plasma was confirmed. Generally, signal enhancement (that is, {the net intensity for organic solvent} / {the net intensity for aqueous solution}) was less for the ultrasonic nebulizer than for a pneumatic nebulizer. It was also found, however, that signal intensities for organic solvents with ultrasonic nebulization, without desolvation, were approximately two to five times superior to those found for pneumatic nebulization. This is close to the difference typically found for aqueous sample introduction. The greatest enhancement found for the ultrasonic nebulizer, without a desolvation system, was a 10 times improvement over that found with a pneumatic nebulizer. In this study, an intermediate-powered (<1.6 kW) ICP was used. The optimum organic solvents for use with the ultrasonic nebulizer were found to be amyl acetate and diisobutylketone.

Further details of ultrasonic nebulization are given in Chapter 8.

6.5.4. Aerosol Vapor Thermostatting

From the discussion of Sections 6.2.4 and 6.4, it is clear that solvent vapor loading in the plasma will vary significantly, depending on both the volatility of the solvent and the rate at which aerosol drops evaporate in the spray chamber on their path to the plasma. The total solvent loading in the plasma will therefore be directly related to the extent of evaporation of the solvent in its passage from nebulizer to plasma, and so will be directly influenced by the temperature of the spray chamber. In contrast to the use of a heated spray chamber described by Greenfield et al. [58] for the high-power nitrogen–argon plasma, in general it has been found to be advantageous for intermediate-power plasmas to use some form of vapor condensation device. This approach was first described by Hausler and Taylor [24] as part of an interfacing study for size exclusion chromatography with an ICP.

After studying a number of spray chamber designs, these workers determined that a water-cooled spray chamber provided a number of advantages over the usual uncooled type, by reducing solvent vapor loading in the plasma. The appearance of the plasma was visibly changed by the use of the cooled spray chamber. The two solvents studied in this work were toluene and pyridine. Detection limits with and without water cooling of the spray chamber are shown in Table 6.5. In a further study by Boorn and Browner [6], the effect of vapor condensation, using a water-jacketed condenser column, was studied for a number of atom and ion lines of Cu, Fe, and Cr. Solvents covering a wide volatility range were examined. Signal enhancements observed are shown in Table 6.8.

Table 6.5. Detection Limits (mg/mL) with Cooled Spray Chamber [24]

Element	Without Cooling[a]	With Cooling (0 °C)[a]	Aqueous
Ag	0.318	0.0246	0.007
Al	54.2	0.137	1.045
Ba	1.27	0.0029	0.0013
Cd	0.055	0.0050	0.0034
Cu	0.243	0.017	0.0054
Fe	2.47	0.069	0.0062
Mg	0.029	0.0005	0.00015
Mn	0.083	0.0018	0.0014
Ni	5.19	0.0346	0.015
Pb	4.02	0.091	0.042
Si	3.05	0.197	0.027
Sn	4.29	0.0414	0.025
Ti	3.55	0.0033	0.0038
V		0.040	0.0075
Zn	0.063	0.0048	0.0040

[a]Toluene solvent.

It is clear from these data that the use of temperature control for either the spray chamber, or the aerosol–vapor mixture leaving the spray chamber, can be very beneficial in reducing the solvent vapor loading in the plasma. This will directly reduce the cooling effect of high solvent vapor loading, which is normally found with organic solvent introduction. As a general rule, the use of a thermostatted, cooled spray chamber, or condenser for more efficient vapor removal, is therefore highly desirable in routine practice.

A study by Maessen et al. [44] and the more recent work [59] emphasize this point (cf. Section 4.7.8 in Part 1). These authors made a thorough study of solvents under a wide range of conditions, and determined that the temperature at which the spray chamber was thermostatted was a critical function of the volatility of the solvent. For example, the maximum tolerable aerosol temperatures were determined to be 7°C for methanol, 12°C for ethanol, 16°C for 2-propanol, and 18°C for toluene, at a forward power of 1.9 kW. At lower forward powers, the critical temperatures were reduced by approximately 10°C. The net signal intensities were shown to decrease very rapidly for a number of elements as the aerosol temperature rose above a critical value. Below that value, the signal was relatively insensitive to temperature, presumably indicating that the solvent vapor loading had been reduced to a limiting value. It was determined that the ion lines were most sensitive to aerosol cooling, in confirmation of the results of Boorn and Browner's study [6].

6.6. ORGANIC COMPOUND SPECTRA IN PLASMA

The introduction of organic solvents has a profound influence on the background spectra emitted by the ICP. Molecular band emission systems have been studied in both the high- and intermediate-powered plasmas (cf. Sections 4.7.7, 7.3.5, and 7.3.6 in Part 1). In general, carbon-containing molecular band spectra are substantially reduced as power in the plasma is raised. For example [58], in a nitrogen–argon plasma, in the region 400–800 nm, with methanol introduced at 2 mL · min^{-1}, it can be observed that C_2 Swan bands at 592.3–619.1, 547.0–563.6, 509.8–562.5, and 467.8–473.7 nm are essentially totally removed as the power is raised from 2 to 7 kW in the plasma. CO bands in the region 711.7–721.0 nm are also largely removed. Similar behavior is observed for all-argon plasmas as the power is to similar levels. In high-powered plasmas, as the power is raised above 5 kW, the H_α line at 656.3 nm, and the H_β line at 486.1 nm also appear. CN bands are less tractable to removal by increased rf power, but may be reduced by the incorporation of oxygen into the outer gas flow. However, the oxygen introduces its own fairly complex molecular spectrum, consisting of O_2^+, O_2, and OH bands. Figure 6.5 shows band spectra for a nitrogen–argon plasma running at powers of 2 and 5 kW.

In general, for intermediate-power plasmas, it is found that the most intense carbon band systems are CN, C_2, and CH. Considering a wide range of solvents [1, 6, 49], including nitrogen-, sulfur- and chlorine-containing solvents, the order of band intensity is: CN (violet) $\geq$ C_2 Swan $\gg$ CS $>$ OH $>$ NO_γ $>$ CH $>$ NH $\geq$ CCl. The wavelengths of the major band heads are recorded in Table 6.6, based upon several reports [1, 30, 49, 54]. Atomic hydrogen and atomic carbon lines are also present, the latter being fairly intense for all solvents investigated. The region 190–300 nm is generally found to be relatively free of molecular bands, except for CS bands from sulfur-containing solvents. The major possibility of spectral overlap in this region is presented by the NO_γ band system, and broadened atomic carbon lines, which may be found at 193.09, 193.36, and 247.86 nm. These could all present a possibility for direct spectral or wing spectral overlap with appropriate lines. Direct spectral overlap requires either the application of correction factors to the measured intensity or, if the overlap is severe, the selection of an alternative analysis line. Wing overlap cases can be corrected for by off-peak background correction at wavelengths selected on preferably both sides of the analysis line. The atomic carbon lines and the NO_γ band system may also be observed when aqueous solvents are used, although their intensities are generally much greater with organic solvents. Generally they cause little spectral interference.

In the spectral region above 300 nm, the potential for spectral overlap is more serious. While the values listed in Table 6.6 refer to the major band heads, vibrational and rotational fine structure of these bands may cause direct spectral

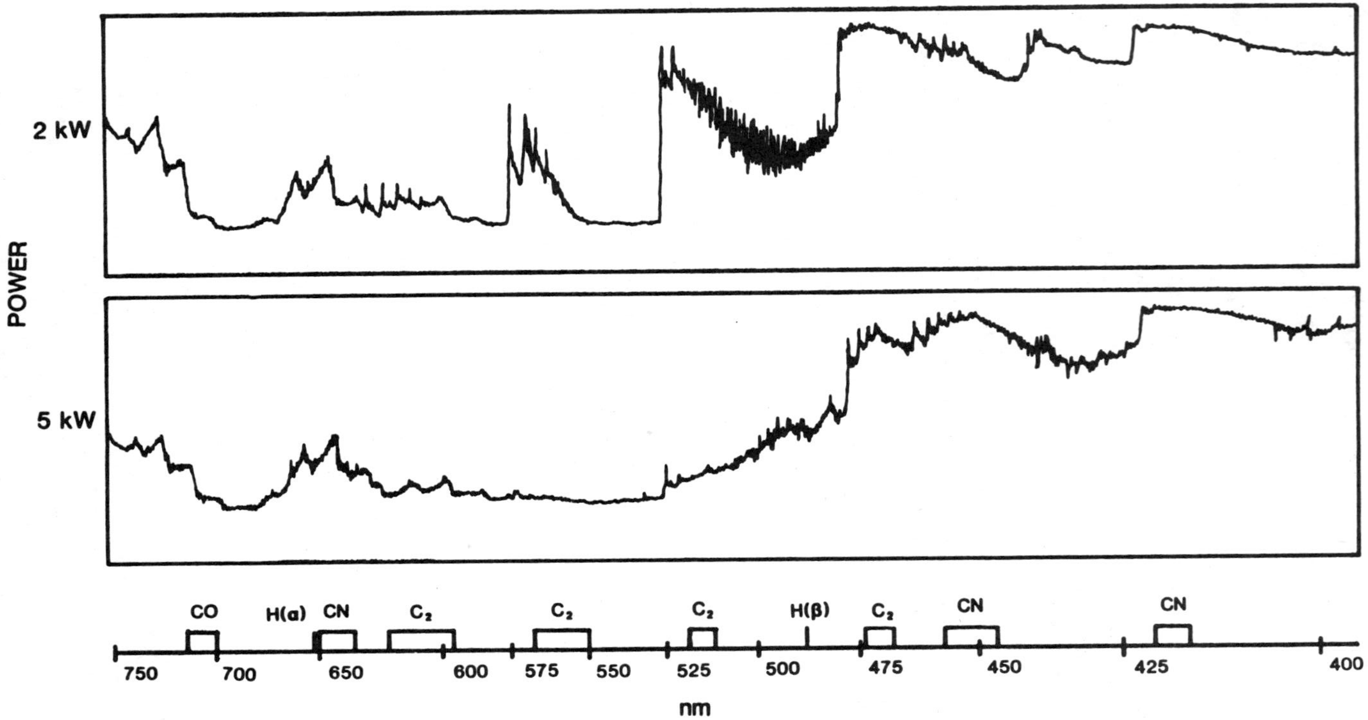

Figure 6.5. Plasma background emission spectra for N_2–Ar plasma at 2 and 5 kW [58]. [Reprinted by permission of ICP Information Newsletter from Greenfield, H., McD. McGeachin, and P. B. Smith, *ICP Information Newslett.* **2,** 167 (1976)].

interferences, even several nanometers from the band head when working at trace levels. Table 6.6 lists some of the potentially more serious spectral interferences that might arise due to the presence of organic solvents. The analysis lines listed are those that lie among the five most intense lines of their individual ICP emission spectra. While it is not certain that all the lines listed will suffer significant spectral interference, they are potentially subject to spectral overlap. In fact, few spectral interferences with organic emission bands have been investigated to date. As can be observed in Table 6.6, the interferences due to spectral features of CN and C_2 bands are most likely to cause problems with less commonly determined elements such as the lanthanides. Xu et al., [52] investigated this specific problem, and determined that certain lanthanide lines did indeed lie directly under minor CN bands. For example, there were direct spectral interferences of both the 371.03-nm Y line, and the 359.26-nm Sm line. In addition, background shift interferences were observed for other lanthanide elements.

The OH band intensity is greatly reduced in the presence of organic solvents, as compared with aqueous solvents, and is still relatively weak even with alcohol introduction. Little direct interference with the 309.37-nm Al and 309.31-nm V lines is observed, except for trace level determinations [6]. In general, it may be concluded that solvent-induced spectral interferences will not normally be a limiting factor for trace element determinations in organic solvents. In general, it is likely that spectral interferences due to overlap of other element lines are liable to present a more serious problem.

6.6.1. Spectral Height Profiles

The three strongest organic spectral features, namely the atomic C line at 193.09 nm, the C_2 line at 516.52 nm, and the CN bands at 358.38 nm, show a substantial variation in intensity with height above the plasma. Height profiles for CN 358.38 nm, C 193.09 nm, and C_2 516.52 nm are shown in Figs. 6.6–6.8. The C atomic and C_2 lines both show a rapid drop in emission intensity with increasing height above the load coil. The CN band emissions show a maximum value in the range of 15–25 mm above the plasma coil. Nitrogen-containing solvents, such as pyridine, show an initial peak low in the plasma, approximately 10 mm above the load coil, and a subsidiary peak higher in the plasma at approximately 22 mm above the coil. This accords with the concept of nitrogen diffusion from the surrounding air being the source of CN band emission for nonnitrogen-containing compounds. In the case of pyridine, the internal source of nitrogen causes the initial peak low in the plasma.

In general, the effect of power on CN, C atomic, and C_2 bands shows a rapid increase of C atomic emission with increasing power, a general reduction of C_2 emission over the same power range, and relatively little response change from

Table 6.6. Major Band Heads for Organic Species Emission and Potential Line Interferences [6]

Species	Wavelength (nm)	Analysis Lines Suffering from Potential Spectral Interferences
C_2	436.52	
	437.14	
	438.25	
	467.86	Cs II 452.67; Cs I 455.53
	468.48	Ba II 455.40
	469.76	Li I 460.29
	471.52	
	473.71	
	512.93	Ba II 493.41; Li I 497.17
	516.52	
	547.03	
	550.19	
	554.07	
	558.55	
	563.55	
CH	431.25	
	431.50	
CN	358.38	Tb II 350.92
	358.59	Dy I 353.17; 353.60
	359.04	Tb II 356.33; Zr II 357.25
	385.09	Sc II 357.25; Sm II 359.26
	385.47	Eu II 381.97; Tm II 384.80
	386.19	Tb II 384.87; U II 385.96
	387.14	Tb II 387.42; Ho II 389.10
	388.34	
	415.34	Nd II 415.61; Ga I 417.21
	416.78	Pr II 417.94; Ce II 418.66
	418.10	Rb I 420.19; Eu II 420.51
	419.70	Sr II 421.55; Rb I 421.56
	421.60	
CS	253.87	P I 253.57; Hg I 253.65
	255.58	Li I 256.23; Zr II 257.14
	257.27	Mn II 257.61
	257.56	
	258.96	Mn II 259.37; Sb I 259.81
	260.59	Fe II 259.94; Mn II 260.57
	262.16	Pb I 261.42; Lu II 261.54
	266.26	Ru II 266.16
	267.70	Au I 267.60; Cu II 267.72
	269.32	Ru II 267.881; Ta II 268.52

Table 6.6. (*Continued*)

Species	Wavelength (nm)	Analysis Lines Suffering from Potential Spectral Interferences
	270.89	Ru II 269.21; Nb II 269.71
	272.67	Ru II 269.21; Nb II 269.71
OH	306.72	Very weak; Al I, 309.271; 309.284
		V II 309.311; 310.230
NH	336.01	Very weak
	337.0	
NO	215.49	Zn I 213.856; Cd II 214.438
	226.94	Cd II 226.50
	237.02	Cd II 228.802
	247.87	
	259.57	Mn II 257.610; Mn II 259.373; Mn II 260.569
	272.22	
CCl	277.76	Very weak
	278.83	
C I	193.09	As I 193.70
	199.36	Se I 199.51
	247.86	

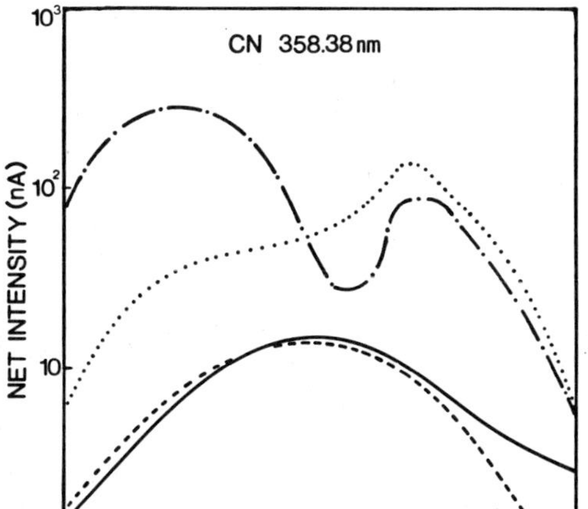

Figure 6.6. Height emission profile of CN emission in argon plasma. 1.75 kW. [6]. [Reprinted by permission from A. W. Boorn and R. F. Browner, *Anal. Chem.* **54,** 1402 (1982). Copyright (1982), American Chemical Society.]

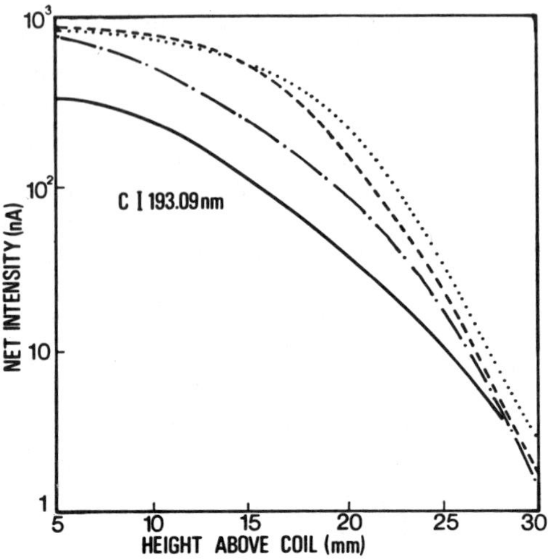

Figure 6.7. Height emission profile of C emission in argon plasma. 1.75 kW. [6]. [Reprinted by permission from A. W. Boorn and R. F. Browner, *Anal. Chem.* **54,** 1402 (1982). Copyright (1982), American Chemical Society.]

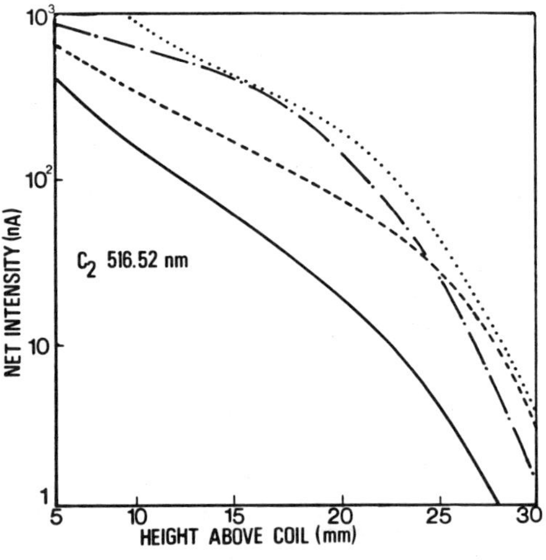

Figure 6.8. Height profile of C_2 emission in argon plasma. 1.75 kW. [6]. [Reprinted by permission from A. W. Boorn and R. F. Browner, *Anal. Chem.* **54,** 1402 (1982). Copyright (1982), American Chemical Society.]

CN [49]. The effect of varying nebulizer gas flow [49], is to cause a relatively sharp peak for the C atomic line at a gas flow of about 0.5 L · min^{-1}, and a steady rise in C_2 emission up to the limit of 1 L · min^{-1} nebulizer gas flow.

6.7. SIGNAL ENHANCEMENTS

Experience with atomic absorption spectrometry leads to the anticipation of a general enhancement of signal when organic solvents are used for sample introduction. However, this experience does not translate directly into the field of ICP spectrometry. In general, without solvent vapor removal, little or no enhancement is observed with organic solvents in comparison to aqueous solvents. For intermediate-power plasmas, a decrease in signal is generally observed, in the absence of vapor removal procedures.

Signal changes produced by the introduction of organic solvents have been generally attributed to changes in sample aspiration rate, and median drop diameter [3]. Uptake rate will vary with changes in surface tension and viscosity of the solvent, and median diameter is predicted to vary with surface tension, viscosity, density, and solvent volumetric flow for a given nebulizer and gas flow, as predicted by Eq. [6.1]. The effect of organic solvents on flame temperature in AAS appears to be minor, in comparison with the effect of such solvents on plasma temperature of intermediate-powered plasmas. With high-powered ICPs, the influence of organic solvents appears to be rather similar to that found in flame AAS [3, 22]. The intermediate-power ICP is very sensitive in its excitation conditions to the influence of organic solvents, and significant excitation temperature drops may be observed (see Section 6.4.2). Low-powered, low-flow plasmas appear closer in their behavior to intermediate-power plasmas, in spite of their similarity to high-powered plasmas in their organic solvent tolerance [48].

6.7.1. Basic Concepts

The process of signal enhancement is a function of two primary factors: analyte mass transport rate to the plasma and plasma excitation temperature. The balance between the effect of these two parameters on the population of excited-state atoms or ions in plasma is what determines the net enhancement or decrease of emission signal observed with any element present in any particular solvent. Consequently, the nature (positive or negative) and the extent of any enhancement observed will vary depending on the precise excitation process involved for a particular atomic or ionic line. Consequently, uniform enhancements are not observed for all lines of the same element in the plasma [6].

6.7.2. High-Powered Plasmas

With high-powered plasmas, one would anticipate that because of their greater tolerance to solvent loading, that signal changes with organic solvents would be largely due to changes in transport properties of the analyte, in other words, that a more volatile solvent would evaporate faster in the spray chamber, increase the analyte mass transport rate to the plasma, and give rise to a higher net emission signal. In their studies with solvent transport properties, Greenfield et al. [3] basically arrived at the same conclusion. In this work, they derived an empirical equation relating the signal with organic solvent to the signal with aqueous solvent. The variables in the relationship were surface tension, viscosity, and density of the solvent:

$$I = I_w \frac{\eta}{(\sigma\rho)^{0.5}} \left(\frac{7.055}{\eta} + 1.245 \right) \tag{6.4}$$

This equation indicates a direct correlation between enhancement and drop size. However, such a correlation suggests that increasing primary drop size gives rise to positive signal enhancements. Typically one would expect the reverse correlation to apply. The production of a primary aerosol with a larger median diameter can only give rise to a reduced analyte mass transport rate W, in the absence of changes in gas dynamics. It is only possible to account for the correlation reported by Greenfield et al. by noting that the plasma was closc to its limit of solvent tolerance, and that the larger drops produced by the nebulizer caused a net reduction in analyte and solvent mass transport to the plasma, with a net gain in excitation power. However, these conclusions must remain speculative in the absence of a more extensive data base regarding the high-powered plasma. The study of Greenfield et al. was based on a limited range of organic solvents, considering specifically several aqueous solutions of organic acids, together with methanol/water mixtures. Ohls and Sommer [22], working with a nitrogen–argon plasma running at 3 kW of forward power studied the signal change for an 80% methanol solution, compared to an aqueous standard. Their studies showed a 2 times signal increase for the Ni 231.6-nm line.

6.7.3. Intermediate-Powered Plasmas

With intermediate-powered plasmas, the situation is much more complex, for the reasons discussed above. Typical signal enhancements for a wide range of organic solvents for several atom and ion lines, measured over a range of observation heights in the plasma, are shown in Table 6.7, and may be compared with enhancements observed for the same elements when a vapor condensation process is used which are shown in Table 6.8. The calculated primary drop diameters and the enhancement factors calculated from Eq. (6.2) are also shown.

Table 6.7. **Signal Enhancements for Organic Solvents [6][a]**

Solvent	Aspiration Rate (mL · min^{-1})	Primary Median Droplet Diameter (μm)	Greenfield Enhancement Factor	Is/Iw Measured[b] Cr II 205.55 nm 10 mm	Cr II 205.55 nm 15 mm	Cr II 205.55 nm 25 mm	Fe II 259.94 nm 10 mm	Fe II 259.94 nm 15 mm	Fe II 259.94 nm 25 mm	Cu I 324.75 nm 10 mm	Cu I 324.75 nm 15 mm	Cu I 324.75 nm 25 mm
Chloroform	2.20	117	1.22	0.004	0.009	0.170	0.079	0.112	0.428	0.897	1.10	0.989
Carbon tetrachloride	1.65	75	1.26	0.008	0.025	0.375	0.093	0.154	0.644	0.911	1.04	1.04
4-methyl-2-pentanone	2.60	133	1.78	0.085	0.159	0.931	0.173	0.317	1.93	1.66	1.80	1.58
Xylenes	2.50	130	1.57	0.140	0.630	2.09	0.250	0.810	2.20	1.70	2.53	1.87
Butanol	0.90	59	2.40	0.950	1.27	1.50	0.862	1.13	1.50	1.69	2.23	1.48
Nitrobenzene	1.30	71	2.40	0.950	1.27	1.50	0.862	1.13	1.50	1.69	2.23	1.48
Water	2.20	105										

[a]Meinhard concentric nebulizer at 1.00-L · min^{-1} gas flow, 1.75-kW rf power, 12-L · min^{-1} outer gas, and 1.0–L · min^{-1} intermediate gas.
[b]Is is the signal for organic solvent. Iw is the signal for aqueous solution.

Table 6.8. Signal Enhancements for Organic Solvents with Vapor Condensation, Natural Aspiration Rate [6]

		Is/IW Measured								
	Aspiration Rate (mL $\cdot$ min^{-1})	Cr II 205.55 nm			Fe II 259.94 nm			Cu I 324.75 nm		
Solvent		10 mm	15 mm	25 mm	10 mm	15 mm	25 mm	10 mm	15 mm	25 mm
Chloroform	2.20	0.12	0.24	1.90	0.24	0.43	1.86	1.75	2.23	2.94
Carbon tetrachloride	1.70	0.81	0.91	2.19	0.69	0.99	2.15	2.24	2.84	3.14
4-Methyl-2-pentanone	2.60	2.60	3.10	3.73	2.29	2.76	3.18	3.01	3.24	3.36
Xylenes	2.50	3.10	3.31	3.67	3.04	2.96	3.11	3.17	3.16	3.18
Nitrobenzene	1.25	0.41	0.29	1.07	0.44	0.46	0.97	0.89	0.99	1.07

The calculated primary drop diameters show the effect of variation in natural aspiration rates on median drop size. This leads to larger primary aerosol drops for some solvents and hence to a predicted reduction in analyte mass transport rate W. The data clearly indicate that significant signal reductions are often observed, but are of far greater magnitude than would be expected if analyte transport were the only determining factor. The reduced signals observed are clearly a direct consequence of the reduction in excitation temperature in the plasma. Furthermore, the height profiles observed correlate well with the idea [6, 46] that with an organic solvent, the region low in the plasma possesses an excitation temperature markedly reduced from that found with a water-containing plasma, whereas higher in the plasma, excitation conditions once more approximate those of a water-containing plasma. In this situation, signals may increase significantly as a function of observation height for most elements.

Influence of Solvent Volatility. The effect of vapor condensation supports the conclusions regarding the effect of solvent loading on excitation conditions. By reducing solvent loading [6, 24, 44], it is clear that significant enhancements may be obtained from organic solvents, provided that measurement is made higher in the plasma than the normal 15-mm observation height. The negative aspect of such a shift is the higher noise level found at greater measurement heights. The effect of increasing rf forward power is to increase the ratio of ion to atom line intensities. Additionally, by counteracting the reduced power loading in the plasma caused by the presence of the organic solvent, there is a net increase in enhancement for most element lines.

6.8. APPLICATIONS

Table 6.9 summarizes the organic sample types that have been analyzed by ICP–AES. As may be seen, a wide range of sample types have been examined, but the great preponderance of samples are of oil analysis for wear metal purposes. The elements determined, and the solvents used, are shown in this table.

6.8.1. Standards

A key factor in the successful application of organic sample introduction to the ICP is the availability of suitable standards. Unfortunately, the situation is much more complex than that of aqueous sample introduction, where highly pure elements and salts are readily available for all elements. The situation with organic samples is much more restricted. The available organic standard materials are shown in Table 6.10. Of these, the Conostan oil-based standards (alkylbenzenesulfonates in base oil), both single and multielement, have prob-

Table 6.9. Organic Samples Analyzed by ICP Spectrometry

Sample	Solvent (Dilution)	Elements	References
Engine oils	Xylene (1:2)	Al, Cu, Cr, Fe, Mg, Mn, Ni, Ti	2
Organophosphates	Methanol (1 + 9) Acetic, propionic, and formic acids Methanol (10–60% v/v H_2O)	Al, Mg Co, Cr, Cu, Fe, Ni	3
Lubricating oils	4-Methyl-2-pentanone (1:10)	Ag, Al, Ca, Cr, Cu, Fe, Mg, Mn, Ni, Pb, Si, Sn, Ti, V, Zn	4
Organophosphates	4-Methyl-2-pentanone, MIBK (1:10)	Ag, Al, B, Ba, Be, Cd, Cr, Cu, Fe, Mg, Mn, Ni, Pb, Si, Ti, V, Zn	11
"Monomers," e.g., maleic anhydride, pentachlorophenol	Butyl acetate (1:10) Isopropyl-alcohol (1:10)	Ag, Al, B, Ba, Be, Cd, Cr, Cu, Fe, Mg, Mn, Ni, P, Pb, Si, Sn, Ti, V, Zn Na and above except P	11
"Polymers," e.g., acrylonitrile	Isooctane (1:10)	As above except P and B	11
"Monomers and polymers"	Tetrahydrofuran (1:10)	As above except P and B	11
Gas oils, bunker oils, petroleum oils	Xylenes (1:10)	All above and Ca, Na	11
Phenolic resins	Ethanol (1:10)	Ca, Cr, Cu, Fe, Mg, Mn, Ni, Zn	11
Diethylammonium diethyldithiocarbamate (DDDC) chelates extracted from acid digested foodstuffs	2-Heptanone (saturated with H_2O + 2% w/v DDDC)	Fe, Cu, Zn, Cd, Co, Pb, Ni	12
NBS SRM 1631 fuel oil "wear oil"	Xylene (1:10)	Al, Cu, Cr, Fe, Mn, Ni, Si, V, Zn	64

Wear metals in used truck engine oil	MIBK (1:10)	Al, B, Ba, Cr, Cu, Fe, Mn, Na, Ni, Pb, Si, Sn	13
Wort, beer, wine, and brewing materials	None 10% aqueous ethanol for standards	Ag, Al, B, Ba, Be, Ca, Cd, Co, Cr, Cu, Fe, K, Li, Mg, Mn, Mo, Na, Ni, Pb, Se, Si, Sn, Sr, Ti, V, Zn, Zr	15
NBS residual fuel oil thiourea	Xylenes	P, S,	16
organic C, I	Water	C, I	
Wear metals in lubricating oil	Xylene	Al, B, Ba, Ca, Cr, Cu, Fe, Na, P, Pb, Si, Sn, Zn	17
HPLC–ICP interfacing: organometallics, gasoline	M/H_2O E/H_2O THF/H_2O ACN/H_2O Hex, xyl, TMP, Tol[a]	As, Cu, Fe, Hg, Mo, Pb, Sn	18
Jet engine oils	Kerosene (1:4)	Mo	65
Unspecified oils	Xylene (1:10)	Al, Ag, B, Ba, Ca, Cr, Cu, Fe, K, Mg, Mn, Mo, Na, Ni, P, Pb, Si, Sn, Ti, V, Zn	19
Geological materials: solvent extracted	10% Tricaprylmethylammonium chloride MIBK	Ag, Au, Bi, Cd, Cu, Pb, Zn	20
Wear metals in lubricating oil	Xylene (1:10)	Al, B, Ca, Cd, Cr, Cu, Fe, Mg, Mn, Ni, Si	
HPLC eluants: organometallics, solvent refined coal	Toluene (1 mmol/10 mL) Pyridine (0.5 g/4 mL)	Ag, Al, Ba, Cd, Cu, Fe, Mg, Mn, Ni, Pb, Si, Sn, Ti, V, Zn	24,25

Table 6.9. (*Continued*)

Sample	Solvent (Dilution)	Elements	References
Metal chelates: extracts of biological materials	Xylene (various)	Al, Cr, Cu, Fe, Mn, Zn	26
Residual and distillate fuel oils	Xylene (1 : 100) MIBK (1 : 100)	S	27
Lubricating oils	Xylene (1 : 10)	Al, Cu, Fe, Mg, Mn, Pb, Si, Zn	28
APDC chelates extracts of wastewaters, plating baths	MIBK (40 mL extracted with 100 mL)	Cd, Co, Cr, Cu, Fe, Mn	
Unused and used lubricating oils; gasoline	Xylenes (1 : 10)	Ag, Al, Ba, Ca, Cd, Cr, Cu, Fe, Mg, Mn, Mo, Na, Ni, P, Pb, S, Si, V, Zn	66
$Cu/CuCrO_4$ thermal battery mixture; $B/CaCrO_4$ pyrotechnic mixtures; 2-theonyltrifluoro-acetone extracts	Xylenes (5 mg extracted with 25 mL)	Cr III	30
Crude, lubricating, axle, and cooking oils	Xylenes MIBK (1 : 10)	Cu, Fe, Ni, P, Si, V	67
Organometallics	Methanol methanol–water	Co, Cu, Fe, Zn	31

Used lubricating oils	None	Fe, Ni	33
Alcoholic beverages	None	Ba, Be, Cd, Co, Cr, Mg, Mn, Mo, Ni, Pb, Ti, V, Zn	34
Lubricating oils (new and used)	Xylenes (1 : 10)	Ag, Al, B, Ba, Ca, Cr, Cu, Fe, K, Mg, Mn, Mo, Na, Ni, P, Pb, Si, Sn, Ti, V, Zn	36
Fatty acid still bottom residues	Isopropyl alcohol–HNO_3 (99 : 1) (0.01–2.5%)	As, Cd, Cu, Hg, Mg, Ni, Pb, Se, Zn	37
Waste, river, and sea waters: (extraction of reduced molybdo-antimonylphosphoric acid)	Diisobutylketone	P; P indirectly via Mo, Sb	38,39
Extraction of various chelating agents	Diisobutylketone	Cd, Cu, Fe, Mo, Ni, P, Pb, Sb, V, Zn	40
Airborne particulates	Xylenes	Fe	41
Oils	Toluene kerosene xylene MIBK (1 : 10)	Ag, Al, Ca, Cd, Cu, Fe, Mg, Mn, Na, Ni, Pb, Si, Zn	42
Fuel oil (NBS SRM No. 1634a)	Xylenes	Mn, Ni, V, Cr	48
Fuel oil (NBS SRM No. 1634a)	Xylenes (1 : 100)	Zn, Ni, Fe, Ca, V	49
Fuel oil (NBS SRM No. 1634)	Xylenes (1 : 100)	Ni, Fe, V	
Wear metals in oil (NBS SRM No. 1084)	Xylenes, kerosene, MIBK	Al, Cr, Cu, Fe, Mo, Ni, Pb, Mg, Ag, Ti	

Table 6.9. (*Continued*)

Sample	Solvent (Dilution)	Elements	References
Wear metals in oil (NBS SRM No. 1085)	Xylenes, kerosene, MIBK	Al, Cr, Cu, Fe, Mo, Ni, Pb, Mg, Ag	
Motor oil		Zn	53
Gasoline		Pb	
Methylcyclopentadienyl-manganese tricarbonyl	Toluene	Mn	68
Various organic compounds	DMSO, CCl_4, H_2O, C_6H_6[b]	B, C, H, I, P, S, Si	54
Various organic compounds		C, H	56
Various organic solvents	Heptane THF MEK[c] 1-propanol	O	57
River and sea water	Diisobutyl ketone	P	69

[a]M = methanol; E = ethanol; THF = tetrahydrofuran; ACN = acetonitrile; Hex = hexane; xyl = xylene, TMP = 2,2,4-trimethylpentane; Tol = toluene.
[b]DMSO = dimethylsulfoxide.
[c]MEK = methylethylketone.

ably achieved the most widespread use, particularly in applications for oil analysis. Oil-based standards are also available from Jarrell-Ash, and from National Spectrographic Laboratories. The other alternative sample type is a powdered organometallic sample, which requires dissolution in an appropriate solvent. In practice, many of these samples are difficult to dissolve completely. However, they offer a greater selection of elements, and also a greater choice of solvents for the production of relatively stable dilute solutions. Long-term stability of diluted solutions may vary significantly for different element–solvent mixtures.

With the organometallic standards in base oil, all that is required for preparation of diluted standards is dilution with appropriate solvent, such as xylenes. The powdered organometallic standards, which are typically metal cyclohexanebutyrates, require dissolution after drying in typically a 1 : 1 : 1 mixture of 2-ethylhexanoic acid–2-ethylhexylamines–xylenes with gentle heating, and then further dilution with the appropriate solvent.

6.8.2. Detection Limits

Detection limits based on the 2σ values are shown in Table 6.11. The wavelength of measurement, detection limit, and solvent are all presented. In general, the conclusions that have been reached on signal enhancements also apply to the detection limits.

6.8.3. Organic Vapor Introduction

In addition to the direct nebulization of organic solutions into the ICP, two groups of workers have examined the possibility of direct introduction of volatile metal chelate vapors into the plasma [55, 70]. Black and Browner [70] showed that it is possible to prepare volatile β-diketonates for a range of elements, including Fe, Zn, Mn, Cr, and Co directly from a range of sample matrices, including metal salts National Bureau of Standards (NBS), Standard Reference Material (SRM) bovine liver, and human blood serum, and to obtain quantitative results for those elements. The process consisted of initially forming the volatile chelate by heating the sample in a closed reaction vessel in the presence of β-diketonate vapor, then flushing the metal β-diketonate vapor directly into the plasma through a conventional torch. Of the range of possible β-diketonates, trifluoroacetylacetone h(tfa), and hexafluoroacetylacetone h(hfa) were tested, and h(tfa) proved the more successful. To maintain a stable plasma, plasma ignition was accomplished with a continuous stream of β-diketone vapor passing into the plasma. The metal β-diketonate vapor was passed as a plug into this stream, in the manner of a GC injection. Detection limits for the elements examined, in comparison with ICP continuous nebulization of aqueous standards, are shown in Table 6.12. The values indicate that the preconcentra-

Table 6.10. Organic Standards Suitable for ICP Spectrometry

Manufacturer	Standard Type	Elements	Concentration (ppm)
Conostan	Single element	Ag, Al, B, Ba, Be, Bi, Ca, Cd, Co, Cr, Cu, Fe, K, La, Li, Mg, Mn, Mo, Na, Ni, P, Pb, Si, Sn, Ti, V, Zn	5000
	Single element	As, Hg, Se	100
	Base oil (blank)		
	12-element mixture	Ag, Al, Cr, Cu, Fe, Mg, Na, Ti, Ni, Pb, Si, Sn	900, 500, 300, 100, 50, 30, 10
	20-element mixture	Ag, Al, B, Ba, Be, Cd, Cr, Cu, Fe, Mg, Mn, Mo, Na, Ni, Pb, Sn, Si, Ti, V, Zn	900, 500, 300, 100, 50, 30, 10
Jarrell-Ash Division	9-element mixture	Al, Cr, Cu, Fe, Pb, Ag, Mg, Ni, Sn	100, 50, 20, 15, 10, 5, 3, 2, 1
Fisher Scientific	Base oil (blank)		
	Single element	Si	200, 100, 50, 20,f 15, 10, 5, 3, 2, 1
	Single element	Ag	500, 200, 100, 50
	8-element mixture	Al, Cr, Cu, Fe, Pb, Mg, Ni, Sn	

National Spectrographic Laboratories	5-Element mixture	Cu, Fe, Cr, Al, Si	4.7–385
	5-element mixture	Mo, Pb, Na, Sn, Mg	4.7–290
	9-element mixture for diesel oil	Fe, Pb, Cu, Cr, Al, Ag, Sn, Si, B	0.5–200
	Blank oil for diesel		
	5-element mixture for diesel oil	Ca, Ba, Zn, P, Mg	24–1980
National Bureau of Standards	Powdered organometallics	Al, Ba, B, Cd, Ca, Cr, Co, Cu, Fe, Pb, Li, Mg, Mn, Hg, Ni, P, Si, K, Ag, Na, Sr, Sn, V, Zn	
Spex	Base powdered organometallics metallics	Al, Ba, Cd, Ca, Co, Cu, Fe, Pb, Li, Mg, Mn, Hg, Ni, K, Ag, Na, Sr, Sn, V, Zn	
Eastman Kodak	Powdered organometallics	Al, Ba, B, Cd, Ca, Cr, Co, Cu, Fe, Pb, Li, Mg, Mn, Hg, Ni, P, K, Si, Ag, Na, Sr, Sn, V, Zn	

Table 6.11. **ICP Detection Limits in Organic Samples 2σ (mg/L) in Solution**

Element	Wavelength (nm)	Detection Limit	Solvent	References
Ag	328.068	0.002	MIBK	4
		0.01	MIBK	11
		0.005	BA	
		0.009	IPA	
		0.01	IO	
		0.03	THF	
		0.004	XYL	
	328.068	0.002	XYL	14
		0.003	MIBK	
		0.003	CB	
		0.003	MIBK	20
		0.037	PY	25
		0.025	T	24
		0.0013	MIBK	29
	328.07	0.008	XYL	66
	328.07	0.020	XYL	36
	338.2	0.062	T	42
		0.004	K	
		0.014	XYL	
Al	308.216	0.009	MIBK	4
	396.152	0.2	MIBK	11
		0.02	BA	
		0.07	IPA	
		0.07	IO	
		0.07	THF	
		0.005	XYL	
	308.216	0.01	XYL	14
		0.02	MIBK	
		0.008	CB	
	396.152	0.008	XYL	22
		0.023	PY	25
		0.014	T	24
	308.216	0.003	XYL	26
	309.3	0.009	XYL	28
	226.916	0.040	MIBK	29
	237.324	0.020	MIBK	
	308.216	0.013	MIBK	
	308.216	0.03	XYL	66
	308.216	0.150	XYL	36
	396.152	0.25	T	42
		0.017	K	
	396.152	0.047	MIBK	

Table 6.11. (*Continued*)

Element	Wavelength (nm)	Detection Limit	Solvent	References
		0.050	XYL	
	—	0.015	XYL	48
	309.3	0.53 ng		55
As	193.69	0.004 (hydride generation)	(9:1) IPA/HNO_3	37
Au	242.8	0.008	MIBK	20
B		0.04	MIBK	11
		0.09	BA	
		0.05	IPA	
		0.1	XYL	
	249.678	0.0008	XYL	22
	208.959	0.020		
	249.773	0.003	MIBK	29
	249.68	0.050	XYL	36
	249.8	1 ng	DMSO	54
Ba	455.40	0.001	MIBK	11
		0.002	BA	
		0.003	IPA	
		0.02	IO	
		0.006	THF	
		0.0015	XYL	
		0.003	T	24
		0.004	PY	25
	233.527	0.002	MIBK	29
	455.40	0.001	XYL	66
	455.40	0.0006	15	34
		0.0013	30% E/H_2O	
		0.0007	60	
	493.41	0.005	XYL	36
Be		0.003	MIBK	11
		0.001	BA	
		0.003	IPA	
		0.003	IO	
		0.003	THF	
		0.0007	XYL	
	234.861	0.00017	MIBK	29
	249.473	0.013	MIBK	
	313.042	0.00010	MIBK	
	313.1	0.0004	30% E/H_2O	34
	313.1	0.0067 ng		55

Table 6.11. (*Continued*)

Element	Wavelength (nm)	Detection Limit	Solvent	References
Bi	306.7	0.120	MIBK	20
C	247.9	12 ng	H_2O	54
Ca	393.367	0.00004	MIBK	4
		0.002	XYL	11
		0.01	E	
	317.9	0.002	XYL	14
		0.003	MIBK	
		0.002	CB	
	315.887	0.020	MIBK	29
	317.93	0.007	XYL	66
	317.93	0.015	XYL	36
	393.3	0.0013	T	42
		0.00013	K	
		0.00027	MIBK	
		0.00033	XYL	
		0.0064	XYL	48
Cd		0.025	MIBK	11
		0.025	BA	
		0.09	IPA	
		0.04	IO	
		0.02	THF	
		0.004	XYL	
	226.5	0.004	XYL	14
		0.003	MIBK	
		0.005	CB	
	226.5	0.015	MIBK	20
	228.8	0.0008	XYL	22
		0.005	T	24
		0.005	PY	25
		0.08	MIBK	28
	214.438	0.004		
	226.502	0.003	MIBK	29
	228.802	0.003		
	226.50	0.004	XYL	66
	214.44	0.009	XYL	6
		0.029	E	
		0.013	MIBK	
		0.046	CT	
	226.5	0.027	15% E/H_2O	34
	228.80	0.033	(99:1) IPA/H_2O	37
	226.50	0.015	DIBK	40

Table 6.11. (*Continued*)

Element	Wavelength (nm)	Detection Limit	Solvent	References
	226.50	0.012	T	42
		0.002	K	
		0.005	MIBK	
		0.009	XYL	
	228.80	0.021	60% M/H_2O	47
	—	0.0024	XYL	48
Co	238.9	0.06	MIBK	28
	228.616	0.006	MIBK	29
	238.892	0.005		
	238.892	0.010	XYL	
		0.010	E	6
		0.010	MIBK	
		0.022	CT	
	237.8	0.006	15% E/H_2O	34
Cr	283.563	0.003	MIBK	4
	357.87	0.15	MIBK	11
		0.008	BA	
		0.02	IPA	
		0.04	IO	
		0.02	THF	
		0.006	XYL	
		0.009	E	
	283.6	0.06	XYL	28
	425.435	0.0008	XYL	22
	267.716	0.005	XYL	26
	267.71	0.005	XYL	66
	205.552	0.033	XYL	6
		0.026	E	
		0.026	MIBK	
		0.046	CT	
		1.7	CHL	
		0.67	CT	
		0.12	MIBK	
		0.036	XYL	
		0.017	BUT	
		0.019	NB	
	205.552, 283.563	0.004	XYL	30, 34
	283.563			
		0.013	15	
		0.013	30% E/H_2O	
		0.033	60	

Table 6.11. (*Continued*)

Element	Wavelength (nm)	Detection Limit	Solvent	References
	267.72	0.030	XYL	36
	—	0.004	XYL	48
	267.7	0.061 ng		55
Cu	325.754	0.0006	MIBK	4
		0.015	MIBK	11
		0.003	BA	
		0.003	IPA	
		0.02	IO	
		0.006	THF	
		0.0007	XYL	
		0.002	E	
		0.001	XYL	14
		0.002	MIBK	
		0.002	CB	
		0.002	MIBK	20
	324.754	0.0004	XYL	22
		0.017	T	24
		0.02	PY	25
		0.0005	XYL	26
		0.1	MIBK	28
		0.009	XYL	
	223.008	0.013	MIBK	29
	224.700	0.013	MIBK	
	324.754	0.0007	MIBK	
	324.75	0.005	XYL	66
	324.754	0.002	XYL	6
		0.002	E	
		0.002	MIBK	
		0.002	CT	
		0.003	CHL	
		0.002	CT	
		0.0009	MIBK	
		0.0005	XYL	6
		0.001	BUT	
		0.002	NB	
	324.754	0.015	XYL	36
	324.754	0.047	(99:1) IPA/HNO_3	37
	324.754	0.034	DIBK	40
	324.754	0.019	T	42
		0.001	K	
		0.002	MIBK	

Table 6.11. (*Continued*)

Element	Wavelength (nm)	Detection Limit	Solvent	References
		0.0023	XYL	
		0.004	XYL	48
	324.8	0.30 ng		55
Fe	259.940	0.004	MIBK	4
		0.015	MIBK	11
		0.004	BA	
	259.940	0.01	IPA	
		0.02	IO	
		0.02	THF	
		0.004	XYL	
		0.004	E	
		0.002	XYL	14
	259.940	0.003	MIBK	
		0.008	CB	
		0.0061	20%	18
	259.940	0.0008	50% E/H_2O	18
		0.001	80%	
		0.004	XYL	22
	not given	0.070	T	24
	not given	0.02	PY	25
	259.940	0.001	XYL	26
	259.940	0.02	MIBK	28
		0.023	XYL	
	238.204	0.003	MIBK	29
	249.940	0.002	MIBK	
	238.20	0.008	XYL	66
		0.010	XYL	6
		0.018	E	
		0.009	MIBK	
		0.026	CT	
		0.019	CHL	
		0.010	CT	
		0.003	MIBK	
	238.20	0.012	XYL	
		0.004	BUT	
		0.004	NB	36
	259.940	0.015	XYL	
	259.940	0.050	DIBK	40
	259.940	0.015	T	42
		0.0017	K	
		0.0027	MIBK	
		0.0047	XYL	

Table 6.11. (*Continued*)

Element	Wavelength (nm)	Detection Limit	Solvent	References
		0.0068	XYL	48
	259.94	0.50 ng		55
H	656.3	27 ng	CCl_4	54
Hg	253.652	0.010 (cold vapor)	(99:1) IPA/HNO_3	37
I	206.2	4 ng	DMSO	54
K	766.49	0.200	XYL	36
Mg	279.553	0.0007	MIBK	4
		0.015	MIBK	11
		0.001	BA	
		0.003	IPA	
		0.001	IO	
		0.01	THF	
		0.001	XYL	
		0.0002	E	
		<0.001	XYL,MIBK,CB	14
		0.0004	XYL	22
		0.003	XYL	28
		0.00004	MIBK	29
	280.270	0.00010		
	279.553	0.001	XYL	66
	279.553	0.0005	XYL	6
		0.0001	E	
		0.002	MIBK	
		0.005	CT	
	279.8	0.0007	15 %E/H_2O	34
		0.0013	30 %E/H_2O	
		0.0013	60 %E/H_2O	
	279.08	0.050	XYL	36
	275.553	0.013	(9:1) IPA/HNO_3	37
		<0.0007	K	42
		<0.0007	MIBK	
		<0.0005	XYL	
		0.001	XYL	48
Mn	257.610	0.001	MIBK	4
		0.007	MIBK	11
		0.002	BA	
		0.004	IPA	
		0.005	IO	
		0.05	THF	

Table 6.11. (*Continued*)

Element	Wavelength (nm)	Detection Limit	Solvent	References
		0.0001	XYL	
		0.0015	E	
	357.6	<0.001	XYL,MIBK,CB	14
		0.002	T	24
		0.005	PY	25
	257.610	0.0008	XYL	22
		0.0005	XYL	26
		0.01	MIBK	
		0.008	XYL	28
	257.610	0.0002	MIBK	29
	259.373	0.0003	MIBK	29
	257.61	0.002	XYL	66
	257.610	0.002	XYL	6
		0.002	E	
		0.002	MIBK	
		0.001	CT	
	257.610	0.0007	15 %E/H_2O	34
		0.0013	30 %E/H_2O	
		0.002	60 %E/H_2O	
	257.610	0.015	XYL	36
		0.0027	T	42
	257.610	0.00033	K	
		0.0006	MIBK	
		0.0011	XYL	
	—	0.0004	XYL	48
	257.610	—	XYL (emulsion)	5
Mo	202.030	0.020	MIBK	29
	281.615	0.005	MIBK	
	202.030	0.008	XYL	66
	281.615	0.013	15 %E/H_2O	34
		0.02	30 %E/H_2O	
		0.02	60 %E/H_2O	
	202.030	0.070	XYL	36
	202.030	0.14	DIBK	40
			B	45
Na	589.00	1.0	IPA	11
		0.06	IO	
		0.1	THF	
		0.06	XYL	
	589.59	0.06	XYL	66
	589.00	0.040	XYL	6
		0.049	E	

Table 6.11. (*Continued*)

Element	Wavelength (nm)	Detection Limit	Solvent	References
		0.064	MIBK	
		0.066	CT	
	589.59	0.090	XYL	36
	588.9	0.02	K	42
		0.087	XYL	
Ni	341.476	0.01	MIBK	4
		0.1	MIBK	
		0.014	BA	
		0.015	IPA	
		0.05	IO	
		0.06	THF	
		0.0025	XYL	
		0.015	E	
	231.6	0.008	XYL	14
		0.014	MIBK	
		0.030	CB	
		0.0008	XYL	22
		0.035	T	24
		0.061	PY	25
	221.647	0.010	MIBK	29
	231.604	0.013	MIBK	29
	232.003	0.020	MIBK	29
	231.604	0.02	XYL	66
	221.647	0.12	XYL	6
		0.088	E	
		0.17	MIBK	
		0.58	CT	
	305.0	0.04	15 %E/H_2O	34
		0.033	30 %E/H_2O	
		0.067	60 %E/H_2O	
	231.604	0.090	XYL	36
	341.476 221.647	0.06	(99:1) IPA/HNO_3	37
		0.073	DIBK	40
	341.476	0.22	T	42
		0.080	K	
		0.02	MIBK	
		0.023	XYL	
		0.0072	XYL	48
O	777.19	650 ng	heptane, THF, MEK,1-propanol	57

Table 6.11. (*Continued*)

Element	Wavelength (nm)	Detection Limit	Solvent	References
P	214.914	1.0	BA	11
		1.3	XYL	
	214.914	0.05	XYL	16
	214.914	0.054	MIBK	29
	214.91	0.06	XYL	66
	178.29	0.200	XYL	36
			DIBK	38
	Indirectly at			
	Mo 202.030	0.000035	DIBK	39
	Sb 206.83	0.00030	DIBK	39
	213.6	0.6 ng	DMSO	54
	214.91	0.00037	DIBK	69
Pb	283.306	0.03	MIBK	4
	220.353	1.0	MIBK	
		0.6	BA	
		3.0	IPA	
		0.6	IO	
		0.3	THF	
	220.353	0.4	XYL	11
		0.025	XYL	14
		0.03	MIBK	
		0.04	CB	
		0.340	MIBK	19
		0.091	T	24
		0.48	PY	25
	283.3	0.03	XYL	28
	216.999	0.060	MIBK	29
	220.353	0.047	MIBK	29
	220.353	0.02	XYL	66
	220.353	0.15	XYL	6
		0.20	E	
		0.14	MIBK	
		0.37	CT	
	280.2	0.067	15% E/H_2O	34
		0.10	30% E/H_2O	
		0.13	60% E/H_2O	
	220.353	0.050	XYL	36
	220.353	5.3	(99:1) IPA/NHO_3	37
	220.353	0.40	DIBK	40
		0.17	T	42
	220.353	0.04	K	
		0.10	MIBK	

Table 6.11. (*Continued*)

Element	Wavelength (nm)	Detection Limit	Solvent	References
		0.15	XYL	
		0.033	K	
	283.3	0.10	XYL	
		0.028	XYL	48
S	190.02	100.0	XYL	14
		10	XYL	16
	180.73	2.0	XYL	27
		5.0	MIBK	
	182.04	3.2	XYL	
		12.3	MIBK	
	182.64	5.2,7.2	XYL	
		19,12	MIBK	
	180.73	0.08	XYL	29
	190.0	250 ng	BE	54
Sb	206.833	0.054	MIBK	29
	217.581	0.034	MIBK	29
Se	196.02	0.0047 (hydride generation)	(99:1) IPA/HNO_3	37
		5.3		
Si	251.6	0.02	XYL	28
		0.20	T	24
		0.53	PY	25
	288.160	0.007	MIBK	4
		0.03	MIBK	11
		0.01	BA	
		0.04	IPA	
		0.03	IO	
		0.04	THF	
		0.01	XYL	
		0.002	XYL	22
	251.611	0.034	MIBK	29
	251.61	0.01	XYL	66
	288.16	0.040	XYL	36
		0.04	T	
	251.611	0.006	K	
		0.0067	MIBK	42
		0.013	XYL	
		0.012	XYL	48
	251.6	0.8 ng	BE	54
Sn	283.999	0.003	MIBK	4
		0.09	BA	11

Table 6.11. (*Continued*)

Element	Wavelength (nm)	Detection Limit	Solvent	References
		0.14	IPA	
		0.3	IO	
		0.07	THF	
		0.1	XYL	
		0.041	T	24
		0.22	PY	25
	235.484	0.040	MIBK	29
	242.949	0.040	MIBK	29
	189.98	0.100	XYL	36
Ti	334.941	0.003	MIBK	4
		0.015	MIBK	11
		0.005	BA	
		0.009	IPA	
		0.02	IO	
		0.02	THF	
		0.0002	XYL	
		0.0008	XYL	22
		0.003	T	24
		0.01	PY	25
	308.8	0.0067	30% E/H_2O	34
	337.28	0.020	XYL	36
V	309.311	0.003	MIBK	4
		0.015	MIBK	11
		0.005	BA	
		0.015	IPA	
		0.02	IO	
		0.02	THF	
		0.0009	XYL	
	292.4	0.004	XYL	14
		0.003	MIBK	
		0.002	CB	
		0.040	T	24
		0.05	PY	25
	268.796	0.004	MIBK	29
	292.40	0.005	XYL	66
	309.31	0.020	XYL	6
		0.040	E	
		0.010	MIBK	
		0.007	CT	
	311.07	0.0033	15 $\%E/H_2O$	34
		0.0047	30 $\%E/H_2O$	
		0.0067	60 $\%E/H_2O$	

Table 6.11. (*Continued*)

Element	Wavelength (nm)	Detection Limit	Solvent	References
	292.40	0.010	XYL	36
	311.07	0.023	DIBK	40
		0.0044	XYL	48
Zn	213.856	0.004	MIBK	4
		1.0	MIBK	11
		0.025	BA	
		0.07	IPA	
		0.02	IO	
		0.02	THF	
		0.07	XYL	
		0.01	E	
		0.002	XYL	14
		0.006	MIBK	
		0.006	CB	
	202.548	0.100	MIBK	20
	213.856	0.002	XYL	22
		0.0048	T	24
		0.008	PY	25
		0.002	XYL	26
		0.004	XYL	28
	202.548	0.013	MIBK	29
	213.856	0.003	MIBK	29
	213.856	0.002	XYL	66
		0.010	XYL	6
		0.013	E	
		0.008	MIBK	
		0.014	CT	
	213.856	0.067	15% E/H_2O	34
	213.856	0.015	XYL	36
	213.856	0.033	(9:1) IPA/HNO_3	37
	202.548	0.056	DIBK	40
		0.0093	T	42
	213.856	0.002	K	
		0.005	MIBK	
		0.008	XYL	
	—	0.003	XYL	48

[a]*Solvents:* MIBK, methylisobutylketone; BA, butyl acetate; IPA, isopropyl alcohol; IO, isooctane; THF, tetrahydrofuran; XYL, xylenes; CB, chlorobenzene; PY, pyridene; T, toluene; K, kerosene; DMSO, dimethylsulfoxide; DIBK, diisobutylketone; CT, carbon tetrachloride; CL, chloroform; E, ethanol; M, methanol; B, butanol; NB, nitrobenzene; BE, benzene.

Table 6.12. Detection Limits for Chelate Vapor Introduction [70]

		Detection Limits		
		for ICP–Chelate		for ICP–Continuous
Metal	Wavelength (nm)	Absolute (ng)[a]	Relative $(ng \cdot mL^{-1})$[b]	Nebulization $(ng \cdot mL^{-1})$
Fe	259.9	0.086	0.086	5
Zn	213.9	0.42	0.42	2
Mn	257.6	0.085	0.085	0.7
Co	238.9	0.67	0.67	3
Cr	267.7	0.95	0.95	1

[a]Experimentally determined from pure trifluoroacetylacetonate complexes.
[b]Based on one-mL sample size and quantitative extraction from matrix.

tion process used, together with the approximately 100% efficient transfer of β-diketonate vapor, give absolute detection limits comparable to those found using graphite furnace ICP emission spectrometry.

Similar reaction chemistry was described by Fujinawa et al. [55] for the elements Al, Be, Cr, Cu, and Fe. These authors examined the thermal stability of the metal chelates, and developed a batch injection system for introduction to the ICP. They found that it was necessary to maintain injection temperatures typically below 100°C. Detection limits were generally good, and excellent linearity of calibration curves over three orders of magnitude dynamic range were found.

6.8.4. Gas Chromatography Interfacing

Relatively little work has been published in the field of gas chromatography (GC)–ICP interfacing, with the majority of publications concerning GC–plasma emission referring to microwave plasma detection. The primary interest in this area seems to be the use of ICP as an efficient atomization and excitation device, to provide elemental analysis of organic compounds. Work in this area has been carried out primarily by Windsor and Denton [54, 56, 71] and Brown and Fry [57, 72]. Using this approach, elemental lines for B, C, H, I, P, S, and Si in the UV and visible range have been readily observed, in addition to the usual background spectra (for example, C, C_2, CH, CN, CS, NH, and OH bands). Typical detection limits and dynamic ranges for these elements are shown in Table 6.13 [54]. The interface between the gas chromatograph and the ICP is very simple, requiring only the addition of makeup Ar to the GC effluent. The effectiveness of the technique for determining empirical formulas was tested for

Table 6.13. Detection Limits and Dynamic Ranges for GC/ICP [54]

Element	Wavelength (nm)	Detection Limit (ng)	Dynamic Range
B	249.77	1	1×10^3
C	247.86	12	1×10^3
H	656.28	27	1×10^2
I	206.16	4	1×10^5
P	213.62	0.6	2×10^4
S	190.03	250	1×10^2
Si	251.61	0.8	5×10^2

a number of elements, and generally a high level of success was found. It was necessary to use only one compound to determine the calibration constants for a range of other compounds. The procedure also provides an approximate carbon number, which may then be used to calculate a molecular formula from the empirical formula for a series of hydrocarbons. The work of Fry et al. [57, 72] has been to take this approach and apply it for N-containing compounds in the ICP. By using near-IR lines, O may also be determined using the 777.194 nm O I line [57], with a quoted detection limit of 25 ng. To avoid atmospheric oxygen entrainment, an extended torch was necessary.

6.8.5. Liquid Chromatography Interfacing

The effectiveness with which a liquid chromatograph (LC) may be coupled to an ICP spectrometer is basically limited by the efficiency of the interface. Peaks leaving a chromatographic column may contain several nanograms of material, and be readily detectable with a conventional UV absorption cell. However, the typically high analyte loss experienced in the LC–ICP interface can reduce the analyte mass transport rate W to the plasma to undetectable levels. The transport efficiency of the interface, therefore, becomes an even more critically limiting factor in LC–ICP interfacing than in conventional liquid sample introduction.

While the majority of work with LC–ICP interfacing has involved aqueous solvents [for example, 73–75], several studies have been made specifically with organic solvents [18, 24, 25, 31]. The increased emphasis on the use of reversed-phase separations, with solvents such as ethanol–water and acetonitrile–water, places stringent requirements on the interface to be able to accommodate these organic components. Further, the almost universal use of gradient elution requires the interface to have a response with is solvent independent over a wide range of compositions of the solvent mixture. The common use of mixed organic–aqueous solvents, rather than pure organic solvents, has the benefit that it results in a reduced organic solvent vapor loading in the plasma (see Section 6.4.2).

The only published work with organic sample introduction for LC–ICP interfacing discussed intermediate-powered plasmas. The studies of Gast et al. [18] established several important points. Among these were difficulties they experienced with pure normal-phase solvents. Further studies, along with those of Hausler and Taylor [24, 25] later addressed this problem through the use of solvent vapor thermostatting (see Section 6.5.4). In the initial studies with organic phase LC interfacing [18], water–methanol, water–ethanol, water–acetonitrile, water–tetrahydrofuran, and tetrahydrofuran–ethanol mixtures are used. A number of organometallic complexes were studied, including iron and molybdenum carbonyls, ferrocene, organoarsenic, organomercury, and organolead compounds. An important point was that the response of the plasma was somewhat compound specific. For example, in ethanol solvent the response of ferrocene was only 83% that of Fe (acetylacetone), whereas the response of 1,1'-diacetylferrocene was 113% that of Fe (acetylacetone). While gradient conditions presented no problems in this study, it should be noted that only low percentages of organic solvents were used (for example, 10–50%), and it is likely that the higher solvent loadings would have produced the typical response changes anticipated as the solvent composition moves to 100% organic solvent.

Hausler and Taylor [24, 25], worked with size-exclusion chromatography. The column used was 100-μm microstyragel, and both toluene and pyridene solvents were used. They were able to separate and determine a number of organosilicon, organolead, and organotin compounds with their system, together with dialkylbenzenesulfonates of 14 other elements, and some organometallic complexes in coal-derived materials. The use of microbore LC for ICP coupling has been described [31], and the lower solvent flow rates used in this system (10–25 $\mu L \cdot min^{-1}$), place a substantially lower solvent burden on the plasma. Cu, Zn, Fe, and Co diethyldithiocarbamate complexes were successfully separated and detected in this study at levels from 80–400 ng injected.

6.9. CONCLUSIONS

The use of the ICP for organic sample analysis now approaches the same general level of simplicity as aqueous sample introduction. There are, however, a number of specific points that still require attention to ensure accurate and reliable data. When working with organic solvents, the physical processes that are associated with the sample introduction process should be remembered at all times. The possibility for practical difficulties occurring with organic samples still remain greater than with aqueous samples, if for no other reason that such a wide range of solvents can be used successfully with the ICP. Within this range of solvents there exists a substantial variation in the degree of interaction with the plasma. In one instance (for example, xylenes), this may mean that very little difference from aqueous sample operating procedures are required, whereas in another (for example, methanol), special precautions may be necessary to ensure plasma stability (cf. Section 4.7.8 in Part 1 and [59]).

The importance of the ICP for organic sample analysis is clear from the wide range of samples already studied by various groups of workers. The most likely area for future study will probably be in the all important area of LC–ICP interfacing, where aspects of the interface still require serious development work. The outstanding problem is the need for improved analyte transport efficiency, coupled with effective utilization by the plasma of the increased mass transport rate W of analyte. In this situation, effective solvent removal is a prerequisite for improved performance.

REFERENCES

1. D. Truitt and J. W. Robinson, *Anal. Chim. Acta* **51,** 61(1970).
2. S. Greenfield and P. B. Smith, *Anal. Chim. Acta* **59,** 341(1972).
3. S. Greenfield, H., McD. McGeachin, and P. B. Smith, *Anal. Chim. Acta* **84,** 67(1976).
4. V. A. Fassel, C. A. Peterson, F. N. Abercrombie, and R. N. Kniseley, *Anal. Chem.* **48,** 519(1976).
5. S. Nukiyama and Y. Tanasawa, *Trans. Soc. Mech. Eng.*, Tokyo, 1938–1940, 4, 5, 6 (translated by E. Hope, "Experiments on the Atomization of Liquids in an Air Stream," Defense Research Board, Dept. of National Defense, Ottawa, Canada, 1950).
6. A. W. Boorn and R. F. Browner, *Anal. Chem.* **54,** 1402(1982).
7. R. F. Browner, A. W. Boorn, and D. D. Smith, *Anal. Chem.* **54,** 1411(1982).
8. M. S. Cresser, *Progr. Anal. Atom. Spectrosc* **5,** 35(1982).
9. A. W. Boorn, M. S. Cresser and R. F. Browner, *Spectrochim. Acta* **35B,** 823(1980).
10. D. D. Smith and R. F. Browner, *Anal. Chem.* **54,** 533(1982).
11. A. Ward and F. Brech, *ICP Information Newslett.* **1,** 171(1976).
12. J. Warren, *ICP Information Newslett.* **2,** 262(1977).
13. A. Varnes and T. E. Andrews, *Jarrell-Ash Plasma Newslett.* **1**(1), 12(1978).
14. A. F. Ward and L. Marciello, *Jarrell-Ash Plasma Newslett.* **1**(4), 10(1978).
15. G. Charalambous, *Jarrell-Ash Plasma Newslett.* **1**(4), 4(1978).
16. A. F. Ward and L. Marciello, *Jarrell-Ash Plasma Newslett.* **2**(2), (1979).
17. J. C. Smith, *Jarrell-Ash Plasma Newslett.* **2**(3), (1979).

18. C. H. Gast, J. C. Kraak, H. Poppe, and F. J. M. J. Maessen, *N. Chromatogr.* **185,** 549(1979).
19. R. N. Merryfield and R. C. Loyd, *Anal. Chem.* **51,** 1965(1979).
20. J. M. Motooka, E. L. Mosier, S. J. Sutley, and J. G. Viets, *Appl. Spectrosc.* **33,** 456(1979).
21. T. Ito, K. Kawaguchi and A. Mizuike, *Bunseki Kagaku* **28,** 648(1979).
22. K. Ohls and D. Sommer, in R. M. Barnes, ed., *Developments in Atomic Spectrochemical Analysis; Proceedings of the International Winter Conference, 1980,* Heyden, London (1981), p. 321.
23. C. T. Apel, T. M. Bieniewski, L. E. Cox, and D. W. Steinhaus, *ICP Information Newslett.* **3,** 1(1977).
24. D. W. Hausler and L. T. Taylor, *Anal. Chem.* **53,** 1223(1981).
25. D. W. Hausler and L. T. Taylor, *Anal. Chem.* **53,** 1227(1981).
26. M. S. Black and R. F. Browner, *Anal. Chem.* **53,** 2224(1981).
27. G. F. Wallace and R. D. Ediger, *Atomic Spectrosc.* **2,** 169(1981).
28. J. A. C. Broekaert, F. Leis, and K. Laqua, *Talanta* **28,** 745(1981).
29. P. W. J. M. Boumans and M. C. Lux-Steiner, *Spectrochim. Acta* **37B,** 97(1982).
30. R. V. Whiteley and R. M. Merrill, *Fresenius Z. Anal. Chem.* **311,** 7(1982).
31. K. Jinno and H. Tsuchida, *Anal. Lett.* **15,** 427(1982).
32. A. Nobile, R. G. Shuler, and J. E. Smith, *Atomic Spectrosc.* **3,** 73(1982).
33. J. D. Algeo, D. R. Heine, H. A. Phillips, and M. B. Denton, Office of Naval Research, USA, Technical Report No. 27, Task NR 051-549, NTIS No: AD A113490, 1982.
34. H. Benli, *Spectrochim. Acta* **38B,** 81(1983).
35. S. B. Smith, R. G. Schleicher, A. G. Dennison, and G. A. McLean, *Spectrochim. Acta* **38B,** 157(1983).
36. R. J. Brown, *Spectrochim. Acta* **38B,** 283(1983).
37. R. M. Barnes, R. P. Khosah, H. S. Mahanti, and A. H. Ullman, *Spectrochim. Acta* **38B,** 291(1983).
38. R. L. Dahlquist and J. W. Knoll, *Appl. Spectrosc.* **32,** 1(1978).
39. A. Miyazaki, A. Kimura, and Y. Umezaki, *Anal. Chim. Acta* **138,** 121(1982).
40. A. Miyazaki, A. Kimura, K. Bansho, and Y. Umezaki, *Anal. Chim. Acta* **144,** 213(1982).
41. A. Sugimae and T. Mizoguchi, *Anal. Chim. Acta* **144,** 205(1982).
42. J. J. Brocas, *Analusis* **10,** 387(1982).
43. B. Magyar, P. Lienemann, and S. Wunderli, *GIT Fachz. Lab.* **26,** 541(1982).
44. F. J. M. J. Maessen, P. J. H. Seeverens, and G. Kreuning, *Spectrochim. Acta* **39B,** 1171(1984).
45. A. Sanz-Medal, J. E. Sanchez Uria, and S. Arribas Jimeno, *Analyst* **110,** 563(1985).
46. M. W. Blades and B. L. Caughlin, *Spectrochim. Acta* **40B,** 579(1985).

47. W. Nisamaneepong, D. L. Haas, and J. A. Caruso, *Spectrochim. Acta* **40B,** 3(1985).
48. R. C. Ng, H. Kaiser, and B. Medding, *Spectrochim. Acta* **40B,** 63(1985).
49. P. Barrett and E. Pruszkowska, *Anal. Chem.* **56,** 1927(1984).
50. A. J. J. Schleisman, W. G. Fateley, and R. C. Fry, *J. Phys. Chem.* **88,** 398(1984).
51. A. Miyazaki, A. Kimura, H. Tao, K. Bansho, and Y. Umezaki, *Bunseki Kagaku* **32,** 746(1983).
52. J. Xu, H. Kawaguchi, and A. Mizuike, *Anal. Chim. Acta.* **152,** 133(1983).
53. K. C. Ng and J. A. Caruso, *Anal. Chem.* **55,** 2032(1983).
54. D. L. Windsor and M. B. Denton, *Appl. Spectrosc.* **32,** 366(1978).
55. T. Fujinaga, T. Kuwamoto, K. Isshiki, and N. Matsubara, *Spectrochim. Acta* **38B,** 1011(1983).
56. D. L. Windsor and M. B. Denton, *Anal. Chem.* **51,** 1116(1979).
57. R. M. Brown and R. C. Fry, *Anal. Chem.* **53,** 532(1981).
58. S. Greenfield, H. McD. McGeachin, and P. B. Smith, *ICP Information Newslett.* **2,** 167(1976).
59. F. J. M. J. Maessen, G. Kreuning, and J. Balke, Proceedings 1985 European Winter Conference, Plasma Spectrochemistry, Leysin, Switzerland, *Spectrochim. Acta* **35B,** 3(1986).
60. H. Anderson, H. Kaiser, and B. Meddings, in R. M. Barnes, ed., *Developments in Atomic Spectrochemical Analysis; Proceedings of the International Winter Conference, 1980*, Heyden, London (1981), p. 251.
61. J. M. Mermet and C. Trassy, *Appl. Spectrosc.* **31,** 237(1977).
62. P. Ch. Hauser and M. W. Blades, *Appl. Spectrosc.* **39,** 872(1985).
63. T. Ito, H. Kawaguchi, and A. Mizuike, *Bunseki Kagaku,* **28,** 648(1979).
64. C. D. Carr and J. E. Borst, *29th Pittsburgh Conference on Applied Spectroscopy and Analytical Chemistry*, Cleveland, OH, 1978, Abstract No. 388.
65. C. S. Saba and K. J. Eisentraut, *Anal. Chem.* **51,** 1927(1979).
66. Anon., "Oil Analysis," Baird Plasma Update, Feb. 1982.
67. H. Uchida, *ICP Information Newslett.* **8,** 155(1982).
68. M. De la Guardia-Cirugeda, G. Legrand, M. Druon, and J. L. Louvrier, *Analusis* **10,** 476(1982).
69. A. Miyazaki, A. Kimura, and Y. Umezaki, *Anal. Chim. Acta* **127,** 93(1981).
70. M. S. Black and R. F. Browner, *Anal. Chem.* **53,** 249(1981).
71. D. L. Windsor and M. B. Denton, *J. Chromatogr. Sci.* **17,** 492(1979).
72. R. M. Brown, S. J. Northway, and R. C. Fry, *Anal. Chem.* **53,** 934(1981).
73. D. M. Fraley, D. A. Yates, S. E. Manahan, D. Stalling and J. Petty, *Appl. Spectrosc.* **35,** 525(1981).
74. B. S. Whaley, K. R. Snable, and R. F. Browner, *Anal. Chem.* **54,** 162(1982).
75. K. E. Lawrence, G. W. Rice, and V. A. Fassel, *Anal. Chem.* **56,** 289(1984).

CHAPTER

7

DIRECT ELEMENTAL ANALYSIS OF SOLIDS BY INDUCTIVELY COUPLED PLASMA EMISSION SPECTROMETRY

JON C. VAN LOON

Departments of Geology and Chemistry
and Institute for Environmental Studies
University of Toronto
Toronto, Ontario, Canada

HIROKI HARAGUCHI and K. FUWA

Department of Chemistry
Faculty of Science
University of Tokyo
Bunkyo, Tokyo, Japan

7.1. Introduction
7.2. Methods Not Involving the Use of Plasmas
7.3. Plasma Methods
- 7.3.1. Introduction
- 7.3.2. Mechanical Agitation of Powders
- 7.3.3. Fluidized Bed Chamber
- 7.3.4. Graphite Cup–DC Plasma
- 7.3.5. Graphite Cupped Graphite Rod–ICP
- 7.3.6. Slurry of Solids
- 7.3.7. Graphite Platform
- 7.3.8. Spark
- 7.3.9. Interrupted Arc
- 7.3.10. Capillary Arc
- 7.3.11. Spark Elutriation
- 7.3.12. Laser Ablation
- 7.3.13. Induction Furnace–Microwave Plasma
- 7.3.14. Volatile Metal Chelates–Microwave Plasma
- 7.3.15. Volatile Chlorides–Microwave Plasma
- 7.3.16. Volatile Chelate–ICP
- 7.3.17. Tantalum Filament
- 7.3.18. Graphite Filament
- 7.3.19. Carbon Rod

7.4. Conclusions
References

7.1. INTRODUCTION

The direct elemental analysis of solids (methods not requiring chemical pretreatments) is an important priority in analytical methods development for use with plasma emission spectrometry. Approaches that have been proposed for this purpose (up to and including 1982) are outlined in this chapter. Although the main emphasis is on methods that have been applied to inductively coupled plasmas (ICP), related work for other types of plasmas (cf. Part 1, Chapter 2) are included when there is a possibility that the technique might be applicable to ICPs.

There is as yet no widely accepted approach to direct solid sample trace analysis by any of the techniques of optical atomic spectrometry: atomic emission (AES), atomic absorption (AAS), and atomic fluorescence (AFS) spectrometry. This is a serious deficiency in the field of analytical methods development.

Table 7.1 presents a summary of methods proposed for the direct trace analysis of solids by the three techniques of atomic spectrometry. None of these

Table 7.1. Summary of Methods of Direct Solid-Sample Trace Analysis by Atomic Spectrometry

	Conventional	Hybrid Techniques
Emission	Solid-in-flame	Furnace-arc
	Solid-in-plasma	Chloride generator
	Arc	Laser-plasma
	Spark	Furnace-plasma
	Electrothermal	Spark-plasma
	Laser	
	Glow discharge	
Absorption	Solid-in-flame	Capsule-in-flame
	Arc	Hollow graphite "T" tube
	Spark	Furnace-flame
	Electrothermal	Arc/spark-flame
	Laser	Chloride generator
	Cathodic sputtering	
Fluorescence	Laser	Furnace-flame
	Laser-spark	
	Electrothermal	

techniques has yet been shown to be entirely satisfactory for direct solid sample analysis.

Ideally a method for direct trace element solid sample analysis should have the following attributes:

1. It should be applicable to a wide range of sample compositions.
2. The method should be relatively fast.
3. Standardization should be simple.
4. To avoid inhomogeneity problems, the method should be capable of handling fairly large samples.
5. Simultaneous multielement analysis is desirable.
6. Cost per analysis should be as low as possible.
7. Repeatability and accuracy should be suitable for the particular application.

Spark AES comes closest to the ideal but is deficient because it is applicable mainly to conducting samples. It is the commonest method used by industry for routine analysis of metals. Arc AES, although applicable to a wide range of sample compositions, gives poor repeatability and accuracy. Both these methods require that standards be used that closely approximate the chemical composition and physical characteristics of the samples.

Graphite furnace AAS has been widely publicized as a method for the direct analysis of solids. In fact L'vov [1], in his pioneering paper proposing the use of furnace atomization for AAS, suggested that such an atomizer would be useful for the analysis of solids directly. However, real progress in this direction has been slow. Most early commercial furnace atomizers suffered from a variety of technical problems, the foremost of which was the failure to permit atomization under isothermal conditions. As a result interference problems were complex and great. Use of the L'vov platform together with the rapid heating rates of recent commercial furnaces has, to a large extent, obviated this problem. Also Zeeman background correction, now available with a few furnaces, in many cases, allows compensation for the larger magnitude of nonspecific interferences that are commonly encountered during solid sample analysis.

However, in spite of this progress, a number of drawbacks to direct solid sample analysis by furnace AAS still exist. These include, failure of even Zeeman background correction to compensate for nonspecific absorption in many instances, the requirement to use a small sample weight (that is, <10 mg), standards must be used that closely approximate the chemical composition and the physical characteristics of the samples and skeletal remains of samples (usually organic samples) that block the light beam.

7.2. METHODS NOT INVOLVING THE USE OF PLASMAS

A few of the other methods (not involving plasma atomizers) that have been proposed for the direct analysis of solids by AES, AAS, and AFS are discussed briefly in this section so that the reader can determine which method may, in the future, be useful in work with ICPs or direct current plasmas (DCP).

In AAS there have been many proposals for the direct analysis of solids, as is evident from Table 7.1. A good review of this subject was written by Langmyhr [2]. Solids have been placed directly in the flame for AAS analysis. For example, sample mixed with NaCl was deposited on an iron screw and placed in a flame [3]. Rubidium was determined in rocks by this approach. Because of the relatively low temperatures of flames, such an approach is restricted to the most volatile elements.

The most common method for AAS direct solids analysis is electrothermal atomization, which was discussed in the preceding section.

Interesting variations on conventional electrothermal atomization have been proposed by L'vov and co-workers [4, 5]. These are the capsule–in–flame atomizer and the circular cavity furnace. In the capsule–in–flame atomizer, the powdered sample, which was mixed with graphite powder (to prevent melting or sintering into a bead), was placed into a cavity in the center of a porous graphite rod. The rod was mounted horizontally between two water-cooled cylindrical holders. Current was passed through thc device for heating. The rod was also placed in a flame with the optical beam passing just over the surface. An air–acetylene or nitrous oxide–acetylene flame was chosen, depending on the volatility of the elements being determined. Atomization occurs in the flame and thus problems from analyte–matrix interactions should be greatly reduced compared to atomization in a conventional graphite furnace atomizer.

In the circular cavity furnace the powdered sample (mixed with graphite powder) was placed in a cavity between the casing (pyrolytical graphite lined) and a porous graphite inner tube. The assembled furnace was electrically heated. As heating occurs, vapors from the sample pass through the walls of the porous inner tube into the optical beam. This gives optimum thermal conditions for dissociation of the more thermally stable oxide compounds.

By comparing these two atomizers the following was concluded.

The range of elements that could be determined was appreciably greater with the capsule-in-flame atomizer (that is, 50 with the latter compared to 35 with the former).

A wider range of sample types could be handled with the capsule-in-flame atomizer.

Standardization was relatively simple, requiring only a mixture of the analyte elements with graphite powder.

The integral method of absorbance measurement must be used.

Electrothermal atomizers have been employed out of the optical beam to produce an aerosol of a solid sample which in turn is swept into the flame. Koop et al. [6] (AAS), Kantor et al. [7] (AAS), and Ip et al. [8] (AFS) used commercial graphite furnaces interfaced with an air–acetylene flame. The flame is a superior device for atomization and as a result problems due to molecular absorption or fluorescence, and light-scatter by particles were greatly reduced. In addition, because the sensitivity is appreciably less than for conventional electrothermal atomization, larger sample sizes can be used. This is important in reducing problems due to sample inhomogeneity.

In AES, as indicated previously, arc and spark atomizers are widely used for the direct analysis of nonconducting and conducting solid samples, respectively. Other methods have also been employed in more restricted applications, as follows.

Powdered samples rolled in filter paper were introduced into the flame for flame emission analysis [9]. This method was improved and automated by Roach [10] and by Stewart and Harrison [11]. A furnace mounted below an arc was used by Preuss to inject sample into the atomizer [12]. The aerosol thus produced was entrained in a stream of inert gas.

Other methods of injecting solids into the arc or spark include the forcing of a sample up through a hollow electrode using a piston [13] and the spreading of the sample out on a cellulose tape which is fed between the gap of the electrodes. Both these approaches have been modified and subsequently used with plasma atomizers.

For the high-precision analysis of conducting samples, Grimm-type discharge lamps are receiving increased attention. This approach is applicable to major and minor element concentrations. Using such an approach, time resolution of analyte and background signals is possible.

7.3. PLASMA METHODS

7.3.1. Introduction

ICPs and DCPs have properties that make them potentially attractive devices for the direct trace analysis of solids. The main one is the high temperature attained in the zone of sample introduction, which may result in excellent breakdown of the sample, atomization, and excitation. In this regard, ICPs are likely somewhat superior to DCPs. Also ICPs and DCPs are very stable and if the method for introduction of solids does not greatly disturb the plasma, then good precision of measurements should be possible. Microwave plasmas, on the other hand, are fairly intolerant to the introduction of all but dilute gaseous samples, and for this reason less will be said about their use with solid samples.

In Table 7.1, methods have arbitrarily been classified as conventional or hybrid techniques, because hybrid techniques (those in which sample breakdown and, to a varying extent, vaporization occur in a device separate from the atomizer) have potential advantages. Most importantly, with the majoirty of hybrid techniques, some time resolution of analyte and background signals occurs. This property is much more important with AAS and AFS than with ICP–AES.

For most samples and most elements, absolute detection limits achievable with ICP–AES are poorer than those obtained by furnace AAS by up to three orders of magnitude. Thus, it is important to devise sample delivery systems that give improved detection limits. The direct analysis of solids avoids dilutions that must occur with wet chemical sample preparation. In the case of liquids, methods have been devised for rapid vaporization of discrete liquid sample residues. In the following discussions, several methods for rapid vaporization of residues of liquid samples are included because they are potentially applicable to some types of solids.

One technical problem that must be considered in any method of solid sample injection into the plasma is that high concentrations of air, over any appreciable period of time, must not be injected into a low-power argon ICP. If this occurs, there is the likelihood that the plasma will shrink down onto the torch, which in turn may result in melting of the torch. Thus, provision must be made to have constant Ar flows in each segment of the torch during the complete procedure.

In the following sections the various methods for introducing solids directly into plasmas are discussed.

7.3.2. Mechanical Agitation of Powders

Hoare and Mostyn [14] placed powdered solid sample into a borosilicate cup at the base of the plasma torch as shown in Fig. 7.1. The cup is mechanically agitated. The rate of agitation controls the amount of powder introduction into the plasma. The approach has been applied successfully to materials of widely varying densities. The torch shown in the figure is a modified version of a "normal" torch. The modification consists of shaping the orifice at the outer silica tube at a 45 ° angle to obtain two advantages: 1 the emission from the lower region of the plasma can be viewed more effectively, and 2 the flow properties of the torch are improved.

The ICP was generated by a high-frequency generator, capable of giving maximum power output of 2.5 kW. The apparatus was employed for the determination of trace impurities in powder samples. For best quantitative work, it

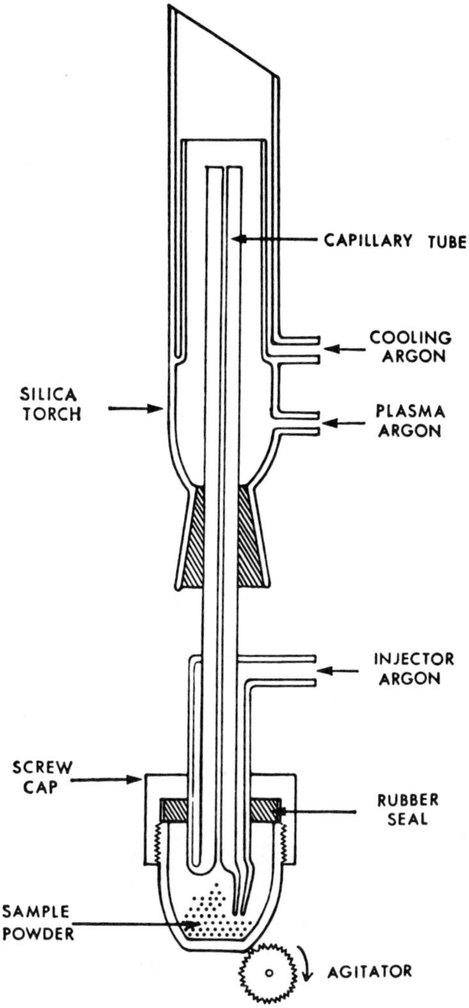

Figure 7.1. Powder injection apparatus [14]. [Reprinted with permission from H. C. Hoare and R. A. Mostyn, "Emission Spectrometry of Solutions and Powders with a High-Frequency Plasma Source," *Anal. Chem.* **39,** 1154 (1967). Copyright (1967), American Chemical Society.]

is necessary to prepare the samples and standards in a closely similar way. Qualitative analysis was done on lithium salts and alumina.

7.3.3. Fluidized Bed Chamber

Dagnall et al. [15] used a fluidized bed chamber (Fig. 7.2) to entrain powders into an Ar carrier gas flowing into the plasma. In this device, powdered sample

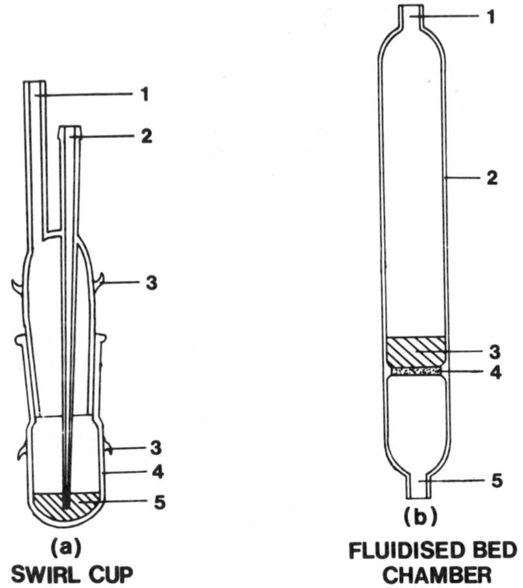

Figure 7.2. (*a*) Swirl cup chamber: (1) gas–powder outlet, (2) gas inlet, (3) securing hook, (4) detachable powder cup, (5) powder. (*b*) Fluidized bed chamber: (1) gas–powder outlet, (2) pyrex tube, (3) powder, (4) sintered glass disc, (5) gas inlet [15]. [Reprinted with permission from R. M. Dagnall, D. J. Smith, T. S. West, and S. Greenfield, "Emission Spectroscopy of Trace Impurities in Powdered Samples with a High-Frequency Argon Plasma Torch," *Anal. Chim. Acta* **54**, 401, (1976). Copyright (1976), Elsevier Science Publishers, Amsterdam.]

was placed on a sintered glass disc through which Ar was flowing. The cyclone chamber above the fluidized bed chamber is to separate large particles. Beryllium and B were determined in magnesium oxide.

7.3.4. Graphite Cup–DC Plasma

The following proposal [16] was developed for qualitative analysis purposes, and is applicable to 4–6 mg of powdered solid samples.

Figure 7.3 shows a diagram of the proposed equipment. A graphite cup, mounted on a metal rod, is inserted into the dc plasma to a position that does not disturb the plasma. This is accomplished by first inserting an empty cup into the plasma and correctly positioning it using a horizontal and vertical adjustment. Then the mechanical stop is activated, and the cup is withdrawn, cooled, and filled with sample. The full cup is then placed back in the plasma at the correct position. The powdered sample must be 200–400 mesh. Using such powders, qualitative analyses of pork, sludge, and algae were done.

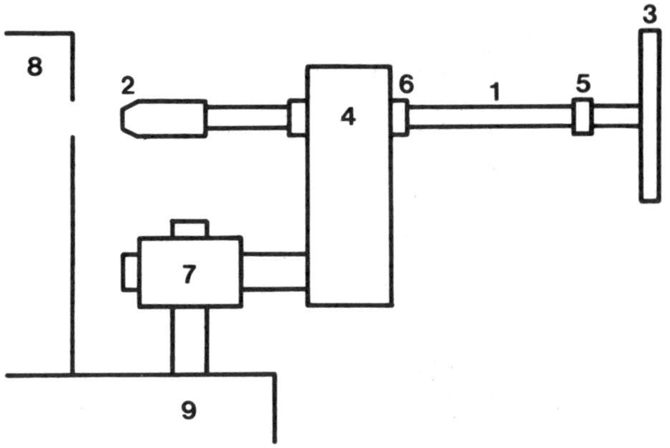

Figure 7.3. Graphite cup positioning assembly: (1) support rod, (2) graphite cup, rod handle (Teflon), (3) support rod, (4) support rod, (5) mechanical stop (Teflon), (6) insulating bushing (Teflon), (7) *x-y* positioner, (8) plasma jet housing, (9) spectrometer base [16]. [Reprinted with permission from T. Yajima, *Plasma Line* **1,** 6, (1980). Copyright (1980), Beckman Instruments, Andover, MA.]

7.3.5. Graphite Cupped Graphite Rod–ICP

Salin and Horlick [17] described the following approach to the introduction of liquids and solids into an ICP. A graphite cup "electrode," such as those commonly used in dc arc emission, was inserted into the central portion of the ICP torch. The sample was placed on the top of the electrode and after the ICP was ignited the sample was moved up into the plasma. The experimental arrangement is shown in Fig. 7.4. The torch is similar to conventional devices, except that the carrier gas tube is replaced by a silica tube through which the cup electrode can be moved. A silica rod supports the graphite electrode and is itself attached to a sliding platform. The rod passes through a Teflon block supporting the torch.

The plasma was ignited with the electrode held at the height of a conventional aerosol injector tube. After ignition the electrode with solid sample was raised into the plasma (positioned just above load coil) using the sliding platform. The electrode, although white hot, is not appreciably consumed because of the Ar atmosphere. No problems were found in sustaining the plasma during the electrode insertion. Almost no adjustment of the matching network was necessary. The plasma was generated at 2.5 kW with an outer Ar flow of 17 L $\cdot$ min^{-1}.

Spex G Standards containing 49 elements but no internal standards were analyzed both by the proposed system and by conventional dc arc. Only the wavelength range of 240–340 nm was used. Results show that equivalent de-

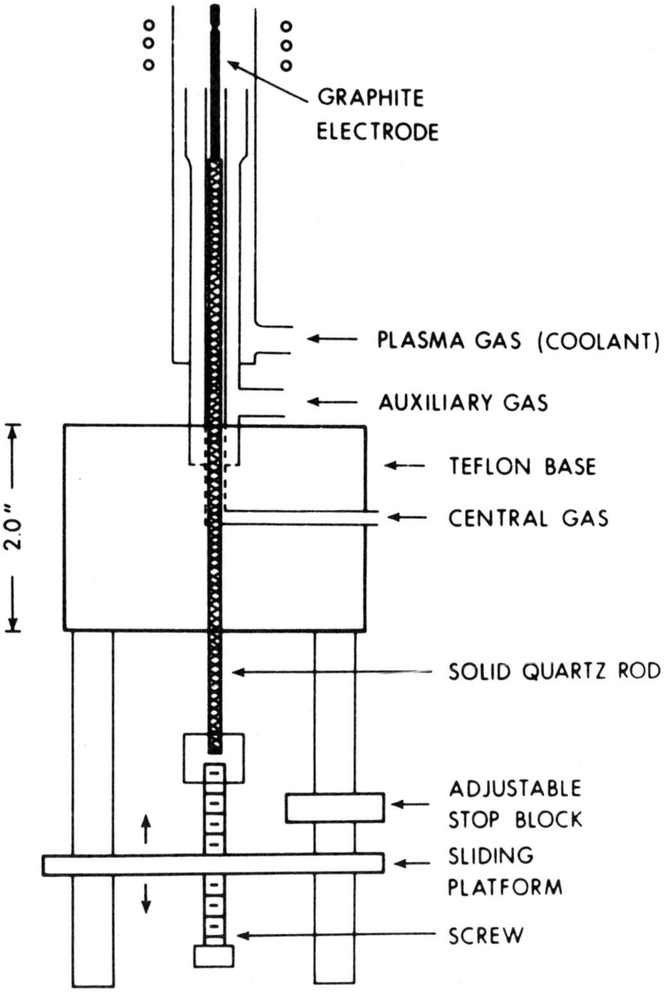

Figure 7.4. Modified torch showing graphite crucible [17]. [Reprinted with permission from E. O. Salin and G. Horlick, "Direct Sample Insertion Device for Inductively Coupled Plasma Emission Spectroscopy," *Anal. Chem.* **51,** 2285, (1979). Copyright (1979), American Chemical Society.]

tection limits were obtained for Ag, Be, Cd, Dy, Fe, Mg, Mn, Sb, and Sn. The solid sample ICP system was best for As, Bi, Cu, Ba, In, Na, Pb, Tl, and Zn. Better results for the dc arc were obtained for B, Ca, Ce, Mo, Ti, and V. The elements Co, Ni, and W were only seen using the dc arc, and Hg, P, and Sr only using the electrode–ICP approach. Preliminary investigations were made of the application of the proposed approach to other samples. In this regard

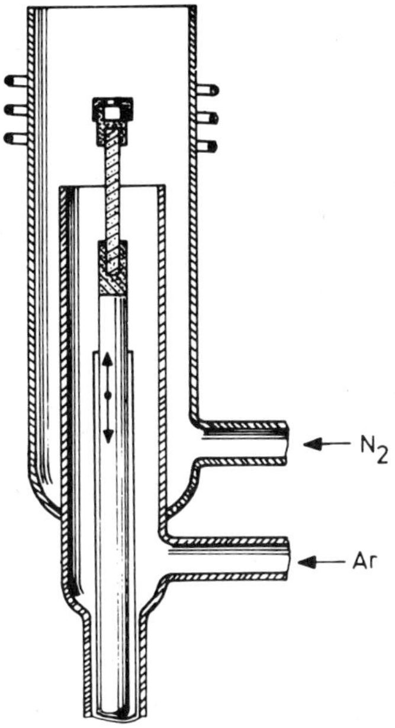

Figure 7.5. Sample insertion torch (SET) [18]. [Reprinted with permission from D. Sommer and K. Ohls, *Fresenius Z. Anal. Chem.* **304,** 98, (1980). Copyright 1980, Springer-Verlag, Berlin, F.R.G.]

qualitative studies were made of National Bureau of Standards (NBS) coal and orchard leaves. The authors caution that careful investigations must now be made, particularly of the temporal behavior of sample emission from the electrode ICP system.

In a related approach Sommer and Ohls [18] used a graphite crucible to insert sample into an N_2-Ar ICP. The modified torch is shown in Fig. 7.5. The crucible is placed into the plasma using a pneumatic elevator device. The outer gas was N_2 flowing at a rate of 28 L $\cdot$ min^{-1}, an Ar was employed as the intermediate gas at a flow rate of 15 L $\cdot$ min^{-1}. The forward power was 4.5 kW or less. Detection limits reported using this method were about one order magnitude better than those obtained with conventional solution nebulization for Mn, Al, Co, Cu, Si, and Ni (cf. Chapter 1).

7.3.6. Slurry of Solids

Attempts have been made to nebulize solids, suspended in a slurry, into the plasma [19] in a similar way as was used in flame AAS. However, the usual problems intrinsic to passing slurries through a nebulizer, which adversely affect precision, are inherent in this approach.

Fuller et al. [20] made a comparison of flame, electrothermal, and ICP atomization of solid samples introduced as a slurry. The solid rock and mineral samples were milled to a size below 6 μm. In the case of the ICP available concentric and cross-flow nebulizers failed to handle the slurries. Thus, a cross-flow, slot type, nebulizer [21] was obtained and fed with a peristaltic pump. In spite of its very high temperature and hence potentially excellent atomization efficiency, the ICP failed to overcome the matrix dependency of the signal. Thus standards had to be used that were similar in composition to the samples in question.

7.3.7. Graphite Platform

Kleinman and Svoboda [22] directly vaporized solid residues into a plasma from a graphite platform support (Fig. 7.6) in the central body of a plasma discharge chamber. The Ar plasma was generated at 40 MHz using 400 W of power. The graphite support must be placed into the induction coil region for best reproducibility of results. Detection limits for Ag, Ba, Co, In, and Ni were better by this approach than by impulse Ar arc. Poorer detection limits by the proposed method were obtained for B, Bi, and Th. The general applicability of this approach to solids is uncertain.

7.3.8. Spark

Solid conducting samples were analyzed directly by ICP emission spectrometry using a spark between an anode and the sample as the cathode to produce metal

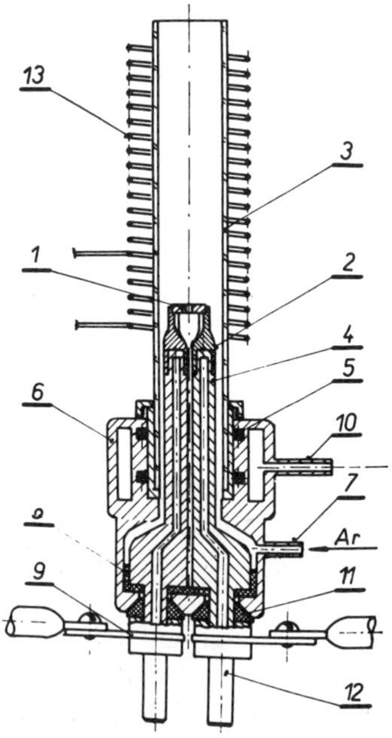

Figure 7.6. Torch with internal electrothermal sampling: (1) graphite support, (2) tantalum holder, (3) quartz tube, (4) copper core, (5) ''O'' ring, (6) brass mantle, (7) gas pot, (8) and (11), Teflon gaskets, (9) cable lub, (10) and (12) cooling water pots, (13) autotransformer [22]. [Reprinted with permission from I. Kleinman and V. Svoboda, ''High-Frequency Excitation of Independently Vaporized Samples in Emission Spectrometry,'' *Anal. Chem.* **41**, 1030, (1969). Copyright (1969), American Chemical Society.]

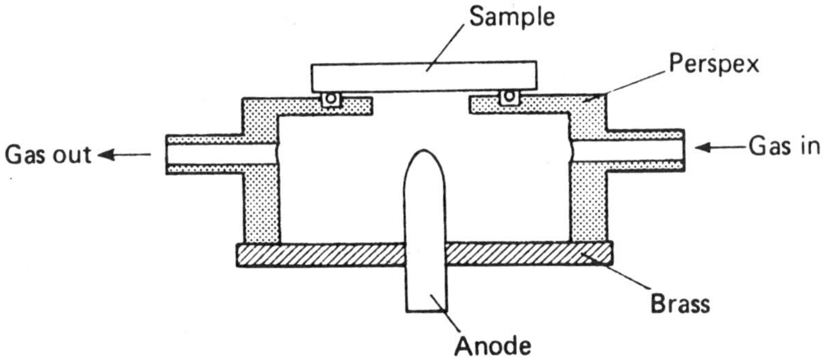

Figure 7.7. Spark chamber [23]. [Reprinted with permission from H. G. C. Human, R. H. Scott, A. R. Oakes, and C. D. West, *The Analyst*, **101,** 266, (1976). Copyright (1976), The Royal Society of Chemistry, London.]

particles. These were then entrained in the carrier gas flow to the plasma [23]; the cell is shown in Fig. 7.7. A Th-treated W rod was used as the anode. The cathode (sample) was sealed to the top of the cell by an O-ring and was pressed down with a pin. The pneumatic nebulizer, used for solution work, was replaced by the spark chamber. An Ar flow of 1 L · min^{-1} was passed through the chamber and into the plasma injection tube.

The device was tested on previously analyzed aluminum and brass samples. Copper, Mg, and Zn were the elements determined. The signal-to-noise ratio was found to increase with increasing inductance, with a value of 75 μH giving best results for this particular cell. Precision varied between 4 and 6% for the elements and samples tried. Detection limits in Al alloys for Cu, Mg, and Zn were 3, 2, and 150 ppm, respectively.

Marks et al. [24] reported on the application of a "Separate Source Excitation Analysis System" for the direct analysis of alloys by ICP–AES. The device employs a spark source, and the authors found that variables affecting spark source analysis also were important in the use of such equipment with the ICP. To obtain best results the alloy surface must be sanded with abrasive. Silicon and B were determined in Ni-based superalloys, but severe spectral interference caused by the major components obviated the determination of P.

7.3.9. Interrupted Arc

Ohls and Sommer [25] used an interrupted arc contained in a sample flow chamber to produce a sample aerosol for injection into the Ar carrier gas flow of the torch. A graphite counter electrode was employed, using a point-to-point technique, for the analysis of metallic samples on briquetted mixtures of Cu powder and oxides. The N_2-Ar plasma had a forward power of 3.3 kW.

7.3.10. Capillary Arc

In a commercial offering described by Dahlquist et al. [26], a capillary arc was used to produce an aerosol of a sample remote from the ICP instrument. Little technical information is available on this device. Conducting samples, only, have been analyzed.

7.3.11. Spark Elutriation

An energetic spark between two electrodes held just above fine-powdered sample (−200 mesh) causes particles of the powder to be elutriated from the sample. Scott [27] used this approach to entrain particles into the Ar carrier gas flow to an ICP as a method of directly analyzing solid samples. The acoustic energy of the spark is thought to be responsible for this phenomenon. In addition, the spark probably causes the break up of particles, thus aiding in the production of a suitable grain size of material for elutriation.

The apparatus for elutriation is shown in Fig. 7.8. Powder (1 g) was weighed into the sample vial. A seal between the electrode assembly and the vial was made with an O-ring. The "T" tube allows an intermediate Ar flow of 0.2 L $\cdot$ min^{-1} to pass through the vial and out the other side, which in turn connects to the main carrier gas flow of the ICP operating at 1 L $\cdot$ min^{-1}. To change samples, the sample vial was replaced by another vial containing the next sample. The carrier gas preferentially flows downward when the sample container is removed, thus the Ar rapidly displaces air from the next vial just prior to its attachment to the electrode assembly and deleterious passage of air into the plasma is prevented.

A Walters high-voltage, controlled waveform spark source was used, but the author states that any spark source used for analytical AES would be suitable.

The signal must be measured at a fixed time (30 s in this study) after the beginning of the elutriation, because there is an exponential drop-off in emission intensity due to the preferential elutriation of the finest powders at the beginning of the process.

This method was tested on geological samples, and Cu was determined in granite, granodionte, dacite, felsic lava, and tuff. Good correspondence of results was obtained compared to values obtained by X-ray fluorescence spectrometry, AAS, and mass spectrometry.

7.3.12. Laser Ablation

Abercrombie et al. [28] used a pulse-transverse, excited CO_2 laser (maximum power 17 kW) to vaporize solid samples. The resultant aerosol was swept in an Ar stream into an ICP. Twenty elements in air particulate could be determined in up to 3400 samples per day. This approach, used for geochemical prospect-

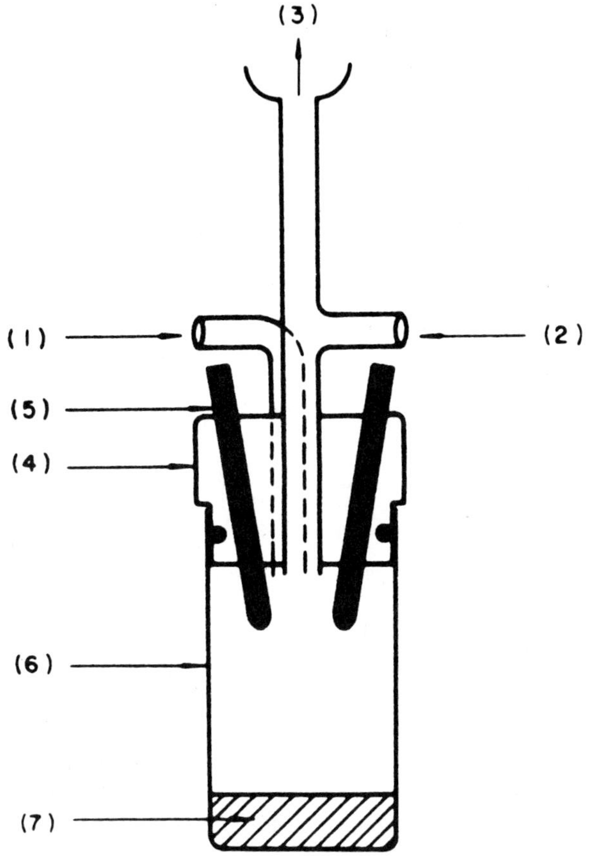

Figure 7.8. Elutriation apparatus: (1) tube for bleeder supply of carrier gas, (2) T-tube side arm for main carrier gas, (3) ball joint to plasma torch, (4) Teflon electrode holder assembly, (5) graphite electrodes, (6) glass or plastic vial, (7) powder sample [27]. [Reprinted with permission from *Spectrochim. Acta*, **33B,** 124, R. H. Scott, "Spark Elutriation of Powders into an Inductively Coupled Plasma," Copyright (1978), Pergamon Journals, Oxford.]

ing, suffers from inhomogeneity problems because of the very small sample being vaporized.

A laser ablation approach was also reported by Thompson et al. [29]. These workers used a commercially available laser microprobe (the LMA 10, Jena). This device consists of a ruby laser capable of producing 1.0-J pulses of energy. The laser pulse was automatically focused using the microscope optics at the center of the field of view. A schematic diagram of the set-up is shown in Fig. 7.9. The ablation chamber was interfaced with an ICP spectrometer.

The approach, although intended for use with geological samples, was first tested on a series of standard steel samples available from the Bureau of Ana-

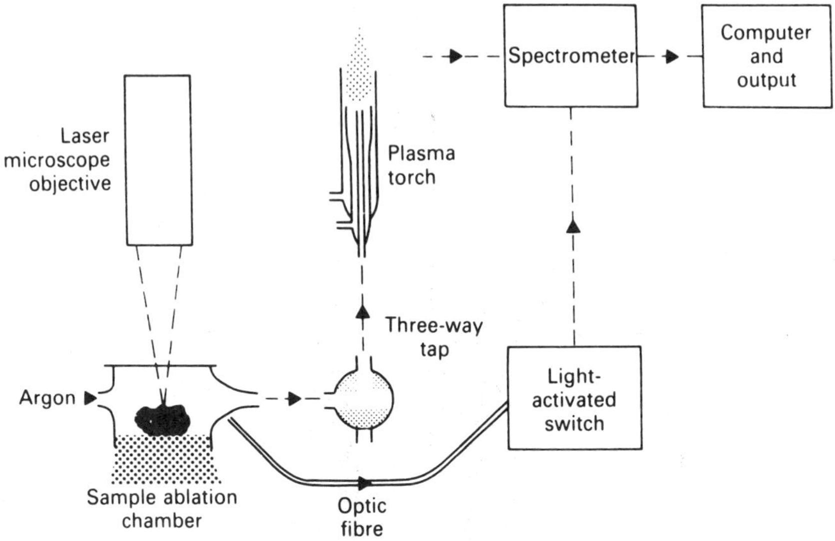

Figure 7.9. Laser–ICP microprobe system [29]. [Reprinted with permission from M. Thompson, J. E. Goulter, and F. Sieper, *The Analyst*, **106**, 33, (1981). Copyright (1981), The Royal Society of Chemistry, London.]

lysed Samples, United Kingdom. Results showed relative standard deviations (RSDs) <2% for Fe and Ni. Copper and Cr gave 4.7 and 9.0% RSDs, respectively. Preliminary work on geological samples was not satisfactory because of problems with inhomogeneity. Attempts to obtain homogeneous samples by fusion with lithium metaborate failed because the bead thus produced was shattered by the laser pulse.

Laser vaporization was used by Carr and Horlick [30] for direct solid sampling of NBS brass and low-Al alloys. Figure 7.10 is a schematic drawing showing the sample chamber-plasma-laser system. A ruby laser (1–2 J) operating mostly in the free-running mode, was employed. The effects of several experimental parameters were closely studied with the following results. The free-running mode was found to release very much more metal from a low-Al alloy sample: 500 mg compared to 25 μg in the Q-switched mode. The signal is very dependent on laser focus, thus it is important that reproducibility of this parameter be excellent. Two laser shots per sample separated by 8 s give best precision (4%). Integration of the signal is employed with signal duration being about 5 s. Calibration curves were plotted for the Al and brass samples. In the case of brass, a log–log plot of Cu-to-Zn ratios of signal and concentration gave good linearity. However, much poorer results were obtained with Al alloys. Detection limits, although not reliably measured, were thought to be in the nanogram range.

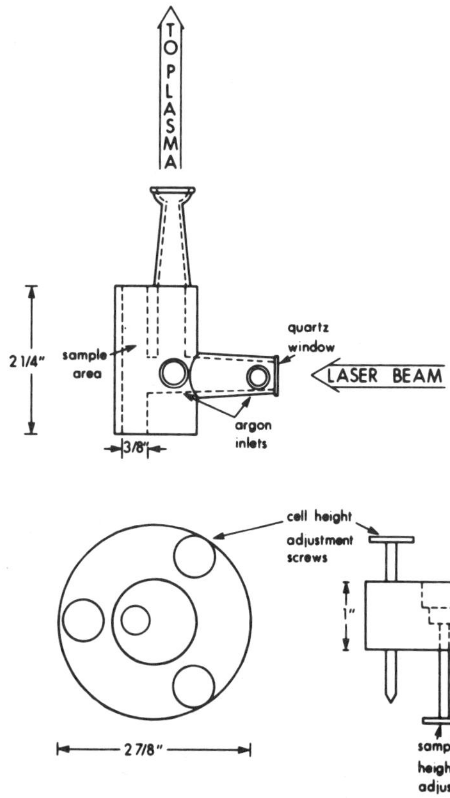

Figure 7.10. Schematic drawing of rod-type sample chamber for laser ablation ICP spectrometry [30]. [Reprinted with permission from *Spectrochim. Acta*, **37B**, 4, J. W. Carr and G. Horlick, "Laser Vaporization of Solid Metal Samples into an Inductively Coupled Plasma," Copyright (1982), Pergamon Journals, Oxford.]

7.3.13. Induction Furnace–Microwave Plasma

Bauer and Natusch [31] describe a modification of a Leco induction furnace for the direct introduction of aerosols from solid samples into a microwave plasma. The modification was a motor-driven, temperature-variable transformer to allow temperature programming. This furnace was used for the determination of the forms of metals and nonmetals in complex samples.

The schematic of the furnace is shown in Fig. 7.11. The transfer line from furnace to plasma must be kept as short at possible; it was heated to 650°C to minimize condensation along its length. Failure to do this can result in spurious peaks, peak-broadening, and memory effects.

Qualitative and quantitative analyses were attempted with this device, and the elements monitored were F, Cl, Br, C, S, N, O, Cd, Hg, Pb, and Zn. The

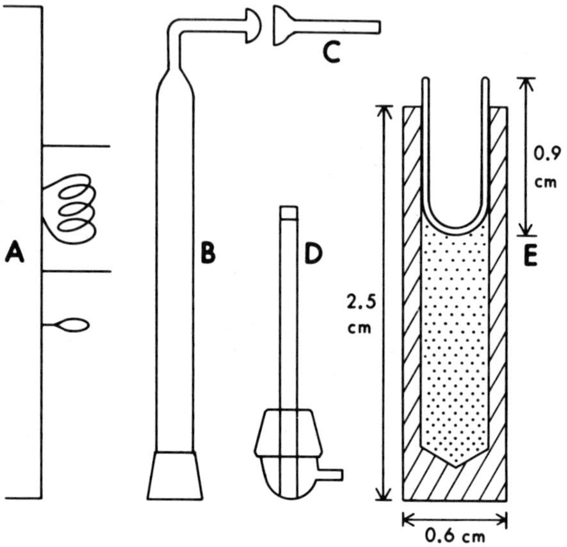

Figure 7.11. Furnace and transfer line components: (A) furnace, showing induction coil above and (B) clamp for furnace below, (C) furnace tube set down through coil and clamped, (D) discharge tube, (E) crucible pedestal, (F) crucible, graphite cup packed with carbon black, quartz cup rests in top [31]. [Reprinted with permission from C. F. Bauer and D. F. S. Natusch, "Speciation at Trace Levels by Helium Microwave-Induced Plasma Emission Spectroscopy," Copyright (1981), American Chemical Society.]

system was used to investigate the chemical forms of elements in solid samples. Both the cation and anion components of a compound could be monitored to make identification more certain. This is a distinct advantage for using a multielement element-specific detector, such as the microwave plasma emission (instead of AAS or AFS) used in this instance. The authors suggest attempting to use this approach with the ICP because of greater freedom from chemical interference with this device.

7.3.14. Volatile Metal Chelates–Microwave Plasma

Runnels and Gibson [32] proposed the introduction of metals into a plasma using volatile metal chelates or volatile inorganic salts such as the halides. In their study a microwave plasma was used, but again such an approach has a potential attraction for ICP–AES.

7.3.15. Volatile Chlorides–Microwave Plasma

Skogerboe et al. [33] proposed the use of a chloride generator to volatilize metals from solid residues into a microwave plasma. The following reaction is applicable:

$$M_nY_{m(s)} + \text{excess } HCl_{(g)} \xrightarrow{\Delta} MCl_{n(g)} + H_mY$$

where Y represents a cation.

The apparatus for chloride generation is shown in Fig. 7.12. The furnace consists of a 10-mm i.d. quartz tube wound with Nicrome wire and insulated with asbestos. The furnace is powered by a lab variac, and sample cuvettes are quartz. Temperatures up to 1000 °C are possible with this system. The above equation goes to completion to the right because of the continuous removal of the chlorides. An Ar microwave plasma was used. The Ar–HCl flow was 200 $mL \cdot min^{-1}$ through the furnace and an additional 2000 $mL \cdot min^{-1}$ flow was needed to sustain the plasma. No details on the cavity or microwave power supply used were given. Samples analyzed included residues of waters, brain, lung, and bone. The dynamic linear working range was about two orders of magnitude. As with the two previous proposals, this system is potentially attractive for application to the ICP.

7.3.16. Volatile Metal Chelate–ICP

Black and Browner [34] reported a volatile metal chelate sample introduction method for the ICP. Metals are converted to volatile β-diketonates for injection

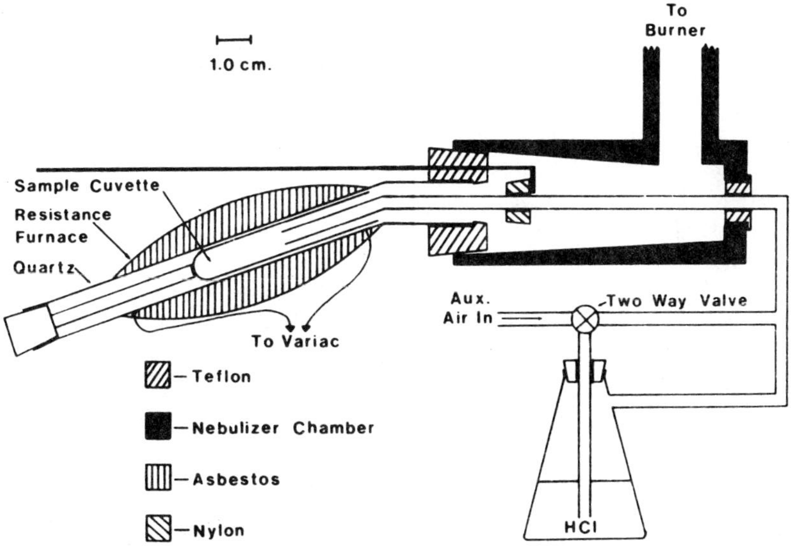

Figure 7.12. Cloride generator and associated manifold [33]. [Reprinted with permission from R. K. Skogerboe, D. L. Dick, D. A. Pavlica, and F. E. Lichte, "Injection of Samples into Flames and Plasmas by Production of Volatile Chlorides," *Anal. Chem.* **47**, 569, (1975). Copyright 1975, American Chemical Society.]

into the Ar sample introduction flow running at 0.9–1.5 L · min^{-1}. A borosilicate glass reaction vessel was used, and was connected through a six-port valve to Ar purge flows and to the center channel of the plasma torch.

Solid metal and NBS bovine liver samples were directly reacted with the chelating reagent using two approaches. In the first method, the sample and chelating reagent were added to the reaction vessel and reacted at 120 °C for 5 min. Alternatively, samples were reacted with chelating reagents in closed ampules in an oven at 110 or 130 °C. The contents were then frozen, and the ampule was opened and placed in the heated reaction vessel. Calibration was done by standard addition. Good results were obtained for the determination of Fe and Zn in bovine liver.

7.3.17. Tantalum Filament

A Ta filament in a quartz dome was used by Nixon et al. [35] to vaporize microliter samples into an ICP. [One of us (JVL) has found this approach to be useful for solid sample vaporization.] The apparatus is shown in Fig. 7.13. Vaporized sample flows directly from the top of the dome into the sample introduction orifice of the plasma torch. Detection limits for Sb, Se, Te, Hg, Be, Mn, and Sn were comparable to or better than the best obtainable by AAS.

7.3.18. Graphite Filament

Another approach, which has been proposed for the introduction of residues from solutions into a plasma, has been described by Gilbert and Hildebrand [36]. The device is a graphite braid transport system interfaced with an echelle plasma emission spectrometer.

The graphite braid filament is fed from a spool which is mounted on the spectrometer chassis. Sample is deposited onto the filament (the filament is precleaned electrically) and desolvation is done. The portion of the filament containing the sample is positioned above the cathode assembly block. There is a hole in the cathode block through which the plasma penetrates to interact with the sample. A blank segment of the filament serves as the cathode when the plasma is activated.

This system has been evaluated for Al, Bi, Cd, Cr, Fe, Mn, Pb, and Ti. The detection limits obtained (ng) were Al 1.5, Bi 4.0, Cd 2.0, Cr 0.2, Fe 4.0, Mn 0.7, Pb 3.0, and Ti 8.0.

7.3.19. Carbon Rod

Gunn et al. [37] described an electrothermal device for the vaporization of discrete samples into the ICP. [Although this device was developed for microliter liquid samples, one of us (JVL), has found it applicable to solids.] The device

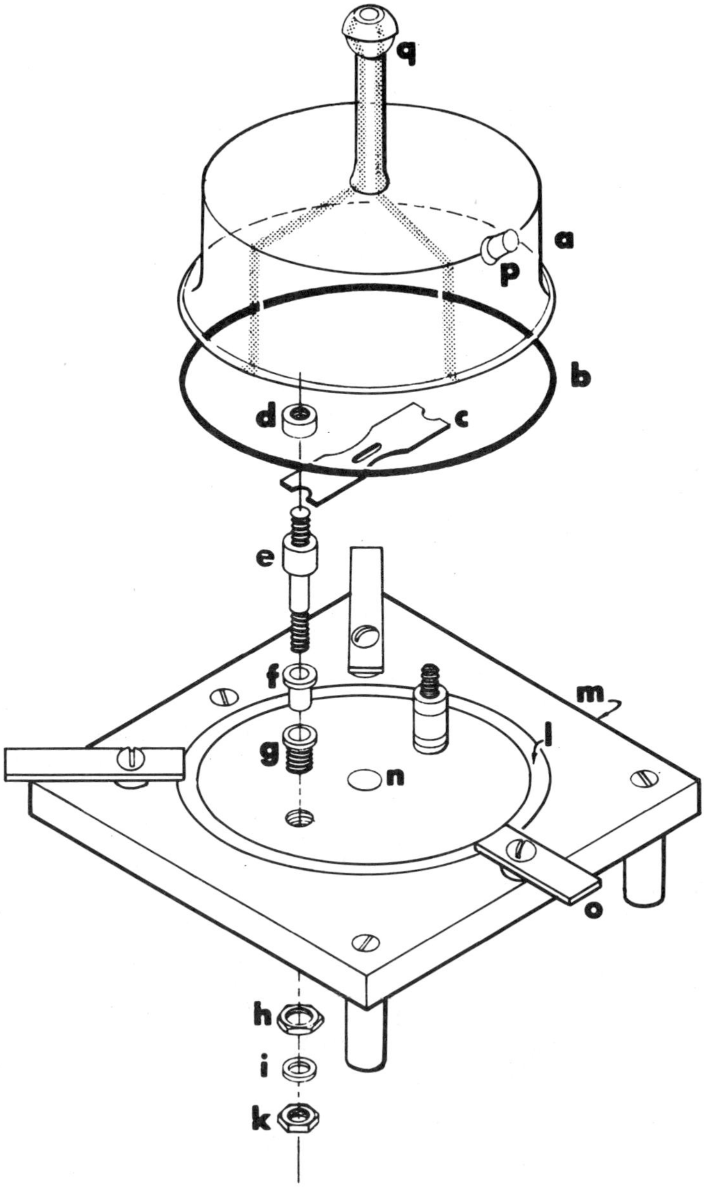

Figure 7.13. Tantalum filament vaporization apparatus: (a) quartz dome, (b) "O" ring, (c) Ta filament, (d–k) Cu post assembly, (l) "O" ring, (m) Al base, (n) Ar gas inlet port, (o) Al sealing tabs, (p) sample injection port, (q) Ar-aerosol flow to torch [35]. [Reprinted with permission from D. A. Nixon, V. A. Fassel, and R. N. Kniseley, "Inductively Coupled Plasma–Optical Emission Analytical Spectroscopy. Tantalum Filament Vaporization of Microliter Samples," *Anal. Chem.*, **46,** 212, (1974). Copyright (1974), American Chemical Society.]

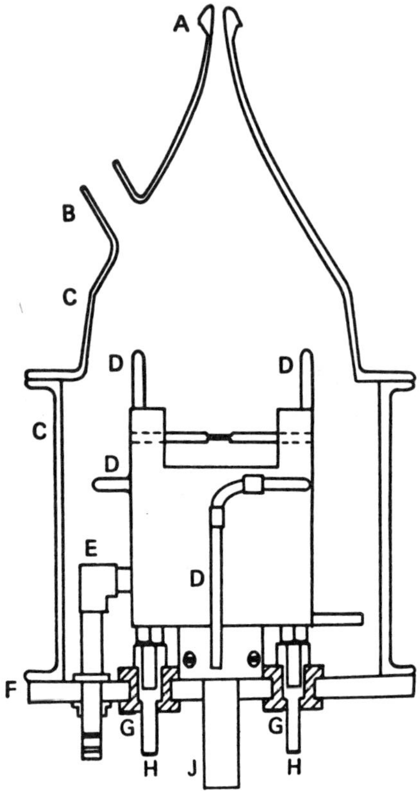

Figure 7.14. Graphite rod vaporization apparatus: (A) ball joint to torch, (B) sample delivery port, (C) cylindrical glass manifold, (D) water cooling line, (E) Ar sample transport gas, (F) circular brass base, (G) insulating blocks, (H) mounting pilar [37]. [Reprinted with permission from A. M. Gunn, D. L. Millard, and G. F. Kirkbright, *The Analyst*, **103,** 1067, (1978). Copyright (1978), The Royal Society of Chemistry, London.]

is shown in Fig. 7.14. A graphite rod is positioned under a glass dome, which is connected at the top to the injector tube of an ICP using 0.5 m of plastic tubing.

The authors point out that, in contrast to use of electrothermal devices for atomic absorption, it is not necessary to produce atoms of the analyte during the heating cycle. It is necessary only that a discrete pulse of analyte be released from the sample and that this be transported to the ICP. The analyte may be in any form (that is, molecular, particulate, and so on). Of course, as in AAS, premature loss of analyte during drying or ashing must be avoided. The appearance time of the analyte signal after the heating pulse was found to be dependent on the length of the connecting tubing. As expected, the longer the tubing the greater the delay in the signal. In addition, as the tube length becomes greater the peak height decreases and the width of the peak increases. Also there is an initial decrease, followed closely by an increase in background intensity caused by the pressure pulse due to heating of the electrothermal device. The tube length must be sufficient to avoid the signal occurring within this period

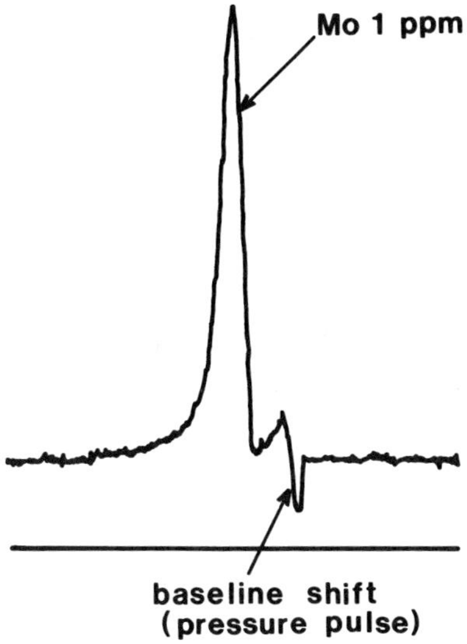

Figure 7.15. Signal obtained during volatilization of Mo from a tungsten filament in an apparatus of the type shown in Fig. 7.14.

of instability. Figure 7.15 presents the signal obtained for the volatilization of Mo from a W filament into an ICP. (This is from recent solid-sample research, now being done cooperatively in Tokyo and Toronto by the present authors.)

Millard et al. [38] studied matrix, interelement, and sample transport effects for electrothermal vaporization (carbon rod) of samples into the ICP. To obtain the optimum signal it is important to have a rapid rate of vaporization from the electrothermal device and then rapid transport of sample into the plasma. Deposition of sample on the atomizer housing and tube walls must be kept to a minimum. In addition, turbulent mixing of aerosol with the gas stream must be minimized.

The authors identify three signal patterns in work with the graphite rod–ICP system:

1. Enhancement of the analyte emission intensity caused by the influence of a concomitant element giving matrix stabilization and thus preventing premature loss of analyte.
2. Variation in emission intensity caused by the effect of concomitants on transport of aerosol from the electrothermal atomizer to the ICP.
3. Depression of analyte emission caused by matrix components decreasing the rate of analyte release.

The elements studied were Cd and As. Aqueous samples were used in the investigation. The presence of Na and Cr(VI) caused a depression in the observed Cd emission, whereas Se(IV) and Se(VI) gave signal enhancements. In the case of As, Na and Se(IV) gave little change in the emission intensity, whereas Cr(VI) and Se(VI) gave enhancements. Se(VI) gave the greatest enhancement for both As and Cd.

The calibration graph for Cd was linear over only 1.5 orders of magnitude of Cd concentration with Cd alone. When 10 μg of Se(VI) was added, linearity over at least four orders of magnitude was obtained. Detection limits were 100 and 1 ng · mL^{-1}, respectively, for these conditions.

Losses of Cd in the atomizer housing and connecting tubing (to ICP) were investigated. In the presence of no concomitants, and for additions of Cr(VI) and Na, approximately 65% of the Cd was lost in the atomizer housing and tubing. When Se(VI) was present (mixed with the sample), only about 10% of the Cd was retained in the atomizer housing and tubing. In the case of As, both Cr(VI) and Se(VI) resulted in good transport of the analyte to the ICP.

This study, while applicable only to the elements Cd and As, is important in outlining problems that may result from other elements by such an approach. When an electrothermal atomizer–ICP approach is used, it is important that similar studies be done for other analytes.

Kirkbright and Snook [39] and Kirkbright [40] employed 0.1% trifluoromethane (Freon 23) in an Ar mixture to improve their procedure when applied to the more refractory elements. This results in the formation of volatile fluorides, which are more readily transported to the ICP. By using Freon 23 these workers were able to avoid the excessive contamination that results from adding fluoride salts as had been done previously. It was necessary to use a compound with the higher bond strength (compared to C—Cl) of the C—F bond to obtain a good detection limit for B. A B detection limit of only 1.0 ng was obtained with a carbon tetrachloride because of prevaporization losses. This is in contrast to a 0.05-ng detection limit with Freon.

The thermal program for the graphite rod was as follows: dry for 45 s at 100 °C and atomized at 2600 °C for 1.6 s. The Freon 23 flow rate through the injector was 0.1 L · min^{-1}. The optimum forward power used was 1.2 kW. The graphite rod sheathing gas flow was connected to the plasma injector using 0.5 m of plastic tubing. Rico and Kirkbright [41] have stated that a long tube is preferable in minimizing the pressure pulse obtained when the atomizer heats. The wavelengths used and detection limits obtained are shown in Table 7.2. Rico and Kirkbright [41] used their electrothermal device to volatilize residues of blood and animal muscle into the ICP, and Ni and Mn were detected. Tetramethyl ammonium hydroxide was used to aid in the pyrolysis of both sample types.

7.4. CONCLUSIONS

As has been indicated previously, there is considerable research being done on the problem of introducing solid samples into the ICP. Detection limits com-

Table 7.2. Wavelengths Used and Detection Limits (2σ) Obtained (Tested on Liquid Samples)

Element	Wavelength (nm)	Argon Only (ng)	Argon and Freon (ng)
Zr	343.8	200	0.01
B	249.7	1.0	0.05
Mo	313.3	1.0	0.10
Cr	357.9	0.5	0.05
W	376.4	5.0	0.06
Ag	328.1	0.01	0.01

pared to conventional methods of nebulizing solutions are often improved by as much as one order of magnitude. This factor is important in trace element analysis because current ICP detection limits are, for the most elements, worse as compared with those obtainable by electrothermal AAS. In addition, errors due to contamination or loss of sample are avoided in direct solid sample analysis. The time taken to analyze each sample is considerably reduced when wet chemical sample preparation is avoided.

Despite a good deal of research into the problem, there is as yet no method that stands out as being superior for direct solid sample introduction into the ICP. In the authors' opinion the two most promising approaches involve direct injection using a crucible (Fig. 7.4) or cup (Fig. 7.3) and the use of an electrothermal atomizer to produce sample aerosol for injection into the sample Ar flow of the plasma. In the case of the latter, a halide-containing material, such as Freon, should be employed to improve the transport of the more refractory elements. One distinct advantage of the electrothermal method is that relatively large sample sizes can be employed. Recent work in the authors' laboratory suggests that up to 0.3 g of sample can be handled. Such sample sizes are essential if problems due to sample inhomogeneity are to be avoided. No matter what method of solid sample introduction is proposed, it is important to do careful studies of the requirements for standardization. Thus methods that show the least matrix dependency must be favored. All proposed methods must be verified through the analysis of standard reference samples, and these materials are now readily available for most sample types.

REFERENCES

1. B. V. L'vov, *Eng. Phys. J. USSR* **2,** 44 (1959).
2. F. J. Langmyhr, *Analyst* **104,** 993 (1976).
3. K. Govindaraju, G. Mevelle, and C. Chouard, *Chem. Geol.* **8,** 131 (1971).
4. B. V. L'vov, *Talanta* **23,** 109 (1976).

5. B. V. L'vov, *Spectrochim. Acta* **33B,** 153, (1978).
6. D. J. Koop, M. D. Silvester, and J. C. Van Loon, *Pittsburgh Conference on Analytical Chemistry and Applied Spectroscopy, Cleveland, Ohio* (1979).
7. T. Kantor, E. Pungor, J. Szatisz, and L. Bezur, *Talanta* **26,** 357 (1979).
8. J. Ip, Y. Tomassen, L. R. P. Butler, and J. C. Van Loon, *Anal. Chim. Acta.* **10,** 1 (1979).
9. H. Ramage, *Nature* **123,** 601 (1929).
10. W. A. Roach, *Nature* **144,** 1047 (1939).
11. F. C. Stewart, and J. A. Harrison, *Ann. Bot.* **3,** 427 (1939).
12. E. Preuss, *Angew. Mineral.* **3,** 8 (1940).
13. J. Noar, *Spectrochim. Acta* **9,** 157 (1957).
14. H. C. Hoare, and R. A. Mostyn, *Anal. Chem.* **39,** 1153 (1967).
15. R. M. Dagnall, D. J. Smith, T. S. West, and S. Greenfield, *Anal. Chim. Acta.* **54,** 397 (1976).
16. T. Yajima, *Plasma Line* **1,** No. 4, Oct. (1980).
17. E. O. Salin, and G. Horlick, *Anal. Chem.* **51,** 2284 (1979).
18. D. Sommer and K. Ohls, *Fresenius Z. Anal. Chem.* **304,** 97 (1980).
19. C. Lerner, Int. Winter Conf. Developments in Atomic Plasma Spectrochemical Analysis, San Juan, Puerto Rico, 7–11 Jan., 1980, Abstract No. 69.
20. C. W. Fuller, R. C. Hutton, and B. Preston, *Analyst* **106,** 1266 (1981).
21. J. F. Wolcott and C. B. Sobel, *Appl. Spectrosc.* **32B,** 591 (1978).
22. T. Kleinman, and V. Svoboda, *Anal. Chem.* **41,** 1029 (1969).
23. H. C. G. Human, R. H. Scott, A. R. Oakes, and C. D. West, *Analyst* **101,** 265 (1976).
24. J. Y. Marks, D. E. Formwalt, and R. E. Yungk, *Spectrochim. Acta* **38B,** 107 (1983).
25. K. Ohls, and D. Sommer, *Fresenius Z. Anal. Chem.* **296,** 241 (1979).
26. R. L. Dahlquist, J. W. Knoll, and R. E. Hoyt. (1975). *Pittsburgh Conference on Analytical Chemistry and Applied Spectroscopy, Cleveland, Ohio*, Abstract No. 49 (1975).
27. R. H. Scott, *Spectrochim. Acta* **33B,** 123 (1978).
28. F. N. Abercrombie, M. D. Silvester, and G. Stoute, *Pittsburgh Conference on Analytical Chemistry and Applied Spectroscopy, Cleveland, Ohio*, Abstract No. 406 (1977).
29. M. Thompson, J. Goulter, and F. Sieper, *Analyst* **106,** 32 (1981).
30. J. W. Carr, and G. Horlick, *Spectrochim. Acta* **37B,** 1 (1982).
31. C. F. Bauer, and D. F. S. Natusch, *Anal. Chem.* **53,** 2020 (1981).
32. J. H. Runnels, and J. H. Gibson, *Anal. Chem.* **39,** 1398 (1967).
33. R. K. Skogerboe, D. L. Dick, D. A. Pavlica, and F. E. Lichte, *Anal. Chem.* **47,** 568 (1975).

34. M. S. Black, and R. F. Browner, *Anal. Chem.* **53,** 249 (1981).

35. D. E. Nixon, V. A. Fassel, and R. N. Kniseley, *Anal. Chem.* **46,** 210 (1974).

36. T. R. Gilbert, and K. J. Hildebrand, *Am. Lab.* **14,** 22 (1982).

37. A. M. Gunn, D. L. Millard, and G. F. Kirkbright, *Analyst* **103,** 1066 (1979).

38. D. L. Millard, H. C. Shan, and G. F. Kirkbright, *Analyst* **105,** 502 (1980).

39. G. F. Kirkbright, and R. D. Snook, *Anal. Chem.* **51,** 1038 (1979).

40. G. F. Kirkbright, R. M. Barnes, ed., in *Developments in Atomic Plasma Spectrochemical Analysis*, Heyden, Philadelphia (1981) p. 223.

41. C. C. Rico, and G. F. Kirkbright, *At. Spectrosc.* **2,** 172 (1981).

CHAPTER 8

FUNDAMENTAL ASPECTS OF AEROSOL GENERATION AND TRANSPORT

RICHARD F. BROWNER

School of Chemistry
Georgia Institute of Technology
Atlanta, Georgia

8.1. Introduction
8.1.1. Figures of Merit for Characterization of Aerosol Properties
8.2. Aerosol Characterization
8.2.1. Definitions of Droplet Diameters and Drop-Size Distributions
8.2.2. Experimental Forms of Distribution Curves
8.3. Experimental Measurement of Particle Size Distributions and Mass Median Diameters
8.3.1. Microscopic Collection Techniques
8.3.2. Collection Techniques Based on Aerodynamic Properties
8.3.3. Optical Scattering Measurements
8.4. Measurement of Mass Transport Rate W and Transport Efficiency ϵ_n
8.4.1. Direct ϵ_n Measurements
8.5. Aerosol Generation
8.5.1. Pneumatic Nebulization
8.5.2. Ultrasonic Nebulization
8.5.3. Secondary Processes of Nebulization
8.6. Factors Modifying Primary and Secondary Aerosols—Gravitational, Impaction, Turbulence, and Centrifugal Loss Processes
8.6.1. Turbulence Loss
8.6.2. Gravitational Aerosol Loss
8.6.3. Centrifugal Loss
8.6.4. Influence of Evaporation on Aerosol Properties

The preparation of this chapter, and part of the work described therein, was supported by the National Science Foundation under Grant No. CHE80-19947.

*Although IUPAC [1] recommends the term efficiency of nebulization for ϵ_n, it has recently been suggested [2] that the important role of components other than the nebulizer should be taken into account in the definition. The term "transport efficiency" has therefore been recommended as a more accurate, and less restrictive, alternative [2].

8.7. Aerosol-Related Interferences
8.7.1. Solution Viscosity Effects
8.7.2. Organic Solvent Effects
8.7.3. Drop Size-Related Phenomena
8.7.4. Aerosol Ionic Redistribution (AIR) Effect
8.7.5. Aerosol Washout from Spray Chambers
8.8. Optimization of Aerosol Characteristics for ICP Analysis
8.8.1. Suggested Additional Sources of Information on Aerosol Generation and Characterization
References

8.1. INTRODUCTION

Solution introduction is by far the most common means of sample introduction in inductively coupled plasma–atomic emission spectrometry (ICP–AES). By and large, the reasons for this popularity parallel those for atomic absorption spectrometry (AAS). From an instrumental viewpoint, these reasons include speed and simplicity of the introduction process, combined with generally excellent measurement accuracy and precision. As with AAS, sample introduction is most readily accomplished when the solution is first converted to an aerosol form before introduction to the plasma. By generating and selecting an aerosol of appropriate drop size, it is possible to ensure that there is rapid solvent loss from the wet aerosol as it enters the plasma. The microscopic salt crystals left after desolvation are also then of appropriate size to vaporize and atomize rapidly in the plasma.

Practical aspects of aerosol production have been considered in Part 1, Chapter 6, and the excitation processes that follow the introduction of aerosol to the plasma are considered in Chapters 9–11. In this chapter, an attempt will be made to discuss certain fundamental aspects of aerosol generation, insofar as they are understood at present, and to indicate the extent to which aerosol properties can influence analytical performance. The list of performance characteristics that are directly influenced by aerosol properties includes all the primary benchmarks of ICP–AES. These are detection limits, interference freedom, and measurement precision. Although precise, quantitative relationships between these variables and aerosol properties are not well understood at present, an attempt will be made in the course of this chapter to provide at least a semiquantitative indication of how they may be interrelated.

8.1.1. Figures of Merit for Characterization of Aerosol Properties

To correlate aerosol properties with analytical performance, it is helpful to have suitable figures of merit which allow the quantitative characterization of critical aerosol properties. Traditionally, the most usual measure of aerosol transport through an atomic spectrometric system has been the transport efficiency, ϵ_n.* Transport efficiency is defined as the ratio of the amount of analyte entering the plasma, to the amount of analyte aspirated [1] expressed as a percent. Experimental techniques suitable for its measurement are discussed in Section 8.4.

Although the transport efficiencies of systems have been widely used as a means of comparing their performance, the value of ϵ_n actually gives relatively little information of direct relevance to analytical spectrometry. For example, probably the most important aerosol transport property is the mass of analyte reaching the plasma per second, W [3]. This will relate directly to the analytical signal, very often in a simple linear manner. Of course, W may be derived from ϵ_n data, provided that the solution aspiration rate, Q_l, is specified, because:

$$W = \epsilon_n Q_l C / 100 \tag{8.1}$$

where C is the analyte concentration. However, W may be measured directly, without the need to calculate ϵ_n, and so in most circumstances it is recommended that the W-parameter is specified for a system, rather than ϵ_n. The use of transport efficiency as a measure of system performance is really only justified when the volume of sample solution to be aspirated is limited, and its efficiency of use becomes important. It must be appreciated, though, when comparing two systems, that the system with the higher transport efficiency may still have the low W value, and so may give rise to the smaller signal from the plasma.

As an aerosol will generally contain a wide range of droplet sizes, it may be important to differentiate between the mass transport properties of different size-range groups. For instance, with very small aerosol droplets, solvent loss and solute vaporization in the plasma may be considered as an essentially instantaneous process. At the other extreme of droplet size, particles may be envisioned that would be sufficiently large, so that there would be inadequate time for efficient atomization to occur during their residence in the plasma.

It has proved convenient, in describing drop-size-related mass transport properties of aerosols, to think directly in terms of the relationship between drop size and analytical signal. In this model, an aerosol containing a wide range of drop sizes is envisioned entering the plasma. Only droplets below a certain size will desolvate and atomize in time to give rise to an appreciable atom or ion population at the height of measurement in the plasma. If the analytical signal is considered as directly proportional to the number of free atoms, or ions, present in the volume of the plasma viewed by the spectrometer, then a simple relationship between drop size and signal can be derived.

If a droplet diameter d_{max} is specified, such that droplets whose diameters exceed this value contribute $< 10\%$ to the analytical signal, then the analyte reaching the plasma can be considered in two distinct groups. The mass contained in droplets of diameter $\leq d_{max}$ can be considered as useful mass (that is, it contributes significantly to the formation of atoms and ions), whereas the mass contained in droplets $> d_{max}$ can be considered as excess mass (that it, it does not contribute usefully to the formation of atoms or ions). From this consideration, a useful-mass-transport rate, W_u, can be defined, such that:

$$W_u = \sum_{0}^{d_{max}} W \tag{8.2}$$

together with an excess-mass-transport rate, W_e, defined by:

$$W_e = \sum_{d_{max}}^{\infty} W \tag{8.3}$$

The total aerosol reaching the plasma is the sum of these two values, hence:

$$W_{tot} = W_u + W_e \tag{8.4}$$

The relationships between the W parameters, in a hypothetical aerosol population, are shown in Fig. 8.1. The ability to determine values for W_{tot}, W_u, and W_e has been facilitated greatly by the recent development of relatively routine and simple aerosol characterization techniques [4–6], which will be discussed in Section 8.3.2.

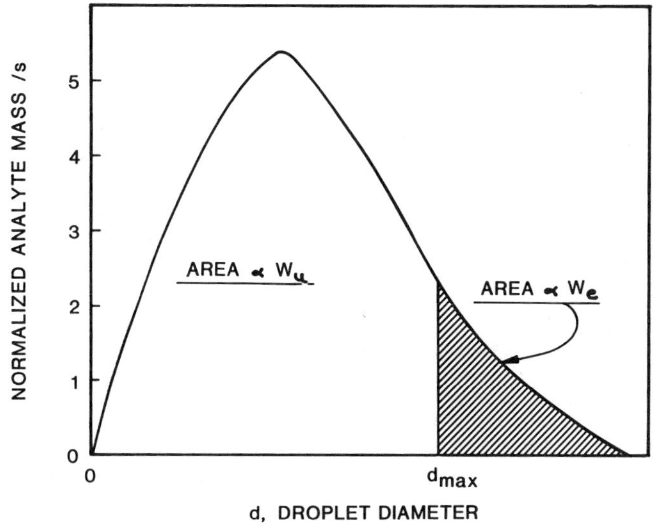

Figure 8.1. Relationship between drop size distribution of aerosol and the mass transport terms W_{tot}, W_u, and W_e. Distribution curve represents complete tertiary aerosol reaching plasma, after passage through spray chamber and torch. Area under curve from $d = 0$ to $d = d_{max} \alpha W_u$; area under curve to right of $d = d_{max} \alpha W_e$; total area under curve αW_{tot}.

An additional transport parameter of extreme importance to the performance of the inductively coupled plasma is the solvent transport. Both aqueous and organic solvents are well known to have a major impact on the excitation properties and the temperature of the ICP [7–9]. Consequently, the rate at which solvent, whether in aerosol or vapor form, reaches the plasma is of crucial importance in determining the excitation properties of the plasma. In extreme cases, it is possible to extinguish the plasma completely by the introduction of certain solvents. For example, high-volatility organic solvents, such as hexane and methanol, are only tolerable in low-power Ar ICPs at low transport rates, and at normal introduction rates may actually quench the plasma [8]. Clearly, therefore, a complete understanding of aerosol sample introduction into the plasma must include a consideration of the influence of the solvent on the plasma characteristics, in addition to the influence of the analyte contained in that solvent. The role of solvent vapor and aerosol on plasma excitation properties is one of the more interesting aspects of aerosol introduction to the inductively coupled plasma, and will be considered in Section 8.6.4.

8.2. AEROSOL CHARACTERIZATION

8.2.1. Definitions of Droplet Diameters and Drop-Size Distributions

There is, unfortunately, no way to define all the relevant properties of an aerosol concisely. Even a parameter, as apparently simple and unequivocal as a droplet diameter, is open to a wide variety of interpretations, depending on the property used actually to characterize the aerosol. For example, if aerosol particles are collected on a microscope slide, the *numbers* of particles of each particular size counted, and the count plotted as a function of particle size, then a distribution curve will be plotted which represents a numerical count distribution. If, by contrast, the *mass* of particles in each particular size range collected is measured directly, then the distribution curve is a mass distribution curve, and has a different form from the numerical count distribution curve. More importantly, if a median droplet diameter is calculated from the two sets of data, then, in general, for aerosols such as those used in ICP–AES, the numerical median diameter d_n will have a smaller value than the mass median diameter d_m, because in the case of the numerical median diameter each particle, however small, is given the same weight in the count. The differences between the two representations of the same aerosol population are illustrated in Fig. 8.2.

The *W*-parameters, which relate directly to the analytical signal, are mass transport terms. Consequently, for analytical purposes, the mass median diameter d_m is most appropriate to describe droplet distributions produced by nebulizers. When aerosol mass collection techniques (such as the cascade impactor described below) are used for characterization purposes, the mass median

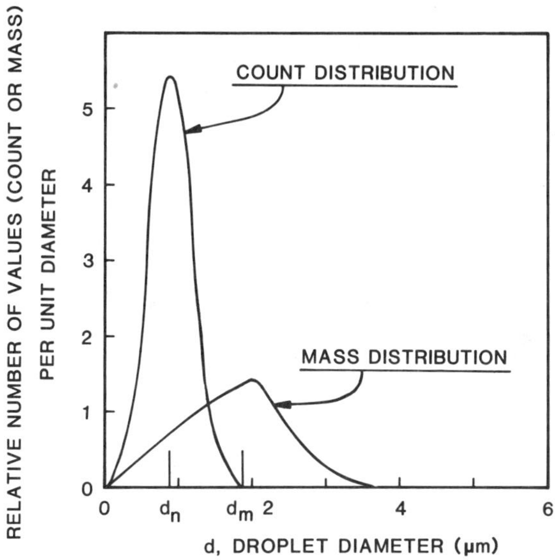

Figure 8.2. Mass and numerical count distributions for typical ICP tertiary aerosol population. d_n = Count mean diameter; d_m = mass median diameter.

diameter of the collected aerosol can be readily determined by plotting cumulative mass percent collected versus particle diameter. The particle diameter, corresponding to a cumulative 50% of collected aerosol mass, is the mass median diameter. A further commonly encountered description of particle diameter is the Sauter mean diameter. The Sauter mean diameter is described by the following equation:

$$d_s = \Sigma d^3 \Delta_n / \Sigma d^2 \Delta_n \tag{8.5}$$

where Δ_n is the number of droplets of diameter d. It can be seen that the Sauter mean diameter is, in fact, a volume-to-surface-area ratio diameter. Statistically, it is also a mean rather than a median value.

Experimentally, however, it has been shown that the value of the Sauter mean diameter and the mass median diameter are very close, at least for pneumatically generated aerosols. Consequently, for practical purposes the two can be taken to be essentially identical. However, both the Sauter mean diameter and the mass median diameter vary considerably, in most instances, from the numerical or count mean diameter d_n, which is defined by:

$$d_n = \Sigma d \cdot \Delta_n / \Sigma \Delta_n \tag{8.6}$$

As an example, for a specific aerosol generated by an AAS nebulizer, values of the count mean diameter, the Sauter mean diameter, and the mass median diameter were found to be, respectively, 0.83, 2.1, and 2.0 μm [3]. It is important to note that experimentally the diameter of an aerosol droplet is rarely the usually perceived geometrical diameter of an ideal, spherical particle. Usually the diameter measured experimentally is some property that follows from the initial property of the aerosol droplet, but from which the original aerosol droplet diameter must be interpolated. The precise means for estimating the particle diameter, and the interpretation to be placed upon that value, are a function of the experimental means used to make the particle diameter determination. For example, the property may be an aerodynamic property, an electrical mobility property or a light-scattering property related to particle size. Only direct microscopic measurement can lead, in principle, to an idealized geometrical diameter. However, as will be shown later, microscopic measurements of small ($<5\ \mu$m) liquid droplets are generally an unreliable means of aerosol characterization.

8.2.2. Experimental Forms of Distribution Curves

A particle size distribution is a mathematical means of describing the variation of some property of the aerosol, such as mass, as a function of particle size. The primary reason to use such a function is that aerosols typically contain a wide range of particle sizes, and so cannot be defined adequately simply by a median diameter. It is particularly helpful to use a particle size distribution curve as a means of predicting properties of aerosols under a variety of experimental conditions. For example, it is in principle possible to predict the influence on transport efficiency of variations in nebulizer operating conditions, spray chamber operating conditions, and so forth, without the need to actually conduct an experimental measurement of the particle size [3].

There have been many attempts made to describe the form of the distribution curves found experimentally for natural aerosols. Equations that have been derived empirically often seem most appropriate to describe the aerosols produced by different specific systems. The distribution that seems most appropriate for pneumatically generated aerosols is the log normal distribution, plotted as a function of mass. Other distribution curves, such as those based on the Rosin-Rammler equation [10], which uses a logarithmic volume distribution, have also been applied to other types of nebulizers.

To consider a log normal distribution, it is convenient first to examine a normal distribution. The normal distribution curve can be considered as follows. Consider an essentially infinite population of particles, each of which possesses a certain characteristic in an amount that can be assigned a numerical value x, which is not the same for all individuals. If the x values are normally distributed among the population, then the relative frequency with which values occur in the interval $x + dx/2$ is given by the equation:

$$f(x)\,dx = \frac{1}{\sigma(2\pi)^{1/2}} \exp\left[-(x-\mu)^2/2\sigma^2\right] dx \tag{8.7}$$

Here σ and μ are, respectively, the standard deviation and mean of the distribution and $f(x)$ is the normal probability density function. The mean μ and the variance (the square of the standard deviation) σ^2 are given by:

$$\mu = \int_{-\infty}^{\infty} x f(x)\, dx \tag{8.8}$$

and

$$\sigma^2 = \int_{-\infty}^{\infty} (x - \mu)^2 f(x)\, dx \tag{8.9}$$

Here, μ and σ are used to identify the parameters of the distribution of an infinite population, in other words, they represent the true mean and standard deviation. Once these parameters are known, the distribution is completely defined. This equation takes the form of a bell-shaped curve, centered over $x = \mu$.

If the quantity $x = \ln d$ is normally distributed, then the distribution of d is said to be lognormal. Figure 8.3 shows typical plots of normal and lognormal distributions.

For a lognormal distribution, the number of particles having logarithms falling in the interval $x + dx/2$ is given by Eq. (8.7), with $\mu = \ln d_m$, where d_m is the median diameter of the population. It is convenient, if a distribution curve follows a lognormal form, to describe it in terms of the median diameter and

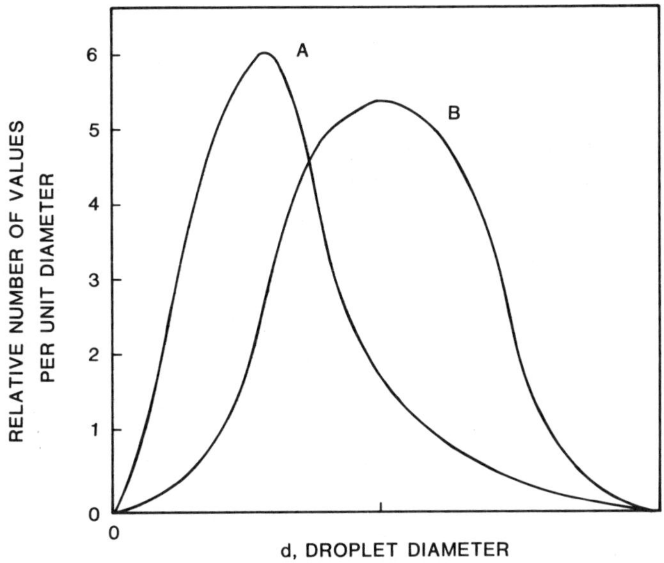

Figure 8.3. Normal and lognormal distributions. A, lognormal; B, normal. The lognormal distribution has a skewed appearance when plotted on a linear scale, as in the figure. If plotted with a logarithmic diameter scale, the lognormal distribution would appear as a symmetric bell shape.

the geometric standard deviation s_g. The value of s_g may be derived from experimental distribution curves, and is a measure of the breadth, or polydispersity, of the drop size distribution. Consequently, large s_g values correspond to highly polydisperse aerosols. The value of s_g may be estimated for a particular aerosol by determining the ratio of the drop diameters corresponding to cumulative mass percentages of 84.13 and 50.0, respectively [11]. However, it should be noted that the use of both d_m and s_g to describe aerosols with distributions other than approximately lognormal can give deceptive results. For example, it is possible for a distribution to be bimodal [4, 5], in which case the mass median diameter of the aerosol could correspond to a value significantly different from either of the two maxima on the curve. Equally, the calculation of s_g, using the method described above, will not yield accurate values for bimodal distributions.

8.3. EXPERIMENTAL MEASUREMENT OF PARTICLE SIZE DISTRIBUTIONS AND MASS MEDIAN DIAMETERS

As discussed in Section 8.2.1, the property most usually envisioned by the term particle diameter, namely, the geometric diameter of a uniform sphere, is rarely capable of direct measurement with aerosols. Even with liquid aerosols, for example, the droplet diameter, as viewed with a microscope, must be corrected to allow for the distortion that inevitably occurs on a microscope slide. The situation is even more complicated with irregular solid particles, which have no "diameter." To describe these, a shape factor must be introduced and the term "equivalent diameter" is often used in place of geometric diameter. The equivalent diameter is defined as the diameter of a sphere that has the same volume as the considered particle. Because of many uncertainties in correlating measured aerosol properties with the precise geometry of aerosol particles, the many terms that are used to describe particle diameters often relate only to a particular measurement system. It can be misleading, therefore, to make comparisons between particle diameters measured with different instruments, unless the meaning of the measured "diameter" is completely understood. The same aerosol, measured with different instruments, will in fact often give different numerical values for particle size. The situation with aerosols formed from dilute aqueous solutions is, fortunately, somewhat simpler. Here, the aerosol consists of droplets of unit density and unit refractive index, which are effectively spherical in shape for most of their existence. Among the many techniques that have been used for particle sizing of aqueous aerosols are the following: 1 microscopic measurement, following collection on glass slides, coated with silicone or other oil film; 2 microscopic measurement after collection on magnesium oxide-coated slides, this time measuring the craters formed by collision of the droplets with the magnesium oxide film; 3 measurements based upon the aerodynamic properties of the droplets, such as gravitational settling or cascade impaction, and 4 particle-scattering techniques based on Mie or Fraunhofer scattering theory.

8.3.1. Microscopic Collection Techniques

The collection of aqueous aerosols on microscope slides was the approach used by Nukiyama and Tanasawa [12] and has since been used by many other groups [13, 14]. In principle, this is the simplest technique. In practice, however, there are a number of experimental difficulties involved with such a technique that are hard to overcome. First of these is the difficulty of collecting small aerosol particles efficiently. To provide uniform collection of particles on the microscope slide, it is necessary that the aerosol should impact on the surface of the slide in an unbiased manner. In other words, there should be no preferential sampling of particles of any particular diameter. In practice it is difficult to achieve efficient sampling of very small aerosol particles, such as those of 5 μm or less, unless very high collection gas velocities are used. For example, to collect efficiently particles of approximately 0.5-μm diameter, gas jet velocities of up to 40 m $\cdot$ s^{-1} are necessary. If the aerosol spray does not travel at such a velocity, small particles will pass around the collection plate and be lost to the measurement. Consequently, the aerosol sample obtained in this manner will have a greater proportion of large droplets than the original aerosol. Impaction processes are generally described in terms of a Stokes number, which is the measure of a particle's ability to follow a gas stream as it flows around an object or surface. The Stokes number, Stk, is given in general by:

$$\mathrm{Stk} = \rho_a K_s d_a^2 U / 9 \eta_g w \tag{8.10}$$

where ρ_a is the aerosol particle density (g $\cdot$ cm^{-3}), d_a the particle diameter (cm), U the gas stream velocity (cm $\cdot$ s^{-1}), η_g the gas viscosity (P), and w the diameter of the tube through which the gas is flowing (cm). K_s is a slip factor that is close to unity for spherical particles >1 μm in diameter, but increases rapidly as the particle diameter approaches the mean free path of the gas molecules. The principle of collection with a microscope slide, coated with a thin film of water-immiscible solvent, is that the droplet should initially penetrate below the surface of the film. However, the densities of the droplet and the film should be sufficiently close, with the density of the droplets slightly exceeding that of the film, so that the droplets rise to just below the surface of the film and remain there for microscopic measurement. The purpose of trapping droplets below the surface of the film is to minimize evaporation. However, it has been shown that small aqueous droplets (2 μm diameter or less) can still evaporate quite rapidly under these conditions. Measurement within a few minutes of collection is therefore essential [15]. There is also, due to the density difference between the two liquid phases, slight distortion of the droplet shape. As the droplet is no longer truly spherical, a correction factor must be applied to the measured diameter. Aerosols measured in this manner are generally sized by counting numbers of particles falling within certain size ranges. This may either be accomplished by the tedious procedure of manual counting with a microscope grid, or alternatively by using more sophisticated computerized counting techniques, based on Vidicon computer-coupled devices. Such techniques, as indicated earlier, will give rise to a numerical count distribution. For this to be useful analytically, it must first be converted to a mass or volume distribution, because a count distribution is heavily weighted toward small particles and is of little use in atomic spectroscopy. The conversion from count to mass or volume can indeed be accomplished, but for good accuracy it is essential that the sampling statistics involve a very large number of particles. A minimum of several thousand particles has been suggested as necessary [16]. Another factor that must be considered when using microscope collection techniques is the high aerosol densities that are often encountered. For example, if one tries to collect directly from a nebulizer, then the plate is rapidly overloaded. Therefore, it is necessary to use a shutter, or other intermittent device, such that aerosol is only sampled for a very small time interval. Overloading of the plate will cause the droplets to coalesce to give multiplets, and so give misleading particle size information.

If the slide is coated with a fine powder of magnesium oxide, instead of a liquid film, then droplets may be sized without direct measurement of the droplet itself. An estimate of the colliding droplet diameter is made by inspection of the crater, left after impaction with the magnesium oxide film. In this case, subsequent evaporation of the droplets becomes unimportant. However, the crater left by the droplet is larger than the diameter of the droplet itself [17], and an empirical correction factor, often quoted as approximately 0.86, must be applied to the droplet diameter measured in this way. The counting approach is similar to that used for liquid aerosols collected on a liquid-coated slide. Microscopic techniques involving collection of dried salt crystals by thermal [18, 19] or electrostatic [20] precipitation have also been widely used.

8.3.2. Collection Techniques Based on Aerodynamic Properties

A number of experimental systems are based on measurement of the aerodynamic properties of aerosols. In this instance, the aerodynamic diameter is measured. The term "aerodynamic diameter" relates to a reference aerosol, composed of spherical particles of unit density. As most aqueous aerosols are indeed composed of spherical particles of unit density, these two terms become experimentally equivalent, which is very convenient for practical measurement purposes. If not, then aerodynamic shape and density correction factors must be introduced. The aerodynamic shape factors of a number of regular solids have been determined experimentally [21], and generally fall within the range 1.0 ±

0.1. This correction would normally be insignificant. The density correction factor is obtained simply by dividing the measured diameter by (particle density)$^{1/2}$. The square root relationship applies because the impaction process balances centrifugal forces [proportional to (particle density)$^{1/2}$] against inertial (Stokes) forces [proportional to particle density]. Over the density range likely to be encountered analytically (for example, from low-density solvents, with $\rho_a \approx 0.7 \text{ g} \cdot \text{cm}^{-3}$, to high-density crystalline solids, with $\rho_a \approx 5 \text{ g} \cdot \text{cm}^{-3}$), the density correction factor will cover the approximate range of 0.8 to 2.2, which can result in an important shift in calculated particle size.

The two most common instruments, based upon aerodynamic properties of aerosols, are 1 horizontal elutriators, in which the gravitational settling of aerosols from a gas stream is used to measure their particle size and 2 cascade impactors, in which an aerosol-laden gas stream is taken successively through stages containing orifices of progressively smaller diameter, such that the average gas velocity constantly increases through the instrument. Of the two types, the cascade impactor has been used most widely.

In a cascade impactor, particles of different size ranges impact in a succession of stages on collection plates placed in the gas stream. This leads to collection of particles over several size ranges, the number of which depends upon the particular design of the cascade impactor. A common design consists of eight stages and then a final absolute filter, which collects particles less than the minimum collection diameter of the bottom stage of the impactor. The collection efficiency of particles of a particular size on the appropriate stage is never 100%. The collection efficiency, for example, on a four-stage impactor is shown in Fig. 8.4. The design of a typical commercial cascade impactor, the Andersen impactor, is shown in Fig. 8.5. Aerosol is collected in eight size ranges, as indicated, plus a final filter for particles less than the bottom range of the final plate, that is, $<0.32\ \mu\text{m}$. Nominal gas impaction velocities, and collection ranges, are shown in Table 8.1. When high aerosol densities of particles greater than 10-μm diameter enter the cascade impactor, there is a distinct possibility of distorted collection in the early stages. For this reason a preimpactor should be used, which ensures a more unbiased collection in the early impactor stages, that is, at the large particle size range of the impactor.

With the cascade impactor, a series of collection plates catches particles in particular size ranges. If the liquid aerosol is very involatile, such as dioctyl phthalate (b p, 384 °C), then the particle size distribution can be obtained directly by weighing the aerosol mass collected on each plate. When more volatile solvents are used, such as aqueous or low-boiling organic solvents, there is always substantial solvent evaporation from the aerosol collected on the plates. This occurs because, generally, the gas flow rate necessary to operate the cascade impactor exceeds the flow rate of the collected gas/aerosol mixture substantially. A typical cascade impactor will operate at $28.4 \text{ L} \cdot \text{min}^{-1}$, compared with the $1 \text{ L} \cdot \text{min}^{-1}$ output of most ICP nebulizers. The additional air sucked

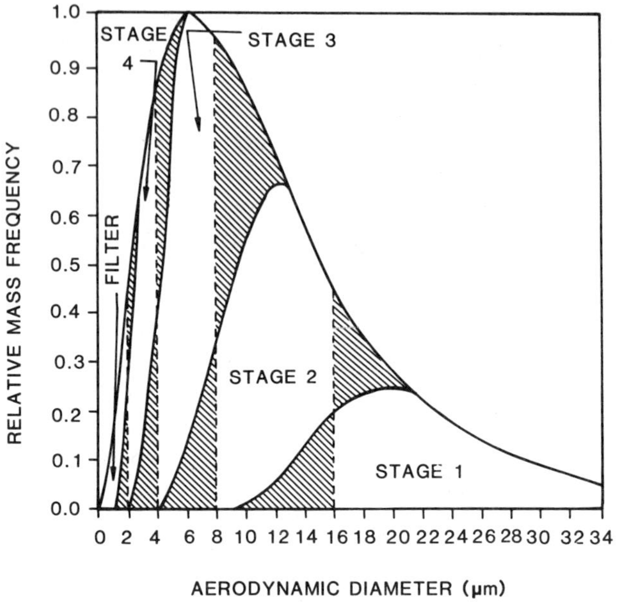

Figure 8.4. Segregation of a mass distribution among the stages of a four-stage cascade impactor. Shaded areas show overlap between successive stages of the impactor.

in by the impactor is generally well below 100% relative humidity, and thus it dries the aerosol collected on the plates. If a pure, aqueous aerosol is collected from a typical ICP system, for example, then when the impactor is opened, the plates will be seen to have no accumulation of water on them at all.

The additional air sucked into the impactor can also cause significant solvent evaporation from the aerosol, as it passes down the various collection stages of the impactor. The result will be a shift of the size distribution of the aerosol passing through the impactor to smaller droplet sizes. This process represents a real shift in the aerosol particle size distribution, and will cause the droplets to be collected on smaller-size-range plates than they would have in the absence of evaporation. As a consequence, the collected aerosol no longer has the same size distribution as the sampled aerosol, and so a significant bias in the measurement toward smaller drop sizes is observed.

It is, fortunately, a simple matter to reduce aerosol evaporation in the impactor to negligible levels. The approach used is to nebulize an aqueous solution containing a relatively high concentration of a dissolved salt. Concentrations in the range of 5 to 10,000 $\mu g \cdot mL^{-1}$ have been found to work well [6]. At concentrations around 200 $\mu g \cdot mL^{-1}$, the shift in apparent drop size distribution resulting from aerosol evaporation is quite pronounced, and can be seen in Fig. 8.6. For this reason, dilute solutions should be avoided.

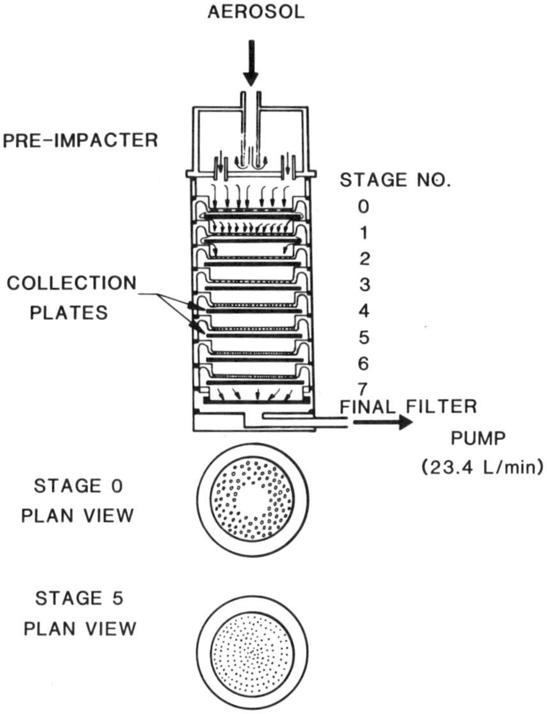

Figure 8.5. Schematic diagram of eight-stage Andersen cascade impactor.

Table 8.1. Collection Ranges and Gas Impaction Velocities of Eight-Stage Cascade Impactor

Stage	Collection Range (μm)	Impaction Velocity (m · s^{-1})
Preimpactor	>10.0	
0	10.0–9.0	0.963
1	9.0–5.8	1.752
2	5.8–4.7	1.814
3	4.7–3.3	2.980
4	3.3–2.1	5.348
5	2.1–1.1	12.996
6	1.1–0.7	24.038
7	0.7–0.32	47.840
Final filter	<0.32	—

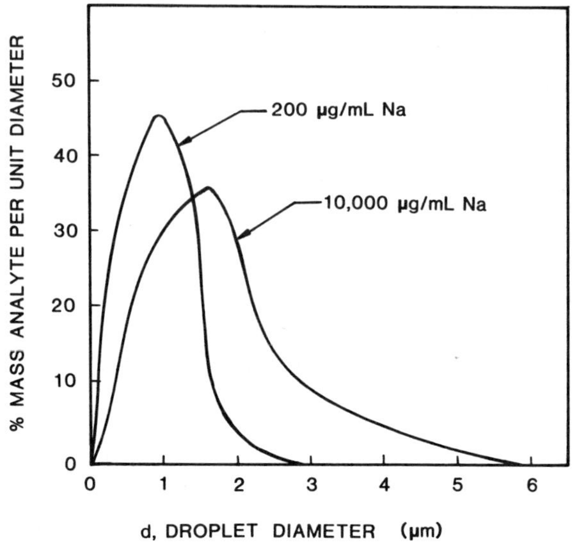

Figure 8.6. Influence of analyte concentration on apparent aerosol drop-size distribution measured with cascade impactor.

Even though *aerosol* evaporation can be reduced to unimportant levels with this approach, evaporation from the plates can still be substantial, and so plate weighing is inappropriate for measuring distributions. The most reliable means to measure the drop size distribution, when collecting aqueous aerosols, is to wash the material collected on each plate into a volumetric flask, and determine the concentration of analyte present. This is conveniently accomplished using AA or ICP–AES. The concentration determined can then be used to calculate the mass of analyte originally present on each plate.

To plot the data in a convenient form for interpretation and comparison, it is necessary to allow for the fact that the size range collected on each plate is unequal. One means of accomplishing this is to simply plot the mass collected on each plate versus drop size, as an unequal size interval histogram. Alternatively, it has been found quite useful to normalize the mass collected on each plate, by dividing this mass by the size range Δd of collection of the plate in micrometers (See Table 8.1). This then allows the plotting of a smooth curve of $W/\Delta d$ versus droplet diameter [3]. The area under this curve is now approximately proportional to the total mass of aerosol collected. Consequently, collection runs made under different conditions can be readily compared, to provide a rapid visual indication of the total mass of aerosol generated under each set of conditions. If the duration of each collection run is measured, this can then be used to normalize further the mass collected, to give a mass / second collected per unit size range. The ordinate axis of the plot now has the dimensions of micrograms of analyte per micrometer · second, and the area under the graph gives the total mass of analyte reaching the ICP per second. This very specific information can be extremely useful in attempting to correlate plasma properties with aerosol introduction properties.

It is important, in any aerosol characterization, that the aerosol actually used for particle sizing is representative of the entire stream. The simplest way to satisfy this condition is to base the measurement on the properties of the entire stream. With cascade impaction techniques used in atomic spectrometry, this is readily achieved by sampling the entire aerosol stream. Unbiased collection is assured by arranging that the gas flow through the impactor exceeds that through the spray chamber by a substantial amount. This usually presents little problem, as typical ICP nebulizers run on ≤ 1.5 L · min^{-1} gas flow and an Andersen impactor draws 28.4 L · min^{-1}. However, sampling problems can still be encountered if the aerosol stream velocity exceeds that of the gas flow into the impactor inlet. This can be true even if the net flow through the impactor is considerably greater than that through the ICP spray chamber. For example, unbiased sampling of the aerosol stream issuing directly from an ICP nebulizer is extremely difficult to achieve, because of complex flow patterns set up in the sampling volume.

If only a portion of the aerosol stream is to be sampled, then isokinetic sampling must be used. This is accomplished by inserting a probe into the aerosol stream, and arranging, by the choice of appropriate probe dimensions, that the gas velocity in the probe exactly matches that in the aerosol stream. If the probe gas velocity exceeds the aerosol stream velocity, collection will be biased to larger particles, and the measured distribution will be correspondingly skewed. When the probe gas velocity is lower than the aerosol stream velocity, the reverse situation applies.

Cascade impaction is a very well-established approach for aerosol characterization studies, and so the theory is well developed. The recent development of techniques suitable for sampling aqueous aerosols, as described previously, further broadens its scope to aerosols of relevance to ICP–AES. Furthermore, cascade impactors are relatively inexpensive, are simple to operate and provide a wealth of information on aerosol properties [2–6, 22, 23]. In consequence, they probably represent the most powerful single tool for analytical aerosol studies currently available. The negative aspects of cascade impactors are that they are necessarily intrusive to the aerosol stream, and, because collection runs take at least a minute or so, are incapable of giving time-resolved information.

8.3.3. Optical Scattering Measurements

By contrast to microscopic and impaction techniques, aerosol characterization by optical scattering measurements, using either Mie or Fraunhofer theory, generally involves extremely expensive and sophisticated equipment. A number of instruments based on Mie scattering theory [24, 25] are available and provide particle size distributions typically in the range of 0.1 to 10 μm. Such a commercial instrument is the Royco particle counter [26]. The Leeds and Northrup Microtrac system is based on Fraunhofer diffraction theory [27, 28]. All optical systems have a great advantage of aerosol measurements using aerodynamic techniques in that they are nonintrusive.

Laser Fraunhofer scattering systems have been used for measuring aerosols from atomic spectrometric systems [29], but unfortunately they only measure in the range of 1.9 to 176 μm. As it has been shown that the mass median diameter of ICP aerosols is typically less than 2 μm [3, 30], this would introduce considerable bias into thc measurement technique. It means, for instance, that the system would not respond to over half of the total aerosol mass. A recent version of the Leeds and Northrup instrument, which measures 90° scattering, instead of the usual small forward angle scattering, is capable of measuring solid suspensions in the range of 0.1 to 20 μm. No applications of this system to aerosols have yet been described, but it would appear to be very suitable for this purpose. Both instruments provide histogram plots of volume % versus particle size, which can be readily corrected, for aqueous aerosols to mass % versus particle size. Mie scattering systems have also been used for characterization of aerosols from ICP spray chambers. However, the Mie scattering systems suitable for use with ICP aerosols do not give particle size distributions, but instead provide only mean particle diameters. The particle diameter measured in this way is the Sauter mean or volume-to-surface-area diameter. The range of this technique is again somewhat limited, but does extend to lower values than the forward angle Fraunhofer system. The range is approximately 0.2–4 μm. This system is based upon the measurement of forward scattering at small angles, such as 5° and 15° [25, 31]. The ratio of these forward scatter angle signals is used according to Mie theory to determine the mean particle size of the aerosol distribution. The limits of the measurement are a consequence of the following: 1 at low particle sizes the ratio of the 5° and 15° beams becomes very close to unity, because the relative magnitude of the scattered signals at those angles becomes similar and 2 at large particle sizes the ratio of the forward scattered angle signals become ambiguous and does not pertain to a unique particle size distribution.

All laser scattering systems have the inherent advantage that they are nonintrusive to the aerosol. However, they have the disadvantage that they have restricted size ranges. Problems can also arise if the aerosol to be measured has a substantial fraction of its mass that falls outside the measurement range of the instrument. In this case, the particle diameter indicated by the instrument may be skewed by the fact that the instrument is not responding to the fraction of the aerosol that falls outside its range. For example, if significant evaporation of an aerosol should take place, such that it should fall below the low particle size range possible with Mie scattering, then this small particle fraction would be ignored by the system, and a measurement biased toward large particles would be obtained.

The measurement of aerosol properties of organic solvent provides a unique difficulty in that evaporation rates can be considerable [8, 32]. Consequently, collection techniques are simply not applicable. Measurements by laser scattering are feasible, although none have yet been reported. In this instance, correction for refractive index would be necessary.

8.4. MEASUREMENT OF MASS TRANSPORT RATE *W* AND TRANSPORT EFFICIENCY ϵ_n

Any assessment of the performance of a nebulizer–spray chamber combination has two facets, as mentioned previously. One of these, the drop size distribution, give important information on the sizes of the droplets entering the plasma, prior to evaporation, decomposition, and atomization. The other, the mass transport term W, describes the mass of analyte reaching the plasma per second as discussed in Section 8.1.1.

Consequently, to measure the amount of aerosol that actually reaches the plasma, it is necessary to make experimental determination for the particular combination of nebulizer–spray chamber and plasma torch, and also for the precise operating conditions selected. All of these may have quite a dramatic effect on the amount of analyte actually reaching the plasma per second. Historically, the parameter used to monitor the transport properties of the system has been the efficiency of nebulization, or transport efficiency ϵ_n.

Methods that have been used to determine transport efficiency fall into two basic categories, namely: 1 direct methods, in which the aerosol itself is collected as it leaves the plasma torch, and 2 indirect methods, in which the liquid going to waste from the system is measured, and the analyte mass passing to the plasma is calculated by difference.

Most published studies have relied on indirect methods to measure ϵ_n. The usual approach is to nebulize a solution of known concentration through the spray chamber, and to collect the solution passing to waste [19, 33]. This is then combined with the washings from the spray chamber, in a volumetric flask, and the concentration of this solution is determined. Both titrimetric acid determinations and AA metal determinations have been used for this purpose. The mass of analyte which does not leave the spray chamber can now be calculated, and by difference from the mass of analyte that was nebulized during the run, the mass assumed to be reaching the plasma can be calculated. If the run is timed, this will also give a value for W. Making a ratio of this mass to the mass nebulized gives the transport efficiency, ϵ_n.

Table 8.2 Transport Efficiency Data for ICP Nebulizer: Comparison of Direct and Indirect Procedures [3]

Method	Transport Efficiency, ϵ_n (%)	RSD (%)
Direct		
Filter	1.4 ± 0.1	6
Cascade impactor	1.4 ± 0.1	4
Silica gel	5.3 ± 0.3	5
Indirect		
Waste Collection[a]	6.9 ± 2	30

System: Concentric all-glass nebulizer, dual concentric spray chamber. Neublizer conditions: 1.0-mL · min^{-1} aspiration rate of 100 μg · mL^{-1} Mn; 1.0-L · min^{-1} gas flow.

[a] Spray chamber washed twice with dilute HNO_3, once with deionized water. Mn in waste determined by AAS.

In view of the simplicity of this approach, it is unfortunate that for ICP nebulizers and spray chambers indirect measurements of W and ϵ_n are generally highly unreliable [2]. This results from the poor efficiency with which solution can be washed from the spray chamber. It seems that 2% recovery loss is not uncommon, in spite of multiple washings [2]. The arithmetic of the calculation converts this small recovery loss into a 400% positive error, as shown in Table 8.2.

Occasionally, the volume of solvent passing to waste has been used to estimate ϵ_n [34]. Unfortunately, this approach can lead to even worse errors than when the previous indirect procedure is used. This is a consequence of the inevitably uneven drainage of typical ICP spray chambers. It has been determined that experimental runs of even 1 h long, under normal circumstances, can give rise to highly nonreproducible data [2]. The tendency with this approach is, once more, to give a strong positive bias to the ϵ_n values.

A final limitation of all indirect procedures for W and ϵ_n, is that aerosol loss on torch surfaces is quite difficult to account for, and measurement precision is sufficiently poor (Table 8.2) that this loss process might be undetected.

8.4.1. Direct ϵ_n Measurement

As a consequence of the limitations of indirect procedures for transport efficiency, it has been suggested that direct measurement procedures be used [2]. These resolve themselves into three categories. The first involves collection on a cascade impactor, in precisely the manner used for measuring droplet size distributions. However, in this instance, it is not necessary to determine the mass of analyte on each of the cascade impactor plates but only to determine the total mass collected. Alternatively, a simpler collection procedure has been shown to be very effective, in which the aerosol leaving the plasma torch is collected on a 45-mm diameter, 0.3-μm pore size, fiberglass filter. By spraying a salt solution of known concentration, and collecting for a known time, it is possible to determine the mass of analyte per second, (W), collected on the filter. The filter is washed with dilute acid solution, and the concentration of analyte in the solution determined. To calculate the transport efficiency, it is now only necessary to determine the mass of analyte aspirated during the run, which can be done most accurately by weight difference. Solutions of any convenient concentration can be used, but typically 100 $\mu m \cdot mL^{-1}$ is a suitable value. Comparisons of transport efficiency measurements, using cascade impaction or filter collection procedures, have shown no significant bias between the two [2]. Both procedures have good accuracy and acceptable precision with relative standard deviations typically 5% or less.

Another approach to direct ϵ_n measurement involves silica gel collection. In this procedure, aerosol issuing from the plasma torch is passed directly into two U-tubes, packed with silica gel and arranged in series. The silica gel traps the water component of the aerosol with high efficiency, so weighing the U tubes before and after the run should give the mass of aerosol reaching the plasma.

It has been suggested that this, therefore, provides a very simple means of determining the mass of analyte transported to the plasma, without the need to determine directly analyte itself [35]. However, it has also been shown that, in fact, silica gel collection procedures do not measure *analyte* transport efficiency, but rather *solvent* transport efficiency [36]. Silica gel will collect both aerosol and water vapor passing to the traps. If the vapor reaching the traps originates only from evaporation of the original aerosol, which would have reached the traps in any case, then this would provide no bias to the measurement. However, it has been shown that with typical ICP nebulizer–spray chambers, the gas issuing from the spray chamber is at approximately 100% relative humidity [36]. The vapor contained in the gas, however, does not apparently originate solely from the aerosol that passes through the spray chamber, but originates to a considerable degree from other sources. Suggested sources include the evaporation of droplets that would otherwise pass to drain [36], and water on the surface of the spray chamber [37]. Consequently, this procedure measures additionally a component of water vapor that is not solely related to the transport properties of the aerosol through the spray chamber. As a result, it has been shown that the use of silica gel collection procedures for aqueous solvent transport efficiency measurements inevitably gives rise to results with a positive bias [2, 36]. With typical ICP nebulizers, both cross-flow and concentric types, operated under normal conditions of 1 $mL \cdot min^{-1}$ aspiration rate and 1 $L \cdot min^{-1}$ Ar flow, results can be positively biased by up to a factor of 3.5. However, the errors resulting when using nebulizers that operate at low gas flows (for example, the design described by Ripson and de Galan [35] (which used only 0.2-L · min^{-1} Ar and 0.1-mL · min^{-1} solution flow rate) are much smaller, giving an estimated 35% positive bias. This may be acceptable in certain circumstances.

It should be noted that the severe errors resulting from indirect measurement of transport efficiency, compared to direct measurement, are primarily a consequence of the difficulty of washing out the ICP spray chamber efficiently. Even a small wash-out loss, when combined with the low values of transport efficiency typical for ICP nebulizer–spray-chambers, arithmetically gives rise to a large error in ϵ_n. If a highly efficient system is used, on the other hand, such as the one described by Layman and Lichte [38], which has ϵ_n values in excess of 30%, then the accuracy and precision of indirect methods will approach those of direct methods, providing that efficient wash-out procedures are provided. Nevertheless, direct methods are still in general to be preferred over indirect methods, because any direct method measures precisely the aerosol reaching the plasma and it is not necessary to take account of other processes that may influence the measured analyte passing to drain, such as losses occurring in the plasma torch. This type of loss alone can account for a 20% reduction in aerosol mass passing to the plasma [2].

When aerosol is collected from a typical ICP torch, it should be noted that it is essential to use the intermediate gas flow through the torch. In the absence of such a flow, aerosol will deposit gravitationally into the torch, and not be collected efficiently by the filter or the cascade impactor system. This will lead to significant negative bias in the transport efficiency data calculated.

8.5. AEROSOL GENERATION

The production of an aerosol, by the disintegration of bulk liquid into small droplets, can be accomplished by two primary means. The first of these, pneumatic nebulization, is a process where a liquid stream interacts with a highly energetic, high-velocity gas stream, in such a manner as to disrupt the liquid surface and produce a stream of fine droplets. Ultrasonic nebulization, on the other hand, is a process where high-frequency acoustic energy is coupled into a bulk liquid, producing surface instability of the liquid and causing the formation of surface waves. These waves collapse, to give rise to fine droplets. Other nebulization techniques, such as those dependent on small-diameter, high-velocity liquid jets [39], or those using high-voltage electrification of liquid streams [40], are primarily of academic interest; to date they have not been applied to inductively coupled plasma systems. Pneumatic nebulization, which is the technique predominantly used in inductively coupled plasma spectrometry liquid sample introduction, will be considered in most detail in this section.

8.5.1. Pneumatic Nebulization

The principles of pneumatic nebulization have been known since the time of Bunsen and Lundegard, and nebulizer design has actually changed little in the interim. There are two basic possible mechanical configurations for the design of a pneumatic nebulizer. One, in which gas and liquid streams interact coaxially, is known generally as a concentric nebulizer. The other, in which gas and liquid streams interact at right angles to one another is known as a cross-flow nebulizer. In the present state of knowledge, there is no clear differention between the mechanisms of aerosol formation with either of the two systems. Aerosol production by both concentric and cross-flow nebulizers has been studied using high-speed photography [12, 41]. Unfortunately, a number of factors militate against the ability of high-speed photography to give precise mechanistic information on droplet formation of direct relevance to ICP nebulizers. First, the droplets of direct importance to aerosol production with the ICP are extremely small, typically with diameters of 10 μm or less. This makes their visualization by photographic means extremely difficult. Second, the time scale of the aerosol production process is sufficently rapid, certainly less than a microsecond [42], so that even high-speed flash photography gives very indistinct images of the primary formation processes. Consequently, although photographic evidence of larger-scale droplet production is helpful in understanding some of the mechanisms of aerosol formation, it does not give a clear picture of the processes that occur during the formation of aerosols of the relevant drop size range. The majority of quantitative information on aerosol production is therefore provided by empirical and indirect means. This typically involves the systematic variation of nebulizer parameters, and the measurement of the drop size distribution of the aerosol so produced.

Photographic studies which have been carried out on liquid streams of larger drops, on the order of a millimeter or so in diameter, show that disruption in a high-velocity gas stream seems to involve various models. For example, in the bursting bag model, a thin film is formed from the primary droplet due to unequal pressure on the surface of the droplet [43]. The thin film, or hollow bag, then bursts to form a string of fine droplets (Fig. 8.7). Likewise, when a liquid stream, emerging from a jet in a nebulizer, interacts with a very-high-velocity gas stream, the process of liquid stream fracture takes place via formation of ligaments [12]. The ligaments then become unstable and move back and forward very rapidly, causing a fragmentation process (Fig. 8.8). Nevertheless, the scale of observation of both of these processes is far in excess of that normally encountered with ICP nebulizers, and consequently it becomes very difficult to extrapolate the observed phenomena to those observed with ICP nebulizers. As a general trend, however, it seems reasonable to assume that the formation of droplets with typical ICP neublizers occurs by essentially a stripping mechanism, where fine ligaments are drawn from an unstable surface of

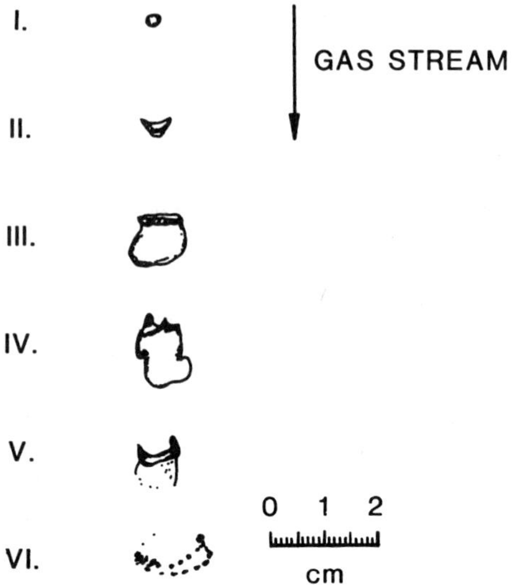

Figure 8.7. "Bursting bag" model, for production of aerosol from large droplets, in gas stream with high relative velocity.

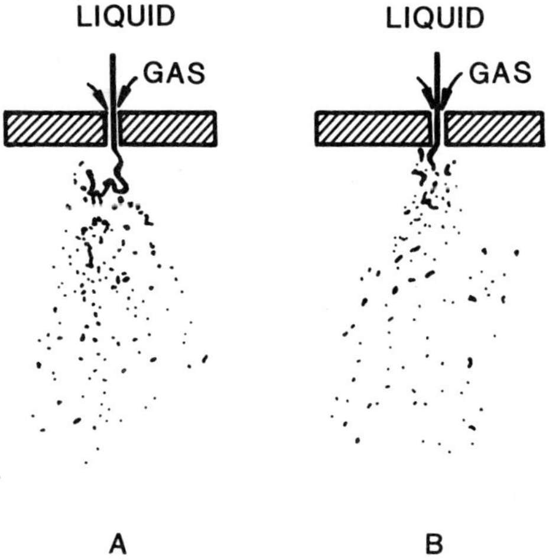

Figure 8.8. Formation of aerosol by "oscillating ligament" model. Water jet interacts with high-velocity gas jet in concentric configuration. (*A*) Air velocity = 30 m · s^{-1}; (*B*) air velocity = 60 m · s^{-1}. Liquid velocity = 0.5 m · s^{-1}; liquid jet diameter = 1.7 mm.

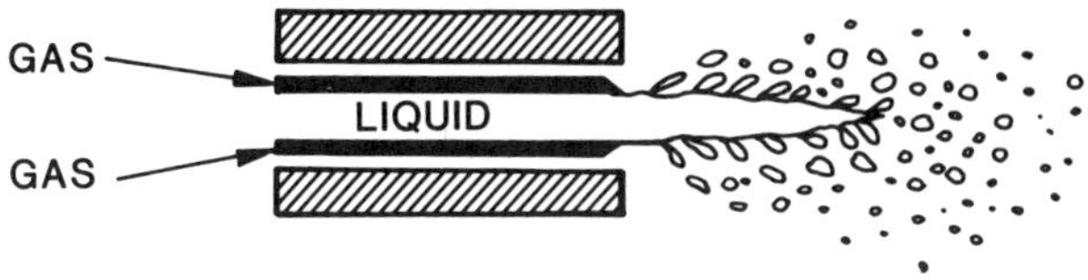

Figure 8.9. "Surface stripping" model for aerosol generation. It requires high relative velocity between gas and liquid streams.

the liquid stream, as it interacts with the high-velocity gas jet [44]. These, in turn, disrupt to give small droplets (Fig. 8.9).

Clearly, it is of great interest to be able to relate the properties of the aerosol droplet size range produced by pneumatic nebulizers to the operating parameters of the nebulizer, namely gas and liquid velocity, surface tension of the liquid, liquid density, liquid viscosity, and volume flow rates of gas and liquid. Such an equation was derived empirically by Nukiyama and Tanasawa [12] in 1939, and has since formed the primary means of estimating aerosol properties produced by pneumatic nebulizers. The equation gives the value of the volume to surface area, or Sauter, mean diameter, d_s, of the droplet distribution produced by such a pneumatic nebulizer. It does not of itself give information on the distribution of particle size within the aerosol spray. The equation is

$$d_s = \frac{585}{V}\left(\frac{s}{\rho}\right)^{0.5} + 597\left[\frac{\eta_l}{(s\rho)^{0.5}}\right]^{0.45}\left[\frac{10^3 Q_l}{Q_g}\right]^{1.5} \tag{8.11}$$

Here, d_s is the Sauter mean diameter (μm), V is the velocity difference between gas and liquid flows to the nebulizer (m · s^{-1}), s is the surface tension of the liquid (dyn · cm^{-1}), ρ is the liquid density (g · cm^{-3}), η_l is the liquid viscosity (P), and Q_l and Q_g are the volume flow rate of liquid and gas (cm^3 · s^{-1}), respectively. Strictly speaking, the use of this equation should be limited to the boundary conditions envisioned by the original workers, namely a particular design of concentric nebulizer, using subsonic gas velocities, and a range of solvents covering a surface tension range 30–70 dyn · cm^{-1} and viscosity 1–30 cP. Typical pneumatic nebulizers used with inductively coupled plasmas work at much higher gas velocities than those used in the experiments of Nukiyama and Tanasawa. However, the work of Bitron [45] has indicated that the Nukiyama and Tanasawa equation may very well be valid at sonic or supersonic gas velocities. Equation (8.11) has been used historically in two ways. First, as an indication of the influence of experimental variables on drop size, and second, as a quantitative indication of the anticipated drop size produced for a particular nebulizer design. It should be emphasized here that the Nukiyama and Tanasawa equation has certain limitations and boundary conditions in addition to those already discussed. First, and foremost, it was derived for, and is only valid for, describing the aerosol formed at the nebulizer. In all instances with ICP–AES, this aerosol will have wholly different properties from that arriving at the plasma. This is because other modifying processes, which take place in the spray chamber and in the transport of aerosol between the nebulizer and the plasma, will act to remove many of the larger droplets from the aerosol stream. This will shift the distribution toward smaller aerosol particles. Also, the measurement technique used in the derivation of this equation involved impaction trapping of aerosol on microscope slides, and subsequent drop counting using microscopic techniques. Because of limitations in this procedure, both in its inability to catch aerosol droplets less than 5 μm efficiently, and also to size them reliably by optical microscopy, this equation does not refer with any degree of certainty to drop sizes below 5 μm. The validity of Eq. (8.11) has not yet been unequivocally established for typical inductively coupled plasma nebulizers of either the concentric or cross-flow configuration. Nevertheless, from previous experience with atomic absorption nebulizers of the concentric variety [3], it would appear that the model is at least of value in a semiquantitative manner for concentric ICP nebulizers. Information is not as yet available to determine its relevance to the description of either AA or ICP cross-flow nebulizers. The equation has been used to predict Sauter mean diameters for a variety of solvents, when nebulized with a concentric pneumatic nebulizer of typical dimensions [8]. Such predictions are shown in Fig. 8.10. These curves indicate the following. First, the Sauter mean drop size of the primary aerosol produced by a typical ICP nebulizer is predicted to be quite large (>20 μm). Second, the lower surface tension exhibited by organic solvents, in comparison to aqueous solutions, by no means always manifests itself as a reduction in the primary mean drop size of the aerosol. Many solvents, such as, for example, nitrobenzene, are indeed predicted to have larger mean drop sizes than water or aqueous solutions. The reason for this behavior is that for ICP nebulizers, with generally subsonic gas velocities, the second term of the Nukiyama and Tanasawa equation exerts a major effect on the calculated drop size for solution flows in excess of approximately 1 mL $\cdot$ min^{-1}. Organic solvents, which generally have lower surface tensions and densities than water, typically give larger values for this second term in the equation, because the surface tension and density terms both appear in the denominator. Organic solvents may have viscosities either higher or lower than water, but the surface tension and density values generally exert the dominant influence on this term.

8.5.2. Ultrasonic Nebulization

Ultrasonic nebulizers operate on a somewhat different principle from pneumatic nebulizers. With ultrasonic nebulizers, the energy that causes disruption of the liquid surface is provided by the oscillation of a transducer, induced by radio frequency (rf) voltages. The mechanism of droplet production with ultrasonic

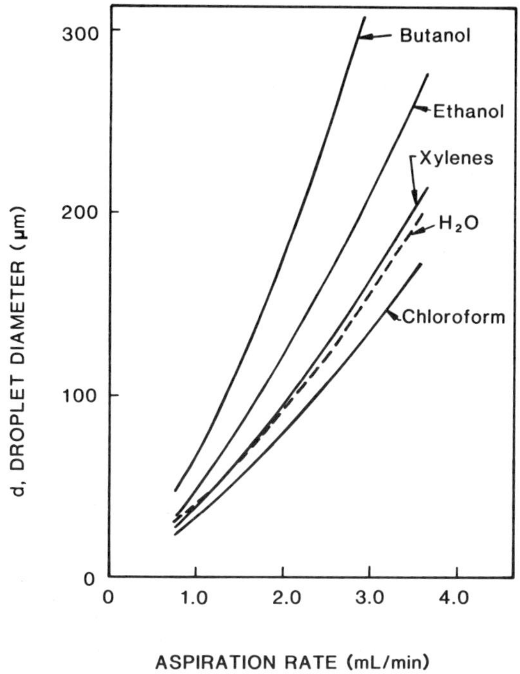

Figure 8.10. Mean droplet sizes for various solvents, as a function of uptake rate to nebulizer. Values calculated for concentric (Meinhard) nebulizer operated at 1.0 L · min^{-1} Ar flow.

nebulization seems to be a function of the frequency of the oscillations. For a fixed level of radiation power in the surface of solution, excited by a rf transducer, the acceleration experienced by the liquid surface increases dramatically as the oscillation frequency rises. Thus, at low exciting frequencies, for example, at 20 kHz, an ultrasonic transducer will produce a cavitation phenomenon [46] in which a large number of oscillating air bubbles are produced within the liquid bulk (see Fig. 8.11). These then collapse, creating large-amplitude shock waves near the surface. This leads to the production and ejection of droplets from the surface of the liquid. It has been suggested [47] that at higher frequencies, for example, in excess of 800 kHz, cavitation bubbles are not formed, but a strong net upward force exists on the liquid surface which causes a column of liquid to be pushed out of the surface. This upward force is supposedly due to radiation, rather than acoustic pressure. The liquid column ejected by the radiation pressure produces a fountain or geyser effect, and from the geyser emanates a fine pulsatory mist of droplets (see Fig. 8.12). However, the nature of aerosol production from the geyser is highly dependent on the power density in the liquid surface, such that at lower powers that are several simultaneous mechanisms active in the geyser. As a result, a fine mist forms in

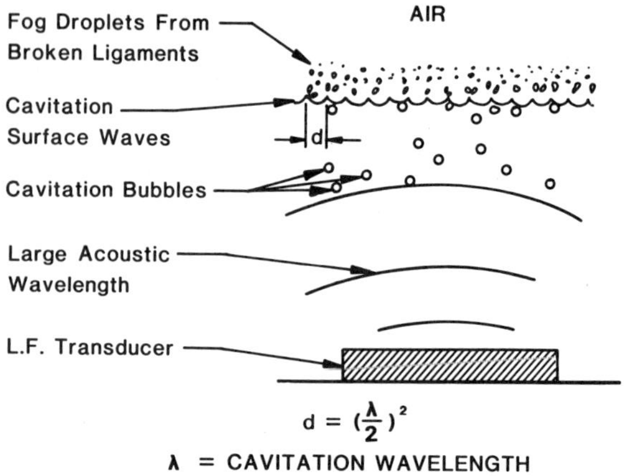

Figure 8.11. "Cavitation" model for ultrasonic aerosol generation at low frequencies (20 kHz) with submersible transducer.

the lower part of the geyser, while larger droplets are produced in the upper part. As the power density in the liquid surface increases, droplet formation becomes more uniform and the geyser becomes more cylindrical in nature. The low-frequency mode of operation gives rise to droplets whose numerical median diameter can be described by a simple equation following the analysis of Lang [48]:

$$d_n = 0.34 \left(\frac{8\pi s}{\rho_1 F^2} \right)^{1/3} \tag{8.12}$$

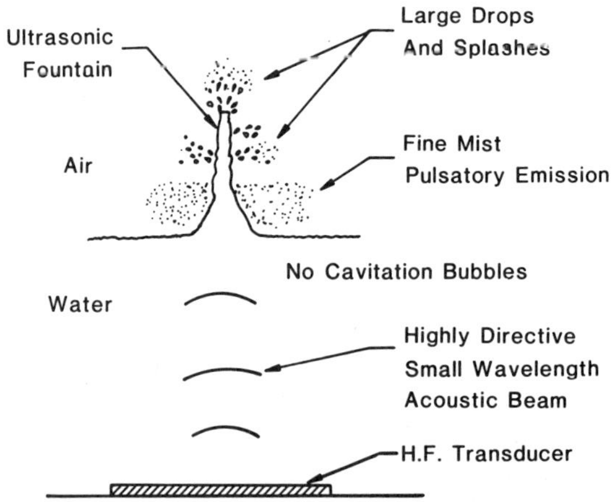

Figure 8.12. "Geyser formation" model for ultrasonic aerosol generation at high frequencies (800 kHz) with submersible transducer.

where s is the liquid surface tension (dyn $\cdot$ cm^{-1}), ρ_l is the liquid density (g $\cdot$ cm^{-3}), and F the excitation frequency (Hz). At high frequencies, on the other hand, characteristics change such that the drop size becomes independent of frequency above approximately 800 kHz. There is some considerable discussion in the literature as to the limiting value of drop size that occurs at high frequencies. However, it would appear that at frequencies above approximately 1 MHz, the majority of droplets produced by ultrasonic nebulizers are below 5 μm diameter [49]. It should be noted, however, that aerosols produced by ICP ultrasonic nebulizers are by no means monodisperse, and the aerosol size range produced may also be a function of power density in the liquid surface [50].

8.5.3. Secondary Processes of Nebulization

In discussing droplet formation by pneumatic nebulizers, it is convenient to consider the processes occurring in three stages. The primary process is that in which, in the case of a pneumatic nebulizer, the initial interaction between gas and liquid occurs. In certain situations, secondary processes may occur immediately after the primary aerosol generation step. These will typically involve the interaction of the high-velocity gas and aerosol stream with some obtruding surface. On striking such a surface, two processes may occur. The first of these will be an impaction process, where larger particles are lost by striking and sticking to the surface of the obtruding surface. Generally, the impaction site takes the form of a spherical, or hemispherical, glass bead placed in the aerosol stream, close to the nebulizer. Consequently, the velocity of the aerosol particles striking this object is considerable, and can be of the order of 100 m $\cdot$ s^{-1}. An empirical equation has been derived for the influence of a spherical impact surface on an aerosol stream. This equation describes a droplet cutoff diameter d_c, in terms of the properties of the aerosol stream and the gas velocity. The cutoff diameter d_c can be considered simply as the value of the droplet diameter on a distribution curve, corresponding to a point where the normalized mass has dropped to half its maximum value. This definition assumes an approximately unimodal distribution. For secondary impaction on a spherical surface, the cutoff diameter is given by [3]:

$$d_c = 1.5\eta_g w/\rho_a v \tag{8.13}$$

where η_g is the absolute gas viscosity (P), w the width of the gas stream (cm), ρ_a the aerosol particle density (g $\cdot$ cm^{-3}), and v the particle velocity (cm $\cdot$ s^{-1}).

Calculations have been made for a typical ICP cross flow nebulizer, using an impact bead, and assuming the aerosol spray diameter approaching the bead to be quite narrow, for example, 2 mm, as might be the case if the bead was placed close to the nebulizer [3]. These gave a value for d_c of approximately 2.6 μm, which is in reasonably good agreement with experimental values that have been measured for aerosol leaving such a nebulizer [51]. This indicates that the model describing the action of an impact bead on an ICP nebulizer as a loss process for large droplets is a reasonable one.

The second mechanism that can, and undoubtedly does, occur at closely spaced impact surfaces is that of secondary aerosol generation. If the aerosol strikes the surface with a sufficiently high velocity, shattering of the droplets will occur at the impact surface. It is not clear which size droplets are most prone to shattering, but presumably the droplet sizes that are most likely to result in secondary shattering are those intermediate in size between the droplets sufficiently large to strike the impact surface and stick to it, and those that are small enough to be carried in the air stream around the impact surface, and avoid any impaction whatsoever. Droplets in the range of 5 to 10 μm could very well be the origin of secondary shattering, although no experimental studies have yet been performed to confirm this. The effect of secondary shattering at an impact bead is the production of an increased proportion of small droplets. However, the droplets produced are not all of one size and typically will cover the range from 0 to 3 μm. As a consequence of the polydispersity of the aerosol produced by secondary processes, the secondary processes of particle shattering, as opposed to those of impaction loss, do not generally result in a significant shift in the droplet size distribution. They do, however, result in a net increase in the density of aerosol reaching the plasma per second, and hence an increase in the W parameter.

8.6. FACTORS MODIFYING PRIMARY AND SECONDARY AEROSOLS—GRAVITATIONAL, IMPACTION, TURBULENCE, AND CENTRIFUGAL LOSS PROCESSES

After the aerosol has passed the nebulizer together with any impact bead or surface placed in its immediate path, it will undergo tertiary modifying processes. Tertiary processes are the major aerosol modifying step in any system that operates without an impact bead, and may result in a significant change in the particle size distribution. They occur because of various interacting forces between the gas flow, the aerosol stream, the spray chamber, and the plasma torch. The net result of these tertiary processes is a preferential loss of large aerosol droplets. Overall, tertiary processes will produce a net reduction in the transport of aerosol and analyte to the plasma. This is inevitably accompanied by a shift in the drop size distribution to smaller droplet diameters. Although detailed measurements have not been made with ICP nebulizers to date, one can envision, semiquantitatively, an overall picture of the aerosol transport process. Substituting typical values into Eq. (8.11), and assuming a lognormal distribution of aerosol, production of an aerosol with a mean diameter between 20 and 100 μm would be predicted at the nebulizer. Precise d_s values will depend on nebulizer design and operating conditions, such as relative gas and liquid flows. It is assumed that no impact bead is present, and so secondary processes can be neglected. The aerosol stream reaching the plasma, having passed through the spray chamber and the plasma torch, will have been modified by the loss of large droplets. As a result, its mass median diameter will be approximately 2 μm, or less. This dramatic shift in droplet distribution is a direct consequence of tertiary processes acting on the aerosol in the spray chamber. In this context, if an impact bead were present, its role as an impaction loss device would achieve essentially the same effect as any subsequent tertiary processes that may occur. This means that if an impact bead is used with an ICP nebulizer, it may not be necessary to design a spray chamber to accomplish any further aerosol modifying processes [51].

The effect on the aerosol of tertiary loss processes is to cause a major part of the primary aerosol droplets to stick to surfaces in the spray chamber, and subsequently run to waste. Experimental measurements carried out on typical ICP nebulizer–spray chamber torch systems have indicated that between 98 and 99.5% of the aerosol is actually lost to waste by tertiary processes [2]. As a result, the transport efficiency of typical ICP systems is only in the range 0.5–2%. These tertiary processes then are clearly of immense importance in determining the transport properties of the ICP aerosol introduction system. They control both the aerosol drop size distribution reaching the plasma, and additionally the total mass per second of analyte (W_{tot}) reaching the plasma. As yet, it is not clear which of these properties is more important in influencing signal magnitude and interference effects in the ICP. This is clearly an area in which further research needs to be carried out.

8.6.1. Turbulence Loss

The most commonly used design of spray chamber for ICP systems is the dual concentric chamber, based on the design of Scott et al. [52]. In this design, the aerosol from the nebulizer enters a chamber initially of approximately 2.5-cm internal diameter. The aerosol-laden gas stream passes through this initial tube, and then back through a concentric tube of larger diameter, before exiting to the plasma injection tube. A typical design is shown in Fig. 8.13. In this type of spray chamber, it can be observed that the predominant site of aerosol loss occurs in the first few centimeters of the inner chamber, following the introduction of aerosol to the chamber. It has been suggested that the loss process occurring here is one due to the interaction of the highly turbulent aerosol gas stream leaving the nebulizer, and the relatively stagnant gas volume in the chamber [3]. This appears to fit both experimental measurements of the drop size dependency of the loss process, and also the influence of gas flow on the loss process. The important practical consequence of a turbulence-induced loss mechanism is that an increase in gas flow rate through the system should result

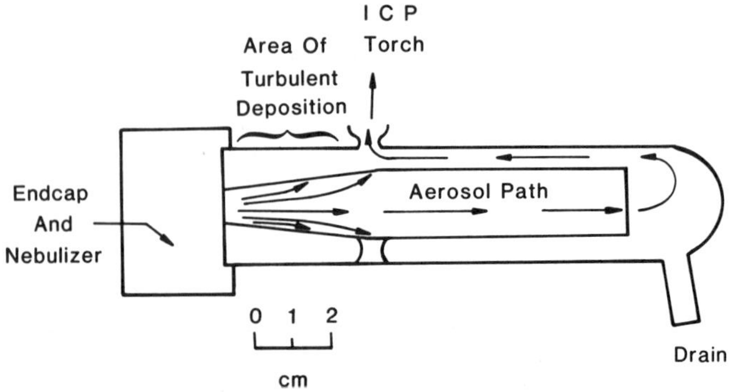

Figure 8.13. Schematic of dual concentric ICP spray chamber, after design of Scott et al. [52]. All-glass construction. Main aerosol path and sites of major aerosol turbulent loss indicated by arrows.

in greater penetration of larger aerosol droplets. Also the shape of the drop size distribution curves should vary somewhat from those expected from other loss processes, such as impaction loss [3]. However, this aspect is not very significant practically. It should be noted that an increased gas flow rate in the ICP gas injection stream is generally accomplished by increasing the gas flow through the nebulizer itself. This will have a tendency to alter both the aerosol production process and the degree of turbulence in the tube, in addition to affecting the turbulence loss process itself. The net result is generally a shift to smaller droplets, accompanied by a small increase in *W*.

8.6.2. Gravitational Aerosol Loss

When an aerosol passes through a tube, each component droplet of the aerosol will rapidly reach its own terminal settling velocity, under the influence of gravity (for example, a 50-μm-diameter droplet will reach its settling velocity in approximately 8 ms). It is possible, therefore, to consider that when an aerosol passes through a horizontally placed tube, the diameter of the tube and the gas velocity impose a boundary condition on the aerosol. As a consequence of this, there will be 50% removal of droplets of diameter d_c, the cutoff diameter, passing through the chamber when

$$d_c = 3(\eta_g D/\rho_a g t)^{1/2} \tag{8.14}$$

where η_g is the gas viscosity (P), D the chamber diameter (cm), ρ_a the aerosol particle density ($g \cdot cm^{-3}$), g the gravitational constant (dyn $cm^2 \cdot s^{-2}$), and t the time spent in the chamber(s). In practice, it has been calculated [3] that the time available for gravitational settling is inadequate for it to be a major factor in influencing aerosol tertiary particle sizes, in all experimentally available spray chamber systems. Nevertheless, if aerosol is passed, subsequent to the spray chamber, down any significant length of tubing, then gravitational settling can cause significant aerosol loss and a noticeable shift to smaller droplet size [3].

8.6.3. Centrifugal Loss

Certain spray chambers, such as the modified atomic absorption spray chamber used by Greenfield et al. [53], impart a centrifugal motion to the aerosol. These devices are used industrially and are known as cyclone chambers. The action of the cyclone is the imparting of a spiral motion to the aerosol. Consequently, particles of larger diameter, on entering the chamber, possess a higher angular momentum, and tend to strike the walls of the spray chamber preferentially and be lost. Due to its industrial importance, this type of chamber has been investigated thoroughly. An expression to describe the cutoff diameter for 50% penetration, d_c, has been derived [54]

$$d_c = \left(\frac{9\eta_g w_i}{2\pi n v_i(\rho_a - \rho_g)} \right)^{1/2} \tag{8.15}$$

where η_g is the gas viscosity (P), w_i the inlet width (cm), n the number of rotations and aerosol path takes (dimensionless), v_i the inlet gas velocity, ρ_a the aerosol particle density ($g \cdot cm^{-3}$), and ρ_g the gas density ($g \cdot cm^{-3}$). For effective removal of large particles, it is clearly desirable to have a relatively high entering gas velocity. This is accomplished, for a fixed gas flow, by designing the chamber with a small-diameter gas inlet. It is also desirable to have a large number of rotations of the aerosol path before it leaves the spray chamber. In predicting cutoff diameters with this design of spray chamber, it is necessary to make an assumption regarding the number of rotations the aerosol path takes before leaving the spray chamber. This design of spray chamber has proved to be quite effective as a means of removing larger droplets from the aerosol spray [53]. However, particle size distribution data on such aerosols has not yet been published.

8.6.4. Influence of Evaporation on Aerosol Properties

The evaporation of aerosol droplets can have a significant effect on both the transport properties of the spray chamber and the solvent vapor loading in the plasma. These effects are most pronounced for organic solvents, when the vapors evolved can be shown to have a dramatic influence on plasma stability. However, specific details regarding organic solvent evaporation have been detailed in Chapter 6 (see also Part 1, Section 4.7.8 and [54]). In this section, general properties of both aqueous and organic aerosol evaporation will be discussed to show some of the basic principles describing aerosol evaporation.

When an aerosol is generated by a pneumatic nebulizer, the process involves the intimate mixing of a high-velocity gas stream and a very high concentration of small droplets. This provides an ideal environment for the rapid attainment of a dynamic equilibrium between liquid and vapor. Consequently, it seems reasonable to assume that significant evaporation of the liquid stream takes place very rapidly under these circumstances. Therefore, it is possible that droplet evaporation may play a major part in the actual primary process of aerosol formation. Unfortunately, detailed experimental measurements of aerosol evaporation properties in ICP spray chambers are lacking at present, except for highly volatile organic solvents. In the case of organic solvents, where the vapor pressure of the solvent is often extremely high, good correlation has been shown between the volatility of the solvent and the transport efficiency [8, 32].

Under controlled conditions of aspiration, it can be shown that volatile organic solvents with very similar properties of surface tension and viscosity have significantly different transport efficiency values [32]. These values can only apparently be correlated with the solvent volatility. For example, at an aspiration rate of 1.8 mL $\cdot$ min^{-1} of various solvents, using a cross-flow nebulizer and dual concentric spray chamber, aqueous solutions show a transport efficiency of 1.5% compared to 6.4% for nitrobenzene, 8% for xylenes, and 22% for benzenes. Here the increase in transport efficiency from water to nitrobenzene may conceivably be accounted for by surface tension and viscosity differences. However, the dramatic increase in transport efficiency between xylenes and benzene can only be related to solvent volatility differences.

With aqueous solutions, the role of solvent evaporation is less clear in its relationship to transport efficiency. Although the droplet of a pure solvent may evaporate totally, given sufficient time, the presence of a dissolved salt will cause a lowering of the saturated vapor pressure of the salt solution. Consequently, for a salt solution, a droplet will tend to evaporate rapidly until it reaches an equilibrium state with the vapor surrounding it. Evaporation will proceed until the droplet vapor pressure is equal to the saturation vapor pressure of the salt solution, this being the vapor pressure in the surrounding gas. The droplet will tend, therefore, to evaporate until it has reached an equilibrium droplet diameter. This diameter, d_x, is given by the following theoretical equation [6, 55]:

$$d_x^3 + \frac{400\, M_s s}{\rho_g RTC\alpha} d_x^2 = d_o^3 \tag{8.16}$$

where d_o is the initial droplet diameter (μm), M_s is the solvent molecular mass, s is the solvent surface tension (dyn $\cdot$ cm^{-1}), ρ_g the density of the gas (g $\cdot$ cm^{-3}), R the gas constant, T the absolute temperature (K), C the mass percent concentration of the solution, and α the degree of dissociation. Plots of theoretical percent decrease in droplet diameter as a function of initial diameter are shown in Fig. 8.14 for various concentrations of sodium chloride solution. It

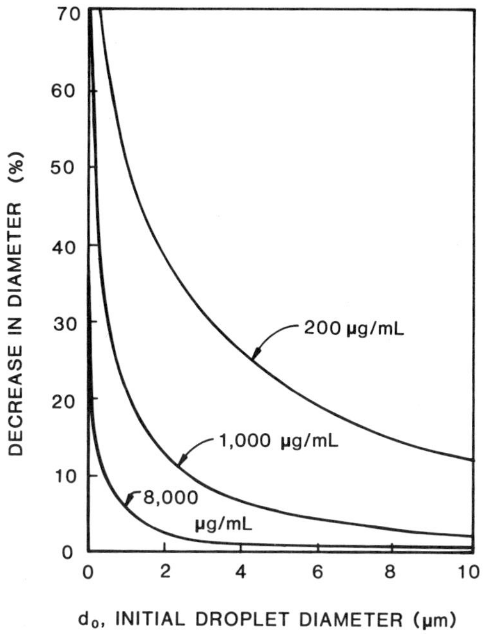

Figure 8.14. Influence of solution concentration and initial droplet diameter d_0 on value of equilibrium droplet diameter after evaporation d_x. Values expressed as percent decrease in diameter from d_0 at equilibrium.

can be seen that at high concentrations, such as in excess of 5000 $\mu g \cdot mL^{-1}$ of Na, the decrease in droplet diameter involved in reaching equilibrium with the vapor is relatively trivial for the majority of aerosol particles observed. In other words, for 2-μm diameter droplets, <5% evaporation is predicted. By contrast, a droplet of 2 μm initial diameter and 200 $\mu g \cdot mL^{-1}$ concentration would be expected to reach an equilibrium diameter at 60% of its initial diameter, in other words, when $d_x \approx 1.2$ μm.

Although the role of droplet evaporation in aqueous aerosol transport has not yet been fully unraveled, solvent evaporation can have a very important effect upon plasma properties. It is well established that both aqueous and organic solvents cause a significant lowering in the ICP temperature [7–9, 56]. If solvent loading is sufficiently high, then the plasma may even be extinguished [8]. This effect is particularly noticeable for volatile organic solvents. The equilibrium between aerosol and vapor can play an important role in the plasma excitation mechanism, and hence in the plasma temperature. It has been demonstrated that an increase in aqueous solvent loading in the plasma, such as provided by the use of ultrasonic nebulizers [57, 58] can cause a significant decrease in plasma excitation temperature. This partially counteracts the increase in analyte loading in the plasma, and as a direct consequence there is a

Table 8.3. Solvent Aerosol and Vapor Transport Data[a]

Temperature (°C)	Aerosol Delivery ($mg \cdot min^{-1}$)	Vapor Delivery ($mg \cdot min^{-1}$)	Theoretical Vapor Delivery, at 100% Relative Humidity	Total Solvent Delivery ($mg \cdot min^{-1}$)
24	5.7	14.9	13.1	20.6
22	5.4	12.9	12.0	18.3

[a] Concentric, all-glass neublizer. Operated at 0.6 $L \cdot min^{-1}$ Ar flow and 1 $mL \cdot min^{-1}$ aspiration rate.

significantly smaller increase in signal to background in the plasma than would be anticipated from the increase in analyte transport rate W. For example, with ultrasonic nebulization followed by desolvation, in which the aerosol is passed first through a heated chamber to evaporate all solvents and then through a condenser to remove the solvent vapor, improvements as much as a factor of 20 over conventional pneumatic nebulization have been observed [57]. However, in the absence of desolvation, improvements in signal to background of only a factor of five maximum have been observed. The deleterious effect of the additional solvent is quite clear.

Recently, information has been published indicating that, with aqueous aerosols, the gas stream flowing with the aerosol reaches approximately 100% relative humidity quite rapidly [36]. Consequently, the solvent vapor loading in the gas stream is a function of the temperature of the gas stream. Typical figures for aerosol and vapor loading, with concentric nebulizer and Scott spray chamber systems, are shown in Table 8.3 at various room temperature values. The full quantitative significance of various aerosol and vapor loadings on plasma performance in terms of both signal to background ratios and interference effects has not yet been fully investigated (see Part 1, Section 4.7.8, and [54], however). It would appear that, at the very least, aqueous solvent vapor loading variation could have a significant effect upon long-term plasma stability. Changes in spray chamber temperature could therefore make a significant difference to the precision of measurements made over a period of several hours.

8.7. AEROSOL-RELATED INTERFERENCES

8.7.1. Solution Viscosity Effects

For a given pressure drop at the nebulizer, and for a given diameter and length of uptake tubing, the natural aspiration rate Q_l of liquid will vary approximately according to the Poiseuille equation, as modified by Kundt and Warburg

[59, 60]

$$Q_l = \frac{\pi(R^4 + 4\eta_l R^3/B)\,\Delta P}{\eta_l l} \tag{8.17}$$

Here, R is the capillary radius (cm), ΔP, the pressure differential (dyn · cm^{-2}), B the coefficient of sliding friction of liquid on the capillary wall (dimensionless), l the capillary length (cm), and η_l the solution viscosity (P). It has been demonstrated [60, 61] that this equation, rather than the simple Poiseuille equation, is most appropriate for describing liquid flow in the narrow capillaries commonly used in ICP spectrometry. Equation (8.17) is able to allow for the fact that the liquid velocity is not zero at the tube walls, and it shows a positive intercept, due to the nonideal flow through the narrow-bore capillary tube. When this equation is considered in conjuction with the Nukiyama and Tanasawa prediction for droplet size [Eq. (8.11)], it can be seen that the predicted effect of an increase in solution viscosity is to cause a decrease in the aspiration rate, which in turn should translate to a decrease in mean droplet size. However, counterbalancing this is the predicted increase in mean droplet size predicted from the increase in the solution viscosity. On balance, the aspiration rate variation due to an increase in viscosity should result in a net reduction in mean droplet size. Clearly, there are two counteracting effects taking place when solution viscosity increases. First, the mean droplet size is predicted to decrease, which should lead to an increase in the transport of analyte to the plasma. However, acting against this is a net decrease in rate of supply of solution to the nebulizer. To predict the net influence of these opposing effects on the mass/second of analyte reaching the plasma, which in turn will relate directly to the analyte emission signal, it is necessary to characterize each nebulizer–spray chamber system individually. Experiments that have been reported, using controlled uptake rates of solution feed to concentric or glass nebulizers and a Scott spray chamber, have shown that the two effects can sometimes counterbalance precisely [3]. It should be noted, however, that in these experiments uptake rate varied not because of solvent viscosity effects, but because of the control of pumping rate. The effect of reducing the uptake rate from 2 to 0.5 mL · min^{-1} resulted in no net change of analytical signal from the plasma. Measurements of the mass of analyte reaching the plasma (W_{tot}) showed that, indeed, the change in uptake caused no net change in W_{tot} at the plasma. In this instance one can only suppose that the increase efficiency of the transport process, resulting from the production of a primary aerosol spray at the nebulizer with a smaller mean droplet size, was exactly counterbalanced by the decreases rate of analyte supplied to the nebulizer. The two effects, being opposed, precisely cancelled one another. Therefore, it is not immediately obvious that changes in aspiration rate to an ICP nebulizer spray chamber system will necessarily cause any change in signal at the plasma.

Nevertheless, several studies of the influence of organic and inorganic acids of varying viscosity, and hence varying natural aspiration rate, have been published which indicate that there can be significant nebulizer effects produced by solutions of varying viscosity. Greenfield et al. [60] studied the influence of a wide range of acids on Ni, Fe, Co, Cr, and Cu solutions aspirated into a high-power, nitrogen-cooled Ar plasma. These workers used a modified atomic absorption cyclone-type nebulizer spray chamber system. With this system, there was in general extremely close correlation between the reduction in intensity found with increasing solution viscosity and the decrease in natural aspiration rate of the solution, compared to pure aqueous solution. The effect of increased viscosity on signal could also be largely removed by using a syringe pump to force concentrated HCl solutions through the nebulizer at approximately constant uptake rate, irrespective of solution viscosity. By this means, HCl solutions varying from 0 to 19 weight % were found to give no more than a 10% average reduction in signal intensity, compared to pure aqueous solutions. Later work by Dahlquist and Knoll [61] showed similar effects over wider concentration ranges of acids, using a low-powered Ar ICP. These workers also found that incorporation of an internal standard into the solution was an effective means of cancelling out viscosity-induced nebulization changes for their system. Alternatively, they recommended matrix matching of samples and standards, to ensure that the viscosity of both sample and standard are effectively identical. This is probably the simplest means to overcome any variations in transport properties resulting from viscosity changes of solutions. Certainly the pumping of concentrated acids using flexible tubing and a peristalic pump is not feasible, as the tubing deteriorates very rapidly under these hostile conditions. It should be noted that, in the work of Dahlquist and Knoll, attempts to compensate the signal change, by normalizing for uptake rate changes due to viscosity changes, provided correction factors that were not satisfactory for practical work. Recoveries, for example, for 10% HCl solutions were as much as 16% reduced from target values. These data are presented in Table 8.4. The difference between the results of Dahlquist and Knoll and Greenfield may well be due to the difference in the types of nebulizer and spray chamber used. Indeed, Dahlquist and Knoll [61] showed that the design of the spray chamber could have a significant effect upon the observed magnitude of the interference effect. They found that a conical spray chamber, with a small internal volume of 40 cm^3, produced lower interference effects with perchloric acid than the normal dual concentric Scott-type spray chamber.

8.7.2. Organic Solvent Effects

Organic solvents, present in varying proportions with aqueous solutions, or in their own, can give rise to significantly different signal intensities for the same analyte concentration, compared with aqueous solutions. Consequently, the

Table 8.4. Influence of Solution Viscosity on Emission Signals [61][a]

	Recovery against Aqueous Standard (%)			
	10% HCl	35% HNO_3	35% $HClO_4$	48% H_2SO_4
Relative nebulizer aspiration rate (% H_2O)	92	73	72	31
Mean recovery against aqueous standards (%)	77	63	63	33
Mean recovery against aqueous standards, normalized for uptake rate (%)	84	87	88	107
Mean recovery against aqueous standards, Co internal standard (%)	100 ± 2	96 ± 9	101 ± 5	102 ± 9

[a] Results calculated for mean of 11 elemental atom and ion lines. Concentric all-glass nebulizer, dual-concentric spray chamber. Results shown as ±, indicating spread of recovery data for the 11 elements.

presence of miscible organic solvents, such as ethanol, in aqueous solution could give rise to significant interference effects if their presence is not allowed for. In the case of pure organic solvents, it is important that the same solvent should be used for the organometallic standard solution as is used for the unknown sample solution. Only by this means can all the effects of transport efficiency and vapor plasma interaction effects be compensated for. If different organic solvents should be used for standards and sample, there will very likely be severe interference problems observed. The reader should refer to Chapter 6 for a more detailed discussion of these organic solvent effects (cf. Part 1, Section 4.7.8).

8.7.3. Drop Size-Related Phenomena

There has been no detailed systematic study of the influence of drop size on plasma signals. In general, it is assumed that the temperature of the plasma is sufficiently high that at the typical viewing height of 15–20 mm above the load coil, even the most involatile species will have been effectively vaporized [62]. While this has not been tested systematically, and there is no experimental evidence for the upper size limit of the particle that should be capable of rapid evaporation in the plasma, nonetheless, experimental evidence of classical volatilization interferences, such as the influence of phosphate on Ca or the influence of Al on Ca, indicates that these interferences are not present to any great extent [61, 63]. This is in contrast to the situation in atomic absorption spectroscopy, using typical commercial nebulizers and spray chambers, and working in the cool air–acetylene flame, where these interferences are quite noticeable. As these classical volatilization interferences disappear in flame AAS, when the higher temperature nitrous oxide–acetylene flame (maximum temperature ~3000 K) is used, it is not surprising that such interference effects are also not present in the far higher temperature ICP sample injection zone (~6000 K) [64]. Furthermore, it has been shown that ICP signals are relatively insensitive to total dissolved solid content. Dilution of solutions containing total dissolved solids in the range of 1 g · 10 mL^{-1} to 1 g · 160 mL^{-1}, showed a maximum mean error for the 10 elements studied (for example, Ca, K, P, Mg, Fe, Mn, Pb, Zn, Cu, and Sr) of only 9% relative [61].

8.7.4. Aerosol Ionic Redistribution (AIR) Effect

One of the more interesting, and unusual, interferences discovered recently in atomic spectroscopy is the aerosol ionic redistribution (AIR) interference. It has been found that when mixtures of certain ions are present in solution, such that one is present at a very much higher concentration than the other, then the concentration ratios of the ions found in the collected aerosol leaving the ICP torch differ from those found in the original solution [65]. Furthermore, the concentration ratio in the collected aerosol is drop-size dependent. The form of this drop-size dependence, and the ratio of ionic concentration found in the different size droplets entering the plasma, are very sensitive to the nature of the ions, their relative concentrations, and the absolute concentration of the solution mixture. The form of the variation of the concentration ratio in the collected aerosol is unpredictable. In certain cases an enrichment of the minor to the major ion may be found in the aerosol compared to the bulk. In other cases a depletion may be found. Furthermore, variations in the concentration ratios of the two ions may cause a change from a net depletion to a net enrichment. The enrichment factor E_d was defined such that:

$$E_d = \frac{(C_{Mi^+}/C_{Ma^+}) \text{ aerosol}}{(C_{Mi^+}/C_{Ma^+}) \text{ bulk}} \tag{8.18}$$

where C_{Mi^+} is a concentration of minor ion and C_{Ma^+} is the concentration of major ion. A value of E_d greater than unity represents an enhancement of the minor ion in the aerosol, and a value less than unity represents a depletion of the minor ion. Analytical curves for Na, in the presence of 10,000 μg · mL^{-1} Mg, resulted in positive enhancements of signal typically greater than a factor of 2 compared with pure Na solutions, at all Na concentrations up to 50 μg · mL^{-1}. Furthermore, the positive trend was maintained at a variety of measurement heights in the plasma (6–25 mm above the coil) and plasma incident powers (0.6–2.5 kW). Measurement of this same aerosol by collections on a cascade impactor, and measurement of the W_{tot} value, compared with the W_{tot} value measured for pure aqueous Na solution, showed a predicted enrichment value of 1.9. This compares with the experimental signal enhancement, determined under these conditions, of 2.1.

Similar studies, for the influence of Li on Na emission, showed somewhat smaller enhancements, and enrichments typically in the range 20–49%. Other observations of AIR have indicated that enhancement or depletion effects may often be found within the range of $\pm 10\%$.

It appears that AIR interferences are very sensitive to operating conditions of the nebulizer. It was found experimentally that variations in gas flow rate, or solution aspiration rate, could cause variations in signal enhancements from 0 to 200% for certain systems, such as the Na–Mg system considered previously. This means that, practically, extreme care must be taken in many analytical sample systems, for example, during the determination of trace or ultratrace elements in the presence of significant concentrations of major ions, such as Ca, Mg, or Na. It may also be that charging nebulizers, or nebulizer operating conditions, could cause a hitherto unmeasured interference effect to appear.

The mechanism of AIR interference processes is not entirely clear at the present time. Thermodynamic considerations, namely the Gibbs' adsorption isotherm, indicate that in a binary solution mixture there will be preferential surface depletion of one species of the mixture. If one imagines the formation of small droplets taking place by the stripping of the surface layer of the liquid stream, as described in Section 8.5.1, then it is possible to imagine that the aerosol droplets reaching the plasma are in fact more representative of the surface properties of the original solution nebulized than of the bulk properties. Consequently, if the surface properties of the solution result in a nonuniform distribution of ions at the surface, it is not unreasonable to assume that if the droplets originate from this surface, they may have different properties from droplets originating from the bulk solution. It is also possible to consider the distributions of ions at a gas–water interface in terms of an electrical double-layer effect [66]. It has been suggested that at a gas–liquid interface, a significant number of water molecules are oriented with the air oxygen atoms facing outward and the hydrogen atoms facing the water bulk. This establishes a polarization charge, which leads to the formation of a double layer. Positive charge at the surface therefore initially attracts a layer of anions, which in turn attracts a looser, more diffuse, layer of cations toward the inner liquid volume. In the rapid droplet formation process, smaller more mobile ions might be anticipated to be drawn more readily to such a double layer than less mobile ions with larger hydrated radii. This could be a major factor determining the magnitude of enrichment. However, at the present time, such considerations are largely conjectural.

8.7.5. Aerosol Washout from Spray Chambers

The gas flow patterns through ICP spray chambers result in far greater stagnant volumes than present in AA spray chambers. This is partly due to the spray chamber design, but also to the considerably lower gas flow, typically 1 L · min^{-1}, compared with approximately 15 L · min^{-1} used with AA systems. Consequently, aerosol passing into the spray chamber will often reach a stagnant region, where its flow is much lower than that of the majority of the aerosol. This results in a substantial washout time for the spray chamber. The washout rate will be a function of the chamber volume and also of its design. Chambers with a smoother gas flow will tend to have less problems resulting from stagnant aerosol than chambers with less favorable gas flow patterns and larger volumes. The decay curve for a particular nebulizer and spray chamber combination will be characteristic and typically exponential in form. The primary problem experienced experimentally occurs when operating over a large linear dynamic concentration range. The ability to operate with solution concentrations covering many orders of magnitude is, of course, one of the major advantages of the ICP. However, if consecutive samples are at high and low concentrations, which may vary by three orders of magnitude or more, then if a high concentration solution sample is followed immediately by a low concentration solution sample, a carry-over problem will result. Unless a long washout time is allowed, this problem can be alleviated somewhat by rapid solvent pumping after each sample, or generally by operating at high (>3 mL · min^{-1}) solution aspiration rates.

8.8. OPTIMIZATION OF AEROSOL CHARACTERISTICS FOR ICP ANALYSIS

On the basis of the detailed discussion of nebulization and aerosol transport phenomena discussed earlier, it is clear that there are certain obvious criteria for the design and operation of nebulizers and spray chambers. For the maximum utilization to be made of the solution, it is important that the nebulizer should generate aerosol with the finest possible droplet size distribution, such that the minimum amount of aerosol will pass to drain. Equally it is important that the spray chamber be designed such as not to permit large aerosol particles to pass through to the plasma torch or the plasma itself. Large droplets could cause the onset of both volatilization interferences, blockage due to impaction on the sample injection tube tip of the torch, and poor precision due to nonuniform vaporization of large droplets in the plasma itself. As yet, optimum characteristics of aerosol size distribution and loading in the plasma have not yet been determined for pneumatic nebulizers. Studies with ultrasonic nebulizers, in which it is possible to produce a much higher aerosol loading in the plasma, have shown that increasing the mass loading in the plasma beyond a certain level causes certain interference effects, such as those due to the presence of easily ionizable elements, to become more severe [67]. It is reasonable to assume that one cannot hope to improve signal-to-background ratios in the plasma indefinitely simply by causing more analyte to enter the plasma. This will particularly be true if the analyte is not desolvated, in which case the increased solvent loading will substantially lower the temperature of the plasma. Adding more solvent to the plasma will require raising ICP power to avoid plasma extinction. This rise in power will adversely effect signal-to-background ratios, and balance the gain achieved by increased analyte loading. If the sample contains major elements that emit line-rich spectra or recombination continua, then, increased sample loading will increase the analyte signal and the background signal in the same proportion once the point has been reached that the background is dominated by the line wings or the recombination continuum (cf. Part 1, Section 7.3.3).

A further consideration in spray chamber design is the minimization of dead volume in the spray chamber, so as to reduce the carry-over problem described in the previous section. It is also of considerable importance to ensure a uniform flow pattern of the aerosol through the spray chamber, in order that fluctuation in the gas pressure, produced by the nebulization process, cannot translate into plasma flicker noise. Other aspects of practical spray chamber design have been considered in Part 1, Chapter 6, and so will not be repeated here.

8.8.1. Suggested Additional Sources of Information on Aerosol Generation and Characterization

Much of the literature on aerosols is highly specialized in nature and difficult to relate to problems of analytical ICP–AES. However, there are certain texts that contain excellent discussions concerning aerosols, which can be used to provide additional references and different approaches to the subject than this one. The monographs of Mavrodineanu and Boiteux [68] and Alkemade and Hermann [69] contain excellent chapters on aerosol introduction for flame spectrometry, and the general subject of aerosol technology is covered in a very accessible manner by Mercer [21].

REFERENCES

1. H. M. N. H. Irving, H. Freiser, and T. S. West, Eds., *Compendium of Analytical Nomenclature*, Chapter 18. Pergamon Press, Oxford (1978).
2. D. D. Smith and R. F. Browner, *Anal. Chem.* **54,** 533 (1982).
3. R. F. Browner, A. W. Boorn, and D. D. Smith, *Anal. Chem.* **54,** 1411 (1982).

4. J. W. Novak and R. F. Browner, *Anal. Chem.* **52,** 287 (1980).
5. J. W. Novak and R. F. Browner, *Anal. Chem.* **52,** 792 (1980).
6. M. S. Cresser and R. F. Browner, *Spectrochim. Acta* **35B,** 73 (1980).
7. J. F. Alder, R. M. Bombelka, and G. F. Kirkbright, *Spectrochim. Acta* **35B,** 163 (1980).
8. A. W. Boorn and R. F. Browner, *Anal. Chem.* **54,** 1402 (1982).
9. P. W. J. M. Boumans and M. C. Lux-Steiner, *Spectrochim. Acta* **37B,** 97 (1982).
10. R. A. Mugele and H. D. Evans, *Ind. Eng. Chem.* **43,** 1317 (1951).
11. N. A. Fuchs, *The Mechanics of Aerosols.* Permagon Press, Oxford (1964).
12. S. Nukiyama and R. Tanasawa, *Experiments on the Atomization of Liquids in an Air Stream*, (E. Hope, Transl.). Defense Research Board, Department of National Defense, Ottawa, Canada (1950).
13. F. Häusser and G. M. Strobl, *Z. Tech. Phys.* **5,** 154 (1924).
14. J. Stupar and J. B. Dawson, *Appl. Opt.* **7,** 1351 (1968).
15. S. Nukiyama and R. Tanasawa, *Experiments on the Atomization of Liquids in an Air Stream*, (E. Hope, Transl.). Defense Research Board, Department of National Defense, Ottawa, Canada (1950), pp. 3–4.
16. T. T. Mercer, *Health Physics* **10,** 873 (1964).
17. K. R. May, *J. Sci. Instrum.* **27,** 128 (1950).
18. W. P. Hendrix and C. Orr, *Rev. Sci. Instrum.* **35,** 1373 (1964).
19. J. B. Willis, *Spectrochim. Acta* **23A,** 811 (1967).
20. P. E. Morrow and T. T. Mercer, *Am. Ind. Hyg. Ass. J.* **25,** 8 (1964).
21. T. T. Mercer, *Aerosol Technology in Hazard Evaluation*, p. 82. Academic, New York (1973).
22. M. S. Cresser and R. F. Browner, *Appl. Spectrosc.* **34,**, 364 (1980).
23. D. D. Smith and R. F. Browner, *Anal. Chem.* **55,** 373 (1983).
24. J. V. Dave, *Appl. Opt.* **10,** 2035 (1971).
25. E. A. Powell, R. A. Cassanova, C. P. Bankston, and B. T. Zinn, in B. T. Zinn, Ed., "Experimental diagnostics in gas phase combustion systems," *Progress in Astronautics and Aeronautics*, vol. 53, American Institute of Aeronautics and Astronautics, New York (1977), p. 449.
26. W. R. Zinky, *Air Pollution Control Assoc. J.* **12,** 578 (1962).
27. A. L. Wertheimer and W. L. Wilcox, *Appl. Opt.* **15,** 1616 (1976).
28. J. Cornillaut, *Appl. Opt.* **11,** 265 (1972).
29. N. Mohamed, R. C. Fry, and D. L. Wetzel, *Anal. Chem.* **53,** 639 (1981).
30. R. F. Browner, M. S. Black, and A. W. Boorn, in R. M. Barnes, Ed., *Developments in Plasma Spectrochemical Analysis*, Heyden, Philadelphia (1981), p. 238.
31. R. Kull and R. F. Browner, *23rd Coll. Spectrosc. Int. and 9th Int. Conf. At. Spectrosc., Amsterdam 1983, Spectrochim. Acta* **38B,** *Supplement*, Abstract No. 51 (1983).

32. A. W. Boorn, M. S. Cresser, and R. F. Browner, *Spectrochim. Acta* **35B,** 823 (1980).

33. C. T. J. Alkenade, Ph.D. Dissertation, University of Utrecht, Utrecht.

34. A. Gustavsson, *ICP Information Newslett.* **5,** 312 (1979).

35. P. A. M. Ripson and L. de Galan, *Spectrochim. Acta* **36B,** 71 (1981).

36. R. F. Browner and D. D. Smith, *Anal. Chem.* **55,** 373 (1983).

37. P. A. M. Ripson and L. de Galan, *Anal. Chem.* **55,** 372 (1983).

38. L. R. Layman and F. E. Lichte, *Anal. Chem.* **54,** 638 (1982).

39. L. Strom, *Rev. Sci. Instrum.* **40,** 778 (1969).

40. B. Vonnegut and R. L. Neubauer, *J. Colloid Sci.* **7,** 616 (1952).

41. M. N. Topp, *Aerosol Sci.* **4,** 17 (1983).

42. J. D. Wilcox and R. June, *J. Franklin Inst.* **271,** 169 (1961).

43. W. R. Lane, *Ind. Eng. Chem.* **43,** 1312 (1951).

44. O. G. Raabe, in B. Y. H. Liu, Ed., "The generation of aerosols of fine particles," *Fine Particles*, Academic, New York (1976), p. 57.

45. M. D. Bitron, *Ind. Eng. Chem.* **47,** 23 (1955).

46. J. D. Bassett and A. W. Bright, *J. Aerosol Sci.* **7,** 47 (1976).

47. R. M. G. Boucher and J. Kreuter, *Ann. Allergy* **26,** 591 (1968).

48. R. J. Lang, *J. Acoust. Soc. Am.* **34,** 6 (1962).

49. T. T. Mercer, R. F. Goddard, and R. L. Flores, *Ann. Allergy* **26,** 18 (1968).

50. A. Faske and R. F. Browner, *Appl. Spectrosc.*, in press (1987).

51. J. W. Novak, D. E. Lillie, A. W. Boorn, and R. F. Browner, *Anal. Chem.* **52,** 579 (1980).

52. R. H. Scott, V. A. Fassel, R. N. Kniseley, and D. E. Dixon, *Anal. Chem.* **46,** 75 (1974).

53. S. Greenfield, I. Ll. Jones, H. McD. McGeachin, and P. B. Smith, *Anal. Chim. Acta* **74,** 225 (1975).

54. F. J. M. J. Maessen, G. Kreuning, and J. Balke, *Spectrochim. Acta* **41B,** 3 (1986).

55. J. Porstendörfer, J. Gebhart, and G. Rösbig, *J. Aerosol Sci.* **8,** 371 (1977).

56. M. W. Blades and B. L. Caughlin, *Spectrochim. Acta* **40B,** 579 (1985).

57. J. M. Mermet and C. Trassy, in R. M. Barnes, Ed., *Developments in Atomic Plasma Spectrochemical Analysis*, Heyden, Philadelphia (1981), p. 245.

58. P. W. J. M. Boumans and F. J. de Boer, *Spectrochim. Acta* **30B**, 309 (1975).

59. A. Kundt and E. Warburg, *Ann. Phys.* **155,** 337 (1975).

60. S. Greenfield, H. McD. McGeachin, and P. B. Smith, *Anal. Chim. Acta* **84,** 67 (1976).

61. R. L. Dahlquist and J. W. Knoll, *Appl. Spectrosc.* **32,** 1 (1978).

62. R. M. Barnes and R. G. Schleicher, *Spectrochim. Acta* **30B,** 109 (1975).

63. G. F. Larson, V. A. Fassel, R. H. Scott, and R. N. Kniseley, *Anal. Chem.* **47,** 238 (1975).

64. D. J. Kalnicky, R. N. Kniseley, and V. A. Fassel, *Spectrochim. Acta* **30B,** 511 (1975).

65. J. A. Borowiec, A. W. Boorn, J. H. Dillard, M. S. Cresser, R. F. Browner, and M. J. Matteson, *Anal. Chem.* **52,** 1054 (1980).

66. L. B. Loeb, *Static Electrification*, pp. 61–80. Springer, Berlin (1958).

67. P. W. J. M. Boumans and F. J. de Boer, *Spectrochim. Acta* **31B**, 355 (1976).

68. R. Maurodineanu and H. Boiteux, *Flame Spectroscopy*, pp. 85–114. Wiley, New York (1965).

69. C. T. J. Alkemade and R. Herrmann, *Fundamentals of Analytical Flame Spectroscopy*, pp. 43–68. Wiley, New York (1979).

CHAPTER

9

PLASMA MODELING AND COMPUTER SIMULATION

MAHER I. BOULOS

Department of Chemical Engineering
University of Sherbrooke
Sherbrooke, Québec,
Canada

RAMON M. BARNES

Department of Chemistry,
University of Massachusetts
Amherst, Massachusetts

9.1. Introduction
9.2. Mathematical Models for the Calculation of the Temperature, Flow, and Concentration Fields
- 9.2.1. One-Dimensional Models
- 9.2.2. Two-Dimensional Models
- 9.2.3. Governing Equations and Boundary Conditions
- 9.2.4. Typical Results

9.3. Single Particle or Aerosol Trajectories and Temperature Histories
- 9.3.1. Governing Equations
- 9.3.2. Typical Results

9.4. Plasma–Particle Interaction Effects
- 9.4.1. Governing Equations
- 9.4.2. Typical Results

9.5. Emission Pattern for a Spectrochemical ICP
- 9.5.1. Calculation Procedure
- 9.5.2. Typical Results

9.6. Nonequilibrium Two-Dimensional, Two-Temperature Models
- 9.6.1. Torch Geometry and Basic Assumptions
- 9.6.2. Governing Equations and Boundary Conditions
- 9.6.3. Calculation Procedure
- 9.6.4. Results and Discussion

9.7. Refinements, Prospects, and Future Developments
References

9.1. INTRODUCTION

Since its development in the early 1960s the inductively coupled plasma (ICP) has been considered as an excellent laboratory tool for such applications as spheroidization, chemical synthesis, nuclear rocket simulation, and extractive metallurgy. One of its widest applications has been as an atom and ion source or atom reservoir in spectrochemical analysis in which it is being increasingly used on a commercial basis. Comprehensive reviews on the subject have been published by Eckert [1] and Barnes [2–6].

The success of these applications, however, depends to some extent on an understanding of the torch characteristics and the operating parameters that control them. Although studies have been carried out to measure the electron and atomic densities, temperature, velocity, and concentration fields in the ICP under different operating conditions, special attention also has been given to the development of mathematical models for the quantitative description of the transport phenomena that occur in the discharge region. These vary from relatively simple one-dimensional models concerned with the calculation of the radial temperature profile in the induction zone to more elaborate, two-dimensional models used to calculate the temperature, velocity, and concentration fields as well as the complete emission pattern from the torch under different operating conditions.

Although the characteristics of the ICP have been to a large extent established experimentally, complete, quantitative models of the atomization, ionization, and excitation processes have only been recently developed owing to the complex nature of the discharge and its deviation from local thermodynamic equilibrium (LTE) [7–19]. Experimental guidelines have been proposed to quantitate departures from equilibrium [10, 11, 15], and the development of accurate models of non-LTE effects remains as a primary, future objective.

The models reviewed in this chapter will be presented under the following five principal groups:

1. One-dimensional and two-dimensional models concerned with the calculation of the temperature, flow, and concentration fields in the discharge.
2. Calculations of single-particle or aerosol trajectories and temperature histories.
3. Plasma-particle interactions under dense loading conditions.
4. Calculations of the complete emission pattern from an ICP spectrochemical torch.
5. Nonequilibrium, two-temperature model.

9.2 MATHEMATICAL MODELS FOR THE CALCULATION OF THE TEMPERATURE, FLOW, AND CONCENTRATION FIELDS

9.2.1. One-Dimensional Models

The first mathematical models proposed for the inductively coupled plasma were one dimensional [20–25]. These were concerned mainly with the calculation of the radial temperature profiles at the center of the discharge by an energy balance between local energy generation and conduction and radiation heat losses. To allow for the analytical solution of the governing equations, the models had to be kept relatively simple, and with few exceptions the following assumptions were adopted:

- LTE: As will be seen later, this assumption has been maintained for most of the one- and two-dimensional models
- Neglected convective heat transfer, with the exception of the model of Keefer et al. [25]

The proposed one-dimensional models differed, however, in the following ways:

- The degree to which the models took into account the variation of the physical properties of the plasma with temperature (electrical and thermal conductivity)
- Whether or not the models took into account in the energy equation radiation heat losses from the plasma.

Freeman and Chase [20] were the first to adopt the channel model, according to which the plasma in the coil region could be replaced by a cylindrical core of constant electrical conductivity, surrounded by an annulus of nonconducting gas. Neglecting axial heat conduction, radiation, and convective heat transfer, they assumed that, under steady-state conditions, the energy dissipation in the core was balanced by radial heat conduction to the wall of the plasma-confining tube. By solving the simplified Elenbaas-Heller equation and the Maxwell equations, they calculated the power density in the discharge for Ar and N_2 induction plasmas at atmospheric pressure as a function of the magnetic field strength for operating frequencies between 250 kHz and 16 MHz.

In an attempt to refine the channel model, Eckert [21] considered the variation of the electrical and thermal conductivity of the plasma across the core region. Assuming that a power–law distribution for the electric field applied and the radius of the plasma column was identical to that of the confining tube, Eckert obtained radial temperature profiles in the discharge that were in general agreement with experimental data. Later, Eckert [23] modified this model further and brought it "one step closer to reality" by including the radiation losses.

Mensing and Boedeker's [22] one-dimensional analysis of the ICP was based on the simultaneous solution of the energy and the simplified Maxwell equations, taking into account both the conduction and radiation energy losses from the plasma. The resulting six ordinary differential equations were solved using a third-order Runge-Kutta method. To avoid having a mixed-boundary value problem, they proceeded with the integration of all of the equations from the centerline of the discharge, which necessitated specifying, a priori, the temperature of the plasma on the centerline.

Pridmore-Brown [24] solved the simplified energy, magnetic, and electric fields equations as a two-point boundary value problem. However, the computed temperature profiles for an Ar plasma at atmospheric pressure were slightly lower than reported experimental data. This could be due to the rather high electrical conductivity values used in the computations.

The only one-dimensional model to include convective heat transfer is that of Keefer et al. [25]. Their analysis was based on the numerical solution of the corresponding energy, electric, and magnetic field equations. Only the radial convective term was maintained and set to an arbitrary value. Their results showed the radial temperature profile in the discharge region to be significantly influenced by the presence of an inward radial gas stream.

Recently serious attempts were made by Eckert [26, 27] and Aeschbach [28, 29] to include nonequilibrium effects in a one-dimensional model for the ICP torch as used in spectrochemical analysis. Their results reveal important differences between the heavy particle and the electron temperatures in the immediate vicinity of the plasma confinement tube and in the center of the discharge in the presence of strong central flow typical of that used in spectrochemical analysis.

9.2.2. Two-Dimensional Models

While the principal advantage of one-dimensional models is that they offer a relatively simple way of estimating the temperature profiles in the center of the discharge, they suffer from two main limitations: First, they provide no information about the temperature field outside the induction zone, and second, they cannot be used for the calculation of the flow field in the torch. A number of two-dimensional models [28–48] was developed. These have the following points in common:

- They take into account conductive, convective, and radiative heat transfer.
- The plasma is assumed to be optically thin.

- They all assume LTE.
- They take into account the variation of the thermodynamic and transport properties of the plasma with temperature.
- They maintain the one-dimensional electric and magnetic fields assumption.

They varied, however, in the following points:

- Whether, or not, they include the full-momentum transfer equations.
- Whether they solve the transient or steady-state equations. In either case they aim at steady-state solutions.

The first two-dimensional model of the ICP was that proposed by Miller and Ayen [30], who calculated the two-dimensional temperature field by solving the corresponding energy equation simultaneously with the one-dimensional electric and magnetic field equations. They treated the radiation heat losses as a volumetric heat sink and assumed the plasma to be optically thin.

Since they did not consider momentum transfer, they assumed the flow to be only in the axial direction and neglected radial convective transfer. At the inlet of the torch the velocity profile was taken as being flat with a step-function increase of the velocity near the wall of the plasma confinement tube. As the plasma gas passed through the discharge zone, the axial velocity profile was allowed to change in such a way as to satisfy the principle of conservation of mass. The numerical technique used was based on the finite difference solution of the transient equations in discrete time steps until steady-state conditions were reached.

Barnes and Schleicher [31] and Barnes and Nikdel [32, 33] modified Miller's model to fit operating conditions used in spectrochemical analysis. They imposed on the flow a central jet stream to represent the aerosol carrier gas. By solving the continuity, energy, electric, and magnetic field equations, they calculated the flow and temperature fields under different operating conditions for Ar and various molecular gases [28, 29, 31–36]. Assuming similarity between the velocity and the concentration distribution of species in the torch, they calculated spectral emittance distributions for elements from an injected sample and the continuum background radiation distribution [31].

Many of the principal features of this model have been verified for Ar [34, 35] and recently for N_2 [36] ICP discharges, and lately the model has been applied to an O_2 ICP. As a consequence of this approach, a number of general features of the spectrochemical ICP have been predicted. These include the following:

1. Improvement in operation and spectrochemical results at higher fre-

quency. Better signal-to-background ratios, lower background continuum, and easy plasma operation were predicted [31] and subsequently verified.

2. Improvement in plasma torch geometry and gas flow design [37, 38]. The importance of the configuration factor and high swirl velocity for practical spectrochemical torches was demonstrated.
3. Operation of molecular gas ICP discharges for spectrochemical analysis. The characteristics, operating parameters, and advantages of molecular gases (for example, nitrogen) for spectrochemical applications were predicted [32], extended to air and oxygen, and subsequently verified [36, 39–42]. High analyte and solvent loading were predicted and achieved with molecular gas ICP discharges, and air ICP discharges now are applied for industrial on-line process control operations.
4. Computation of non-LTE effects in an Ar ICP. Computation of spatial changes in electron temperature and number density in an ICP was performed [28, 29] and has led to further modeling of non-LTE effects.

Although the above models represented an important step forward, it was obvious that a detailed description of the characteristics of the flow and temperature fields in the discharge region could only be achieved through the incorporation of the full-momentum transfer equations. Independently, Delettrez [43] and Boulos [44] calculated the two-dimensional velocity and temperature fields in the induction plasma by solving the corresponding continuity, momentum, and energy equations simultaneously with the one-dimensional electric and magnetic field equations. Their mathematical approach was different, however, since Delettrez [43] solved the transient transport equations in terms of temperature and the axial and radial velocity components, while Boulos et al. [44–46], on the other hand, solved the steady-state, two-dimensional momentum and energy equations in terms of the enthalpy, stream function, and vorticity simultaneously with the one-dimensional electric and magnetic field equations and the phase difference between them.

The results obtained for an Ar plasma at atmospheric pressure assuming laminar flow conditions revealed an important influence of the electromagnetic forces on the flow field in the torch. This was responsible for the formation of two recirculation eddies in the discharge region, with the downstream eddy being swept away with the increase of the plasma gas flow rate.

Recent work by Boulos and his collaborators [47, 48] further refined this model and improved its computational stability by solving the continuity, momentum and energy equations in terms of their primitive variables, that is, velocity, pressure, and enthalpy rather than the stream function, vorticity, and enthalpy used earlier. A different numerical algorithm is also used in this case. This is known as "Semi-Implicit Method for Pressure-Linked Equations, Revised,'' or SIMPLER, and was developed by Patankar and Spalding [49]. A detailed comparison between the performance of the two computer codes is given by Mostaghimi et al. [47].

In a comparison of the refined model employing the SIMPLER algorithm [47, 48] with the modified Miller model, Kovacic and Barnes observed that although the modified Miller model compiled and computed at twice the speed of the model of Mostaghimi et al. for an oxygen ICP and provided similar temperature fields, the latter model provided a more complete description of the fluid dynamics of the plasma.

9.2.3. Governing Equations and Boundary Conditions

For a typical induction plasma arrangement, illustrated by any of the torch geometries or sizes given in Fig. 9.1, the two-dimensional continuity, momentum, energy, and mass-transfer equations can be written in terms of their primitive variables in the axially symmetric cylindrical system of coordinates as follows:

Continuity:

$$\frac{1}{r}\frac{\partial}{\partial r}(r\rho v) + \frac{\partial}{\partial z}(\rho u) = 0 \tag{9.1}$$

Momentum transfer:

$$\begin{aligned} \rho\left(v\frac{\partial u}{\partial r} + u\frac{\partial u}{\partial z}\right) &= -\frac{\partial p}{\partial z} + 2\frac{\partial}{\partial z}\left(\mu\frac{\partial u}{\partial z}\right) \\ &\quad + \frac{1}{r}\frac{\partial}{\partial r}\left[\mu r\left(\frac{\partial u}{\partial r} + \frac{\partial v}{\partial r}\right)\right] + \rho g \end{aligned} \tag{9.2}$$

$$\begin{aligned} \rho\left(v\frac{\partial v}{\partial r} + u\frac{\partial v}{\partial z}\right) &= -\frac{\partial p}{\partial r} + \frac{2}{r}\frac{\partial}{\partial r}\left(\mu r\frac{\partial v}{\partial r}\right) \\ &\quad + \frac{\partial}{\partial z}\left[\mu\left(\frac{\partial v}{\partial z}\right) + \frac{\partial u}{\partial r}\right)\right] - \frac{2\mu v}{r^2} + \xi\sigma E_\theta H_z \cos\chi \end{aligned} \tag{9.3}$$

Energy transfer:

$$\begin{aligned} \rho\left(v\frac{\partial h}{\partial r} + u\frac{\partial h}{\partial z}\right) &= \frac{1}{r}\frac{\partial}{\partial z}\left(r\frac{k}{c_p}\frac{\partial h}{\partial r}\right) \\ &\quad + \frac{\partial}{\partial z}\left(\frac{k}{c_p}\frac{\partial h}{\partial z}\right) + \sigma E_\theta^2 - P_r \end{aligned} \tag{9.4}$$

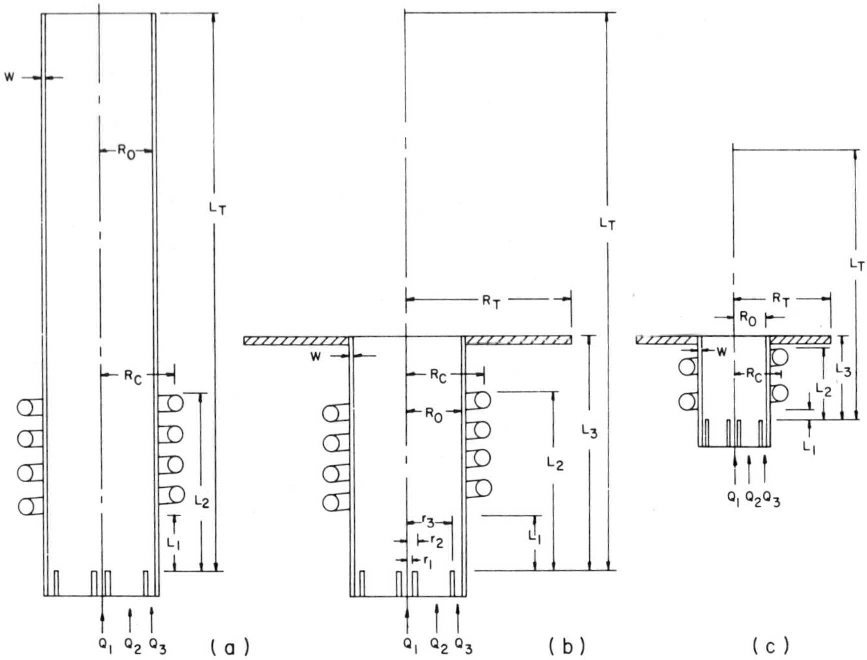

Figure 9.1. ICP torch geometries and dimensions. (*a*) Confined plasma; (*b*) free-discharge plasma; (*c*) spectrochemical torch.

Mass transfer:

$$\rho\left(v\frac{\partial y}{\partial r} + u\frac{\partial y}{\partial z}\right) = \frac{1}{r}\frac{\partial}{\partial r}\left(rD\frac{\partial y}{\partial r}\right) + \frac{\partial}{\partial z}\left(D\frac{\partial y}{\partial z}\right) \tag{9.5}$$

where v and u are the plasma velocity components in the radial and axial directions, respectively; h is the plasma specific enthalpy; y is the mass fraction of a given gaseous component in the mixture; r and z are distances in radial and axial directions; ρ is the density; μ the dynamic viscosity; k the thermal conductivity; c_p the specific heat; D the diffusion coefficient; p is the local pressure; and p_r represents radiation losses per unit volume.

The corresponding one-dimensional electromagnetic field equations are given by eq. (9.6):

$$\begin{aligned} \frac{1}{r}\frac{d}{dr}(rE_\theta) &= -\zeta\omega H_z \sin\chi \\ \frac{dH_z}{dr} &= -\sigma E_\theta \cos\chi \\ \frac{d\chi}{dr} &= \frac{\sigma E_\theta}{H_z}\sin\chi - \frac{\zeta\omega H_z}{E_\theta}\cos\chi \end{aligned} \tag{9.6}$$

where E_θ is the electric field intensity in the θ direction; H_z the axial magnetic field intensity, and χ the phase difference between them; ω is the oscillator angular frequency ($\omega = 2\pi f$); σ the electrical conductivity of the plasma; and ξ the magnetic permeability of free space.

Solution of these equations requires the specification of the appropriate boundary conditions, which will depend on the particular torch configuration and operating conditions used. For a confined plasma, shown in Fig. 9.1*a*, the boundary conditions for Eqs. (9.1–9.4) can be summarized as follows:

Inlet conditions ($z = 0$):

$$v = 0$$

$$u = \begin{cases} Q_1/\pi r_1^2 & r < r_1 \\ 0 & r_1 \le r \le r_2 \\ Q_2/\pi(r_3^2 - r_2^2) & r_2 < r \le r_3 \\ Q_3/\pi(R_0^2 - r_3^2) & r_3 < r \le R_0 \end{cases}$$

$$T = 350 \tag{9.7}$$

Centerline ($r = 0$):

$$v = 0$$

$$\frac{\partial u}{\partial r} = 0$$

$$\frac{\partial h}{\partial r} = 0 \tag{9.8}$$

Wall ($r = R_0$):

$$v = u = 0$$

$$\frac{k}{c_p}\frac{\partial h}{\partial r} = \frac{k_c}{w}(T - T_{w0})$$

$$T_{w0} = 350 \tag{9.9}$$

Exit ($z = L_T$)

$$\frac{\partial h}{\partial z} = 0$$

$$\frac{\partial v}{\partial z} = 0$$

$$\frac{\partial(\rho u)}{\partial z} = 0 \tag{9.10}$$

The boundary conditions for the electromagnetic field, Eq. (9.6), are set at $r = 0$ and $L_1 \leq z \leq L_2$, that is, $\chi = \pi/2$, $E_\theta = 0$.

$$H_z = \frac{H_{c\infty}}{2(L_2 - L_1)}\left[\frac{L_2 - z}{\left\{R_c^2 + (L_2 - z)^2)\right\}^{1/2}} - \frac{L_1 - z}{\left\{R_c^2 + (L_1 - z)^2)\right\}^{1/2}}\right] \tag{9.11}$$

where Q_1, Q_2, Q_3 are the gas flow rates introduced in the different regions of the torch as shown in Fig. 9.1. The corresponding torch dimensions, r_1, r_2, r_3, L_1, L_2, R_0, w, and L_T are also indicated in Fig. 9.1. $H_{c\infty}$ is the magnetic field intensity for an infinite solenoid defined in Eq. (9.12) as

$$H_{c\infty} = \frac{NI}{L_2 - L_1} \tag{9.12}$$

Normally $H_{c\infty}$ is not specified, but rather the total power input to the plasma P_0 is known. This is set as an integral type boundary condition.

$$P_t = \int_{z=L_1}^{L_2} \int_{r=0}^{R_0} 2\pi r\, P dr dz \tag{9.13}$$

were $P_t \neq P_0$, a correction factor to the electric field and magnetic field is found.

$$\alpha_c = \sqrt{\frac{P_0}{P_t}} \tag{9.14}$$

Multiplying the electromagnetic fields by α_c insures that the condition $P_t = P_0$ is satisfied.

9.2.4. Typical Results

As an illustration of the present induction plasma modeling capabilities, typical results will be given in this section for the calculated flow, temperature, and concentration fields obtained using these models. The computations were made mostly for atmospheric pressure Ar and N_2 plasmas, although in other studies they have been achieved for He and O_2. The range of torch dimensions and operating conditions used is summarized in Table 9.1.

One of the obvious applications of these models was to determine the effect of the different operating parameters on the flow and temperature field in the discharge.

Effect of Swirl in the Sheath Gas. Since the presence of a swirl in the plasma or sheath gas has a stabilizing effect on the operation of an induction plasma

Table 9.1. Torch Sizes and Operating Conditions

Tube diameter (mm)	$d_0 = 18 - 50$
Frequency (MHz)	$f = 0.3 - 26.3$
Power in the plasma (kW)	$P_t = 1 - 15$
Total gas flow rate ($L \cdot min^{-1}$)	$Q_0 = 10 - 50$
Powder or aerosol carrier gas flow rate ($L \cdot min^{-1}$)	$Q_1 = 0.0 - 7.0$
Inlet swirl velocity ($m \cdot s^{-1}$)	$v_\theta = 0 - 20$

[38], the flow and temperature fields in the torch were calculated in the presence and absence of a swirl. This required the addition of an equation for the tangential momentum conservation to those listed previously. In this particular study the momentum equations were written in term of the stream function and vorticity, as given by Boulos et al. [45] and Gagné [46]. A schematic of the torch is given in Fig. 9.1*a*. It consisted of a 30-mm i.d. quartz tube with a four-turn induction coil placed at 15 mm from the inlet gas. The intermediate gas (Q_2) and outer gas (Q_3) flow rates were fixed at 2 and 18 $L \cdot min^{-1}$, respectively. No central carrier gas was admitted in this case ($Q_1 = 0$). The oscillator frequency was set to 3 MHz, and the net power dissipated in the discharge was 3 kW.

The computed flow and temperature fields for a torch oriented vertically upward in the absence and presence of a swirl ($\overline{v}_\theta = 13.3\ m \cdot s^{-1}$) are given in Figs. 9.2 and 9.3, respectively. The tangential velocity component (Fig. 9.3*c*), which was limited at the outer gas inlet of the torch, drops rather rapidly from its initial maximum value of 19.8 to $< 12\ m \cdot s^{-1}$ over approximately 15 mm from the gas inlet. However, swirl velocities as high as 19.6 $m \cdot s^{-1}$ reappear downstream around the center of the coil owing to the constriction of the flow by the inward electromagnetic pumping. A comparison of Figs. 9.2*a* and 9.3*a* reveals that the presence of a swirl results in a slight outward displacement of the recirculation eddy in the fireball with a corresponding decrease of the backflow along the centerline of the torch. The effect of the swirl on the corresponding temperature fields (Figs. 9.2*b* and 9.3*b*) is relatively much smaller.

Effect of Confinement. To determine the effect of plasma confinement, computations were made for a free plasma discharge under the same operating conditions as those used for the confined plasma. The geometry and dimensions for the torch used are shown in Fig. 9.1*b*. While having the same diameter as that used for the confined plasma calculations, the plasma confining tube extended only 15 mm beyond the end of the induction coil. At this point the plasma emerged as a free jet in an ambient atmosphere, which was assumed to be the same as the plasma gas (Ar). In both confined and free plasma cases, the dimensions of the induction coil and the gas distributer were identical.

Typical streamlines, temperature, and swirl contours obtained for a plasma-

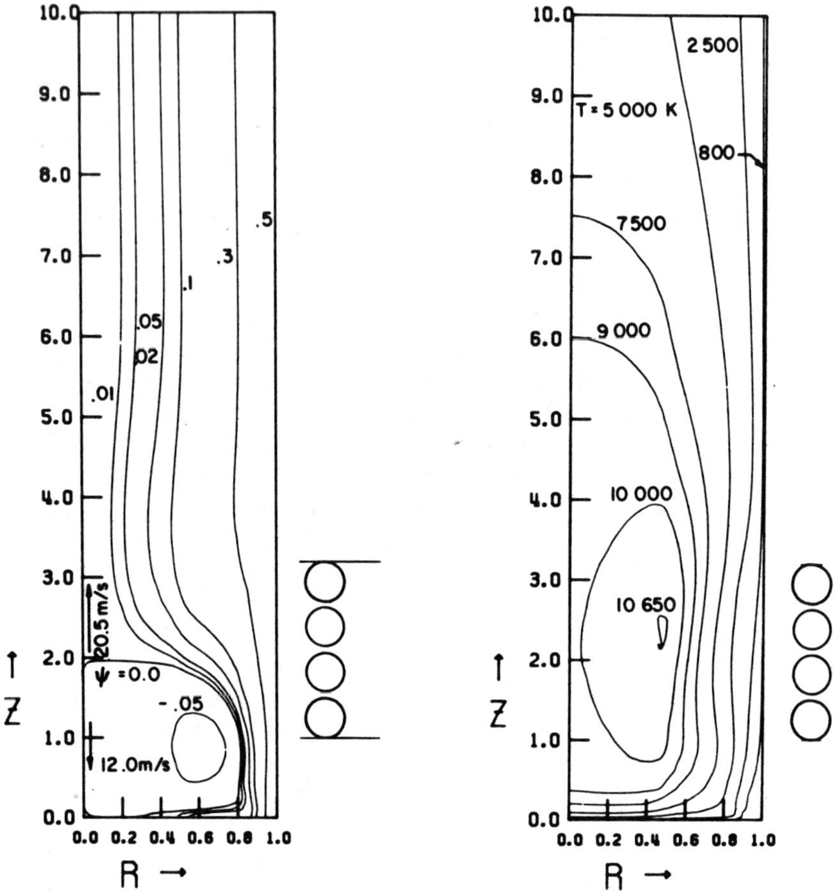

Figure 9.2. Flow (*a*) and temperature (*b*) fields for a confined plasma in the absence of swirl: $f = 3$ MHz, $P_t = 3.0$ kW, $Q_1 = 0.0$, $Q_2 = 2.0$, $Q_3 = 18.0$ L · min^{-1}, and $\overline{v}_\theta = 0.0$. (*a*) Streamlines; (*b*) temperature.

oriented, vertically upward under free discharge conditions and with an inlet average swirl velocity of 13.3 m · s^{-1} are given in Fig. 9.4. The flow field in the induction zone is similar to that obtained for the confined plasma under the same operating conditions. Beyond the confinement region, the hot plasma gas streams vertically upwards entraining a substantial amount of ambient gas. As shown in Fig. 9.4*a*, the mass flow rate of the entrained gas, as indicated by the values of the stream function ψ, can be larger than the plasma gas itself. The entrained gas, however, does not seem to mix with the plasma tail flame and hardly exerts an influence on the centerline axial velocity.

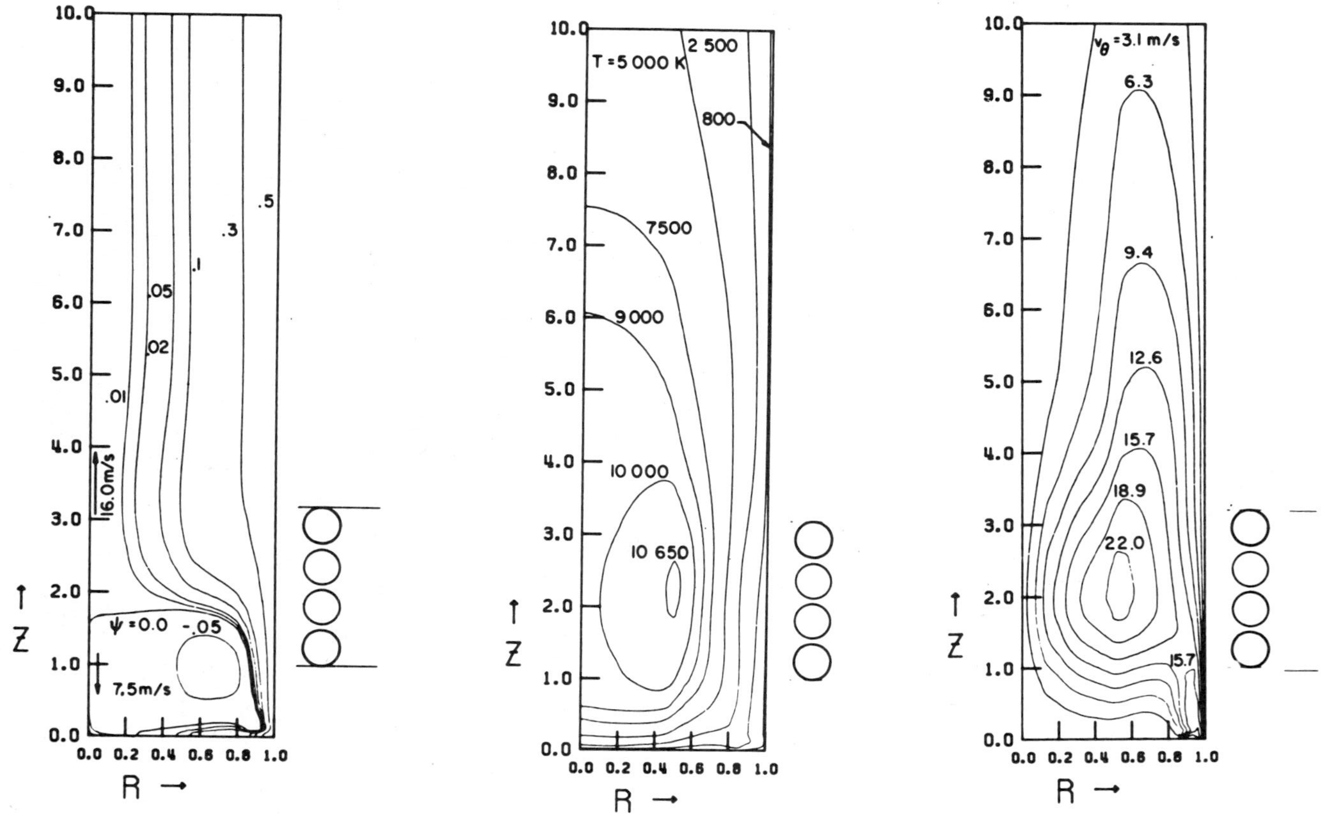

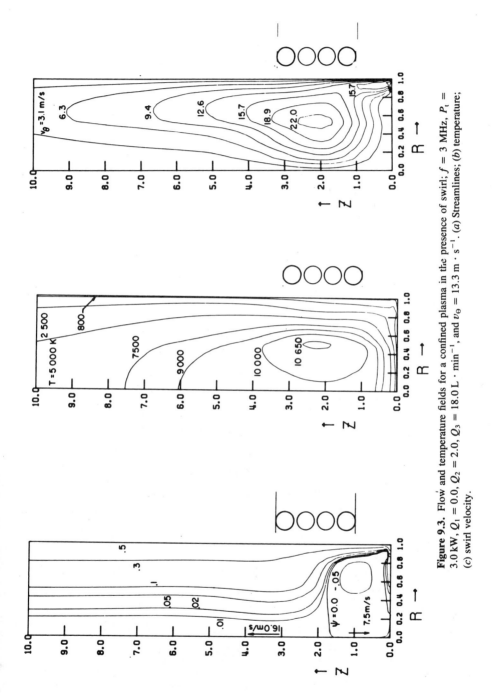

Figure 9.3. Flow and temperature fields for a confined plasma in the presence of swirl; f = 3 MHz, P_t = 3.0 kW, $Q_1 = 0.0$, $Q_2 = 2.0$, $Q_3 = 18.0\ \mathrm{L \cdot min^{-1}}$, and $v_\Theta = 13.3\ \mathrm{m \cdot s^{-1}}$. (*a*) Streamlines; (*b*) temperature; (*c*) swirl velocity.

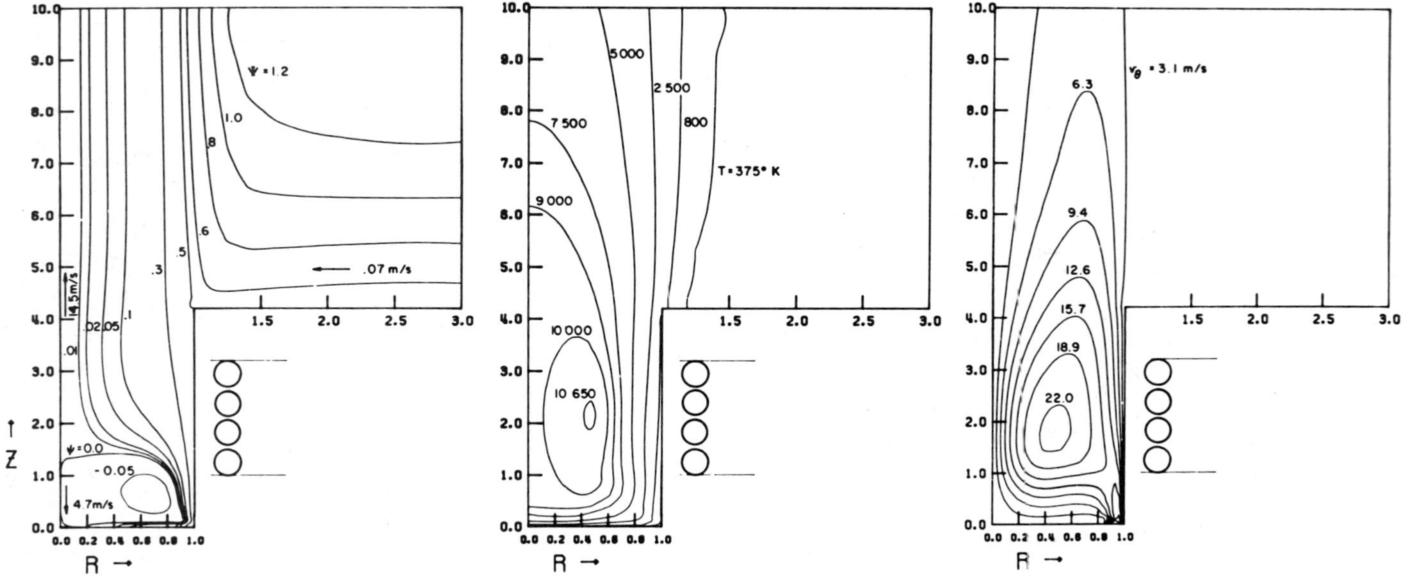

Figure 9.4. Flow and temperature fields for a plasma under free discharge conditions in the presence of swirl: $f = 3$ MHz, $P_t = 3.0$ kW. $Q_1 = 0.0$, $Q_2 = 2.0$, $Q_3 = 18.0$ L · min^{-1}, and $\overline{v}_\Theta = 13.3$ m · s^{-1}. (*a*) Streamlines; (*b*) temperature; (*c*) swirl velocity.

Little change is predicted in the temperature field in the coil region under confined or free discharge conditions (Fig. 9.4*b*). As the gases emerge from the confining tube, however, they heat up the ambient gas giving rise to its typical laminar contours. Calculations made with different plasma gas flow rates showed an increase in the expansion of the discharge at the exit of the torch with decreasing plasma gas flow rate.

In spite of the relatively high swirl velocity in the sheath gas at the inlet of the torch (19.8 $m \cdot s^{-1}$), the tangential velocity field (Fig. 9.4*c*) decays to < 7 $m \cdot s^{-1}$ at the exit of the plasma confinement tube.

In contrast to the assumption of hot Ar passing into ambient Ar, for practical spectrochemical applications, the Ar plasma flows into ambient air, which rapidly cools the fringes of the discharge [34] principally by quenching reactions and heat conduction. As noted by Barnes and Nikdel [50] the mixing of ambient N_2 above and within an Ar ICP differs substantially from that of gas mixing at ambient temperature.

Effect of Frequency. The effect of frequency on the flow and temperature fields in the induction plasma was determined for an arrangement used in spectrochemical analysis, with configuration and principal dimensions given in Fig. 9.1*c*. This torch was rather similar to that used earlier for the free discharge calculations, except for the smaller overall dimensions and the use of a two-turn induction coil. The total plasma gas was again divided between the intermediate gas (Q_2) and outer gas (Q_3), which were set to 1.0 and 9.0 $L \cdot min^{-1}$, respectively, with no central carrier gas, $Q_1 = 0$. The inlet swirl flow was restricted to the outer gas and was set to an average swirl velocity of 13.3 $m \cdot s^{-1}$. The total power dissipated in the plasma was 1.0 kW.

Computations were made for a torch operating at 3, 13, and 26.3 MHz, and typical results obtained at 26.3 MHz are shown in Fig. 9.5. While the increase of frequency resulted in the elimination of the recirculation flow caused by the electromagnetic pumping (Fig. 9.5*a*), the inward radial flow around the middle of the coil region is still observed.

The effect of frequency on the flow field seems to be the result of the increase in the effective power input at the edges of the plasma, with the increase in frequency leading to the outer displacment of the high-temperature, energy dissipation zone toward the plasma confinement tube wall. The temperature contours shown in Fig. 9.5*b* also reveal an overall collapse in the size of the plasma discharge compared with that obtained at lower frequencies. The increase in frequency had a relatively small influence on the tangential velocity field in the torch (Fig. 9.5*c*).

Effects of frequencies between 15 and 70 MHz on the temperature distribution radial input power, energy losses, and central velocities were calculated earlier by Schleicher and Barnes with the modified Miller model [31]. They too

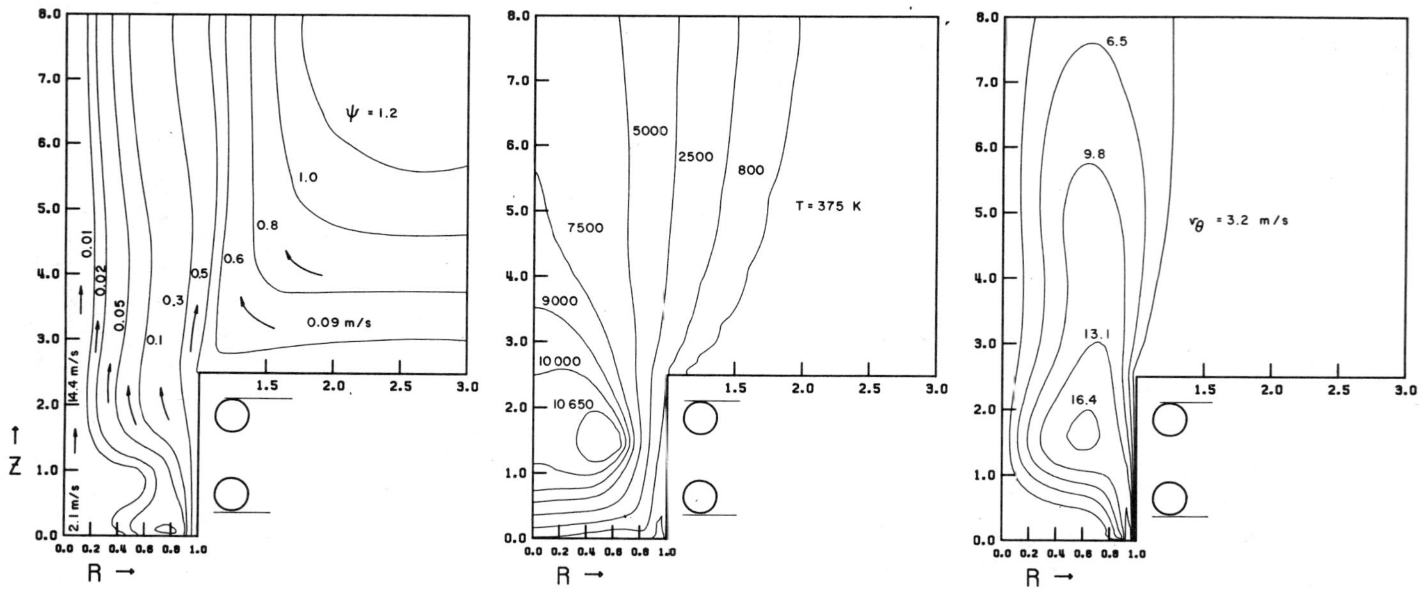

Figure 9.5. Flow and temperature fields for the spectrochemical torch: $f = 26.3$ MHz, $P_t = 1.0$ kW, $Q_1 = 0.0$, $Q_2 = 1.0$, $Q_3 = 9.0$ L · min^{-1}, and $\overline{v}_\Theta = 13.3$ m · s^{-1}. (*a*) Streamfunction; (*b*) temperature; (*c*) swirl velocity; (*d*) concentration.

observed a reduction in the minimum magnetic field, discharge temperature, gas velocity, and radiation energy loss.

Effect of the Central Carrier Gas. To determine the effects of the central carrier gas on the flow and temperature fields in the induction plasma, computations were repeated for the torch used in spectrochemical analysis at a frequency of 26.3 MHz and essentially the same operating conditions as those used in the previous section. The central carrier gas flow rate Q_1 was gradually increased to 0.2, 0.4, 0.6, and 1.0 L · min^{-1}. Results obtained with $Q_1 = 1.0$ L · min^{-1} are given in Fig. 9.6 for the stream function, temperature, swirl velocity, and concentration isocontours. The latter were obtained assuming the presence of a tracer in the carrier gas stream Q_1, which allows it to be identified without having an effect on the thermodynamic and transport properties of the plasma.

In Fig. 9.6*a*, the streamline corresponding to a valve of $\psi = 0.045$ represents the boundary of the central carrier gas. According to the definition of a streamline, no fluid can convectively cross such a line. As will be seen later, however, mass transfer by diffusion can account for the presence of the aerosol tracer elements far beyond a streamline boundary.

In the presence of the central carrier gas (Fig. 9.6*b*), a strong local cooling of the plasma on the upstream side of the discharge zone occurs similar to the effect predicted and measured earlier by Barnes et al. [31, 34, 35]. Even at a central carrier gas flow rate of 1 L · min^{-1}, the central gas still heats to almost 8000 K. The presence of the central carrier gas does not seem to have a noticeable influence on the swirl velocity field in the torch, (Fig. 9.6*c*) or upon the flow in the outer region of the discharge. These effects were confirmed experimentally [35].

The concentration field, expressed in terms of a nondimensional mass fraction y, defined as the ratio of the tracer element concentration to that originally present in the carrier gas at the inlet of the torch, is given in Fig. 9.6*d*. The tracer diffuses far beyond the central channel of the flow and reaches most of the discharge regions, although its maximum concentration still remains on the centerline of the torch.

The effect of central injection on the flow and temperature fields for a confined plasma has also been investigated [48]. The torch geometry in this case was similar to that illustrated in Fig. 9.1*a* with $R_0 = 25$ mm, $R_c = 33$ mm, $r_1 = 1.7$ mm, $r_2 = 3.7$ mm, $r_3 = 18.8$ mm, $L_1 = 10$ mm, $L_2 = 74$ mm, $L_T = 250$ mm, and $w = 2.0$ mm. Computations were carried out with the momentum and energy-transfer equations written in terms of their primitive variables for both Ar and N_2. A summary of the operating conditions is given in Table 9.2.

The calculated isotherms and streamlines for an Ar plasma at a power level of 3 kW and different central injection flow rates ($Q_1 = 1.0$–7.0 L/min) is shown in Fig. 9.7. The corresponding plasma gas flow rate Q_2 is kept constant

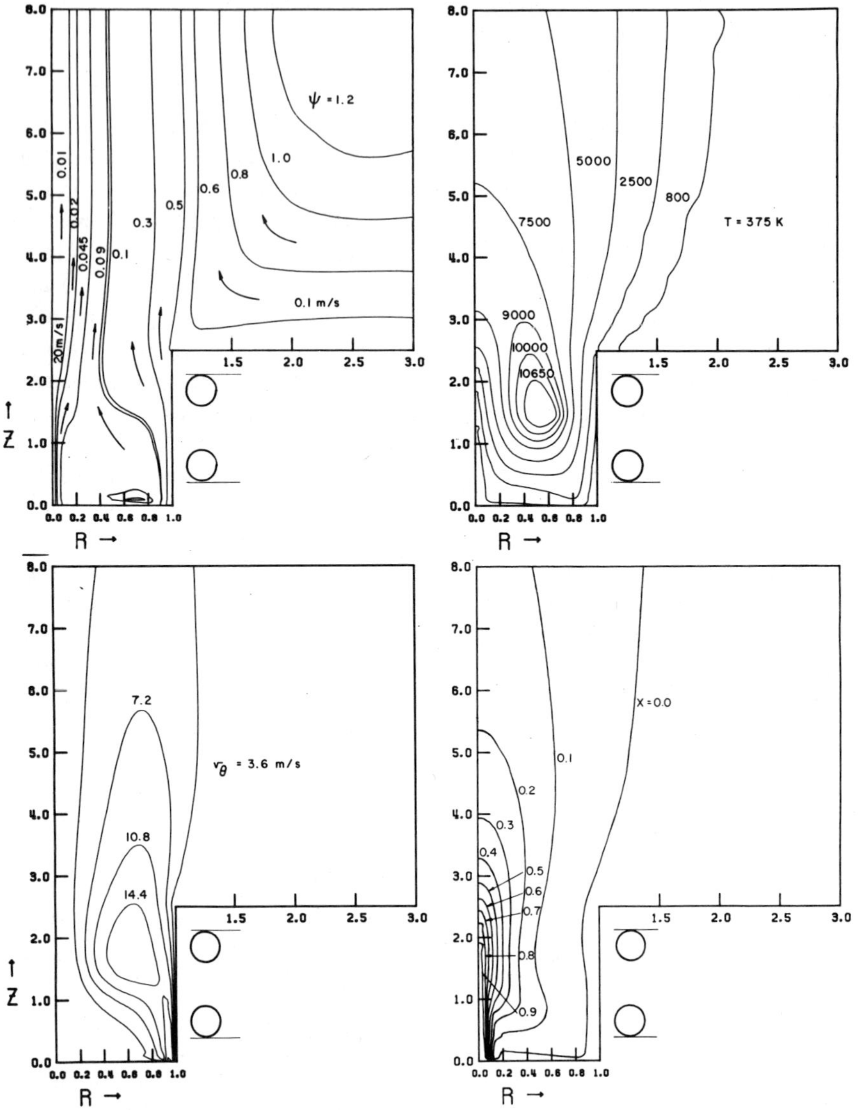

Figure 9.6. Flow, temperature, and concentration fields for the spectrochemical torch in the presence of a central carrier gas: f = 26.3 MHz, P_t = 1.0 kW, Q_1 = 1.0, Q_2 = 1.0, Q_3 = 9.0 L · min^{-1}, and $\bar{v}_\Theta$ = 13.3 m · s^{-1}. (*a*) Streamfunction; (*b*) temperature; (*c*) swirl velocity; (*d*) concentration.

Table 9.2 Torch Operating Conditions [48]

Plasma Gas	f (MHz)	P_0 (kW)	Q_0 ($L \cdot min^{-1}$)	Q_1 ($L \cdot min^{-1}$)	Q_2 ($L \cdot min^{-1}$)	Q_3 ($L \cdot min^{-1}$)
a. Ar	3.0	3.0	20.0	1.0	3.0	16.0
				3.0		14.0
				5.0		12.0
				7.0		10.0
b. N_2	3.0	10.0	20.0	0.0	3.0	17.0
				1.0		16.0
				2.0		15.0
				3.0		14.0
				4.0		13.0
				5.0		12.0
c. Ar	3.0	5.0	20.0	0.0	3.0	17.0
			30.0		4.5	25.5
			40.0		6.0	34.0
			50.0		7.5	42.5
d. Ar	3.0	3.0	20.0	1.0	3.0	16.0
		5.0				
		7.0				
		10.0				

while the outer gas flow rate Q_3 is adjusted so that the total flow rate Q_0 is constant. As the central injection flow is increased, the entrance region close to the centerline cools and the high-temperature region ($T > 9600$ K) is pushed closer to the wall of the plasma-confinement tube. As will be seen later, this results in a decrease of the radiation losses and an increase of the conduction losses to the wall (see Fig. 9.11). The flow fields at the exit of the torch do not change appreciably with the increase of Q_1 over the range investigated although the centerline temperature decreases.

At $Q_1 = 1.0 \text{ L} \cdot \text{min}^{-1}$ a strong circulation appears in the coil region and a small secondary recirculation zone near the wall (Fig. 9.7*a*). As the central injection is increased, the circulation eddy becomes weaker and the backflow on the centerline eventually disappears. By contrast, the recirculation zone on the wall becomes systematically larger.

The effect of the central flow rate on the velocity profile along the centerline is shown in Fig. 9.8. For $Q_1 = 1.0 \text{ L} \cdot \text{min}^{-1}$, the backflow has a maximum velocity of about $7.0 \text{ m} \cdot \text{s}^{-1}$, while the maximum velocity downstream of the coil is about $13 \text{ m} \cdot \text{s}^{-1}$. With an increase of the central flow to $3.0 \text{ L} \cdot \text{min}^{-1}$, the backflow completely disappears, and for $Q_1 = 7.0 \text{ L} \cdot \text{min}^{-1}$ the central jet goes through the coil region before it starts to decay.

The corresponding effect of the central flow rate on the temperature profile

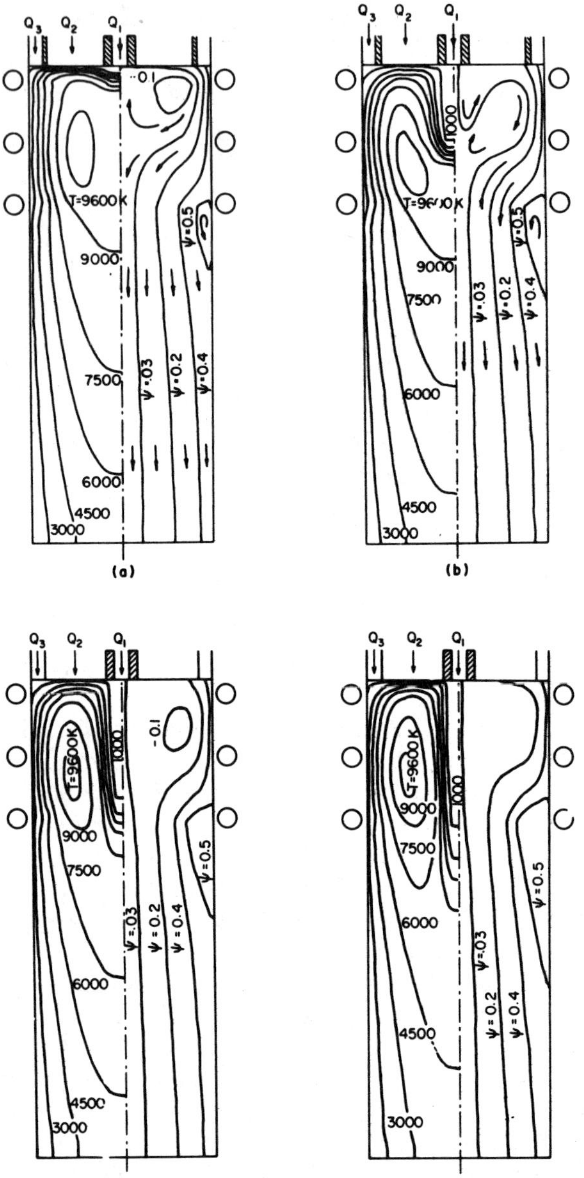

Figure 9.7. Isotherms (*left*) and streamlines (*right*) for Ar plasma, P_0 = 3 kW, Q_0 = 20 L · min^{-1}. (*a*) Q_1 = 1 L · min^{-1}; (*b*) Q_1 = 3 L · min^{-1}; (*c*) Q_1 = 5 L · min^{-1}; (*d*) Q_1 = 7 L · min^{-1}.

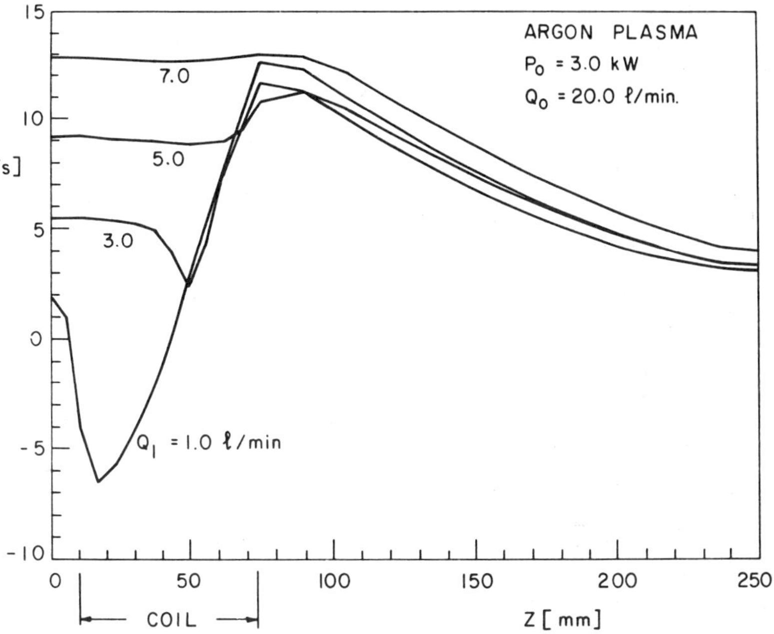

Figure 9.8. Velocity profile along the centerline for an Ar plasma (see Table 9.2*a* for the operating conditions).

along the torch centerline is represented in Fig. 9.9. As expected, a systematic drop in the maximum temperature occurs, and for $Q_1 = 7.0\ \mathrm{L \cdot min^{-1}}$ the gas maintains its inlet temperature for the full length of the coil region.

The radial temperature profiles in the middle of the coil, given in Fig. 9.10, show that increasing the central injection flow rate results in substantial lowering of the temperature along the axis accompanied by a slight increase in the temperature close to the wall of the plasma confinement tube.

The variation of the central injection gas flow rate has virtually no effect on the energy distribution in the torch (Fig. 9.11). Over the range investigated, only 25% of the dissipated energy leaves the torch as enthalpy in the gas, while 15 and 60% of the energy is lost by radiation and conduction to the plasma confinement tube, respectively. These values are dependent on the length of the torch and should be interpreted in absolute terms only for the torch geometry considered. Energy losses as a function of central flow rates also were calculated with the modified Miller model [31], and as the central flow rate increased so did the gas enthalpy.

Similar computations also were carried out for a N_2 plasma at atmospheric pressure. The results showed that in spite of the fact that the power level was considerably higher for the N_2 plasma (10 kW) compared with that for the Ar

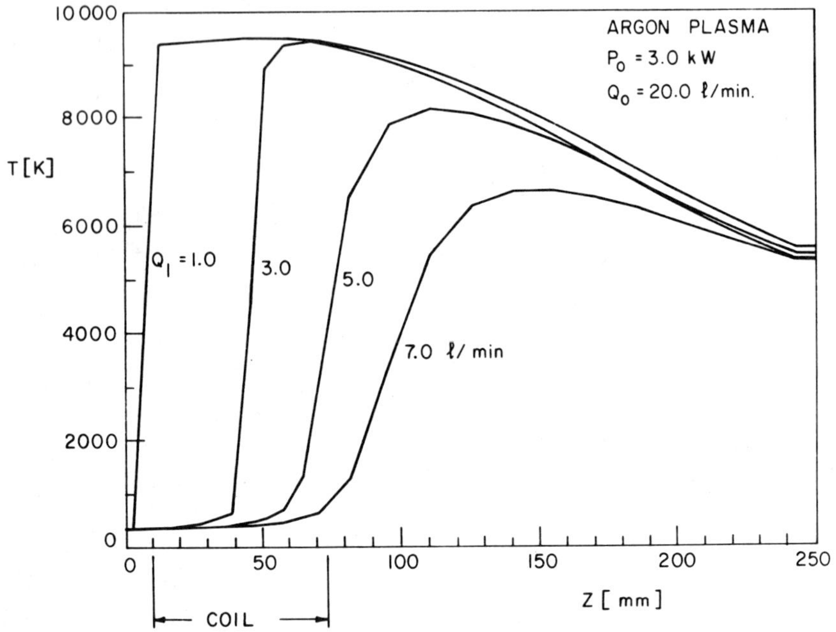

Figure 9.9. Temperature profile along the centerline for an Ar plasma (see Table 9.2*a* for the operating conditions).

plasma (3 kW), the temperature of the plasma was lower, showing a maximum value slightly above 7500 K. The axial velocity along the centerline of the N_2 plasma, on the other hand, was higher than that for the Ar discharge. Recently, calculations for an O_2 plasma in a spectrochemical configuration were completed.

Effect of the Total Gas Flow Rate. To determine the effect of increasing the total gas flow rate Q_0 on the flow and temperature fields in the torch, computations were also carried out. In this case (see Table 9.2*c*), Q_1 is kept at zero while Q_2 and Q_3 are proportionally increased to give values of Q_0 varying between 20 and 50 ($L \cdot min^{-1}$). The Ar plasma gas is taken at atmospheric pressure, and the total power level is 5 kW. The results indicate that increasing the outer gas flow rate considerably cools the regions close to the wall of the plasma confinement tube. Since the comparison is carried out for a constant total power dissipated in the plasma, a corresponding increase of the temperature in the current carrying region is observed.

For a total flow rate of 20 $L \cdot min^{-1}$, the local heat loss through the wall reaches its maximum value close to the downstream end of the coil (Fig. 9.12).

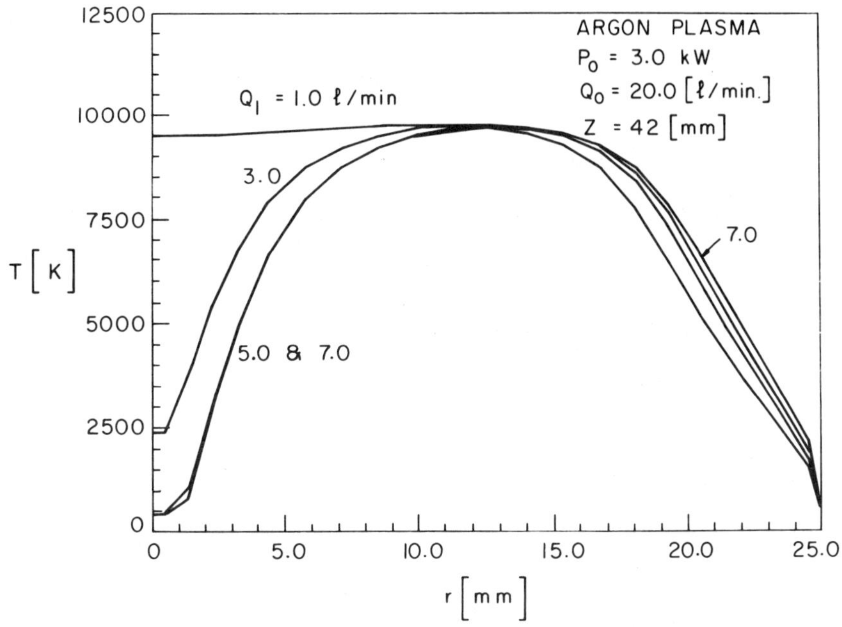

Figure 9.10. Temperature profile in the middle of the coil region for an Ar plasma (see Table 9.2*a* for the operating conditions).

This seems to coincide with the stagnation point where the secondary recirculation zone on the wall starts (cf. Fig. 9.7). As the flow rate is increased, a large reduction in the heat flux to the wall is observed in the coil region. The point of maximum heat flux, however, moves systematically downstream of the induction coil.

The overall energy balance for that particular torch geometry and dimensions is shown in Fig. 9.13. Radiation losses are virtually unchanged, whereas the heat loss by conduction to the wall decreases and the enthalpy of the gas at the exit of the torch increases almost linearly with the increase of the total gas flow rate. The specific enthalpy of the exit gas, on the other hand, does not seem to change substantially.

Effect of the Plasma Power. The effect of the power input on the overall energy balance was investigated also for this case. As shown in Fig. 9.14, increasing the power level over the range $P_0 = 3.0$–10.0 kW results in a rapid increase of the fraction of the total power that is lost as radiation. This is compensated by a slight decrease in the percentage of the total power that is lost to the wall of the plasma confinement tube and that which appears as enthalpy of the gas at

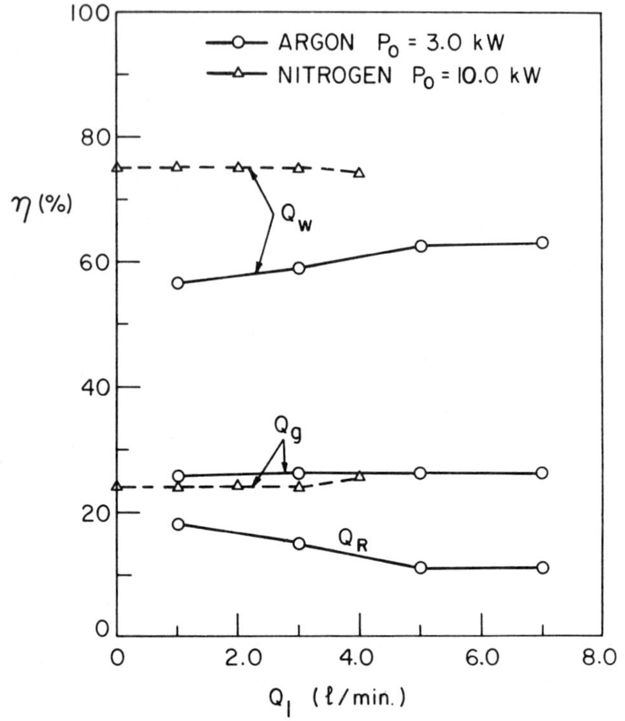

Figure 9.11. Distribution of power for Ar and N_2 plasmas as a function of central gas flow rate. P_w, Energy transferred to the wall of the plasma confinement tube; P_g, energy available as enthalpy in the gas; P_r, energy lost by radiation.

the exit of the torch. The values of η given in Fig. 9.14 strongly depend on the length of the plasma confinement tube, L_T, which was in this case taken as 250 mm. Obviously, for shorter tubes, P_w and P_R would decrease with a corresponding increase of P_g. Similar results were observed with the modified Miller model [31].

9.3. SINGLE PARTICLE OR AEROSOL TRAJECTORIES AND TEMPERATURE HISTORIES

Since the thermal treatment of powders in plasma torches and furnaces represents one of the most important applications of plasma technology, not surprisingly considerable attention has been given to the important problem of plasma-particle heat transfer. A number of mathematical models have been developed

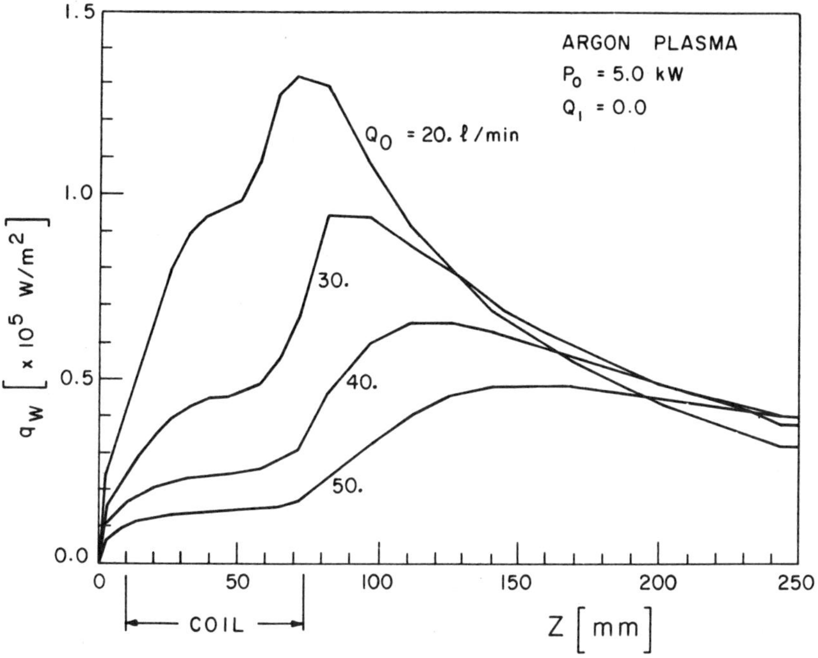

Figure 9.12. Effect of the total gas flow rate, Q_0, on the conduction losses to the wall.

for the thermal treatment of powders in dc plasma torches [51–56], and a few were applied to the induction plasma [31, 33, 57–62]. In either case, the models differed in the assumptions made and whether or not they took into account the effects of internal heat conduction in particles on the overall heat transfer process between the plasma and the particles. Only Fiszdon [54] and Yoshida and Akashi [58] followed the internal heat conduction in the particles, whereas the others assumed the particles to have a uniform temperature.

The specific question of whether, and when, the internal heat conduction in the particles must be considered has been studied by Bourdin et al. [63] and Chen and Pfender [64]. The results reported by Bourdin et al. [63] showed the differences as high as 1000 K could develop between the surface temperature and that of the center of alumina particles as small as 20 μm in diameter when immersed in a N_2 plasma at 10,000 K. The controlling parameter seems to be the Biot number, which is simply the ratio of the thermal conductivity of the plasma to that of the particle (k/k_s). According to Bourdin et al. [63], during the transient heating of a particle under plasma conditions, internal heat conduction in the particle should be taken into account if the Biot number is >0.02. On the other hand, Chen and Pfender [64] indicate that in spite of the differences

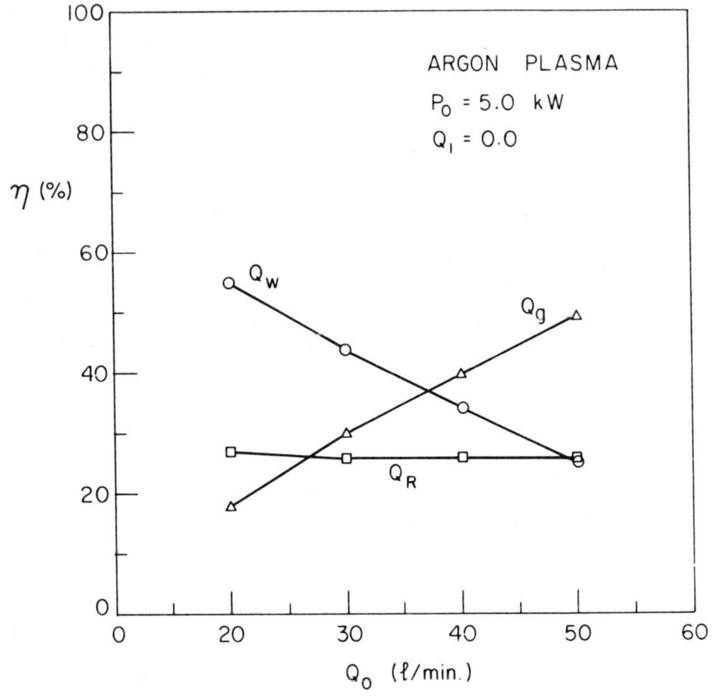

Figure 9.13. Distribution of power for an Ar plasma as a function of the total flow rate.

in initial heating, the analytical expression based on infinite thermal conductivity predicts the correct total time for both heating and evaporation, even for low-conductivity materials such as alumina.

Chen and Pfender [65–67] also studied the evaporation of single particles under plasma conditions and the behavior of small particles in a thermal plasma flow. In the latter case, they propose a Knudsen number correction of the heat transfer rate to the particles to account for deviations from continuum fluid mechanics that becomes increasingly important for submicrometer particles.

In this section an outline will be given of the models proposed for the calculation of the trajectory and temperature history of single particles as they are injected in the discharge region of an ICP. The discussion will be limited to dilute systems, which implies that the particle loading is sufficiently low so as to have neither an influence on the plasma flow and temperature fields, nor any particle–particle interaction effects.

For simplicity, the additional assumption of uniform particle temperature will be maintained throughout. The important problem of plasma–particle interaction will be discussed separately in Section 9.4.

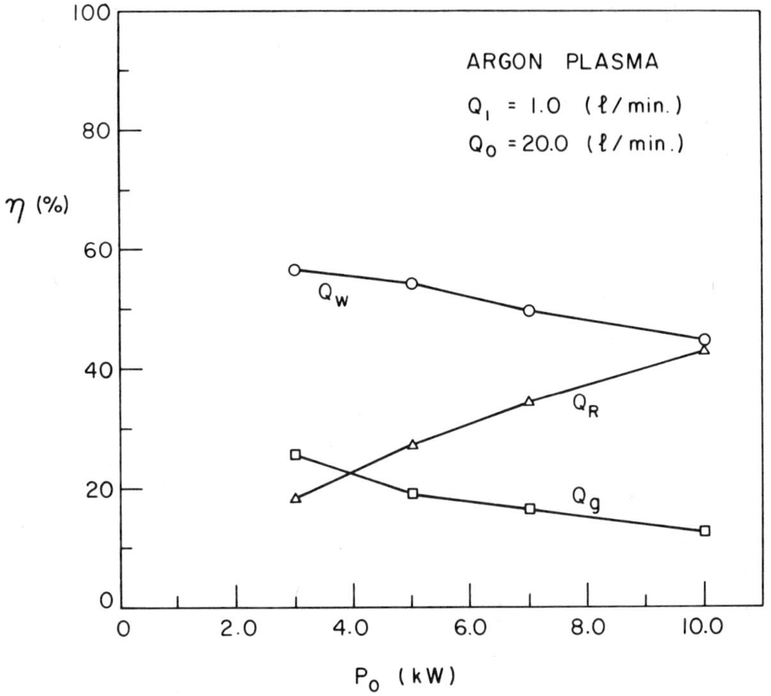

Figure 9.14. Distribution of power losses for an Ar plasma as a function of power input.

9.3.1. Governing Equations

Assuming that the only forces affecting an individual particle trajectory are the drag and the gravity, the momentum equations for a single-particle injected vertically downward into the plasma can be expressed as [57]:

$$\frac{du_p}{dt} = -\frac{3}{4} C_D(u_p - u) U_R \left[\frac{\rho}{\rho_s d_p}\right] + g \tag{9.15}$$

$$\frac{dv_p}{dt} = -\frac{3}{4} C_D(v_p - v) U_R \left[\frac{\rho}{\rho_s d_p}\right] \tag{9.16}$$

with

$$U_R = \sqrt{(u_p - u)^2 + (v_p - v)^2} \tag{9.17}$$

where u_p and v_p are the axial and the radial particle velocities, respectively; U_R is the relative speed between the particle and the plasma gas; d_p is the particle diameter; and C_D is the drag coefficient that can be estimated as a function of the particle Reynolds number ($\mathrm{Re} = \rho U_R d_p/\mu$) using the following relations [57].

$$C_D = \begin{cases} \dfrac{24}{\mathrm{Re}} & \mathrm{Re} \leqslant 0.2 \\ \dfrac{24}{\mathrm{Re}}\left(1 + \dfrac{3}{16}\mathrm{Re}\right) & 0.2 < \mathrm{Re} \leqslant 2.0 \\ \dfrac{24}{\mathrm{Re}}(1 + 0.11\,\mathrm{Re}^{0.81}) & 2.0 < \mathrm{Re} \leqslant 21.0 \\ \dfrac{24}{\mathrm{Re}}(1 + 0.189\,\mathrm{Re}^{0.62}) & 21.0 < \mathrm{Re} \leqslant 200 \end{cases} \tag{9.18}$$

The particle temperature is determined by an energy balance between conductive and convective heat transfer between the particle and the plasma and radiative energy loss from the particle to the surroundings. The energy transfer equation can be written for a single particle as follows:

$$Q_p = \pi\, d_p^2 h_c (T - T_p) - \pi\, d_p^2\, \sigma_{sb}\epsilon (T_p^4 - T_a^4) \tag{9.19}$$

where Q_p is the net heat exchange between the particles and its surroundings,

$$Q_p = \begin{cases} \left(\dfrac{\pi}{6}\rho_s d_p^3 c_{ps}\right)\dfrac{dT_p}{dt} & \text{for } T < T_m \text{ and } T_m < T < T_v \\ \left(\dfrac{\pi}{6}\rho_s d_p^3 H_m\right)\dfrac{dx}{dt} & \text{for } T = T_m \\ -\left(\dfrac{\pi}{2}\rho_s d_p^2 H_v\right)\dfrac{dd_p}{dt} & \text{for } T = T_v \end{cases} \tag{9.20}$$

and

$$h_c = \left(\frac{k}{d_p}\right)(2.0 + 0.515\,\mathrm{Re}^{0.5}) \tag{9.21}$$

where h_c is the heat-transfer coefficient; σ_{sb} is the Stefan–Boltzmann constant; ϵ is the particle emissivity; c_{ps} is the specific heat of the solid or liquid particles; x is the liquid mass fraction of the particle; and T_p, T_m, and T_v are the particle temperature, melting temperature, and boiling temperature, respectively. H_m and H_v are the corresponding latent heat of melting and evaporation, respectively.

9.3.2. Typical Results

As an example, the trajectories of fine alumina particles with a diameter of 10–200 μm injected in the discharge region of a confined Ar induction plasma are given in Fig. 9.15 [57]. A summary of the pertinent physical properties of pure alumina used in these calculations is given in Table 9.3. In this case, the plasma confinement tube had an internal diameter of 28 mm. The central powder carrier

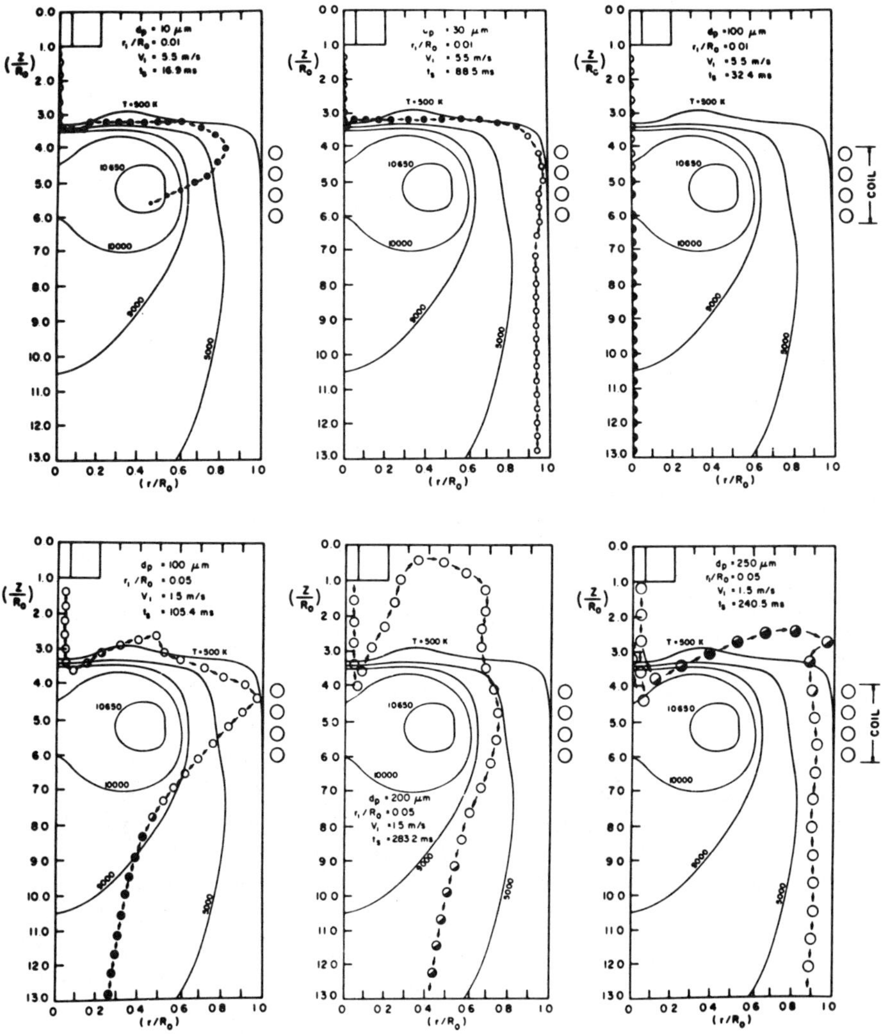

Figure 9.15. Particle trajectories for alumina powder injected in an induction plasma: f = 3 MHz; P_t = 3.77 kW, Q_1 = 0.4, Q_2 = 2.0, Q_3 = 16.0 L · min^{-1}, and $\bar{v}_\Theta$ = 0.

Table 9.3 Physical Properties of Pure Alumina [27]

Density (kg · m^3)	ρ_s	= 3900
Melting point (K)	T_m	= 2323
Latent heat of fusion (kJ · kg^{-1})	H_m	= 1071
Boiling point (K)	T_v	= 3800
Latent heat of evaporation (kJ · kg^{-1})	H_v	= 24660
Emissivity (−)	ϵ	= 0.3
Specific heat (kJ · kg^{-1}K^{-1})	c_{ps}	= 1.038 at 500 K
		= 1.200 at 1000 K
		= 1.405 at 1510 K
		= 1.958 at 2575 K

gas, intermediate, and outer gas flow rates were set to 0.4, 2.0, and 16.0 L · min^{-1}, respectively. The oscillator frequency was 3 MHz, and the net power dissipated in the discharge was 3.8 kW.

In the representation of the trajectories given in Fig. 9.15, the size of the circles used to indicate the position of the particles is also an indication, although not to scale, of their diameter. Moreover, an open circle indicates a solid particle whereas a closed circle represents a liquid droplet.

Predictions that could be made using this model can be useful to optimize the injection condition of a powder or an aerosol to obtain either a physical and/or chemical change in the powder or complete evaporation of the powder or aerosol. As an example, Barnes and Nikdel [32] compared the decomposition of aluminum oxide particles in Ar and N_2 spectrochemical ICP discharges. The higher efficiency for decomposing particles in the N_2 ICP has led to the development of industrial applications of the molecular gas spectrochemical ICP [39–42] (cf. Part 1, Sections 4.1.8 and 4.9).

9.4. PLASMA-PARTICLE INTERACTION EFFECTS

Although the assumption of dilute system has generally been accepted for the calculation of individual particle trajectories and temperature histories under plasma conditions, the interpretation of the results obtained is greatly hindered by the simple fact that any application of plasma technology for the in-flight processing of powders will have to be carried out under sufficiently high loading conditions in order to make efficient use of the thermal energy available in the plasma. With the local cooling of the plasma resulting from the presence of the particles, model predictions using the low-loading assumption can be substantially in error.

In an attempt to take into account the plasma–particle interaction effects, Boulos and his collaborators [59–62] developed a mathematical model which

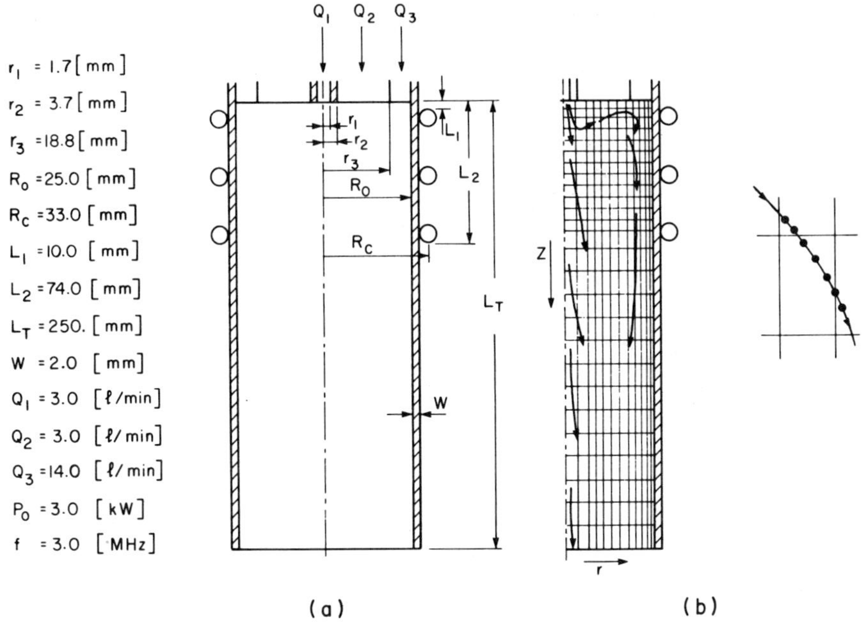

Figure 9.16. Schematic of the torch (*a*) and the computational domain (*b*).

through the iterative procedure refines continuously the computed plasma temperature, velocity, and concentration fields. The interaction between the stochastic single-particle trajectory calculations and those of the continuum flow, temperature, and concentration fields is incorporated through the use of appropriate source-sink terms in the respective continuity, momentum, energy, and mass transfer equations. These are estimated using the so-called particle-source-in-cell model (PSI-Cell) [68], which can be represented schematically as shown in Fig. 9.16.

9.4.1. Governing Equations

According to the PSI-Cell model, the passage of a particle through a finite difference cell (Fig. 9.16) would result in the exchange of momentum, energy, and mass between the particle and the fluid in that cell. To account for this exchange, an appropriate source-sink term defined as $S_{\mathrm{p}}^{\mathrm{c}}$, $S_{\mathrm{p}}^{\mathrm{Mz}}$, $S_{\mathrm{p}}^{\mathrm{Mr}}$, $S_{\mathrm{p}}^{\mathrm{E}}$, and $S_{\mathrm{p}}^{\mathrm{c}}$, respectively, must be added to each of the Eqs. (9.1–9.5). A full listing of the corresponding equations could then be written as follows:

Continuity:

$$\frac{1}{r}\frac{\partial}{\partial r}(r\rho v) + \frac{\partial}{\partial z}(\rho u) = S_{\mathrm{p}}^{\mathrm{c}} \tag{9.22}$$

Momemtum transfer:

$$\rho\left(v\frac{\partial u}{\partial r}+u\frac{\partial u}{\partial z}\right)=-\frac{\partial p}{\partial z}+2\frac{\partial}{\partial z}\left(\mu\frac{\partial u}{\partial z}\right)$$
$$+\frac{1}{r}\frac{\partial}{\partial r}\left[\mu r\left(\frac{\partial v}{\partial r}+\frac{\partial u}{\partial r}\right)\right]+\rho g+S_{\mathrm{p}}^{\mathrm{Mz}} \tag{9.23}$$

$$\rho\left(v\frac{\partial v}{\partial r}+u\frac{\partial v}{\partial z}\right)=-\frac{\partial p}{\partial r}+\frac{2}{r}\frac{\partial}{\partial r}\left(\mu r\frac{\partial v}{\partial r}\right)$$
$$+\frac{\partial}{\partial v}\left[\mu\left(\frac{\partial v}{\partial z}+\frac{\partial u}{\partial r}\right)\right]-\frac{2\mu v}{r^2}+\xi\sigma E_0 H_z\cos\chi+S_{\mathrm{p}}^{\mathrm{Mr}} \tag{9.24}$$

Energy transfer:

$$\rho\left(v\frac{\partial h}{\partial r}+u\frac{\partial h}{\partial z}\right)=\frac{1}{r}\frac{\partial}{\partial r}\left(r\frac{k}{c_{\mathrm{p}}}\frac{\partial h}{\partial r}\right)$$
$$+\frac{\partial}{\partial z}\left(\frac{k}{c_{\mathrm{p}}}\frac{\partial h}{\partial z}\right)+\sigma E_\theta^2-P_{\mathrm{r}}+S_{\mathrm{p}}^{\mathrm{E}} \tag{9.25}$$

Mass transfer:

$$\rho\left(v\frac{\partial y}{\partial r}+u\frac{\partial y}{\partial z}\right)=\frac{1}{r}\frac{\partial}{\partial r}rD\frac{\partial y}{\partial r}$$
$$+\frac{\partial}{\partial z}\left(D\frac{\partial y}{\partial z}\right)+S_{\mathrm{p}}^{\mathrm{c}} \tag{9.26}$$

The corresponding electromagnetic field equations and the boundary conditions remain essentially unchanged [see Eqs. (9.6–9.14)].

The formulation of the source terms in Eqs. (9.22–9.26) deserves special attention. This can be described as follows. Let N_{t}^0 be the total number of particles injected per unit time, n_{d} the particle size distribution, and n_{r} the fraction of N_{t}^0 injected at each point over the central tube radius (Fig. 9.16). The total number of particles per unit time traveling along the trajectory (l, k) corresponding to a particle diameter d_{pl} injected at the point r_{k} is

$$N^{0(l,k)}=n_{dl}n_{rk}N_{\mathrm{t}}^0 \tag{9.27}$$

The particle concentration in a given cell crossed by the trajectory (l, k) is

$$C_{ij}^{(l,k)}=\frac{N^{0(l,k)}\,\tau_{ij}^{(l,k)}}{V_{ij}} \tag{9.28}$$

$\tau_{ij}^{(l,k)}$ is the residence time of the (l, k) particles in the (ij) cell of volume V_{ij}. The mass source term for the (ij) cell, due to all the trajectories with initial diameter d_{pl} and initial injection point r_k, is given as

$$S^{\mathrm{c}}_{\mathrm{p},ij} = \sum_{l,k} C^{(l,k)}_{ij} \frac{\Delta m^{(l,k)}_{\mathrm{p}}}{\tau^{(l,k)}_{ij}} \tag{9.29}$$

$\Delta m^{(l,k)}_{\mathrm{p}}$ is the mass evaporated by a particle with (l, k) trajectory cell (ij). The corresponding source term in the energy equation includes the heat given to the particles $Q^{(l,k)}_{\mathrm{p},ij}$, as well as the superheat needed to bring the particle vapors into thermal equilibrium with the plasma gas $Q^{(l,k)}_{\mathrm{v},ij}$, namely,

$$S^{\mathrm{E}}_{\mathrm{p},ij} = \sum_{l,k} C^{(l,k)}_{ij} \left[Q^{(l,k)}_{\mathrm{p},ij} + Q^{(l,k)}_{\mathrm{v},ij} \right] \tag{9.30}$$

where

$$Q^{(l,k)}_{\mathrm{p},ij} = \frac{1}{\tau^{(l,k)}_{ij}} \int_0^{\tau^{(l,k)}_{ij}} \pi d^2_{\mathrm{p}} h_{\mathrm{c}} (T_{ij} - T^{(l,k)}_{\mathrm{p},ij})\, dt \tag{9.31}$$

and

$$Q^{(l,k)}_{\mathrm{v},ij} = \frac{1}{\tau^{(l,k)}_{ij}} \int_0^{\tau^{(l,k)}_{ij}} \frac{\pi}{2} d^2_{\mathrm{p}} \rho_{\mathrm{s}} \left(\frac{dd_{\mathrm{p}}}{dt} \right) c_{\mathrm{pv}} (T_{ij} - T^{(l,k)}_{\mathrm{p},ij})\, dt \tag{9.32}$$

where c_{pv} is the specific heat of the particle material in vapor form. The source-sink terms for the corresponding momentum transfer equations are

$$S^{\mathrm{Mr}}_{\mathrm{p},ij} = \sum_{(l,k)} C^{(l,k)}_{ij} \frac{\Delta(m_{\mathrm{p}} v_{\mathrm{p}})}{\tau^{(l,k)}_{ij}} \tag{9.33}$$

and

$$S^{\mathrm{Mz}}_{\mathrm{p},ij} = \sum_{(l,k)} C^{(l,k)}_{ij} \frac{\Delta(m_{\mathrm{p}} u_{\mathrm{p}})}{\tau^{(l,k)}_{ij}} \tag{9.34}$$

9.4.2. Typical Results

As an example of possible plasma–particle interaction effects in induction plasma modeling under heavy loading conditions, this section will report results of Proulx et al. [62]. These were obtained for an ICP operated with Ar at atmospheric pressure. Details of the torch dimensions and operating conditions are given in Table (9.4*a*).

Copper powder with a mean particle diameter of 70 μm and a standard deviation of 30 μm is injected through the central tube along the central axis into the coil region of the discharge. The choice of Cu made it possible to take into account the effect of the presence of Cu vapor on the transport properties of the Ar under plasma conditions. For these, the data published by Mostaghimi and

Table 9.4a. Torch Dimension and Summary of the Operating Conditions

$r_1 = 1.7$ mm	$L_1 = 10.0$ mm	$Q_1 = 3.0\ \text{L} \cdot \text{min}^{-1}$
$r_2 = 3.7$ mm	$L_2 = 74.0$ mm	$Q_2 = 3.0\ \text{L} \cdot \text{min}^{-1}$
$r_3 = 18.8$ mm	$L_T = 250.0$ mm	$Q_3 = 14.0\ \text{L} \cdot \text{min}^{-1}$
$R_0 = 25.0$ mm	$w = 2.0$ mm	$P_0 = 3.0\ \text{L} \cdot \text{min}^{-1}$
$R_c = 33.0$ mm		$f = 3.0$ MHz

Table 9.4b. Plasma–Particle, Q_p, and Plasma–Vapor, Q_v, Energy Transfer under Different Loading Conditions

$\dot{m}_p$ [a] $(\text{g} \cdot \text{min}^{-1})$	$\dot{m}_v$ [b] $(\text{g} \cdot \text{min}^{-1})$	Q_p (w)	Q_v (w)	Percent of Total Energy Absorbed
1.0	0.50	71.0	21.0	3.1
5.0	1.40	255.0	42.0	9.9
10.0	1.40	386.0	35.0	14.0
15.0	1.16	460.0	28.0	16.3
20.0	0.94	511.0	23.0	17.8

Material: Copper powder, $d_p = 70\ \mu\text{m}$, $\sigma_s = 30\ \mu\text{m}$, $\rho_s = 8900\ \text{kg} \cdot \text{m}^{-3}$. $T_m = 1356$ K, H_m 204.7 $\text{kJ} \cdot \text{kg}^{-1}$, $c_{ps} = 0.425\ \text{kJ} \cdot \text{kg}^{-1}$, $T_v = 2840$ K, $H_v = 4794.0\ \text{kJ} \cdot \text{kg}^{-1}$, $c_{pv} = 0.480\ \text{kJ} \cdot \text{kg}^{-1}$.

[a] $\dot{m}_p$, Particle feed rate $(\text{g} \cdot \text{min}^{-1})$.

[b] $\dot{m}_v$, Particle evaporation rate $(\text{g} \cdot \text{min}^{-1})$.

Pfender [69] were used, which showed that the presence of even small amounts of Cu vapor can have a pronounced effect on the electrical conductivity of Ar.

By dividing the Gaussian particle size distribution of the Cu powder into n_d discrete fractions and their spatial distribution in the powder injection tube into n_r discrete positions, the problem simplifies to one of calculating $n_d \times n_r$ different possible trajectories. Assuming that the injection velocity of the particles to be equal to that of the carrier gas velocity at the point of injection, 35 such individual trajectories could be calculated corresponding to valves of n_d and n_r of seven particle diameters and five injection points, respectively. These were combined with the interactive solution of Eqs. (9.6 and 9.22–9.26) with the auxiliary relations (Eqs. 9.27–9.34) and the appropriate boundary conditions to determine the effect of particle loading on the flow, temperature, and concentration fields in the discharge.

Results were reported [62] for different mass feed rates of the Cu powder varying between 1.0 and 20.0 $\text{g} \cdot \text{min}^{-1}$. The isotherms, streamlines, and the concentration of Cu vapor in the plasma for a feed rate of 5 $\text{g} \cdot \text{min}^{-1}$ are represented in Fig. 9.17. Because the trajectories of the particles in this case are very close to the axis of the torch, the plasma gas is significantly cooled in

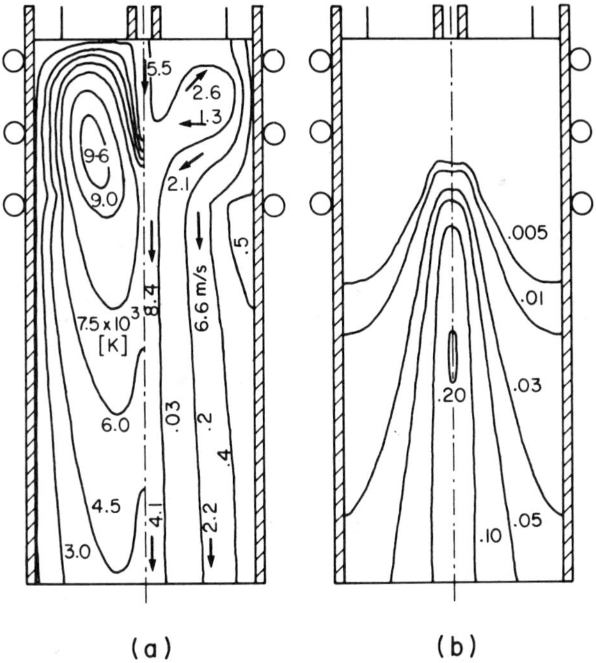

Figure 9.17. (*a*) Isotherms and stream lines for 5.0 (g · min^{-1}) and copper feed rate. (*b*) Iso-concentration contours of Cu vapor for this case.

this region (Fig. 9.18). On the other hand, the outer region of the plasma, where most of the power is dissipated, remains largely unaffected by the increase in the copper feed rate (Fig. 9.19). The result clearly demonstrates that although the overall loading ratio of the Cu powder to plasma gas might be small (0.19 g Cu / g Ar), the local cooling effects are significant. Clearly, the local plasma–particle interactions could be very important under loading conditions, which are generally assumed to be safe to neglect the changes in plasma temperature owing to the presence of the powder.

Momentum transfer between the gas and the particles is found to be negligible, and the flow field is affected only through the local plasma temperature changes. As the particles pass through the plasma, a portion of the powder evaporates and the vapor diffuses into the plasma medium (Fig. 9.17). The heat absorbed by the solid particles Q_p, and the superheat absorbed by the Cu vapor to heat to the plasma temperature Q_v are given in Table 9.4*b*. Q_v is generally much smaller than Q_p. The ratio Q_v/Q_p changes from 0.296 at 1 g · min^{-1} feed rate to 0.045 at 20 g · min^{-1} powder feed rate, and the total energy absorbed by the powder is between 3.1 and 17.8% of the plasma power input.

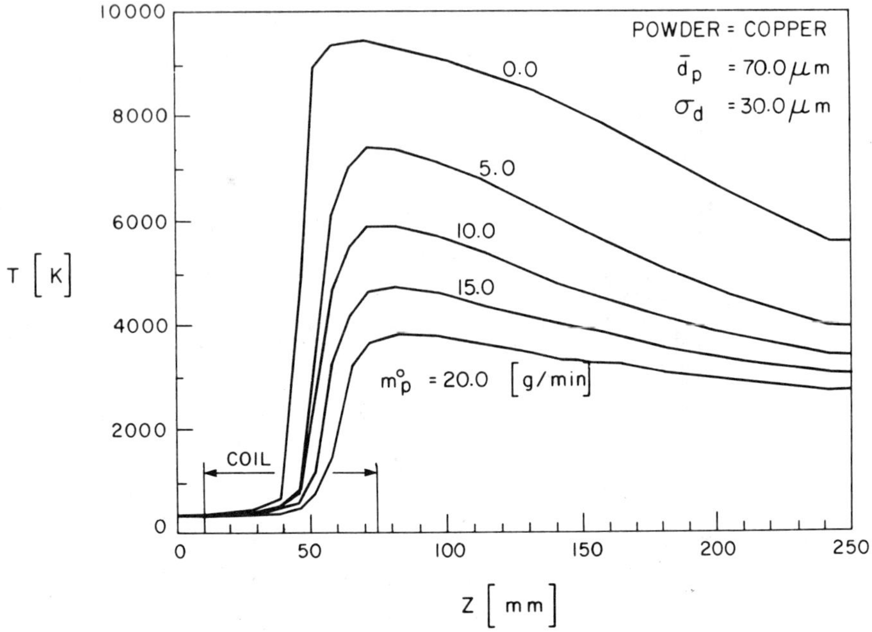

Figure 9.18. Effect of particle feed rate on the temperature profile along the axis of the torch.

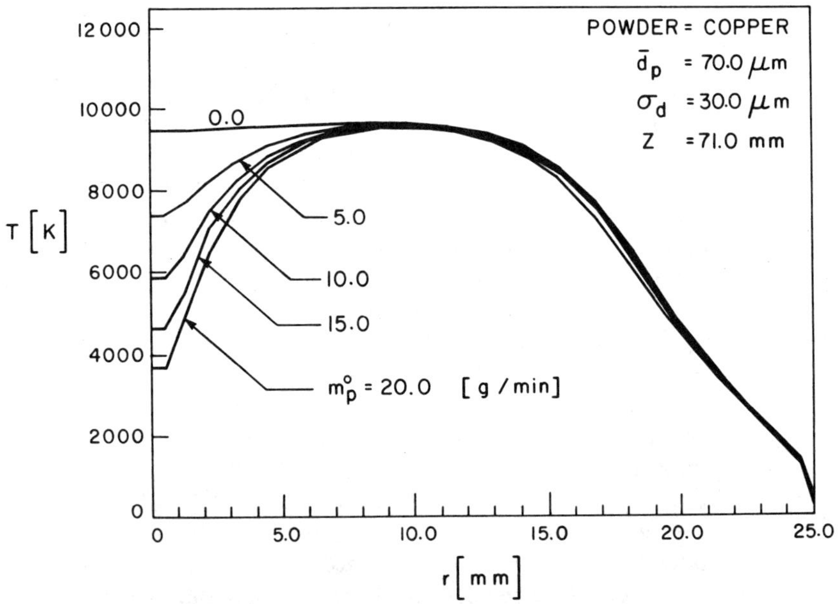

Figure 9.19. Effect of particle feed rate on the radial temperature profile at $z = 71$ mm.

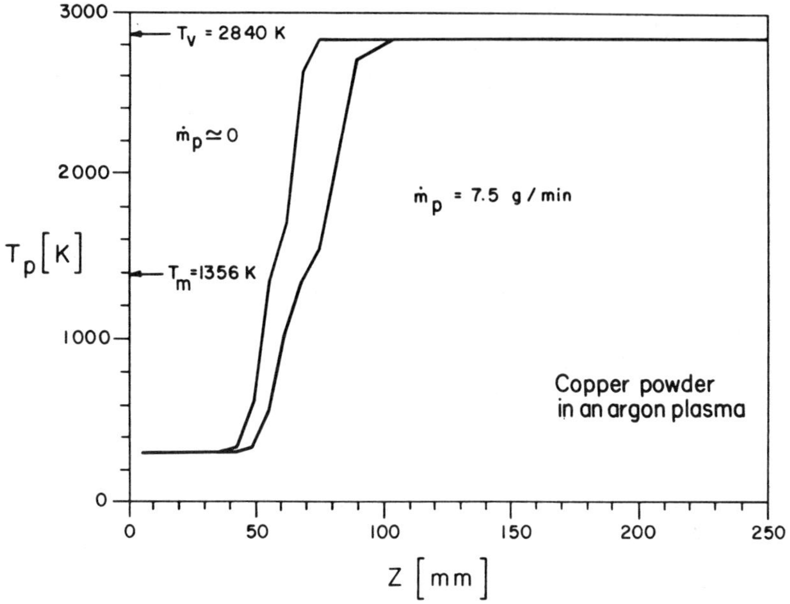

Figure 9.20. Particle temperature history along the axis of the discharge in the presence and absence of particle-loading effects.

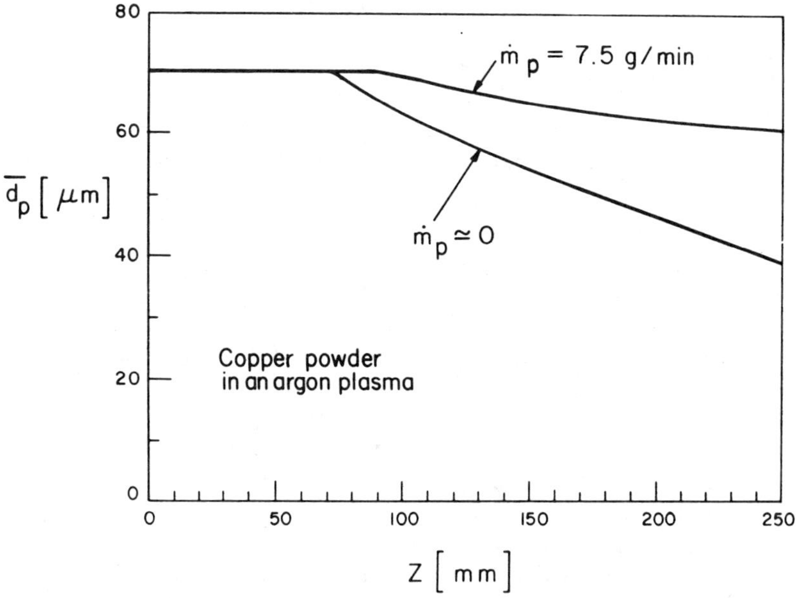

Figure 9.21. Mean particle diameter along the axis of the discharge in the presence and absence of Cu loading effects.

The plasma–particle interaction effect, through its influence on the flow and temperature fields in the discharge, could obviously have an important influence on the predicted particle-processing efficiency. Particle temperature history profiles and average particle size, respectively in the presence and absence of loading effects, are illustrated in Figs. 9.20 and 9.21.

Local cooling of the plasma resulting from the presence of the particles is responsible for the reduction of the heating rate of the particles as they are injected in the torch and the downstream shift of the point at which they reach their boiling point (Fig. 9.20). The effect results in a corresponding reduction of mass fraction of the powder which is evaporated by the time the powder exits the discharge. As shown in Fig. 9.21, for the Cu powder with an initial mcan particle diameter of 70 μm, the mean particle size at the exit of the torch is close to 60 μm for a powder feed rate of only 7.5 g $\cdot$ min^{-1}, compared with a value of 40 μm for the no-loading case ($\dot{m}_p \approx 0$).

9.5. EMISSION PATTERN FOR A SPECTROCHEMICAL ICP

Among the long-term objectives of the mathematical modeling efforts for the spectrochemical ICP is the development of a complete model that could be used to relate the observed emission pattern from the ICP to the concentration of the different elements in the aerosol.

A number of theoretical studies of emission patterns in spectrochemical ICPs and analytical flames have been reported in the literature [31]. Originally Barnes and Schleicher [31] applied the result of the modified Miller model to predict the extent of decomposition of aluminum oxide assumed to flow along the discharge centerline. From this calculation, they obtained the radial emission profiles for Al I and Al II, assuming LTE conditions and congruent diffusion of atomic species with gas flow. In addition, they computed the Ar continuum radiation from the ICP.

Eckert and Danielsson [70] proposed simple expressions for the relative intensity of the analyte atom and ion lines for the two limiting cases of vanishing and full ionization of the analyte. A parabolic temperature profile and a Gaussian distribution of the analyte density is assumed. All the calculations are performed under assumption of LTE. Eckert [71] also calculated the analytical curves and detection limits for a high-pressure electrodeless lamp. The results show a linear dependence between element concentration and the emission intensity over several orders of magnitude. Boss and Hieftje [72] studied, on the other hand, the spatial distribution of atoms created by the vaporization of a single aerosol particle in a laminar flame. The proposed model considers that the vaporization and diffusion rates are of the same time magnitude.

Recently, Mostaghimi et al. [61] proposed a two-dimensional model that could be used to calculate the emission pattern of different elements for a given

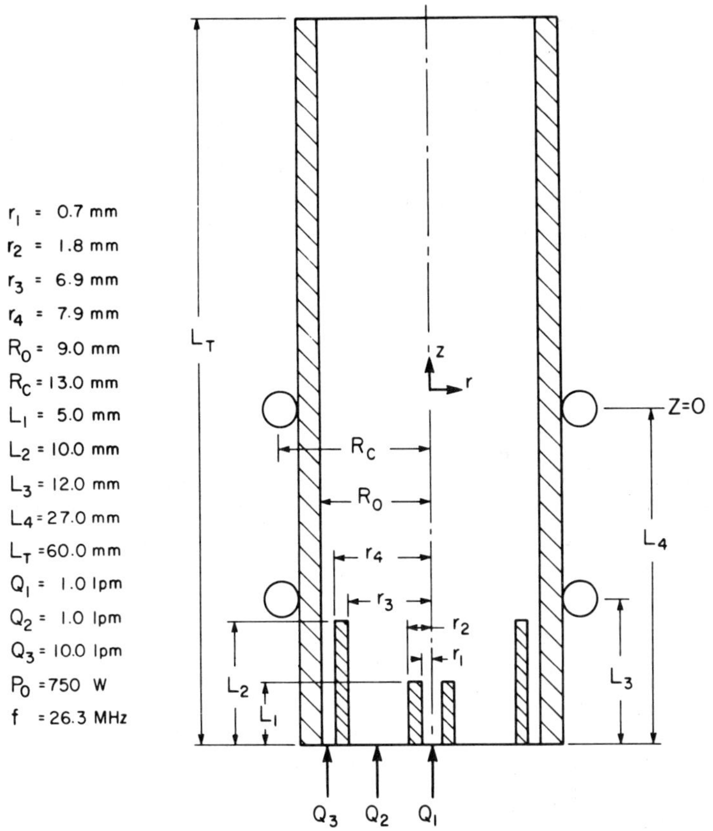

Figure 9.22. Torch geometry and system of coordinates.

ICP under typical operating conditions. The model, developed for the torch geometry given in Fig. 9.22, was based on the following assumptions:

- Axisymmetric system of coordinates with two-dimensional temperature, flow, and concentration fields, and one-dimensional electromagnetic fields
- Steady laminar flow
- Optically thin plasma
- Local thermodynamic equilibrium
- Dilute system, that is, negligible plasma–aerosol interaction

Among these conditions, the last two assumptions are the most serious ones since LTE does not necessarily hold for some regions in the ICP [73].

Two cases were investigated in which the analyte was introduced into the plasma in the following ways.

1. As aerosol droplets with a mean diameter of 25 μm. The analyte concentration in the aerosol was assumed to be 100.0 $\mu g \cdot mL^{-1}$ and the aerosol total feed rate was 0.1 $mg \cdot min^{-1}$. Once the droplets were evaporated completely, the analyte was assumed also to be in vapor form.
2. As analyte vapor. This served to identify the effects caused by the liquid droplet evaporation process on the concentration profiles in the discharge. The mass flow rate of the analyte in this case was the same as in 1.

9.5.1. Calculation Procedure

1. For given torch geometry and operating conditions, the flow, temperature, and concentration fields of the water vapor and the analyte vapor clouds were computed using essentially the same calculation procedure as reported by Proulx et al. [60]. Owing to the low loading ratio, the plasma–particle interaction effects were neglected. The computations were made for Ar as the plasma gas. The diffusion coefficients of the water vapor and the analyte were determined using the Chapman–Enskog kinetic theory [74]. The interaction between the plasma gas and the water or the analyte vapor was assumed to follow Lennard-Jones potential.
2. Neglecting the presence of the water vapor in the plasma, the equilibrium composition of the plasma was computed for Ar and the analyte, X, using the following equations [75]:

$$\begin{aligned} \mathrm{Ar} &\rightleftharpoons \mathrm{Ar}^+ + e^- \\ \mathrm{X} &\rightleftharpoons \mathrm{X}^+ + e^- \end{aligned} \tag{9.35}$$

$$\begin{aligned} \frac{p_{\mathrm{Ar}^+}\, p_{\mathrm{e}}}{p_{\mathrm{Ar}}} &= K_{\mathrm{Ar}}(T) \\ \frac{p_{\mathrm{X}^+}\, p_{\mathrm{e}}}{p_{\mathrm{X}}} &= K_{\mathrm{X}}(T) \\ p_{\mathrm{e}} + p_{\mathrm{Ar}} + p_{\mathrm{Ar}^+} + p_{\mathrm{X}} + p_{\mathrm{X}^+} &= P \\ p_{\mathrm{Ar}^+} + p_{\mathrm{X}^+} &= p_{\mathrm{e}} \\ p_{\mathrm{X}} + p_{\mathrm{X}^+} &= f\,(p_{\mathrm{Ar}} + p_{\mathrm{Ar}^+}) \end{aligned} \tag{9.36}$$

where p_j values are the partial pressures and P is the total pressure assumed to be 1 atmosphere. $K(T)$ is the equilibrium constant in the Saha equation, and f

is the ratio of analyte for argon concentration. $K(T)$ is given by Eq. (9.37)

$$K_X(T) = 2\,\frac{2U_{X^+}}{U_X}\left[\frac{2\Pi m_e}{h^2}\right]^{3/2}(kT)^{5/2}\exp\left[-\frac{E_i - \Delta E_i}{kT}\right] \quad (9.37)$$

U is the partition function, h the Planck constant, k the Boltzmann constant, m_e electron mass, T temperature, E_i ionization energy, and ΔE_i the lowering of the ionization energy which was calculated using Griem's theory. For given T, P, and f Eqs. (9.36) could be solved iteratively. Once p_j values were known, the population of the atoms, ions, and electrons were calculated in terms of number densities using Eq. (9.38).

$$n_j = \frac{p_j}{kT} \quad (9.38)$$

3. Assuming a Boltzmann distribution for the population of the excited state y,

$$\frac{n_{j,y}(r, z)}{n_j(r, z)} = \frac{g_{j,y}}{U_j(r, z)}\exp\left[-\frac{E_{j,y}}{kT(r, z)}\right] \quad (9.39)$$

where $g_{j,y}$ is the statistical weight of the excited state with an energy of $E_{j,y}$. The relative radial distributions of the excited atoms/or ions can be obtained by

$$\begin{aligned}\epsilon_R(r, z) &= \frac{n_{j,y}(r, z)}{n_{j,y}(0, z)} \\ &= \frac{n_j(r, z)}{n_j(0, z)}\frac{U_{\xi}(0, z)}{U_j(r, z)}\exp\left[-\frac{E_{j,y}}{k}\left(\frac{1}{T(r, z)} - \frac{1}{T(0, z)}\right)\right]\end{aligned} \quad (9.40)$$

The relative lateral emission intensity $I_R(x, z)$ was calculated as function of $\epsilon_R(r, z)$ as follows:

$$I_R(x, z) = 2\int_x^{R_0}\frac{\epsilon_R(r, z)}{(r^2 - x^2)^{1/2}}\,dr \quad (9.41)$$

where x is the lateral coordinate and R_0 is the radius of the plasma. The absolute emission intensity of an atom/or ion line can be expressed as [75]

$$\epsilon = n\,\frac{g_q\,A_{qp}\,hc}{4\,\Pi\,U\,\lambda}\exp\left[\frac{-E_q}{kT}\right] \quad (9.42)$$

where ϵ is the emission intensity, A_{qp} the Einstein transition probability, and λ the wavelength of the transiton; p and q refer to the lower and upper energy states, respectively. Values of g, A, λ, and E used were those of reference [76] and the partition functions were obtained from reference [77].

9.5.2. Typical Results

Flow, Temperature, and Concentration Fields. Computations were carried out for the torch dimensions given in Fig. 9.22 and the following operating conditions:

Plasma gas	Ar
Flow rates	$Q_1 = 1\ \mathrm{L \cdot min^{-1}}$
	$Q_2 = 1\ \mathrm{L \cdot min^{-1}}$
	$Q_3 = 10\ \mathrm{L \cdot min^{-1}}$
Operating frequency	$f = 26.3$ MHz
Power	$P_0 = 500$–750 W

Typical results for $P_0 = 750$ (W) are given in Fig. 9.23. A distinct difference

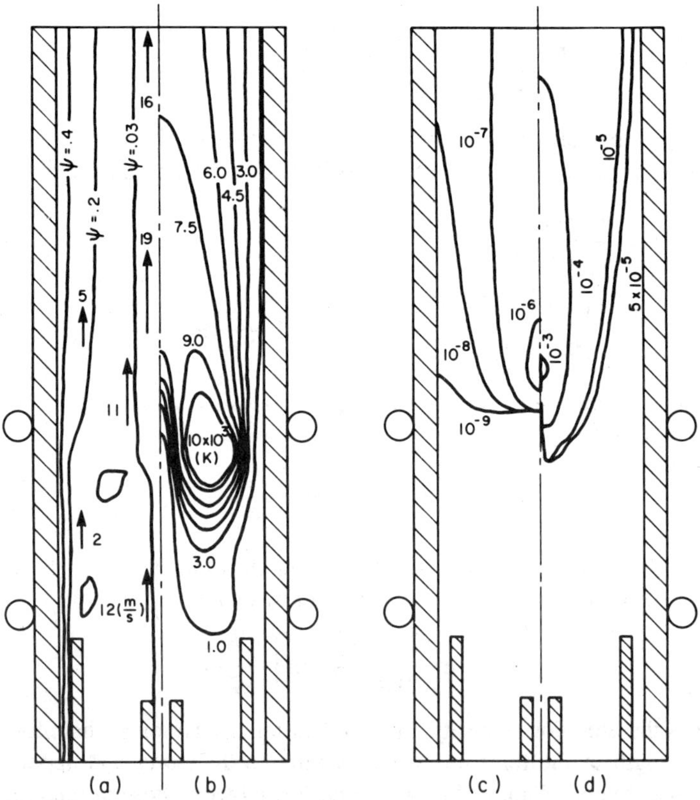

Figure 9.23. Flow, temperature; and concentration fields for $P_0 = 750$ W. (*a*) Streamlines; (*b*) temperature; (*c*) analyte concentration; (*d*) water vapor concentration.

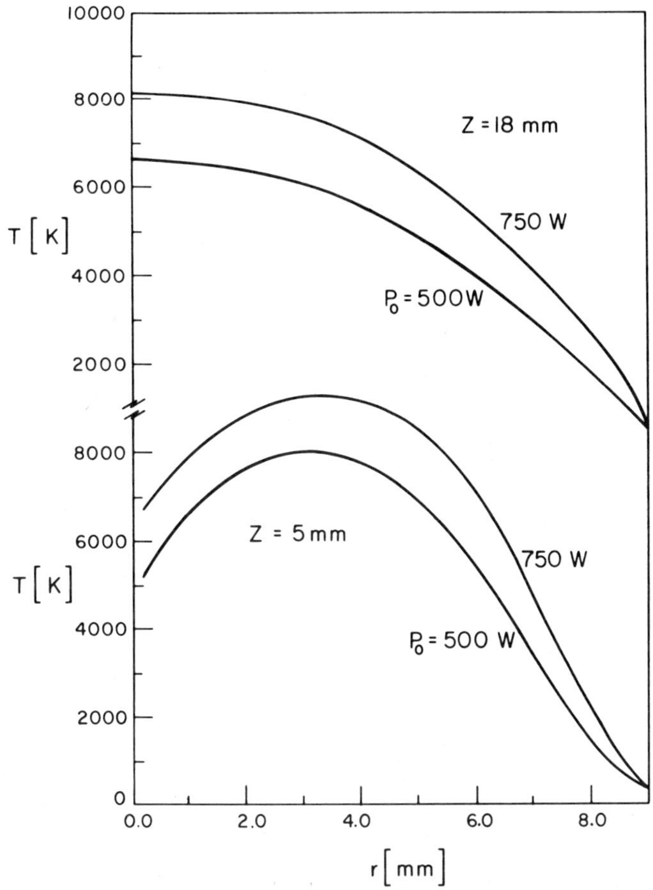

Figure 9.24. Radial temperature profiles at 5 and 18 mm above the coil region.

exists between the concentration fields of the analyte and the water vapor cloud represented in Fig. 9.23*c* and *d*. The effect of the plasma power level on the radial temperature profiles is illustrated in Fig. 9.24 at 5 and 18 mm above the coil. The corresponding electron density distribution under LTE conditions is given in Fig. 9.25. If the LTE assumption is valid in the regions where the electron density is $> 10^{15}$ cm^{-3} [78], it follows from Fig. 9.25 that the central region of the discharge and the wall boundary will exhibit distinct deviations from LTE (see Refs. 10–12 and Chapter 11).

The temperature variations along the axis and at 3 mm off the axis for the two power levels are given in Fig. 9.26. For P_0 = 750 W, the temperature downstream of the coil is about 1500 K higher than that for P_0 = 500 W. These results resemble earlier measured temperatures [34, 35] and electron densities

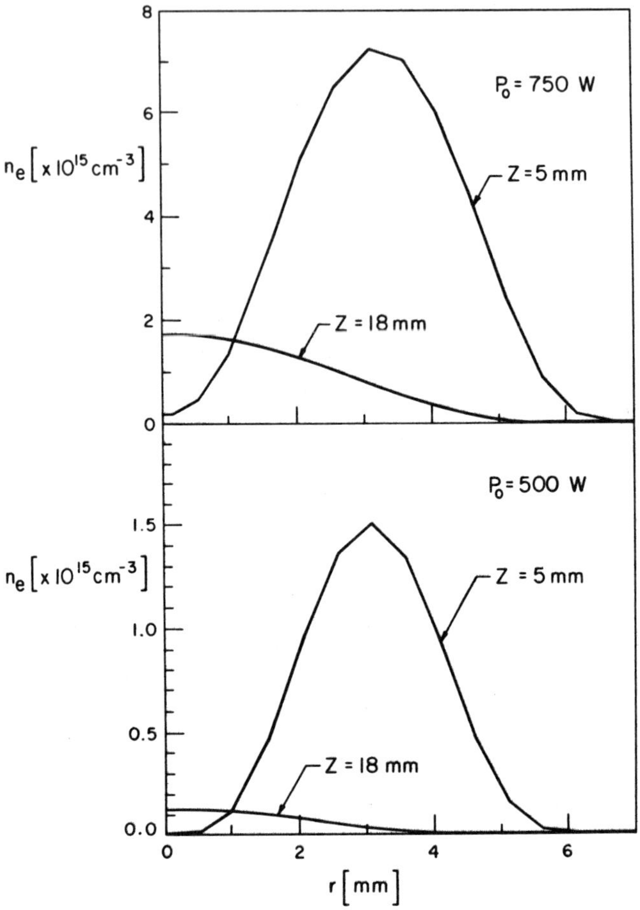

Figure 9.25. Radial profiles of the electron density distribution.

[15, 17], although in most experimental spectrochemical cases $Q_2 < 0.2$ L · min^{-1}, which may account for mismatch between calculated and experimental values. Comparison of computed and experimental values at $Q_2 = 0$ are illustrated in [61].

To calculate the emission intensity profiles, several atom and ion lines were considered: Li I 670.8 nm, Ca I 422.6 nm, Ni I 336.9 nm, Cu I 327.4 nm, Cu I 324.5 nm, Fe I 382.4 nm, and Ca II 393.6 nm, Ca II 397.3 nm, and Fe II 258.9 nm. Typical results are given in Fig. 9.27.

The concentration distribution of the analytes depends on the atomic weight of the element as well as the way in which the analyte is introduced into the torch. The lighter elements, for example, because of their higher diffusion coef-

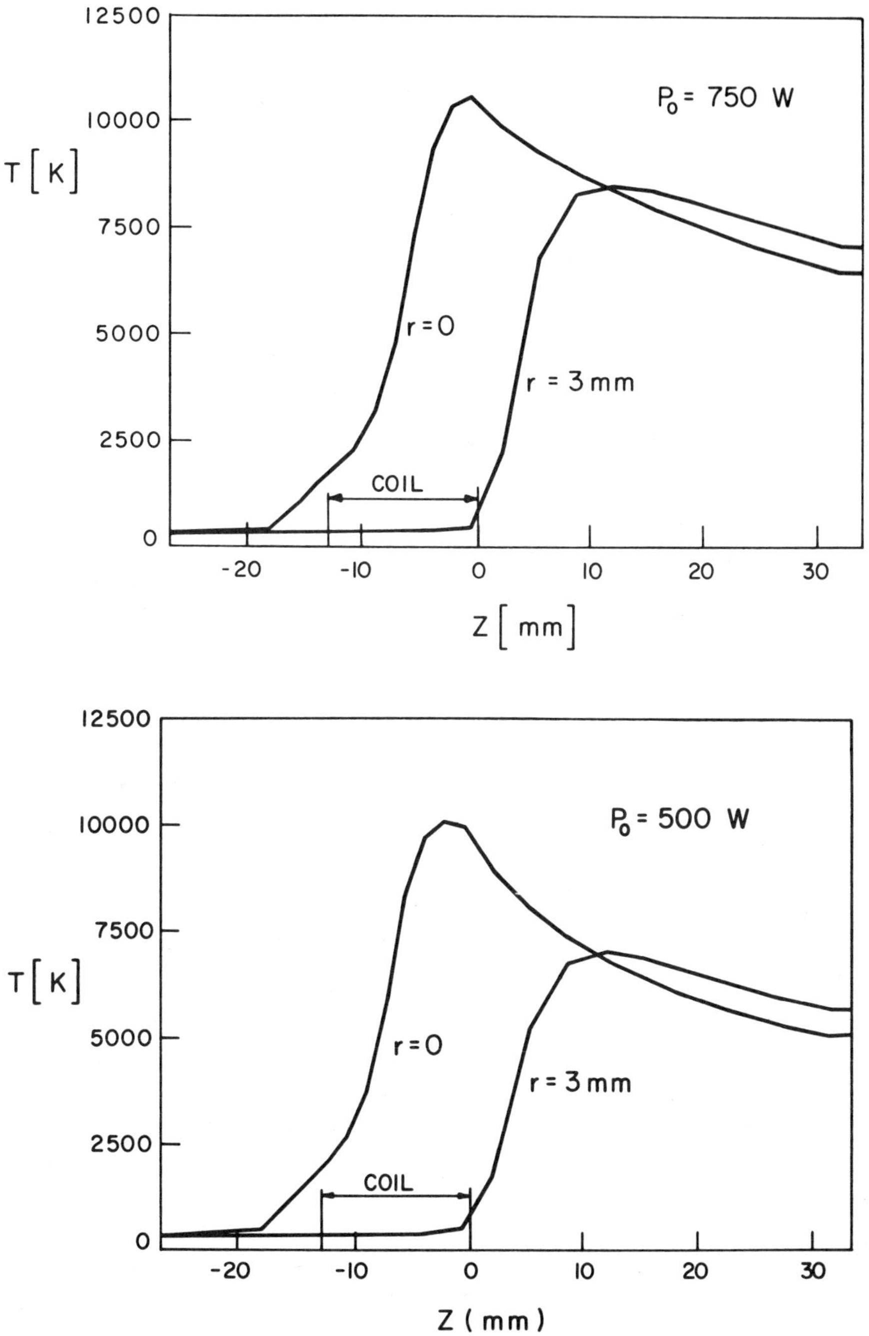

Figure 9.26. Axial temperature profiles for $r = 0$ and $r = 3$ mm. (*a*) $P_0 = 500$ W; (*b*) $P_0 = 750$ W.

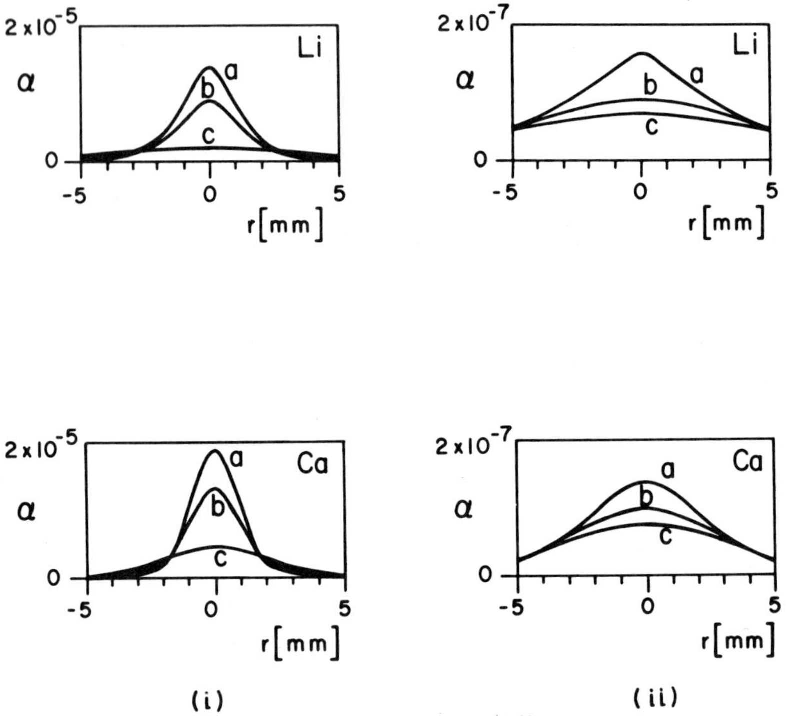

Figure 9.27. Analyte concentration profiles at different levels above the coil. (*a*) $z = 5$ mm; (*b*) $z = 10$ mm; (*c*) $z = 18$ mm. (*i*) $P_0 = 750$ W analyte introduced as aerosol water droplets; (*ii*) $P_0 = 500$ W analyte introduced as vapor.

ficients, tend to exhibit a flat radial distribution, whereas the heavy ones, for example, Cu, show a bell-shaped profile. When the analyte is introduced as a vapor (Fig. 9.27ii), the concentration profiles for the same observation heights are more uniform than the aerosol droplet case (Fig. 9.27i). The diffusion process for the aerosol starts later in the coil only after the water droplets are completely evaporated. The analyte was assumed to be in the vapor phase once the water droplets were completely evaporated. Since vaporization occurs on the same time scale as diffusion [72], this assumption was regarded only as a first approximation. Moreover, the effect of the water vapor on the thermodynamic and transport properties of the plasma was not included in this calculation.

Vertical and Radial Emission Profiles of Atom and Ion Lines. The radial and the lateral emission intensity profiles were calculated at different levels above the load coil (Figs. 9.28 and 9.29). In Fig. 9.30, the axial intensity profiles for Li I 670.8 and Ca I 422.6 are given. The results indicate for the elements

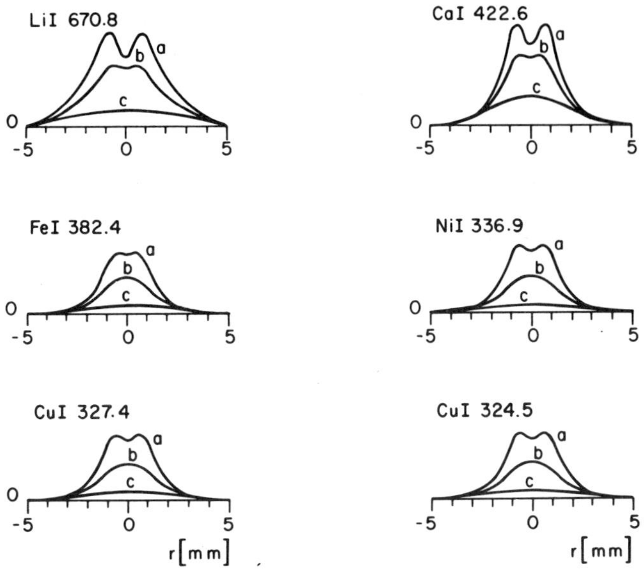

Figure 9.28. Radial emission intensity profiles for neutral atom lines at different levels above the coil, P_0 = 750 W. (*a*) z = 5 mm; (*b*) z = 10 mm; (*c*) z = 18 mm.

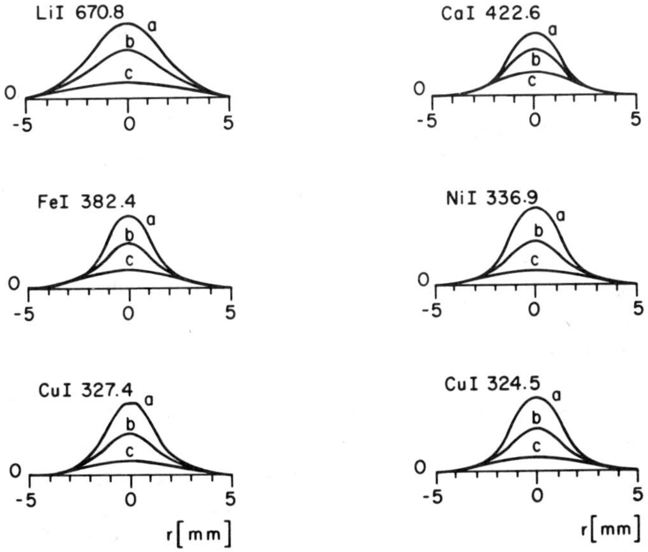

Figure 9.29. Lateral emission intensities at different levels above the coil, P_0 = 750 W. (*a*) z = 5 mm; (*b*) z = 10 mm; (*c*) z = 18 mm.

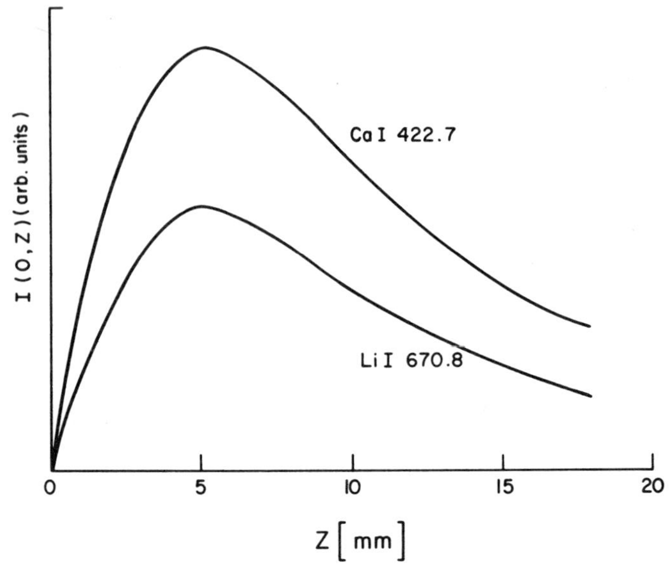

Figure 9.30. Lateral intensity profile along the axis, $P_0 = 750$ W.

considered here, that the profiles have essentially the same characteristic. The maximum intensity occurs at about 5 mm above the load coil, and the off-axis peak separation is close to the boundary between the central jet and the plasma. This is in good agreement with the results reported by Furuta and Horlick [79].

Experimental results show that the ICP discharge can be divided into two regions [7, 80, 81] (cf. Part 1, Section 4.7.2). The first one is a thermal region, and it is approximately between 5 to 15 mm above the load coil. The second one is an upper zone, which is nonthermal in nature. Because of the LTE assumption, the present results must therefore be compared within the thermal region.

Analytical Curves. To obtain a correlation between the intensity of the signal reaching the photomultiplier and the concentration of the analyte in the aerosol droplets, the following calculations were performed. The procedure is essentially similar to that used by Eckert [71]. The radial intensities for several lines of Cu, Fe, Ni, and Ca were found using Eq. (9.42).

The lateral emission intensities were calculated for the side-on observation on the axis by integrating the radial intensity profiles over the whole diameter (Eq. 9.41). The intensities calculated as such were multiplied by an optical response factor of 10^{-5} estimated for a monochromator with an entrance slit area of 10^{-3} cm^2 (that is, 5 mm high and 20 μm wide), a 1 : 1 image of the source, and an $f/9$ aperture, that is, 10^{-2} sr, [71]. No radiation losses were

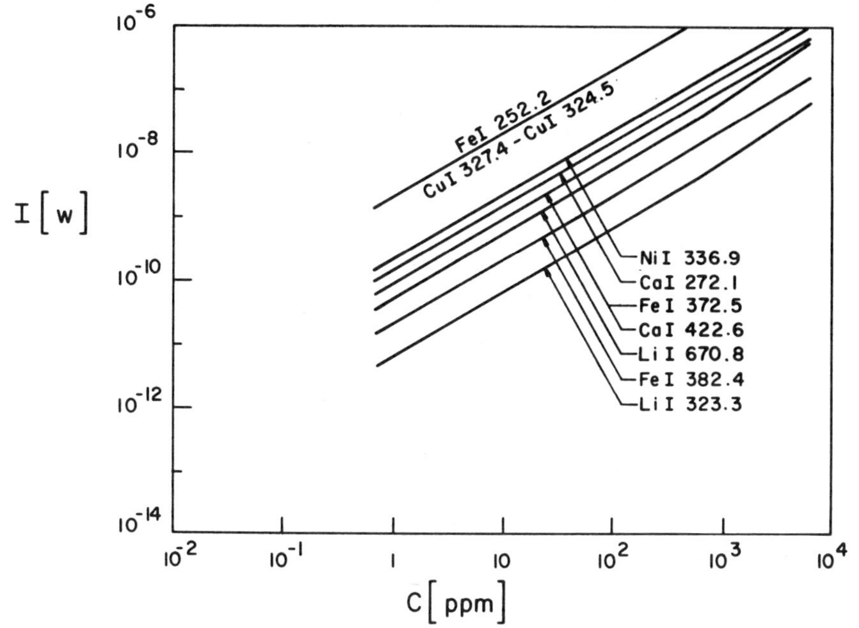

Figure 9.31. Lateral emission intensity as function of the analyte concentration for aerosol feed at P_0 = 750 W and z = 5 mm.

considered. Typical lateral emission intensity functions of concentration for an observation point of 5 mm above the coil are illustrated in Figs. 9.31 and 9.32. The slope of the lines is equal to one. When the concentration is sufficiently high, the equilibrium electron density increases significantly, followed by the analyte signal and its slope. However, this effect is likely to be offset by self-absorption which is not taken into account in this model. At the 500 W power level, a deviation of the slope from 1 occurs at lower concentrations. This is due to the lower temperature fields in the ICP. The effect of the analyte on the electron population is more pronounced at low temperatures. This effect is also observed with the low ionization potential elements, for example, Li, which shows a change in the slope at lower concentrations.

9.6. NONEQUILIBRIUM TWO-DIMENSIONAL, TWO-TEMPERATURE MODELS

As noted earlier, most existing models have assumed that the discharge is in LTE [48, 82]. Only a few authors have considered theoretical analysis of the nonequilibrium effects, and those were limited to one-dimensional models [26, 28, 83–85].

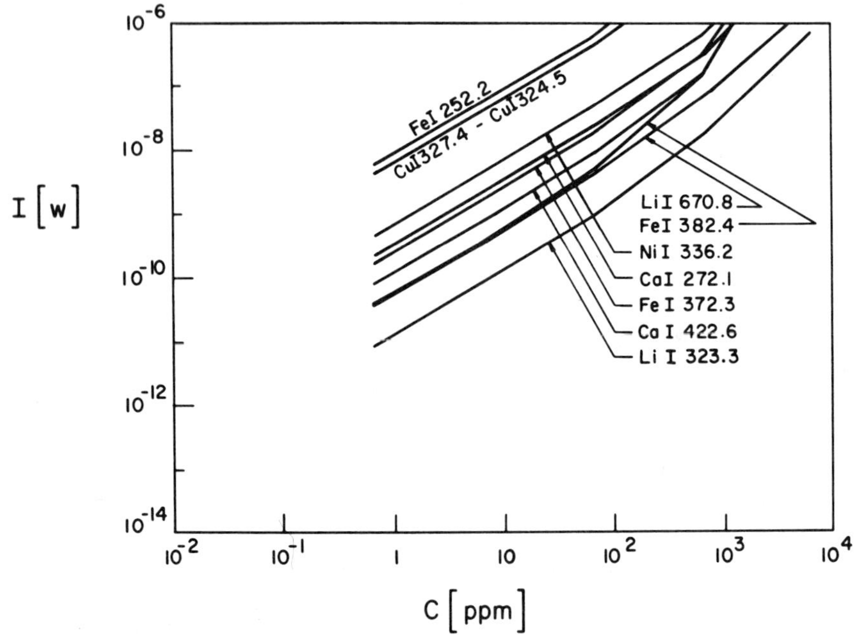

Figure 9.32. Lateral emission intensity as function of the analyate concentration for aerosol feed at $P_0 = 750$ W and $z = 5$ mm.

In this section, a non-LTE mathematical model proposed by Mostaghimi et al. [86] for the calculation of the two-dimensional flow, electron, and heavy-particle temperatures, and the electron density distributions (based on ambipolar diffusion) is presented. Results obtained for an Ar plasma under atmospheric pressure and soft vacuum conditions are compared with those obtained using the earlier LTE model [48].

9.6.1. Torch Geometry and Basic Assumptions

A schematic of the torch geometry, and its principal dimensions and operating conditions is given in Fig. 9.33. The following assumptions are made:

- Axi-symmetric, two-dimensional temperature and flow fields
- One-dimensional electromagnetic fields with negligible displacement currents
- Steady, laminar flow with negligible viscous dissipation
- Optically thin plasma
- Maxwellian distribution of the electron gas

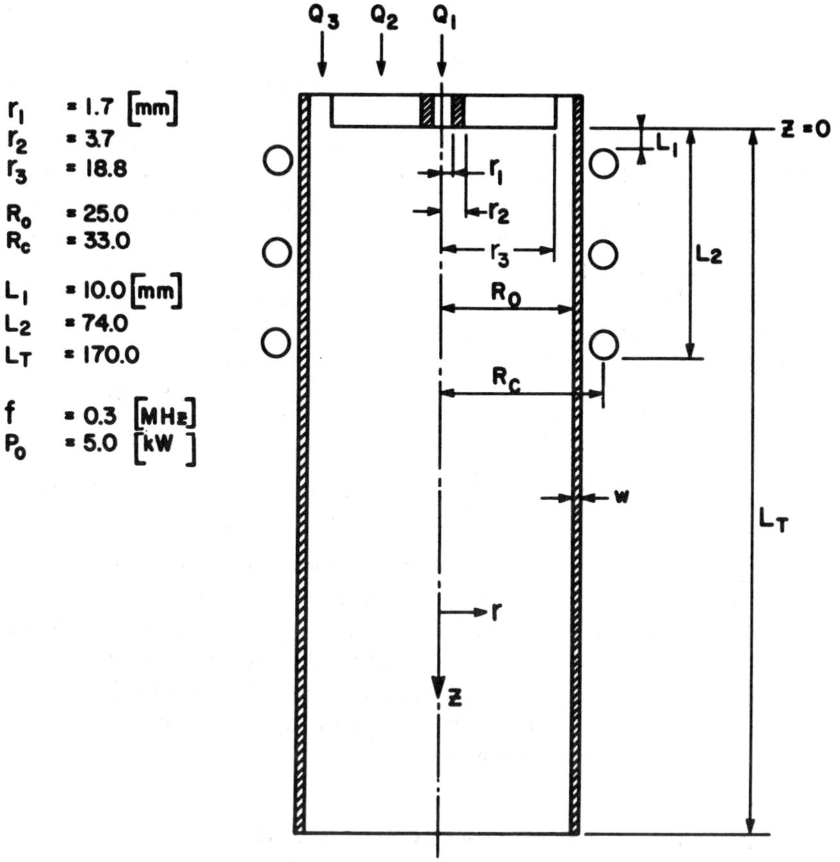

Figure 9.33. Schematic of the torch geometry.

- Collision-dominated plasma gas
- All the electromagnetic energy is picked up by the electrons

9.6.2. Governing Equations and Boundary Conditions

The continuity, momentum, electron, and the heavy particle energy, and the electron continuity equations are as follows:

$$\vec{\nabla} \cdot \rho \vec{U} = 0 \tag{9.43}$$

$$\rho \vec{U} \cdot \vec{\nabla} \vec{U} = -\vec{\nabla} P + \vec{\nabla} \cdot \mu \vec{\nabla} \vec{U} + \vec{J} \times \vec{B} \tag{9.44}$$

$$\vec{\nabla} \cdot \left[\left(\tfrac{5}{2} kT_e + \epsilon_i\right) n_e \vec{U}\right] = \vec{\nabla} \cdot \lambda_e \vec{\nabla} T_e + \sigma_e E^2 - R^0 - E^0_{eh} \tag{9.45}$$

$$\vec{\nabla} \cdot \left[\tfrac{5}{2} k(n_a + n_e) T\vec{U}\right] = \vec{\nabla} \cdot \lambda_h \vec{\nabla} T + E^0_{eh} \tag{9.46}$$

$$\vec{\nabla} \cdot n_e \vec{U} = \vec{\nabla} \cdot D_{amb} \vec{\nabla} n_e + \alpha n_a \left[S(T_e) - (n_e^2/n_a)\right] \tag{9.47}$$

where ρ is the density, $\vec{U}$ is velocity, P is pressure, μ is viscosity, $\vec{J}$ is current density, $\vec{B}$ is magnetic induction, k is the Boltzmann constant, T_e is electron temperature, ϵ_i is ionization energy, n_e is electron density, λ_e is electron thermal conductivity, σ_e is electrical conductivity, E is electric field strength, R^0 is radiation loss, E^0_{eh} is electron-heavy particle energy exchange term, n_a is atom density, T is heavy particle temperature, λ_h is heavy particle thermal conductivity, D_{amb} is ambipolar diffusion coefficient, α is recombination coefficient, and $S(T_e)$ is the Saha equilibrium constant. Electromagnetic field equations are the same as in previous work [48]. The energy exchange term, E^0_{eh}, is given by [87].

$$E^0_{eh} = 3\left(\frac{m_e}{m_h}\right) n_e \bar{\nu}_{eh}(T_e - T) \tag{9.48}$$

where m_e and m_h are the electron and the heavy particle mass, respectively, and $\bar{\nu}_{eh}$ is the average electron-heavy particle collision frequency.

The boundary conditions for the momentum and the heavy particle energy equations, as well as the electric and magnetic field equations, are the same as in our earlier LTE model [48]. The gradients of the electron temperature and the electron density are assumed to be zero on the boundaries at the centerline and the wall of the plasma confinement tube.

9.6.3. Calculation Procedure

The governing equations (9.43–9.47) are solved using a finite difference iterative scheme [49]. The electromagnetic field equations are solved by the Runge-Kutta method. Since the electron and the heavy particle temperatures are different, the transport and the thermodynamic properties will not be the same as in the case of LTE models. The heavy particle thermal conductivity and viscosity are calculated using the Chapmann-Enskog method [69]. For the electrical conductivity and the electron thermal conductivity, Frost formulae [87] are used. The ambipolar diffusion for Ar is given by Devoto [88] and the recombination coefficient is from reference [89]. Since most of the radiation loss occurs at temperatures where LTE exists, the LTE data given in reference [90] are used. The effect of pressure on the radiation losses is assumed to be linear.

Figure 9.34 shows the energy exchange term E^0_{eh} as a function of $\Theta = T_e/T$ for different values of T_e. It is noticed that at constant electron temperature a slight deviation from LTE significantly increases the energy exchanges term. However, after the initial increase in E^0_{eh}, the effect of increase in Θ on E^0_{eh} is relatively small.

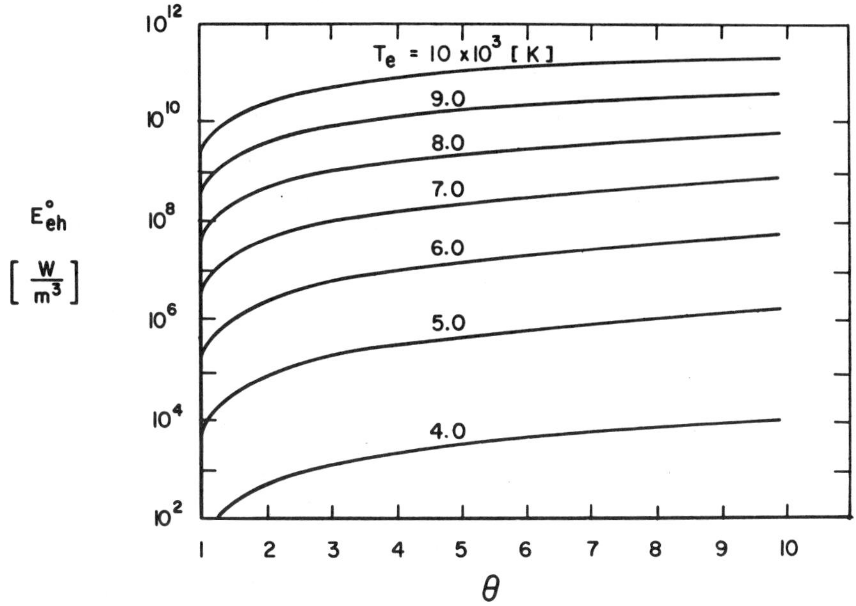

Figure 9.34. Electron–heavy particle energy exchange term in the energy Eqs. (9.45) and (9.46) as function of $\Theta = T_e/T$.

9.6.4. Results and Discussion

Computations were carried out for a power level of 5 kW and, an oscillator frequency of 0.3 MHz. The plasma and outer gas flow rates were as follows:

Case a—Argon under atmospheric pressure conditions:

$$Q_1 \doteq 0$$

$$Q_2 = 3 \text{ L} \cdot \text{min}^{-1}$$

$$Q_3 = 34 \text{ L} \cdot \text{min}^{-1}$$

Case b—Argon at 228 Torr (0.3 atm) with the same gas mass flow rate as in case (a):

$$Q_1 = 0$$

$$Q_2 = 10 \text{ L} \cdot \text{min}^{-1}$$

$$Q_3 = 113.3 \text{ L} \cdot \text{min}^{-1}$$

Figures 9.35 and 9.36 show the flow and temperature isocontours for the atmospheric pressure and a soft vacuum discharge, cases a and b, respectively. It may be noted that the temperature profiles along the centerline of the torch

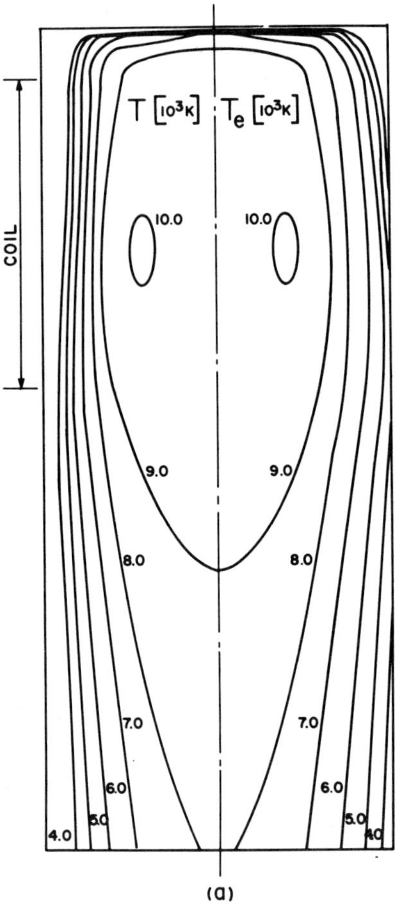

Figure 9.35. Flow and temperature fields in the torch under atmospheric pressure conditions: $P = 760$ torr, $Q_1 = 0$, $Q_2 = 3.0\ \text{L} \cdot \text{min}^{-1}$, $Q_3 = 34.0\ \text{L} \cdot \text{min}^{-1}$, $f = 0.3$ MHz, $P_0 = 5$ kW. (*a,b*) Two-temperature model; (*c*) LTE model.

Fig. 9.37, obtained using the two-temperature and the LTE models, are in good agreement for the atmospheric pressure case. Substantial deviations are noted, however, between the results of the two models under soft vacuum conditions. It is also noted that the entrance region has a predominantly lower temperature under soft vacuum conditions compared with that at atmospheric pressure.

The radial profiles of the electron and heavy particle temperatures, on the other hand, show marked differences between the non-LTE and LTE models, under both atmospheric pressure and soft vacuum conditions in the regions close to the wall of the plasma confinement tube. The combination of the high elec-

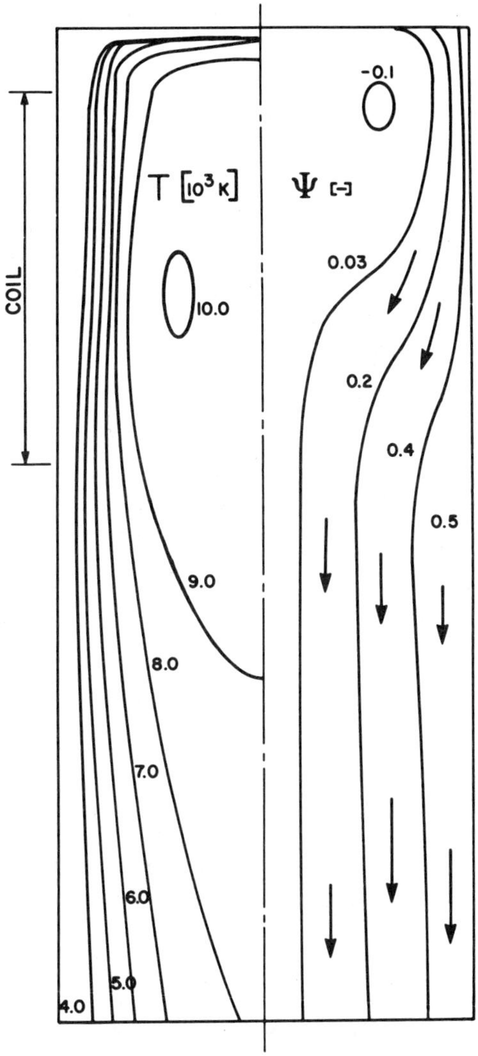

(b)

Figure 9.35. (*Continued*)

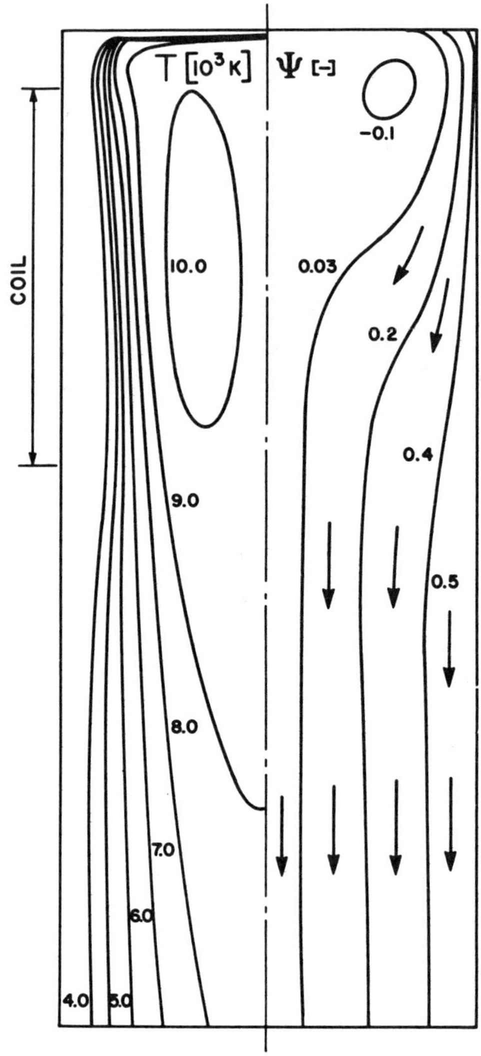

(C)

Figure 9.35. (*Continued*)

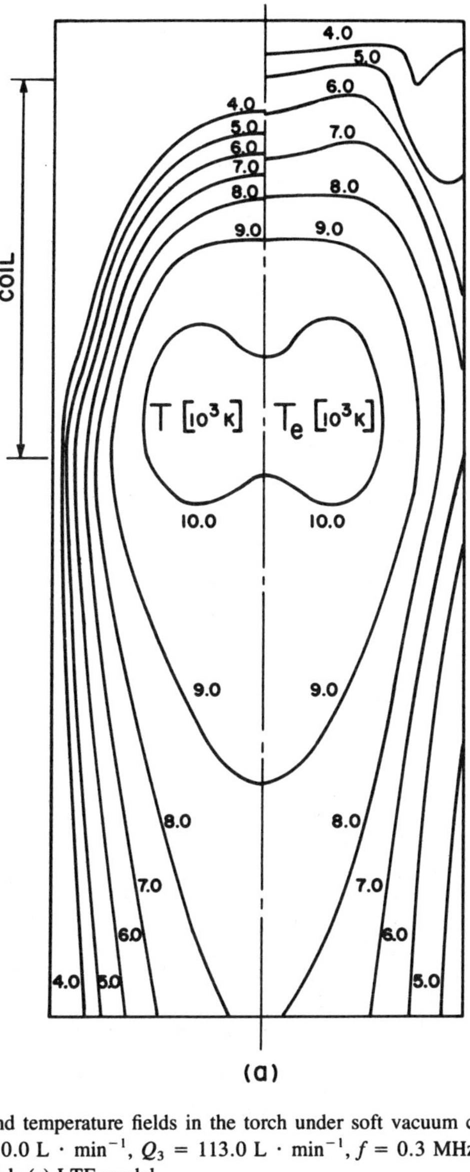

Figure 9.36. Flow and temperature fields in the torch under soft vacuum conditions: $P = 228$ torr, $Q_1 = 0$, $Q_2 = 10.0 \text{ L} \cdot \text{min}^{-1}$, $Q_3 = 113.0 \text{ L} \cdot \text{min}^{-1}$, $f = 0.3$ MHz, $P_0 = 5$ kW. (*a,b*) Two-temperature model; (*c*) LTE model.

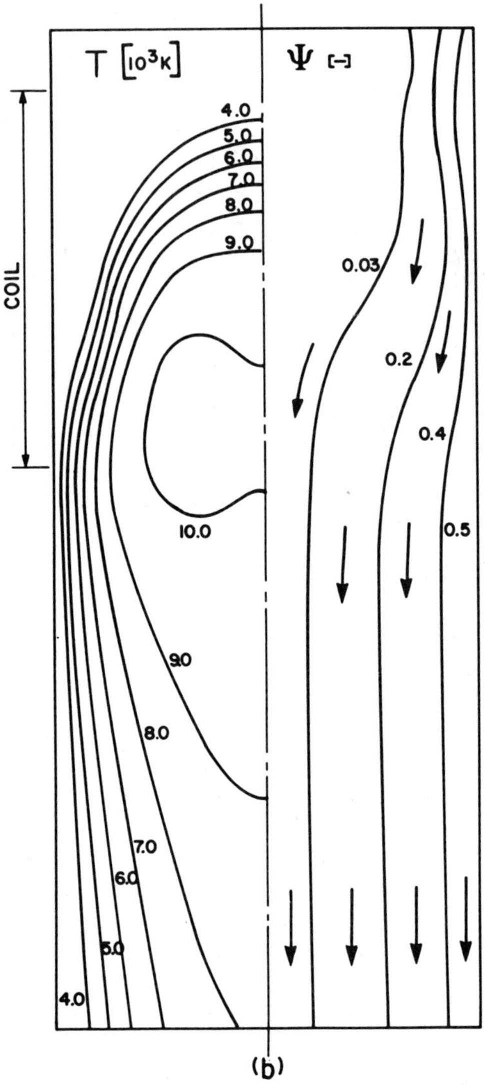

Figure 9.36. (*Continued*)

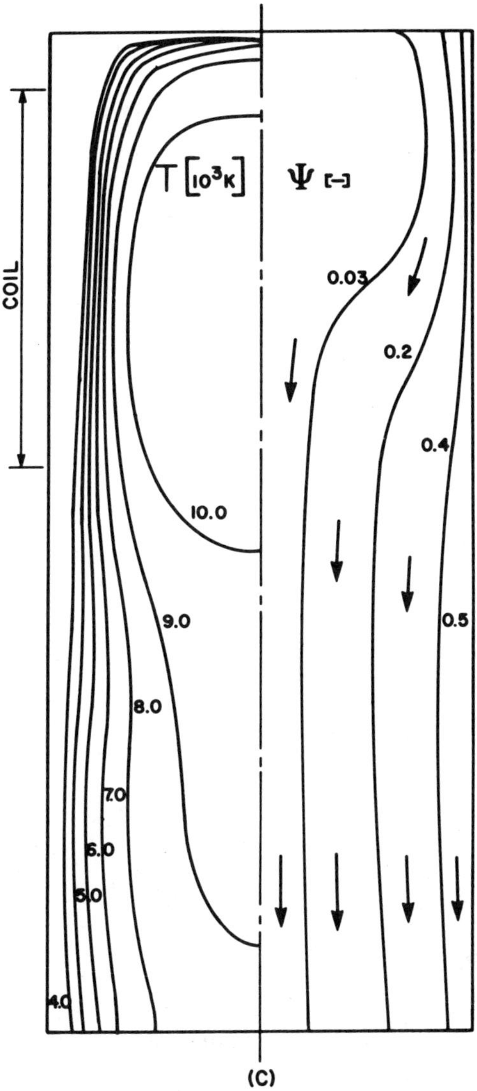

Figure 9.36. (*Continued*)

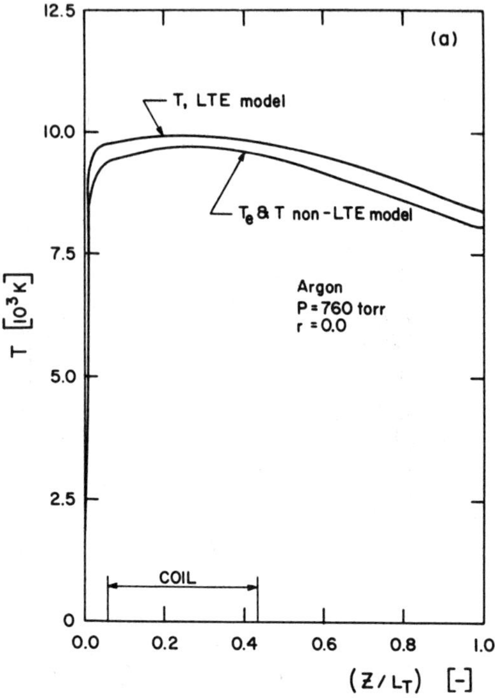

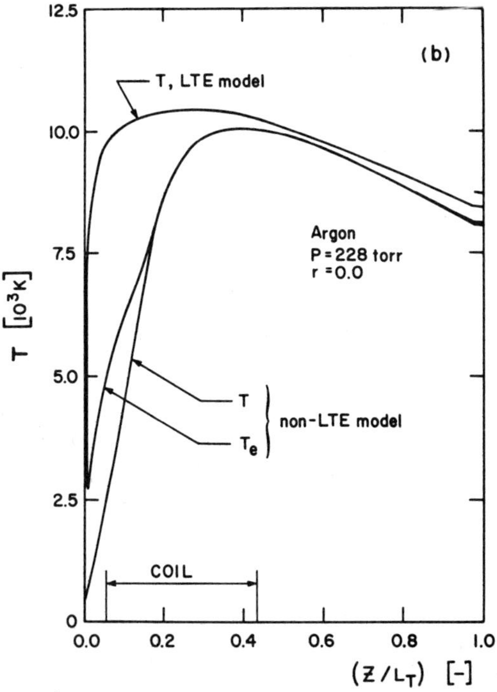

Figure 9.37. Temperature profile along the centerline of the torch under atmospheric pressure (*a*) and soft vacuum conditions (*b*).

tron temperature and electron density close to the wall seems to influence strongly the energy dissipation pattern in the discharge and could account for the high heavy particle temperatures obtained in this region using the non-LTE model.

9.7. REFINEMENTS, PROSPECTS, AND FUTURE DEVELOPMENTS

The models described and results illustrated demonstrate the effectiveness of computer simulations of ICP discharges. Not only can these models be applied to spectrochemical analysis, but their application in engineering and materials processing is significant. The accuracy of predicted temperature, flow, concentration, and radiation fields still requires experimental verification with sophisticated diagnostic tools. The modeling of the spectrochemical ICP is incomplete still, because the introduction of solvent aerosols containing analyte and the evaporation and perturbation of the discharge by the solvent vapor remains to be solved. Furthermore, unconfided configurations in which ambient air mixes with the argon discharge should be computed.

Computation of non-LTE effects, in particular, atom, ion, and electron populations, require extending present capabilities to include appropriate mechanisms [7–19] (cf. Chapter 11). Nevertheless, consideration of molecular gas ICP discharges found nearer to LTE can proceed without the added complication of non-LTE computations. Consideration of ionization interferences and coupling with mathematical descriptions of the rf electrical and impedance matching circuits should result in a truly comprehensive model of the ICP.

REFERENCES

1. H. U. Eckert, *High Temperature Sci.* **6,** 99 (1974).
2. R. M. Barnes, *CRC Crit. Rev. Anal. Chem.* **7,** 203 (1978).
3. R. M. Barnes, *Wiss. Z. Karl-Mark-Univ.* **28,** 383 (1979).
4. R. M. Barnes, *Trends Anal. Chem.* **1,** 51 (1981).
5. R. M. Barnes, *Phil. Trans. R. Soc. (London)* **A305,** 499 (1982).
6. R. M. Barnes, *J. Testing Eval.* **12,** 194 (1984).
7. P. W. J. M. Boumans, *Spectrochim. Acta* **37B,** 75 (1982).
8. M. W. Blades, *Spectrochim. Acta* **37B,** 869 (1982).
9. R. J. Lovett, *Spectrochim. Acta* **37B,** 969 (1982).
10. I. J. M. Raaijmakers, P. W. J. M. Boumans, B. van der Sijde, and D. C. Schram, *Spectrochim. Acta* **38B,** 697 (1983).
11. D. C. Shram, I. J. M. M. Raaymakers, B. van der Sijde, and P. W. J. M. Boumans, *Spectrochim. Acta* **38B,** 1545 (1983).

12. L. de Galan, *Spectrochim. Acta* **39B,** 537 (1984).
13. R. S. Houk and J. A. Olivares, *Spectrochim. Acta* **39B,** 575 (1984).
14. J. W. Mills and G. M. Hieftje, *Spectrochim. Acta* **39B,** 859 (1984).
15. B. L. Caughlin and M. W. Blades, *Spectrochim. Acta* **39B,** 1583 (1984).
16. G. M. Hieftje, G. D. Rayson, and J. W. Olesik, *Spectrochim. Acta* **40B,** 167 (1985).
17. N. Futura, Y. Nojiri, and K. Fuwa, *Spectrochim. Acta* **40B,** 423 (1985).
18. M. W. Blades and B. L. Caughlin, *Spectrochim. Acta* **40B,** 579 (1985).
19. W. Gunter, K. Visser, and P. B. Zeeman, *Spectrochim. Acta* **40B,** 617 (1985).
20. M. P. Freeman and J. D. Chase, *J. Appl. Phys.* **39,** 180 (1968).
21. H. U. Eckert, *J. Appl. Phys.* **41,** 1520 (1970).
22. A. E. Mensing and L. R. Boedeker, Theoretical Investigations of R.F., Induction Heated Plasmas, NASA CR-1312 (1969).
23. H. U. Eckert, *J. Appl. Phys.* **41,** 1529 (1970).
24. D. C. Pridmore-Brown, *J. Appl. Phys.* **41,** 3621 (1970).
25. D. R. Keefer, J. A. Sprouse, and F. C. Leper, *IEEE Trans. Plasma Sci.* **PS-1,** 71 (1973).
26. H. U. Eckert, in R. M. Barnes, ed., *Developments in Atomic Plasma Spectrochemical Analysis*, Heyden, London, Philadelphia (1981), p. 35.
27. H. U. Eckert, 5th Int. Symp. Plasma Chem. (ISPC-5), Edinburgh, UK, (1981) p. 781.
28. F. Aeschbach, *ICP Information Newslett.* **6,** 272–287 (1980).
29. F. Aeschbach, *Spectrochim. Acta* **37B,** 987 (1982).
30. R. J. Miller and J. R. Ayen, *J. Appl. Phys.* **40,** 5260 (1969).
31. R. M. Barnes and R. G. Schleicher, *Spectrochim. Acta* **B30,** 109 (1975).
32. R. M. Barnes and S. Nikdel, *Appl. Spectrosc.* **29,** 477 (1975).
33. R. M. Barnes and S. Nikdel, *J. Appl. Phys.* **47,** 3929 (1976).
34. R. M. Barnes and R. G. Schleicher, *Spectrochim. Acta* **36B,** 81 (1981).
35. R. M. Barnes and J. L. Genna, *Spectrochim Acta* **36B,** 299 (1981).
36. R. M. Barnes, N. Kovacic, and G. A. Meyer, *Spectrochim Acta* **40B,** 907 (1985).
37. C. D. Allemand and R. M. Barnes, *Appl. Spectrosc.* **31,** 434 (1977).
38. J. L. Genna, R. M. Barnes, and C. D. Allemand, *Anal. Chem.* **49,** 1450 (1977).
39. G. A. Meyer and R. M. Barnes, *Spectrochim. Acta* **40B,** 893 (1985).
40. N. Kovacic, G. A. Meyer, K.-L. Lui, and R. M. Barnes, *Spectrochim. Acta* **40B,** 943 (1985).
41. G. A. Meyer and M. D. Thompson, *Spectrochim. Acta* **40B,** 195 (1985).
42. G. A. Meyer and R. M. Barnes, U.S. Patent 4, 482,246, November 13 (1984).
43. J. A. Delettrez, Ph.D. Dissertation, University of California, Davis (1974).
44. M. I. Boulos, *IEEE Trans. Plasma Sci.* **PS-4,** 28 (1976).
45. M. I. Boulos, R. Gagné, and R. M. Barnes, *Can. J. Chem. Eng.* **58,** 367 (1980).

46. R. Gagné, M.Sc.A. Thesis, University of Sherbrooke, Québec (1970).

47. J. Mostaghimi, P. Proulx, and M. I. Boulos, *J. Numerical Heat Transfer* **8,** 187 (1985).

48. J. Mostaghimi, P. Proulx, and M. I. Boulos, *J. Plasma Chem. Plasma Proc.* **4,** 129 (1984).

49. S. V. Patankar, *Numerical Heat Transfer and Fluid Flow.* McGraw-Hill, New York (1980).

50. R. M. Barnes and S. Nikdel, *Appl. Spectrosc.* **29,** 477 (1975).

51. M. I. Boulos and W. H. Gauvin, *Can. J. Chem. Eng.* **52,** 355 (1974).

52. D. Bhattacharya, and W. H. Gauvin, *Am. Inst. Chem. Eng. J.* **21,** 879 (1975).

53. B. Gal-Or, *J. Eng. Power* **102,** 589 (1980).

54. J. K. Fiszdon, *Int. J. Heat Mass Trans.* **22,** 749 (1979).

55. M. Vardelle, A. Vardelle, P. Fauchais, and M. I. Boulos, *Am. Inst. Chem. Eng. J.* **29,** 236 (1983).

56. D. Wei, S. M. Correa, D. Apelian, and M. Paliwal, *6th Int. Symp. Plasma Chem. (ISPC-6), Montreal, Canada* **1,** 83 (1983).

57. M. I. Boulos, *IEEE Trans. Plasma Sci.* **PS-6,** 93 (1978).

58. T. Yoshida and K. Akashi, *J. Appl. Phys.* **48,** 2252 (1977).

59. P. Proulx, M.Sc.A., University of Sherbrooke, Québec. (1984).

60. P. Proulx, J. Mostaghimi, and M. I. Boulos, *6th Int. Symp. Plasma Chem., (ISPC-6), Montreal, Canada,* **1,** 59 (1983).

61. J. Mostaghimi, P. Proulx, M. I. Boulos, and R. M. Barnes, *Spectrochim. Acta* **40B,** 153 (1985).

62. P. Proulx, J. Mostaghimi, and M. I. Boulos, *Int. J. Heat Mass Transfer* **28,** 1327 (1985).

63. E. Bourdin, P. Fauchais, and M. I. Boulos, *Int. J. Heat Mass Transfer* **26,** 567 (1983).

64. Xi Chen and E. Pfender, *J. Plasma Chem. Plasma Proc.* **2,** 293 (1982).

65. Xi Chen and E. Pfender, *J. Plasma Chem. Plasma Proc.* **2,** 185 (1982).

66. Xi Chen and E. Pfender, *J. Plasma Chem. Plasma Proc.* **3,** 97 (1983).

67. Xi Chen and E. Pfender, *J. Plasma Chem. Plasma Proc.* **3,** 351 (1983).

68. C. T. Crowe, M. P. Sharma, and D. E. Stock, *J. Fluids Eng.* **99,** 325 (1977).

69. J. Mostaghimi and E. Pfender, *J. Plasma Chem. Plasma Proc.* **4,** 199 (1984).

70. H. U. Eckert and A. Danielsson, *Spectrochim. Acta* **38B,** 15 (1983).

71. H. U. Eckert, *Spectrochim. Acta* **33B,** 591 (1978).

72. C. B. Boss and G. M. Hieftje, *Anal. Chem.* **51,** 895 (1979).

73. N. Furuta and G. Horlick, *Spectrochim. Acta* **37,** 53 (1982).

74. J. O. Hirschfelder, C. F. Curtiss, and R. B. Bird, *Molecular Theory of Gases and Liquids*, Wiley, New York (1954).

75. P. W. J. M. Boumans, *Theory of Spectrochemical Excitation.* Hilger, London/ Plenum, New York (1966).

76. D. C. Morton and W. H. Smith, *Astrophys. J. Suppl.* **26,** 333 (1973).
77. H. W. Drawin and P. Felenbok, *Data for Plasmas in Local Thermodynamic Equilibrium.* Gauthier-Villars, Paris (1965).
78. P. W. J. M. Boumans and J. F. de Boer, *Spectrochim. Acta* **32B,** 365 (1977).
79. N. Furuta and G. Horlick, *Spectrochim. Acta* **37,** 53 (1982).
80. T. E. Edmonds and G. Horlick, *Appl. Spectrosc.* **31,** 536 (1977).
81. H. Kawaguchi, T. Ito, and A. Mizuike, *Spectrochim. Acta* **36B,** 615 (1981).
82. T. Yoshida, K. Nakagawa, T. Narada, and K. Akashi, *J. Plasma Chem. Plasma Proc.* **1,** 113 (1981).
83. S. V. Dresvin and V. S. Klubnikin, *High Temp.* **9,** 3, 475 (1971).
84. S. V. Dresvin (Ed.), *Physics and Technology of Low Temperature Plasmas.* Iowa State University Press, Ames, IA (1977).
85. V. M. Goldfarb, in R. M. Barnes, ed., Developments in Atomic Plasma Spectrochemical Analysis Heyden, London, Philadelphia (1981), p. 725.
86. J. Mostaghimi, P. Proulx, and M. I. Boulos, *7th Int. Symp. Plasma Chem., (ISPC-7), Eindhoven, The Netherlands,* **3,** 865 (1985).
87. M. Mitchner and C. H. Kruger, *Partially Ionized Gases*, Wiley, New York (1973).
88. R. S. Devoto, *Phys. Fluids* **10,** 354 (1967).
89. S. V. Desai and W. H. Corcoran, *J. Quant. Spectrosc. Radiat. Transfer* **9,** 1371 (1969).
90. D. L. Evans and R. S. Tankin, *Phys. Fluids* **10,** 1137 (1967).

CHAPTER

10

SPECTROSCOPIC DIAGNOSTICS: BASIC CONCEPTS

J. M. MERMET

Laboratoire des Sciences Analytiques
Université Claude Bernard-Lyon I
Villeurbanne, France

10.1. Introduction
10.2. Definition of Line Parameters
10.3. Intensity Gradient
10.4. Various Species in a Plasma
10.5. Definition of Temperatures
10.5.1. Translational or Kinetic Temperature T_{kin}
10.5.2. Excitation Temperature or Electronic Excitation Temperature T_{exc}
10.5.3. Rotational Temperature T_{rot}
10.5.4. Ionization–Recombination Temperature T_{ion}
10.5.5. Electron Temperature T_e
10.5.6. Radiation Temperature T_{rad}
10.5.7. Norm Temperature
10.6. Classification of Temperature Measurements in the ICP
10.7. Temperature Determination Using Absolute Line Intensities
10.8. Temperature Determination Using Relative Line Intensities
10.8.1. Boltzmann Plot
10.8.2. Line Pair Intensity Ratio Method
10.9. Linewidth Method
10.10. Temperature Determination from the Line-to-Continuum Ratio
10.11. Classification of Electron Number Density Measurements
10.12. Electron Number Density Measurement Using the Stark Effect
10.12.1. Stark Effect on Hydrogen Lines
10.12.2. Stark Effect on Argon lines
10.13. Electron Number Density Determination from the Continuum Intensity
10.14. Inglis–Teller Method
References

10.1. INTRODUCTION

Only little information can be directly obtained from spectrocopic observation of a plasma discharge. Parameters such as temperatures or species densities can be deduced if some hypothesis is assumed on equilibrium or mechanisms in the discharge. Although not sufficient to explain every process in an inductively coupled plasma (ICP), knowledge of the various parameters of the plasma is necessary to describe some properties of the discharge, in particular with analytical applications in mind, and also to compare several ICP systems. Several types of ICPs are presently available and they can differ by the frequency, oscillator type, input power, coil, and torch geometries. For the sake of comparison and for a better understanding of processes, spectroscopic diagnostic measurements have been carried out since the beginning of ICP research. Such measurements are still necessary to explain line intensities, line widths, and background intensity for analytical purposes.

Several books or review papers [1–11] have been devoted to plasma diagnostics, although none of them has been specialized in the ICP field. The purpose of this chapter is to provide information about conventional measurements of temperatures and electron number densities in the discharge. More recent advanced diagnostics are dealt with in Chapter 11.

10.2 DEFINITION OF LINE PARAMETERS

If we observe the variation of the intensity as a function of the frequency, it is possible to define the total intensity I of a line, the line profile $I(\nu)$, which is the spectral intensity distribution within the line, and the intensity I_c of the continuum (continuous background). Thus, we have

$$I_\nu = \int_0^\infty I(\nu)\, d\nu \tag{10.1}$$

For practical spectroscopy and in this wavelength range [ultraviolet (UV) to visible], wavelengths λ are used rather than frequencies ν. Wavelength and frequency are interrelated by

$$\nu = c/\lambda \tag{10.2}$$

and

$$d\nu = c/\lambda^2\, d\lambda \tag{10.3}$$

where c is the speed of the light. Sometimes, especially when dealing with line-

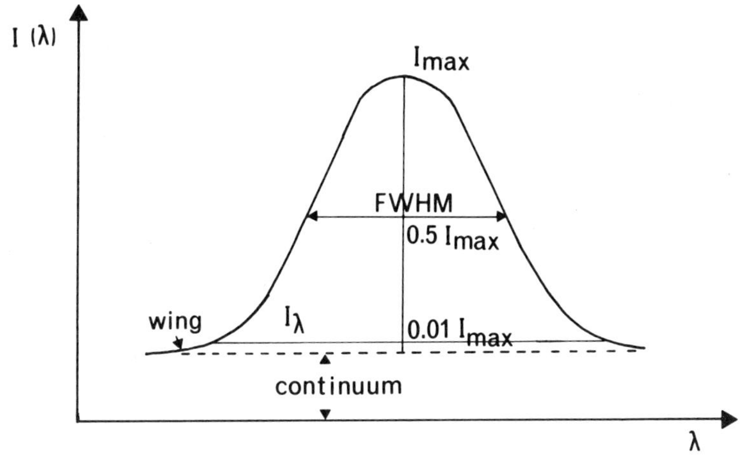

Figure 10.1. Line profile with definitions of full width at half-maximum (FWHM) and line wing.

broadening theories, the angular frequency ω can be used:

$$\omega = 2\pi\nu \tag{10.4}$$

In contrast with I_λ, which is represented by an area, the maximum intensity (peak intensity) I_{max} is defined as the maximum value of the line profile function. The central wavelength corresponds to the peak intensity.

Usually, the shape of a line profile (Fig. 10.1) is represented by the full width at half-maximum (FWHM), which corresponds to the width measured at $I_{max}/2$. In some books, it is also called "half-width" and half of its value "semi-half-width." When the value of the intensity is below 1% of the peak intensity, the corresponding part of the line profile is called the "wing of the line."

The continuum or continuous radiation arises from various causes; among them are free–free and free–bound electron transitions, incandescent radiation, and unresolved band spectra. The term "background" is often improperly used to indicate the continuum. It should be used for all the radiation that originates from the plasma and reaches the spectrometer when no analyte is present (cf. Part 1, Sections 4.6 and 7.3).

10.3. INTENSITY GRADIENT

Because of the gradients of temperatures and species densities within the discharge, only an average value of the intensity—the experimental or lateral value—is measured directly. To obtain a value in a single point (radial value), a mathematical transformation called the Abel inversion has to be carried out (see, for example, [1, 4]). This is possible because of the circular symmetry of the discharge. The experimental intensity is defined as the energy radiated per wavelength unit per unit area in unit time within a solid angle perpendicular to the discharge. After Abel inversion, a local volume emission coefficient ϵ_λ is obtained. It may be defined as the energy emitted by a unit volume in unit time per unit solid angle and per unit wavelength ($J \cdot m^{-3} \cdot s^{-1} \cdot sr^{-1}$). Thus, we have

$$I_\lambda = K \int_l \epsilon_\lambda(l)\, dl \tag{10.5}$$

where l is the length of the zone where ϵ_λ can be considered as constant.

10.4. VARIOUS SPECIES IN A PLASMA

Although a plasma, especially a rare gas discharge, seems to be a simple medium, many types of neutral or ionized species are present, either excited or in the ground state. Knowledge of their role and density may be necessary. Besides electrons (e), neutral (Ar) and ionized (Ar^+) argon are present. Neutral argon can be excited (Ar*), and the lowest levels are either resonant levels (direct optical transitions with the neutral ground state) or metastable levels (forbidden optical transitions to lower states). Molecular excited Ar (Ar_2^*) and molecular Ar ion (Ar_2^+) may also be considered. Usually, injected species are observed in several forms (X, X^+, X^{2+}, X*, X^{+*} . . .), depending on the respective ionization and excitation energies. Some negative ions (X^-) are also present in the discharge. Although the kinetic temperature of the plasma can be high enough to atomize every species, some radicals still exist, at least at trace levels (OH, N_2, N_2^+, CH, CN, . . .), especially with organic compounds. Some of them (NO) are due to the surrounding air.

Obviously, a crucial species in the discharge is formed by the photons originating from radiative deexcitation of excited levels and from radiative recombination.

10.5. DEFINITION OF TEMPERATURES

Usually, in classical gas thermodynamics, a temperature T can be defined and related to the mean translational energy (or kinetic energy) E_{kin} of the particles, namely,

$$E_{kin} = \tfrac{3}{2}(kT) \tag{10.6}$$

where k is the Boltzmann constant. In this case, all the species have the same mean translational energy and therefore the same temperature T. The physical state of such media can be described by a finite number of parameters, temperature, pressure, and concentration of each species in the case of a mixture.

However, when considering laboratory plasmas such as the ICP, it is difficult to define a single temperature. The different temperatures will be defined according to the physical phenomena that are observed. A statement of temperatures is a way to describe the energies of the various particles.

10.5.1. Translational or Kinetic Temperature T_{kin}

The Maxwell distribution function $f(v)$ for the random velocity v is given by:

$$f(v) = 4\pi v^2 (m/2\pi kT_{\text{kin}})^{3/2} \exp(-mv^2/2kT_{\text{kin}}) \tag{10.7}$$

where m is the mass of the particle considered (heavy particles and electrons). This function defines the kinetic temperature T_{kin}, which represents the temperature of the system for the state of complete thermodynamic equilibrium (CTE).

Because of the motion of the emitting particles, the central wavelength λ_0 appears to be shifted by $\lambda - \lambda_0$. This is the Doppler effect:

$$\lambda - \lambda_0 = \lambda_0 v/c \tag{10.8}$$

where v is the velocity component parallel to the direction of observation. Using the Maxwellian distribution, the fraction dn/n of the particles with velocities between v and $v + dv$ is found to be

$$dn/n = (1/\pi^{1/2})(m/2kT)^{1/2} \exp(-mv^2/2kT) \tag{10.9}$$

If we use the most probable velocity $\langle v \rangle$:

$$\langle v \rangle = (2kT/m)^{1/2} \tag{10.10}$$

then

$$dn/n = (1/\pi^{1/2}) \exp\left[-(v/\langle v \rangle)^2\right](dv/\langle v \rangle) \tag{10.11}$$

If we define:

$$\Delta\lambda = \lambda_0 \langle v \rangle / c \tag{10.12}$$

we have

$$v/\langle v \rangle = (\lambda - \lambda_0)/\Delta\lambda \tag{10.13}$$

and

$$dv/\langle v \rangle = d\lambda/\Delta\lambda \tag{10.14}$$

Substituting Eqs. (10.13) and (10.14) into Eq. (10.11) we obtain:

$$dn/n = [1/(\pi^{1/2}\Delta\lambda)] \exp\left[-((\lambda - \lambda_0)/\Delta\lambda)^2\right] d\lambda \qquad (10.15)$$

The line intensity is proportional to the concentration dn of emitting particles in the interval $d\lambda$, and the intensity distribution is

$$I_\lambda = (I_0/\pi^{1/2}\Delta\lambda) \exp\left[-((\lambda - \lambda_0)/\Delta\lambda)^2\right] \qquad (10.16)$$

The line profile exhibits a Gaussian shape. The intensity reaches one half of its maximum when the value of the exponential factor is equal to one half. Therefore

$$\lambda - \lambda_0 = (\ln 2)^{1/2}\Delta\lambda \qquad (10.17)$$

The FWHM Doppler width then is

$$\Delta\lambda = 2(\ln 2)^{1/2}\Delta\lambda \qquad (10.18a)$$

Using Eqs. (10.12) and (10.10) we thus find:

$$\Delta\lambda = 7.10 \times 10^{-7}\lambda_0(T/M)^{1/2} \qquad (10.18b)$$

where T is expressed in K and M in atomic mass units.

10.5.2. Excitation Temperature or Electronic Excitation Temperature T_{exc}

The Boltzmann distribution allows us to define the ratio of the populations or number densities n_j and n_k of bound electrons in the states of energy E_j and E_k with $k > j$.

$$n_k/n_j = (g_k/g_j) \exp\left[-(E_k - E_j)/kT_{exc}\right] \qquad (10.19)$$

where g_k and g_j are the statistical weights of the two levels ($g = 2J + 1$). If we consider the total number density n of the ionization state in question, we have

$$n_k/n = \left[g_k \exp(-E_k/kT_{exc})\right]/Q(T) \qquad (10.20)$$

T_{exc} is then called the excitation temperature and $Q(T)$ the internal partition function defined as:

$$Q(T) = \sum_k g_k \exp(-E_k/kT_{exc}) \qquad (10.21)$$

For low temperatures (in chemical flames, for instance), the value of $Q(T)$ is most of the time very close to the statistical weight g_0 of the ground state. But for a temperature greater than 4000 K, the partition function may vary with the temperature and the electron number density [4, 12, 13]. Practical formulas in the form of polynomials have been published [14] for the calculation of the partition function of most elements.

The line intensity formula is deduced from the Boltzmann equilibrium

$$I = (h\nu/4\pi)\,An_k \tag{10.22}$$

where A is the transition probability (s^{-1}) for spontaneous emission. Using Eqs. (10.2) and (10.20) we obtain from eq. (10.22)

$$I = (hc/4\pi)\,\lambda g_k A(n/Q(T_{exc})\,\exp\,(-E_k/kT_{exc}) \tag{10.23}$$

In this formula, A is sometimes replaced by the oscillator strength f using the relationship:

$$g_k A = C g_j f/\lambda^2 \tag{10.24}$$

where $C = 6.67 \times 10^{15}$, if λ is expressed in Ångströms and $C = 6.67 \times 10^{13}$ if λ is in nanometers.

10.5.3. Rotational Temperature T_{rot}

Similarly, for a transition $J' - J''$, a rotational line intensity can be calculated:

$$I = D\nu^4 S\,\exp\,(-E_r/kT_{rot}) \tag{10.25}$$

where coefficient D contains the rotational partition function, the statistical weight $(2J' + 1)$ and universal constants, and S is the oscillator strength and E_r the rotational energy:

$$E_r = B_v\,hc\,J'(J' + 1) \tag{10.26}$$

where B_v is a rotational constant belonging to the vibrational quantum number v.

Very often, the rotational temperature T_{rot} is assumed to be of the same magnitude as the kinetic temperature T_{kin}, because of the low energies involved in the rotation process.

10.5.4. Ionization–Recombination Temperature T_{ion}

The distribution of the different species in the plasma can be described by a mass action law that is a function of the temperature. If we consider the case of the ionization–recombination process, this law is given by the Saha–Eggert relation:

$$S(T_{ion}) = n_{i+1}(n_e/n_i) \tag{10.27}$$

$$S(T_{ion}) = 2(Q_{i+1}/Q_i)\,(2\pi mk/h^3)^{3/2}\,T_{ion}^{3/2}\,\exp\,(-E_{ion}/kT_{ion}) \tag{10.28}$$

where n_i and n_{i+1} are the total number densities of the ith and $(i + 1)$th ionization states. Q_i and Q_{i+1} are the respective partition functions and the coefficient 2 is the equivalent partition function for the electron; m is the electron mass, h the Planck constant, and E_{ion} the ionization energy between states i and $i + 1$.

T_{ion} is then the ionization–recombination temperature, usually called the ionization temperature. Although this temperature does not have a physical meaning, it describes the ionization equilibrium between two successive ionization states.

10.5.5. Electron Temperature T_e

After ionization processes, electrons are no longer bound, which means that they are in nonquantisized levels. Many transitions, either free–free or free–bound can occur. In the UV–visible part of the spectrum emitted by an Ar ICP, radiative recombination is the predominant process:

$$Ar^+ + e \rightarrow Ar^* + h\nu$$

This process leads to the emission of a continuum of which the intensity is a function of the wavelength, namely,

$$\epsilon(\lambda, T) = C_1 \frac{n_e n^+}{T_e^{1/2}} \frac{Z^2}{\lambda^2} \xi(\lambda, T)\left[1 - \exp\left(-\frac{hc}{kT}\right)\right] \tag{10.29}$$

where Z is the effective charge number and $\xi(\lambda, T)$ a correction factor (see Section 10.10 and [77]).

10.5.6. Radiation Temperature T_{rad}

The radiation field provides a radiation density $U(T)$ which depends only upon the temperature and which is given by the Planck function:

$$U(T)\, d\lambda = (8\pi hc/\lambda^5) \cdot \left[\frac{d\lambda}{\exp\left(\dfrac{h\lambda}{kT_{\text{rad}}}\right) - 1}\right] \tag{10.30}$$

where T_{rad} is the radiation temperature.

10.5.7. Norm Temperature

The norm temperature is the value of the excitation temperature for which a spectral line has a maximum intensity due to competing effects of ionization and excitation processes. It is a function of the total pressure and of the plasma composition (cf. Part 1, Section 4.7.2).

10.6. CLASSIFICATION OF TEMPERATURE MEASUREMENTS IN THE ICP

A great many papers [15–98] have dealt with temperature measurements in ICPs. The various methods can be classified into various categories depending on the use of

- Absolute line intensities
- Relative line intensities
- Linewidths
- The line-to-continuum intensity ratio

10.7. TEMPERATURE DETERMINATION USING ABSOLUTE LINE INTENSITIES

This method directly uses Eq. (10.23) and provides the excitation temperature. Application of the formula is not easy because knowledge of the absolute value of the intensity requires a calibration of the optical system. Moreover, high accuracy is required for the values of the transition probabilities and it is necessary to calculate the number density. Nevertheless, several authors [15, 16, 18, 19, 23, 27, 39, 58] have determined the excitation temperature with this method, and, in particular, it was the first temperature determination in an ICP similar to those at present used. Reed [15] used the absolute intensity of the ArI 763.5-nm line a 4-MHz, 10-kW ICP. A maximum value of 16,000 K was found in the center of the discharge. Since then, it has been demonstrated that a large error was involved because Reed used a wrong value of the transition probability, which illustrates the importance of such a value. Many authors [17, 18, 27, 31, 99–126] have published data on neutral Ar transition probabilities in the 340- to 700-nm range. Among the lines usable for temperature measurements, the line ArI 430.0 nm has one of the most accurately known transition probabilities [50, 123]. Values for this line are given in Table 10.1.

10.8. TEMPERATURE DETERMINATION USING RELATIVE LINE INTENSITIES

Two methods have been mainly used: the Boltzmann plot and line pair methods. In these methods, it is not necessary to measure the absolute line intensity and to know the concentration of emitting species, which explains their wide use among ICP explorers.

Table 10.1. *A* Values for Ar I 430.0 nm (in 10^8 s^{-1})

A Values	Reference
0.0358	[99]
0.0325	[100]
0.13	[103]
0.0411	[104]
0.0317	[106]
0.0318	[17]
0.0330	[107]
0.0098	[108]
0.14	[109]
0.0331	[27]
0.031	[110]
0.0314	[111]
0.0366	[112]
0.028	[113]
0.031	[114]
0.0394	[115]
0.0308	[117]
0.0277	[121]
0.034	[122]
0.372	[123]
0.030	[124]
0.032	[125]
0.015	[126]

10.8.1. Boltzmann Plot

From Eq. (10.23), we can see that the value log $(I\lambda/gA)$ is a linear function of the excitation energy.

$$\log\,(I\lambda/gA) \propto E_{\mathrm{exc}} \tag{10.31}$$

The slope is related to the excitation temperature and is equal to $-0.625/T_{\mathrm{exc}}$ when E is in reciprocal centimeters (cm^{-1}) and equal to $-5040/T_{\mathrm{exc}}$ when E is in electron volts (eV). The oscillator strength can also be used, and using Eq. (10.24) we can write:

$$\log\,(I\lambda^3/gf) \propto E_{\mathrm{exc}} \tag{10.32}$$

These two formulae can be applied only when a Boltzmann equilibrium exists. A more accurate temperature determination is obtained when a large range of excitation energies is covered. This method has been applied to Ar lines and to lines of injected species, including H.

10.8.1.1. Argon

In the case of Ar, several authors [21, 27, 28, 30, 31, 33, 39, 41–43, 50, 57, 59, 64, 74] have used the Boltzmann plot method with lines between 340 and 700 nm, which corresponds to an energy range between 107,500 and 124,750 cm^{-1}. The energy difference is rather small, which is explained by the unusually high energies for the first excited levels. Table 10.2 lists lines used for the Boltzmann plot as described in the literature. Several sets of A values have been tested for this determination, which can provide a difference of about 1000 K in the result [50]. The A values of Malone and Corcoran [17] and the compilation of Wiese [115] are the most used data for this purpose. Note that the A values of Malone and Corcoran have been determined with an ICP. The main set of values [17, 110, 112, 115, 124] is given in Table 10.3.

10.8.1.2. Elements

In the case of injected elements, the choice is not very large, because of the following requirements: large energy range, good knowledge of A or f values, and lines located at closely spaced wavelengths to avoid a calibration of the wavelength response. For these reasons, lines of neutral Fe or ionized Ti are used most of the time.

With neutral Fe, most measurements have been carried out at wavelengths between 367 and 382 nm. For these lines, several papers [127–150] have been published. Traditionally for Fe, data are usually presented in the form of log gf values. Results before 1966 are summarized in a review paper of Corliss and Warner [134]. More recent work [142, 144, 148, 149] indicates that previous results provided wrong values when energies were above 34,000 cm^{-1}. These last results are mutually very consistent. They are normalized on an absolute scale, using the log gf value of FeI 371.994-nm line (log $gf = -0.43$) selected from Wagner and Otten [151] and Bells and Tubbs [152]. In Table 10.4, log gf values are given from [131, 134, 142, 144, 148, 149] after normalization using the FeI 371.994 nm line. The Boltzmann plot with Fe lines has been applied by several authors [50, 57, 63, 70–72, 74, 80, 83, 91, 94]. The use of various sets of log gf values can provide a variation of 2000 K between the results [50, 63, 91]. The lowest values of the temperature were obtained with Allen [129] and Corliss and Bozman data [131], the highest values with Huber and Parkinson [142] and Bridges and Kornblith data [144]. Because of their better accuracy, it is preferable to use values from [142, 144, 148, 149].

With ionized Ti, measurements have been carried out between 321 and 324 nm [42, 50, 63]. Several log gf values are available [1, 131, 153–157] and they are listed in Table 10.5.

Table 10.2. Argon Line Selection for the Boltzmann Plot Excitation Temperature Determination, after the literature

λ (nm)	Reference [27]	Reference [28]	Reference [30]	Reference [31]	Reference [33]	Reference [50]	Reference [57]	Reference [59]	Reference [63]	Reference [74]
340.6		x						x		
383.4		x	x					x		
404.4										
419.8		x		x						
420.0		x		x						
425.1		x	x	x	x	x	x		x	x
425.9	x				x	x	x		x	x
426.6						x	x		x	x
427.2					x	x	x	x	x	
430.0	x		x			x	x		x	x
433.3						x	x		x	x
433.5						x	x		x	
434.5	x	x	x	x		x	x		x	
451.0					x			x		
516.2										
555.8		x	x	x						
557.2	x	x	x	x						
603.2										
667.7										
695.5			x							

Table 10.3. Selection of Argon A Values from the Literature for the Excitation Temperature Determination Using the Boltzmann Plot Method (in 10^8 s^{-1})

λ (nm)	g_j	g_k	E (cm^{-1})	Reference [17]	Reference [110]	Reference [112]	Reference [115]	Reference [124]
340.618	3	1	124,750	0.0353			0.041	
357.230	3	1	123,385	0.0467			0.054	
364.983	3	1	122,791	0.0736			0.085	
383.468	3	1	121,470	0.0691			0.080	
394.898	5	3	118,460	0.0445	0.035		0.0467	0.317
404.442	3	5	118,469	0.0325	0.028	0.0435	0.0346	0.0241
418.188	1	3	118,460	0.0487	0.046	0.0511	0.058	0.045
419.832	3	1	117,563	0.242	0.22	0.228	0.276	0.266
420.067	5	7	116,943	0.0801	0.082	0.0865	0.103	0.092
425.118	5	3	116,660	0.0089	0.0076	0.0075	0.0113	
425.936	3	1	118,871	0.366	0.32	0.364	0.415	0.289
426.629	3	5	117,184	0.0265	0.023	0.0294	0.0333	0.0254
427.217	3	3	117,151	0.0688	0.063	0.0769	0.084	0.0645
430.010	3	5	116,999	0.0318	0.031	0.0366	0.0394	0.030
433.356	3	5	118,469	0.0506	0.048	0.0551	0.060	
433.534	3	3	118,460	0.0308	0.029	0.0385	0.0387	
434.517	3	3	118,407	0.0273	0.022	0.0278	0.0313	
451.073	3	1	117,563	0.102	0.10	0.0946	0.123	0.083
515.139	3	1	123,509		0.14		0.249	0.257
516.229	3	3	123,468		0.12	0.178	0.198	0.192
522.127	7	9	124,610				0.092	0.117
549.587	7	9	123,653		0.16	0.170	0.176	0.197
555.870	3	5	122,087				0.148	
557.254	5	7	123,557		0.0036		0.069	0.091
603.213	7	9	122,036		0.18		0.246	0.30
614.544	5	7	123,557				0.079	
667.728	3	1	108,723				0.0241	
696.543	5	3	107,496				0.674	

Table 10.4. Selection of Log gf Values for Fe from the Literature, for the Excitation Temperature Determination Using the Boltzmann Plot Method

λ(nm)	g_j	g_k	E (cm^{-1})	Reference [131]	Reference [134]	Reference [142]	Reference [144]	Reference [148]	Reference [149]
367.991	9	9	27,167	−1.37	−1.41	−1.51	−1.55	−1.58	
368.223	5	5	55,754	1.07	1.07		0.28		0.24
368.411	9	7	49,135	0.50	0.54		−0.28		−0.31
370.109	7	9	51,192	1.10	0.90		−0.01		−0.04
370.446	11	9	48,703	0.33	0.30		−0.57		−0.59
370.557	7	7	27,395	−1.24	−1.21	−1.27	−1.30	−1.31	
370.925	9	7	34,329	−0.37	−0.41	−0.58	−0.57		−0.65
371.994	9	11	26,875	−0.43	−0.43	−0.43	−0.43	−0.43	−0.43
372.256	5	5	27,560	−1.22	−1.13	−1.22	−1.28	−1.31	
372.438	5	7	45,221	−0.05	0.11		−0.70		−0.72
372.762	7	5	34,547	−0.39	−0.41	−0.53	−0.52		−0.63
373.240	5	5	44,512	0.04	0.21		−0.50		−0.54
373.332	3	3	27,666	−1.26	−1.29		−1.36		
373.487	11	11	33,695	0.48	0.46	0.30	0.31		0.32
373.713	7	9	27,167	−0.63	−0.61	−0.57	−0.57	−0.58	−0.58
373.831	11	13	53,094	0.60	0.71		0.05		0.01
374.826	3	5	27,560	−0.97	−0.98	−0.98	−1.01	−1.00	
374.949	9	9	34,040	0.29	0.33	0.18	0.17		0.16
375.361	7	5	44,184	−0.23	−0.18		−0.89		−0.93
375.823	7	7	34,329	0.20	0.20	0.00	0.00		−0.03
376.005	13	15	45,978	−0.07	0.04		−0.69		−0.74
376.053	3	5	44,512	−0.83	−0.58		−1.19		
376.379	5	5	34,547	−0.02	0.00	−0.18	−0.19		−0.24
376.554	13	15	52,655	0.88	0.97	0.53	0.53		0.50
376.719	3	3	34,692	−0.15	−0.13	−0.35	−0.34		−0.39
381.584	9	7	38,175	0.42	0.49	0.37	0.25		0.30

Table 10.5. Selection of Log gf Values for Ti from the Literature, for the Excitation Temperature Determination Using the Boltzmann Plot

λ (nm)	g_j	g_k	E (cm^{-1})	Reference [153]	Reference [154]	Reference [131]	Reference [155]	Reference [156]	Reference [157]
321.706	8	10	31,301	−0.43	−0.42	−0.48	−0.45	−0.54	−0.54
321.827	10	8	43,741		0.23	0.43	0.42	−0.06	−0.06
322.284	6	8	31,114	−0.32	−0.40	−0.41	−0.40	−0.49	−0.49
322.424	12	10	43,781		0.26	0.40	0.50	0.04	0.04
322.860	4	2	39,675	0.03		0.24	0.10	−0.20	−0.20
323.228	8	6	39,927	0.03	0.04	−0.07	0.08	−0.25	−0.25
323.452	10	10	31,301	0.28	0.14	0.31	0.26	0.31	0.31
323.657	8	8	31,114	0.15	−0.01	0.19	0.14		0.16
323.904	6	6	30,959	0.02		0.06	−0.04	−0.02	−0.02
323.966	6	4	39,603	0.03		−0.15	0.13		−0.24
324.198	4	4	30,837	−0.08	−0.30	−0.14	−0.20	−0.12	−0.12

Table 10.6. *A* Values for the Balmer Series Lines of H_2

λ (nm)	g	g	E (cm^{-1})	A ($10^8\ s^{-1}$)
656.280	8	18	97,492	0.4410
486.132	8	32	102,824	0.08419
434.046	8	50	105,292	0.02530
410.173	8	72	106,632	0.009732
397.007	8	98	107,440	0.004389

Other measurements have been described with V [63, 88, 91] with transition probability values from [131, 155, 158], Cd [72] with values from [159], W [160] with values from [131, 161], and Mn [160] with values from [131, 162].

Finally, H can be used [54, 82] since the values of the transition probabilities are accurately known [163]. They are given in Table 10.6. The only difficulty is the strong line broadening due to Stark effect. As a consequence, the line intensity determination may be difficult.

10.8.1.3. Molecular Species

Similarly to the excitation temperature, a rotational temperature can be deduced from relation (10.25). Several species have been used for this purpose. Among them, the molecular ion of N_2 (N_2^+) [25, 62, 83] and the OH radical [59, 83] are the most common.

In the case of N_2^+, the first negative system $B^2\Sigma_u^+ - X^2\Sigma_g^+$ corresponds to the emission of two branches R and Q with $K' = K'' + 1$ for the R branch and $K' = K'' - 1$ for the Q branch, K'' being the assignment of the lower state. We have:

	S_J	$K'(K'+1)$
R branch	$2(K''+1)$ or $(K'+K''+1)$	$(K''+1)(K''+2)$
Q branch	$2K$ or $(K'+K''+1)$	$K''(K''+1)$

It is possible to determine the temperature from the slope of the curve $\log [I/(K' + K'' + 1)]$ versus $K'(K'+1)$. The value of the slope is

$$-Bhc/kT = -2.983/T \tag{10.33}$$

When a decadic logarithm is used, the value of the slope becomes $-1.296/T$. The assignments of the rotational lines as a function of K'' are based on the works of Childs [164] and Coster and Brons [165]. Due to the even–odd alternation [166], line intensities with odd K'' must be multiplied by 2. Moreover,

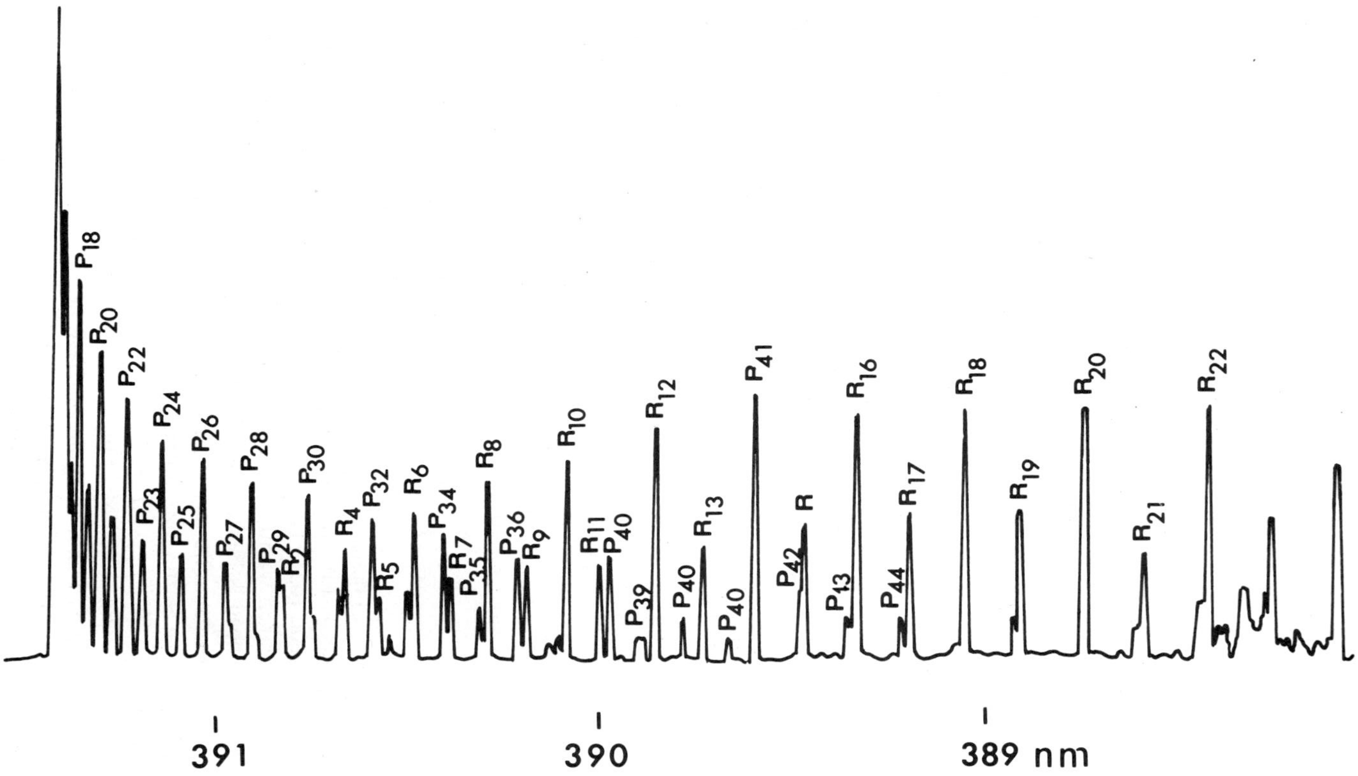

Figure 10.2. Spectral scan of (0–0) band of N_2^+ with bandhead at 391.4 nm. The figure includes the designations of the overlapping components of the *P* and *R* branches (cf. Table 10.7).

Table 10.7. Wavelengths and $K'(K'+1)$ Values for the (0–0) Band of the First Negative System of N_2^+

	P Branch		*R* Branch	
K''	λ (nm)	$K''(K''+1)$	λ (nm)	$(K''+1)(K''+2)$
6			390.49	56
7			390.40	72
8			390.29	90
9			390.19	110
10			390.08	132
11			389.97	156
12			389.85	182
13			389.73	210
16			389.33	306
18	391.35	306	389.04	380
20	391.30	380	388.74	462
21	391.29	420	388.58	506
22	391.25	462		
23	391.20	506		
24	391.15	552		
25	391.10	600		
26	391.04	650		
27	390.97	702		
28	390.91	756		
29	390.84	812		
30	390.76	870		
31	390.68	930		
32	390.60	992		
33	390.51	1056		
34	390.41	1122		
35	390.31	1190		

the line K'' from the *R* branch is closely overlapped (0.02 nm) by the line $(K'' + 27)$ of the *P* branch. An effective resolving power of at least 30,000 is necessary for this method [62]. The *R* branch with $K'' = 6$–10, 12, 13, 16, 18, and 20 and the *P* branch with $K'' = 21$–35 have been used [62] from the (0-0) band, whose band head is located at 391.4 nm (Fig. 10.2). Parameters of the lines used for the temperature measurement are given in Table 10.7.

With OH bands, it is similarly possible to determine the rotational temperature. Because of a $A^2\Sigma^+ - X^2\Pi$ transition, five main branches (*O*, *P*, *Q*, *R*, *S*) can be observed with a total of 12 branches: O_{12}, P_1, P_2, P_{12}, Q_1, Q_{21}, Q_2, Q_{12}, R_1, R_{21}, R_2 and S_{21}. Line assignments have been dealt with in detail by Dieke and Crosswhite [167] and transition probabilities have been given in [167, 168]. Using the (0-0) band (Fig. 10.3), R_2 and Q_1 can be selected. The rela-

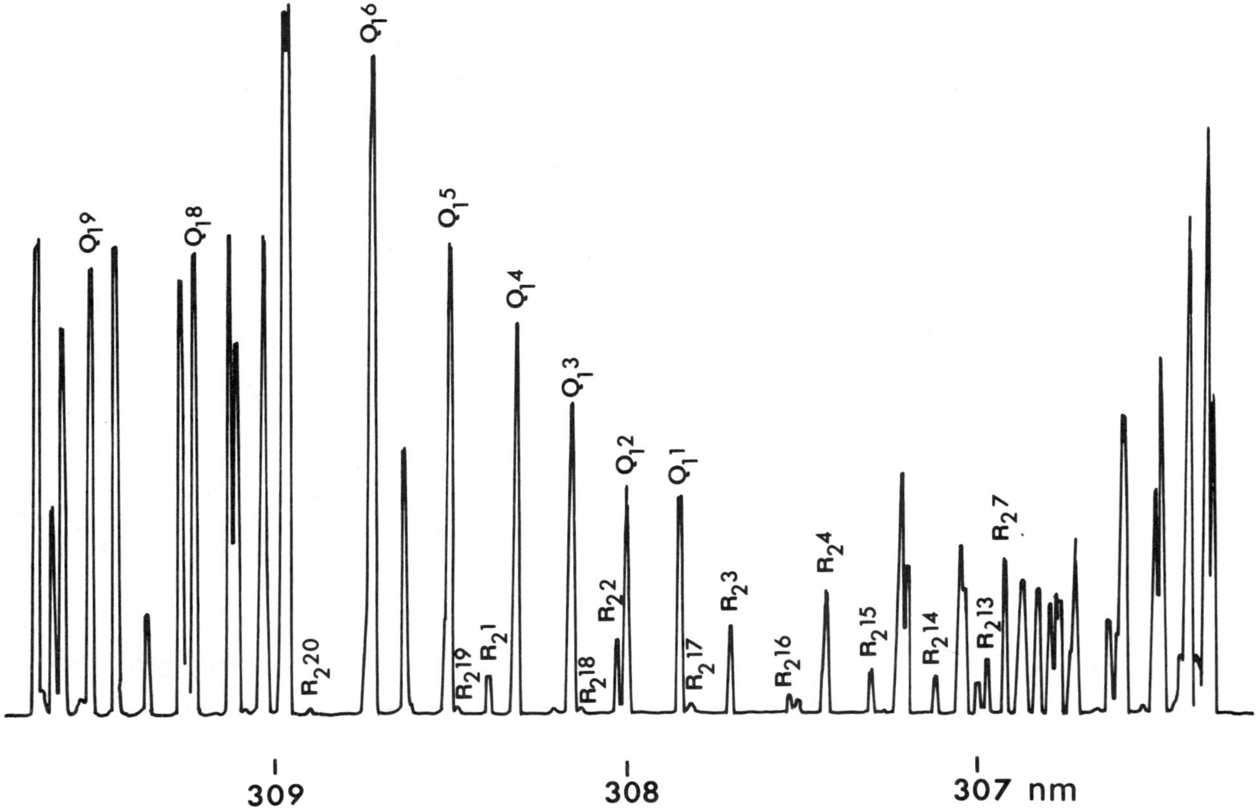

Figure 10.3. Spectral scan of the (0–0) band of OH. Either the R_2 or Q_1 branch can be selected for the determination of rotational temperatures (cf. Table 10.8).

tionship between line intensity and energy is given in the form of log $(I\lambda/A)$ versus E, with a reference level of 32,440.6 cm^{-1} for the energy. Line assignment, wavelengths, transition probabilities, and energies are given in Table 10.8. The temperature is deduced from the slope $-0.625/T$.

Other molecular species can be used for this purpose, such as CN [33], BO [38], and C_2 [41, 43].

Table 10.8 Assignment, Wavelengths, Energies, and A Values for the R_2 and Q_1 Branches of the OH (0–0) Band

K	λ (nm)	E (cm^{-1})	A (10^8 s^{-1})
		R_2 branch	
1	308.405	32,542	2.7
2	308.023	32,643	5.7
3	307.703	32,778	8.9
4	307.437	32,947	12.8
7	306.918	33,650	24.8
13	306.967	35,912	49.1
14	307.114	36,393	53.2
15	307.303	36,903	57.2
16	307.553	37,440	61.3
17	307.807	38,004	65.3
18	308.125	38,594	69.3
19	308.489	39,209	73.4
20	308.901	39,847	77.4
		Q_1 branch	
1	307.844	32,475	0.0
2	307.995	32,543	17.0
4	308.328	32,779	33.7
5	308.520	32,948	42.2
6	308.734	33,150	50.6
8	309.239	33,652	67.5
9	309.534	33,952	75.8
10	309.859	34,283	84.1
13	311.022	35,462	100.6
14	311.477	35,915	108.8
15	311.967	36,397	125.2
16	312.493	36,906	133.3
17	313.057	37,444	141.5
18	313.689	38,008	149.6
19	314.301	38,598	157.7

10.8.2. Line Pair Intensity Ratio Method

Another way to avoid the use of absolute line intensity measurements is to determine the excitation temperature with the ratio of two line intensities I_1 and I_2 of an element in the same ionization state. A or f values can be used and T can be deduced from formula (10.23), namely,

$$T = \frac{0.625(E_1 - E_2)}{\log \dfrac{g_1 A_1 \lambda_2}{g_2 A_2 \lambda_1} - \log \dfrac{I_1}{I_2}} \qquad (10.34a)$$

Several test elements have been used: titanium with Ti I 390.48/392.45 [22], Ti II 310.62/313.08 and 333.21/334.034 [37], Ti II 322.28/322.42 [42, 50, 87], iron with Fe I 391.67/391.72 [32], Fe I 381.58/382.44 [59], Fe I 382.59/382.44 [84], zinc with Zn I 307.59/328.23 [40], strontium with Sr II 407.77/430.54 [86], calcium with Ca II 393.36/373.69 [90, 95].

Titanium is a convenient element with Ti II 322.284 and 322.424, because these two lines are very close and have a large energy difference between the upper levels (31,114 and 43,781 cm^{-1}). Using the f values of Warner [155] and Roberts and Voigt [156] or Wiese and Furh [157], we have

$$T = 7920/[C - (\log I\, 322.4/\log I\, 322.2)] \qquad (10.34b)$$

with $C = 1.9$ with Warner data and $C = 1.53$ with Roberts or Wiese data. The last values provide higher temperatures than the Warner values.

The line pair method is less accurate than the Boltzmann plot method, and a large energy difference is necessary to obtain an acceptable precision:

$$\Delta T = \frac{T^2}{0.625(E_1 - E_2)} \left(\frac{\Delta A_1}{A_1} + \frac{\Delta A_2}{A_2} + \frac{\Delta I_1}{I_1} + \frac{\Delta I_2}{I_2} \right) \qquad (10.35)$$

The main source of error is caused by the transition probabilities. An uncertainty of at least 20% must be allowed for. With $T = 5000$ K, an error of 1000 K corresponds to an energy difference of 16,000 cm^{-1}. An error of only 100 K would correspond to a difference of 160,000 cm^{-1}, which is impossible to obtain practically. On the other hand, this method is very useful to follow the relative variation of the excitation temperature when one or several parameters of the plasma are modified.

10.9. LINEWIDTH METHOD

Since the line broadening due to the Stark effect (see Section 10.12) is not very sensitive to temperature variations, it is mainly the Doppler effect that is used. The FWHM measurement allows the determination of the kinetic temperature after Eq. (10.18). With an ICP, FWHMs due to the Doppler effect are generally between 0.0010 and 0.0060 nm. A high practical resolving power is required for this measurement and a deconvolution must be carried out (see Part 1, Section 7.7.2.) to eliminate the line broadening due to the foreign gas broadening (or pressure broadening) and due to the optical system (resultant spectral slit width and optical aberrations). For this purpose, echelle monochromators [92, 169, 170] or interferometry [33, 55, 93] can be used with Ar lines [33, 55] or injected element lines. The possibility of using a Fourier transform spectrometer allows a direct determination of the physical linewidth of the elements [93] because of the high resolution used. However, echelle spectrometers too, provide the possibility of measuring physical linewidths if the instrument function is approximately accounted for [92, 169, 170]. Thus, Boumans and Vrakking [170] published the results of such measurements for 350 prominent ICP lines of 65 elements. That work also covered the measurement of the Doppler temperature using a Be line.

10.10. TEMPERATURE DETERMINATION FROM THE LINE-TO-CONTINUUM INTENSITY RATIO

Below 500 nm, the radiative recombination process is predominant. The intensity is given by Eq. (10.29) and its measurement should give the value of the electron temperature. Unfortunately, knowledge of the electron number density is required and, moreover, the continuum intensity is an insensitive function of the temperature. Therefore, this method is not accurate enough to be applied. That is why several authors [23, 77, 96] preferred the use of the ratio between the Ar line and the adjacent continuum. The formula is rather complex and is a function of the ξ factor and the bandpass of the spectrometer. Ar I 430.0 nm line is generally used [23, 77, 96] because of the good knowledge of its *A* value (see Section 10.7) and because the ξ factor is independent of the temperature at this wavelength. Ar I 518.7 [77] and Ar I 451.5 nm [23] have been also used for this purpose. Bastiaans and Mangold [96] have given a plot of Ar I 430.0 nm to the adjacent continuum ratio as a function of the electron temperature, for several values of the bandpass of the spectrometer (0.1, 1.0, and 10 nm).

10.11. CLASSIFICATION OF ELECTRON NUMBER DENSITY MEASUREMENTS

Many papers have dealt with electron number density measurements in ICPs [16, 19, 22, 23, 25–27, 29, 33, 35, 36, 39, 46, 49, 57–59, 63, 71, 74, 77, 82, 90, 94, 96–98, 171–173] using several methods, classified as follows:

- Stark effect on H or Ar lines
- Continuum intensity
- Inglis–Teller method

10.12. ELECTRON NUMBER DENSITY MEASUREMENT USING THE STARK EFFECT

Because of the strong Stark effect on Balmer series lines of H, these lines are commonly used to determine the electron number density, in particular the H_β 486.1 nm line. But the Stark effect is also noticeable on some Ar lines, particularly those between 500 and 600 nm.

Stark broadening is caused by the action of charged particles, that is, electrons and ions. According to the kind of particles, two extreme theories have been used to describe the experimental line profile. In the quasistatic theory, the pertubers are almost stationary because of their mass (that is, ions). The pertubation is constant over the whole emission process. The electrical field causes a symmetrical splitting of the line into several components. This theory was first described by Holtsmark [175]. In the impact theory, fast-moving pertubers (that is, electrons) are considered. In contrast with the previous theory, the duration of the pertubation is very small between collisions. The electron broadening provides an overall profile averaging the intensity of each component. The intensity distribution of the profile is Lorentzian, thus termed after Lorentz [176], who described the theory. More recent papers [177, 178] described the impact theory. For a long time, the electron broadening was considered negligible, but several experiments indicated their role. Therefore a more refined Stark broadening for hydrogen-like atoms was developed which takes simultaneously the electrons and the ions into account. Several groups of workers have been involved: Griem, Kolb, and Shen (referred to as GKS) [2, 8, 179], Kepple and Griem [180], and Vidal, Cooper, and Smith (referred to as VCS) [181].

More recent theories (especially in the case of H_β) consider that the dynamic effect of ions must be taken into account, because of the thermal motions of ions [182, 183]. The state of the art of Stark broadening has been recently described in [184].

Several Stark effects are observed. From the central frequency ν_0 several components are located at a distance depending on the electrical field F.

$$\Delta \nu = aF + bF^2 + \cdots \tag{10.36}$$

When $\Delta \nu = aF$, linear Stark effect is observed; when $\Delta \nu = bF^2$, quadratic Stark effect is observed. The full width at half-maximum $\Delta \lambda$ is related to the electron number density n_e by

$$\Delta \lambda = C(n_e, T_e) n_e^a \tag{10.37}$$

With the linear Stark effect, $a = \frac{2}{3}$, and with the quadratic effect $a = 1$. The linear effect applies to H, in contrast to Ar, where the quadratic effect should apply. Usually, the parameter C depends only weakly on the electron temperature.

10.12.1. Stark Effect on Hydrogen Lines

The H_β 486.1-nm line is used very often for electron number density measurements, because its FWHM is very large compared with the contributions from other line broadening effects (Doppler).

A very efficient measurement would be to determine the whole profile and to compare it with the theoretical profile. Because of the various gradients in the ICP, this measurement is actually time consuming, and, most of the time, only the FWHM is measured.

For the determination of the H_β line broadening, a small amount (around 1% or less) of H must be added (either mixed with the outer gas or with the carrier gas), which modifies the appearance and the properties of the discharge. The resistivity and the thermal conductivity of H are different from those of Ar, which results in a variation of the coupling. Therefore, results obtained with H must be extrapolated with care to describe the properties of the discharge in the absence of foreign gas.

Because of the absence of the unshifted Stark component at high electron concentration, H_β exhibits a central dip. For $n_e = 2.4 \times 10^{15}$ cm^{-3}, the dip theoretically should be 40%. But practically, the value is smaller. This discrepancy along with the asymmetry of the lines (the lower wavelength side shows a higher maximum) are explained in part by self-absorption, Doppler effect, and ion dynamics. This asymmetry is the cause of the difficulty of determining the actual FWHM. Moreover, because of the Lorentzian shape, the true value of the background can also be difficult to estimate. Usually, this determination has to be made far from the central wavelength.

Note that the central dip no longer exists for FWHMs below 0.2 nm, which corresponds to an electron concentration of 10^{15} cm^{-3}.

Although the electron temperature is a parameter that has a small action on the FWHM, for the following calculations we have assumed a value of 10,000 K, based on the ratio between an Ar line and the adjacent continuum [77].

From Eq. (10.37), we can calculate n_e from the value of $\Delta\lambda$:

$$n_e = C(n_e, T_e)\Delta\lambda^{3/2}\, 10^{13} \tag{10.38}$$

Griem et al. [2, 179] have given the value of C for n_e equal to 10^{14} and 10^{15} cm^{-3}. For $T_e = 10{,}000$ K and λ in angstroms (Å), C is equal to 38.0 and 35.8, respectively.

Since C is also a function of n_e, interpolation between values of the electron number densities is not easy. That is why Hill [185, 186] has proposed a formula to avoid the interpolation, based on Griem results:

$$n_e = (C_0 + C_1 \log \Delta\lambda)\Delta\lambda^{3/2}\, 10^{13} \tag{10.39}$$

where log is the decadic logarithm and λ is expressed in angstroms. After Hill, C_0 is equal to 36.57 and C_1 to -1.72. Calculations for $\Delta\lambda = 0.1$–1.0 nm (which cover the range of electron number densities usually observed in the ICP) are given in Table 10.9.

In 1968, Kepple and Griem [180] have improved the GKS theories and given the semi-half-width $\alpha_{1/2}$ of the reduced Stark profile, related to the electrical field:

$$\left.\begin{aligned} \alpha_{1/2} &= \Delta\lambda/F \\ F &= 2.61 e n_e^{2/3} \\ F &= 1.25 \times 10^{-9}\, n_e^{2/3} \\ \Delta\lambda &= 2.50 \times 10^{-9}\, \alpha_{1/2}\, n_e^{2/3} \end{aligned}\right\} \tag{10.40}$$

For $T = 10{,}000$ K, $n_e = 10^{15}$ cm^{-3}, $\alpha_{1/2} = 0.0803$, which provides a value of equal to 0.2 nm. In each case, the FWHM is determined by averaging the two maximum intensities to obtain the peak intensity.

Greig et al. [187] have given a more sophisticated formula with four parameters:

$$n_e = \left[C_0 + C_1 \ln \Delta\lambda + C_2(\ln \Delta\lambda)^2 + C_3(\ln \Delta\lambda)^3\, \Delta\lambda^{3/2}\, 10^{13}\right. \tag{10.41}$$

ln is the Neperian logarithm.

$$\left.\begin{aligned} C_0 &= 36.84 \qquad & C_1 &= -1.430 \\ C_2 &= -0.133 \qquad & C_3 &= 0.0089 \end{aligned}\right\} \tag{10.42}$$

Results are presented similarly in Table 10.9.

More recently, based on the VCS theory, Czernichowski and Chapelle [188] have proposed a formula including the variation of T.

$$\left.\begin{aligned} &\log n_e = C_0 + C_1 \log \Delta\lambda + C_2(\log \Delta\lambda)^2 + C_3 \log T \\ &C_0 = 22.578 \qquad C_1 = 1.478 \\ &C_2 = -0.144 \qquad C_3 = -0.1265 \end{aligned}\right\} \tag{10.43}$$

In this case, $\Delta\lambda$ is in nanometers and the FWHM is determined by doubling the semi-half-width taken on the lower wavelength side of the line profile. Calculation is possible between 3×10^{14} and 3×10^{16} cm^{-3}. Results are given in Table 10.9.

Table 10.9. Values of the Electron Number Densities (10^{15} cm^{-3}) Versus the FWHM of the H_β Line, According to Various Theories

λ (nm)	Reference [180]	Reference [186]	Reference [187]	Reference [188]
0.1		0.366	0.368	
0.2	1.00	1.02	1.01	0.93
0.3		1.86	1.82	1.82
0.4		2.84	2.77	2.89
0.5		3.95	3.82	4.11
0.6		5.18	4.98	5.46
0.7		6.50	6.23	6.91
0.8		7.98	7.55	8.46
0.9		9.43	8.95	10.10
0.974	10.00			
1.0		11.02	10.42	11.80

The VCS theory provides slightly higher values, which seems to correspond to more accurate values [123]. Diagnostics using H_β line-broadening have been applied very often to the ICP [16, 19, 22, 25, 26, 29, 33, 39, 46, 57, 63, 71, 74, 94, 98, 171, 172, 174].

10.12.2. Stark Effect on Argon Lines

To avoid the use of hydrogen, electron number densities can be determined directly from some Ar lines such as Ar I 549.5 and 565.0 nm, although the line-broadening is not so large compared to the values obtained with H_β. Griem

Table 10.10. Parameters and FWHMs for Ar I 549.5 and Ar I 565.0 nm Broadened by Stark Effect, at Different Electron Number Densities, and Comparison with Doppler Broadening

	n_e	10^{14}	10^{15}	10^{16}
	$1 - 0.75r$	0.854	0.787	0.688
549.5	ω	0.00246	0.0246	0.246
	α	0.0278	0.0494	0.088
	Δλ (nm)	0.0051	0.0524	0.542
565.0	ω	0.00094	0.00945	0.0945
	α	0.00158	0.00281	0.005
	Δλ (nm)	0.0019	0.0190	0.190
Doppler			0.0044	

[2] has given a formula:

$$\omega = 2[1 + 1.75\alpha(1 - 0.75r)]\omega_t \tag{10.44}$$

where ω_t is the semi-half-width due to electron impact, proportional to n_e, α accounts for ion-broadening, and is proportional to $n_e^{1/4}$, and r is the ratio of the mean distance between ions and the Debye radius.

Griem [2] has given values of ω_t and α for $T = 10{,}000$ K and $n_e = 10^{16}$ cm^{-3}. Extrapolated values and corresponding Stark broadenings are given in Table 10.10. From this table, it may be seen that the value of the Doppler effect is no longer negligible below values of the electron number density of 10^{15} cm^{-3}.

10.13. ELECTRON NUMBER DENSITY DETERMINATION FROM THE CONTINUUM INTENSITY

Once the electron temperature has been determined (see Section 10.10), the electron number density can be deduced from Eq. (10.29), if the absolute intensity of the continuum is measured at a given wavelength (usually 430 nm). The value of the coefficient C_1 is

$$C_1 = \frac{16\pi e^6}{3c^3(6\pi m^3 k)^{1/2}(4\pi\epsilon_0)^3} = 5.46 \times 10^{-52} \tag{10.45a}$$

when the frequency unit is used in the relation (10.29). If wavelength is used, $1/\lambda^2$ appears in the formula and c^3 becomes c^2 in the coefficient C_1; therefore

$$C_1 = 1.63 \times 10^{-43} \tag{10.45b}$$

At $T_e = 10{,}000$ K and $\lambda = 430$ nm, the value of ξ is about 1.8, and the value of the exponent about 0.035. It is to be noted that relation (10.29) involves an emission coefficient ϵ. Practically, the continuum intensity I_c implies a spectral interval and is related to the emission coefficient by the spectral bandpass:

$$I_c = \epsilon\Delta\nu \quad \text{or} \quad I_c = \epsilon\Delta\lambda \tag{10.46}$$

Bastiaans and Mangold [96] have given a theoretical plot of the continuum coefficient versus electron number density, for several electron temperatures.

10.14. INGLIS-TELLER METHOD

The electron number density can be estimated from the last discernible line in a series with a principal quantum number m [189] using the relation:

$$\log n = 23.26 - 7.5 \log m \tag{10.47}$$

The density *n* is equal to the sum of the charged particles (ions and electrons). This method has been applied to H [22] and to the alkalis [172, 173].

REFERENCES

1. W. J. Pearce, "Plasma Jet Temperature Study," in P. J. Dickermann, ed., *Optical Spectroscopic Measurements in High Temperature*, University of Chicago Press, Chicago, IL (1960); p. 125.
2. H. R. Griem, *Plasma Spectroscopy*, McGraw-Hill, New York (1964).
3. R. H. Huddlestone and S. L. Leonard, eds., *Plasma Diagnostics Techniques*, Academic, New York (1965).
4. P. W. J. M. Boumans, *Theory of Spectrochemical Excitation*, Adam Hilger/London, Plenum/New York (1966).
5. W. Lochte-Holtgreven ed., *Plasma Diagnostics*, North-Holland Company, Amsterdam (1968).
6. H. W. Drawin, *High Pressures-High Temperatures* **2,** 359 (1970).
7. F. Cabannes and J. Chapelle, in M. Venugopalan, ed., *Reactions under Plasma Conditions*, vol. 1, Wiley-Interscience, New York (1971).
8. H. R. Griem, *Spectral Line Broadening by Plasmas*, Academic, New York (1974).
9. R. Rompe and M. Steenbeck, eds., *Progress in Plasmas and Gas Electronics*, vol. 1, Akademie-Verlag, Berlin (1975).
10. P. Fauchais, K. Lapworth, and J. M. Baronnet, eds., First Report on Measurement of Temperature and Concentration of Excited Species in Optically Thin Plasmas, IUPAC, Subcommittee on Plasma Chemistry (1979).
11. J. Chapelle, in *L'arc electrique et ses applications*, vol. 1, Editions du Centre National de la Recherche Scientifique, Paris (1984).
12. C. H. Corliss, *Astrophys. J.* **136,** 916 (1962).
13. H. W. Drawin and P. Felenbok, *Data for Plasmas in Local Thermal Equilibrium*, Gauthiers-Villars, Paris (1965).
14. L. de Galan, R. Smith, and J. D. Winefordner, *Spectrochim. Acta* **23B,** 521 (1968).
15. T. B. Reed, *J. Appl. Phys.* **32,** 821 (1961).
16. V. M. Goldfarb and S. V. Dresvin, *High Temp.* **3,** 303 (1965).
17. B. S. Malone and W. H. Corcoran, *J. Quant. Spectrosc. Radiat. Transfer* **6,** 443 (1966).
18. F. Molinet, *C.R. Acad. Sc. Paris, Serie B* **262,** 1377 (1966).
19. P. D. Johnston, *Phys. Lett.* **20,** 499 (1966).
20. G. Abravanel, *Methodes Phys. Anal.* **10,** 289 (1966).
21. D. W. Hughes and E. R. Wooding, *Phys. Lett.* **24A,** 70 (1967).
22. J. Besombes-Vailhe, *J. Chim. Phys. Phys. Chim. Biol.* **64,** 370 (1967).

23. R. E. Rovinskii, V. A. Gruzdev, T. M. Gutenmakher, and A. P. Sobolev, *High Temp. (USSR)* **5,** 502 (1967).
24. M. E. Britske, V. M. Borisov, and Y. S. Sukach, *Zavod. Lab.* **33,** 252 (1967).
25. H. U. Eckert, F. L. Kelly, and H. N. Olsen, *J. Appl. Phys.* **39,** 1846 (1968).
26. R. Assous, *J. Phys. Paris* **29,** 877 (1968).
27. P. D. Scholz and T. P. Anderson, *J. Quant. Spectrosc. Radiat. Transfer* **8,** 1411 (1968).
28. S. V. Desai and W. H. Corcoran, *J. Quant. Spectrosc. Radiat. Transfer* **8,** 1721 (1968).
29. V. M. Goldfarb, V. H. Goikhman, and S. V. Dresvin, *Proc. 14th Coll. Spectrosc. Intern.*, Debrecen 1967, vol. 2, Hilger, London (1968) p. 751.
30. S. V. Desai and W. H. Corcoran, *J. Quant. Spectrosc. Radiat. Transfer* **9,** 1371 (1969).
31. S. V. Desai and W. H. Corcoran, *J. Quant. Spectrosc. Radiat. Transfer* **9,** 1489 (1969).
32. W. B. Barnett, V. A. Fassel, and R. N. Kniseley, *Spectrochim. Acta* **25B,** 139 (1970).
33. I. Kleinmann and J. Cajko, *Spectrochim. Acta* **25B,** 657 (1970).
34. S. V. Dresvin and V. S. Klubnikin, *Teplof. Vys. Temp.* **9,** 475 (1971).
35. A. D. Stokes, *J. Phys. D: Appl. Phys.* **4,** 916 (1971).
36. V. N. Soshnikov, E. S. Trekhov, A. F. Fomenko, and Y. M. Khoshev, *High Temp. (USSR)* **9,** 446 (1971).
37. H. Triche, A. Saadate, B. Talayrach, and J. Besombes-Vailhe, *Analusis* **1,** 413 (1972).
38. B. Talayrach, J. Besombes-Vailhe, and H. Triche, *Analusis* **1,** 135 (1972).
39. S. L. Leonard, *J. Quant. Spectrosc. Radiat. Transfer* **12,** 619 (1972).
40. P. W. J. M. Boumans, H. Wagenaar, and F. J. de Boer, Preprints 17th Coll. Spectrosc. Intern., Florence, 1973, vol. 1, (1973) p. 114.
41. J. M. Mermet and J. Robin, *Rev. Intern Htes Temp. Refract.* **10,** 133 (1973).
42. J. M. Mermet, J. Jarosz, and J. Robin, Preprints 17th Coll. Spectrosc. Intern., Florence, 1973, vol. 1, (1973) p. 101.
43. J. F. Alder and J. M. Mermet, *Spectrochim. Acta* **28B,** 421 (1973).
44. H. U. Eckert, *High Temp. Sci.* **6,** 99 (1974).
45. G. R. Kornblum and L. de Galan, *Spectrochim. Acta* **29B,** 249 (1974).
46. J. Jarosz, J. M. Mermet, and J. Robin, *C.R. Acad. Sc. Paris, Serie B* **278,** 885 (1974).
47. G. F. Kirkbright and A. F. Ward, *Talanta* **21,** 1145 (1974).
48. D. J. Kalnicky, R. N. Kniseley, and V. A. Fassel, *Spectrochim. Acta* **30B,** 511 (1975).
49. V. S. Klubnikin, *High Temp. (USSR)* **13,** 439 (1975).
50. J. M. Mermet, *Spectrochim. Acta* **30B,** 383 (1975).

51. M. H. Abdallah, J. Jarosz, J. M. Mermet, C. Trassy, and J. Robin, Preprints 18th Colloq. Intern. Spectrosc., Grenoble, 1975, vol. 3, (1975) p. 758.
52. G. R. Kornblum and L. de Galan, Preprints 18th Colloq. Intern. Spectrosc., Grenoble, 1975, vol. 2, (1975) p. 370.
53. P. W. J. M. Boumans and F. J. de Boer, *Spectrochim. Acta* **31B,** 355 (1976).
54. K. Visser, F. M. Hamm, and P. B. Zeeman, *Appl. Spectrosc.* **30,** 34 (1976).
55. H. G. C. Human and R. H. Scott, *Spectrochim. Acta* **31B,** 459 (1976).
56. P. W. J. M. Boumans and F. J. de Boer, *Spectrochim. Acta* **32B,** 365 (1977).
57. D. J. Kalnicky, V. A. Fassel, and R. N. Kniseley, *Appl. Spectrosc.* **31,** 137 (1977).
58. K. G. Lapworth and L. A. Allnutt, *J. Phys. E: Sci. Instr.* **10,** 727 (1977).
59. G. R. Kornblum and L. de Galan, *Spectrochim. Acta* **32B,** 71 (1977).
60. G. R. Kornblum and L. de Galan, *Spectrochim. Acta* **32B,** 455 (1977).
61. P. B. Zeeman, S. P. Terblanche, K. Visser, and F. H. Hamm, *Appl. Spectrosc.* **32,** 572 (1978).
62. M. H. Abdallah and J. M. Mermet, *J. Quant. Spectrosc. Radiat. Transfer* **19,** 83 (1978).
63. J. Jarosz, J. M. Mermet, and J. P. Robin, *Spectrochim. Acta* **33B,** 55 (1978).
64. J. Jarosz, J. M. Mermet, and J. Robin, *Spectrochim. Acta* **33B,** 365 (1978).
65. R. M. Barnes, *CRC Crit. Rev. Anal. Chem.* **7,** 203 (1978).
66. J. A. C. Broekaert, F. Leis, and K. Laqua, *Spectrochim. Acta* **34B,** 167 (1979).
67. J. F. Alder, R. M. Bombelka, and G. F. Kirkbright, *Proc. Analyt. Div. Chem. Soc.* **16,** 21 (1979).
68. R. N. Savage and G. M. Hieftje, *Anal. Chem.* **51,** 408 (1979).
69. V. K. Goykhman and V. M. Goldfarb, *ICP Information Newslett.* **4,** 537 (1979); **5,** 15 (1979).
70. H. Uchida, R. Negishi, R. Yamazaki, and Y. Imai, *Bunseki Kagaku* **28,** 244 (1979).
71. J. F. Alder, R. M. Bombelka, and G. F. Kirkbright, *Spectrochim. Acta* **35B,** 163 (1980).
72. B. Cheron, J. Jarosz, and P. Vervisch, *J. Phys. B: Atom. Mol. Phys.* **13,** 2413 (1980).
73. H. Uchida, *Spectrosc. Lett.* **14,** 665 (1981).
74. H. Uchida, K. Tanabe, Y. Nojiri, H. Haraguchi, and K. Fuwa, *Spectrochim. Acta* **36B,** 711 (1981).
75. M. W. Blades and G. Horlick, *Spectrochim. Acta* **36B,** 861 (1981).
76. H. Kawaguchi, T. Ito, and A. Mizuike, *Spectrochim. Acta* **36B,** 615 (1981).
77. A. Batal, J. Jarosz, and J. M. Mermet, *Spectrochim. Acta* **36B,** 983 (1981).
78. J. A. C. Broekaert, F. Leis, and K. Laqua, in R. M. Barnes, ed., *Developments in Atomic Plasma Spectrochemical Analysis*, Heyden, London, Philadelphia, (1981) p. 84.

79. N. Furuta and G. Horlick, *Spectrochim. Acta* **37B,** 53 (1982).
80. B. Capelle, J. M. Mermet, and J. Robin, *Appl. Spectrosc.* **36,** 102 (1982).
81. G. R. Kornblum and J. Smeyers-Verbake, *Spectrochim. Acta* **37B,** 83 (1982).
82. A. Batal, J. Jarosz, and J. M. Mermet, *Spectrochim. Acta* **37B,** 511 (1982).
83. M. H. Abdallah and J. M. Mermet, *Spectrochim. Acta* **37B,** 391 (1982).
84. J. P. Rybarczyk, C. P. Jester, D. A. Yates, and S. R. Koirtyohann, *Anal. Chem.* **54,** 2162 (1982).
85. J. E. Roederer, G. J. Bastiaans, M. A. Fernandez, and K. J. Fredeen, *Appl. Spectrosc.* **36,** 383 (1982).
86. W. H. Gunter, K. Visser, and P. B. Zeeman, *Spectrochim. Acta* **37B,** 571 (1982).
87. P. A. M. Ripson, L. de Galan, and J. W. de Ruiter, *Spectrochim. Acta* **37B,** 733 (1982).
88. A. Batal and J. M. Mermet, *Can. J. Spectrosc.* **27,** 37 (1982).
89. H. Kawaguchi, Y. Oshio, and A. Mizuike, *Spectrochim. Acta* **37B,** 809 (1982).
90. W. H. Gunter, K. Visser, and P. B. Zeeman, *Spectrochim. Acta* **38B,** 949 (1983).
91. L. M. Faires, B. A. Palmer, R. Engleman, and T. M. Niemczyk, *Spectrochim. Acta* **39B,** 819 (1984).
92. T. Hasegawa and H. Haraguchi, *Spectrochim. Acta* **40B,** 123 (1985).
93. L. M. Faires, B. A. Palmer, and J. W. Brault, *Spectrochim. Acta* **40B,** 135 (1985).
94. M. W. Blades and B. L. Caughlin, *Spectrochim. Acta* **40B,** 579 (1985).
95. W. Gunter, K. Visser, and P. B. Zeeman, *Spectrochim. Acta* **40B,** 617 (1985).
96. G. J. Bastiaans and R. A. Mangold, *Spectrochim. Acta* **40B,** 885 (1985).
97. N. Kovacic, G. A. Meyer, Liu Ke-Ling, and R. M. Barnes, *Spectrochim. Acta* **40B,** 943 (1985).
98. B. L. Caughlin and M. W. Blades, *Spectrochim. Acta* **40B,** 987 (1985).
99. H. W. Drawin, *Z. Physik* **146,** 295 (1956).
100. W. E. Gericke, *Z. Astrophysik* **53,** 68 (1961).
101. H. N. Olsen, *J. Quant. Spectrosc. Radiat. Transfer* **3,** 59 (1963); **3,** 305 (1963).
102. A. Pery-Thorne and J. E. Chamberlain, *Proc. Phys. Soc. (London)* **82,** 133 (1963).
103. R. H. Garstang and J. Van Blerkom, *J. Opt. Soc. Am.* **55,** 1054 (1965).
104. C. H. Popenoe and J. B. Shumaker, *J. Res. Natl. Bur. Stand. Sect. A* **69,** 495 (1965).
105. J. Richter, *Z. Astrophysik* **61,** 57 (1965).
106. P. B. Coates and A. G. Gaydon, *Proc. R. Soc. London, Ser. A* **293,** 452 (1966).
107. J. F. Bott, *Phys. Fluids* **9,** 1540 (1966).
108. M. S. Frish, *Opt. Spectrosc.* **22,** 9 (1967).
109. P. D. Johnston, *Proc. Phys. Soc. London* **92,** 896 (1967).
110. B. Wende, *Z. Physik* **213,** 341 (1968).

111. J. Chapelle, A. Sy, F. Cabannes, and J. Blondin, *J. Quant. Spectrosc. Radiat. Transfer* **8,** 1201 (1968).

112. B. T. Wujec, *Acta Phys. Polon.* **36,** 269 (1969).

113. H. M. Meiners, *J. Quant. Spectrosc. Radiat. Transfer* **9,** 1493 (1969).

114. I. Bues, T. Haag, and J. Richter, *Astron. Astrophys.* **2,** 249 (1969).

115. W. L. Wiese, M. W. Smith, and B. M. Miles, *Atomic Transition Probabilities*, vol. 2, *Sodium through Calcium*, NSRDS-NBS 22, The Superintendent of Documents, Washington, DC (1969).

116. R. A. Nodwell, J. Meyer, and T. Jacobsen, *J. Quant. Spectrosc. Radiat. Transfer* **10,** 335 (1970).

117. D. Van Houwelingen and A. A. Kruithof, *J. Quant. Spectrosc. Radiat. Transfer* **11,** 1235 (1971).

118. A. D. Stokes, *J. Phys. D: Appl. Phys.* **4,** 930 (1971).

119. D. Garz, *Z. Naturforsch.* **28A,** 1459 (1973).

120. J. V. Specht, *Astron. Astrophys.* **34,** 363 (1974).

121. P. Ranson and J. Chapelle, *J. Quant. Spectrosc. Radiat. Transfer* **14,** 1 (1974).

122. H. Nubbemeyer *J. Quant. Spectrosc. Radiat. Transfer* **16,** 395 (1976).

123. R. C. Preston, *J. Phys. B: Atom Mol. Phys.* **10,** 1377 (1977).

124. P. Ranson, Thesis, University of Orléans (1978).

125. P. Baessler and M. Kock, *J. Phys. B: Atom. Mol. Phys.* **13,** 1351 (1980).

126. K. Katsonis and H. W. Drawin, *J. Quant. Spectrosc. Radiat. Transfer* **23,** 1 (1980).

127. R. B. King and A. S. King, *Astrophys. J.* **87,** 24 (1938).

128. G. Jurgens, *Z. Physik* **138,** 613 (1954).

129. C. W. Allen and A. S. Asaad, *Mon. Not. R. Astron. Soc.* **117,** 36 (1957).

130. H. M. Crosswhite, *The Spectrum of Fe I*, The Johns Hopkins University, Baltimore (1958).

131. C. H. Corliss and W. R. Bozman, *Experimental Transition Probabilities for Spectral Lines of Seventy Elements*, NBS Mongraph 53, The Superintendent of Documents, Washington, DC (1962).

132. M. Margoshes and B. F. Scribner, *J. Res. Natl. Bur. Stand. Sect. A* **67,** 561 (1963).

133. A. K. Valters and G. P. Startsev, *Opt. Spectrosc.* **17,** 262 (1964).

134. C. H. Corliss and B. Warner, *J. Res. Natl. Bur. Stand. Sect. A* **70,** 325 (1966).

135. C. H. Corliss and J. L. Tech, *Transition Probabilities for 3228 Lines of Fe I*, NBS Monograph 108, The Superintendent of Documents, Washington, DC (1968).

136. T. Garz and M. Kock, *Astron. Astrophys.* **2,** 274 (1969).

137. J. M. Bridges and W. L. Wiese, *Astrophysics. J. Lett.* **131,** 71 (1970).

138. S. J. Wolnik, R. O. Berthel, and G. W. Wares, *Astrophys. J.* **162,** 1037 (1970).

139. J. M. Bridges and J. Richter, *Proc. 10th Intern. Conf. Ionized Gases*, p. 374. (1971).

140. I. G. Reif, V. A. Fassel, and R. N. Kniseley, Preprints 16th Colloq. Intern. Spectrosc., Heidelberg, 1971, vol. 2, Adam Higler, London, (1971) p. 317.

141. M. Martinez-Garcia, W. Whaling, D. L. Mickey, and G. M. Lawrence, *Astrophys. J.* **165,** 213 (1971).

142. M. C. E. Huber and W. H. Parkinson, *Astrophys. J.* **172,** 229 (1972).

143. F. P. Banfield and M. C. E. Huber, *Astrophys. J.* **187,** 335 (1973).

144. J. M. Bridges and R. L. Kornblith, *Astrophys. J.* **192,** 793 (1974).

145. C. H. Corliss and J. L. Tech., *J. Res. Natl. Bur. Stand. Sect. A* **80,** 787 (1976).

146. D. E. Blackwell, P. A. Ibbetson, A. D. Petford, and R. B. Willis, *Mon. Not. R. Astron. Soc.* **177,** 219 (1976).

147. I. Reif, V. A. Fassel, R. N. Kniseley, and D. J. Kalnicky, *Spectrochim. Acta* **33B,** 807 (1978).

148. D. E. Blackwell, A. D. Petford, M. J. Shallis, and G. J. Simmons, *Mon. Not. R. Astron. Soc.* **191,** 445 (1980).

149. W. L. Wiese and G. A. Martin, *Wavelengths and Transition Probabilities for Atoms and Atomic Ions, Part II, Transition Probabilities*, NSRDS-NBS 68, The Superintendent of Documents, Washington, DC (1980).

150. J. R. Furh, G. A. Martin, W. L. Wiese, and S. M. Younger, *J. Phys. Chem. Ref. Data* **10,** 305 (1981).

151. R. Wagner and E. W. Otten, *Z. Physik* **220,** 349 (1969).

152. G. D. Belles and E. F. Tubbs, *Astrophys. J.* **159,** 1093 (1970).

153. R. B. King, *Astrophys. J.* **94,** 27 (1941).

154. J. B. Tatum, *Mon. Not. R. Astron. Soc.* **122,** 311 (1961).

155. B. Warner, *Mem. R. Astron. Soc.* **70,** 165 (1967).

156. J. R. Roberts and P. A. Voight, *Astrophys. J.* **197,** 791 (1975).

157. W. L. Wiese and J. R. Furh, *J. Chem. Ref. Data* **4,** 263 (1975).

158. J. R. Roberts, T. A. Andersen, and G. Sorensen, *Astrophys. J.* **181,** 587 (1973).

159. B. Laniepce, *J. Physique* **31,** 439 (1970).

160. A. Goldwasser and J. M. Mermet, *Spectrochim. Acta* **41B,** 725 (1986).

161. J. E. Clawson and M. H. Miller, *J. Opt. Soc. Am.* **63,** 1598 (1973).

162. R. L. Kurucz and E. Peytremann, *A Table of Semiempirical gf Values*, Smiths. Astrophys. Obs. Special Rep. 362, Cambridge, MA (1975).

163. W. L. Wiese, M. W. Smith, and B. M. Glennon, *Atomic Transition Probabilities*, vol. 1, *Hydrogen through Neon*, NSRDS-NBS-4, The Superintendent of Documents, Washington, DC (1966).

164. W. H. J. Childs, *Proc. R. Soc. London Ser. A* **137,** 641 (1932).

165. D. Coster and F. Brons, *Z. Physik* **73,** 747 (1932).

166. G. Herzberg, *Spectra of Diatomic Molecules*, Van Nostrand, Princeton, NJ (1950).

167. G. H. Dieke and H. M. Crosswhite, *J. Quant. Spectrosc. Radiat. Transfer* **2,** 97 (1962).

168. A. K. Hui, M. R. McKeever, and J. Tellinghuisen, *J. Quant. Spectrosc. Radiat. Transfer* **21,** 387 (1979).

169. P. W. J. M. Boumans and J. J. A. M. Vrakking, *Spectrochim. Acta* **39B,** 1239 (1984).

170. P. W. J. M. Boumans and J. J. A. M. Vrakking, *Spectrochim. Acta* **41B,** 1235 (1986).

171. J. Jarosz and J. M. Mermet, *J. Quant. Spectrosc. Radiat. Transfer* **17,** 237 (1977).

172. A. Montaser, V. A. Fassel, and G. Larson, *Appl. Spectrosc.* **35,** 385 (1981).

173. A. Montaser and V. A. Fassel, *Appl. Spectrosc.* **36,** 613 (1982).

174. N. Furuta, Y. Nojiri, and K. Fuwa, *Spectrochim. Acta* **40B,** 423 (1985).

175. J. Holtsmark, *Ann. Physik, 4th Series* **58,** 577 (1919).

176. H. A. Lorentz, *Proc. Acad. Sci. Amsterdam* **8,** 591 (1906).

177. M. Baranger, *Phys. Rev.* **111,** 494 (1958).

178. A. C. Kolb and H. R. Griem, *Phys. Rev.* **111,** 514 (1958).

179. H. R. Griem, A. C. Kolb, and K. Y. Shen, *Astrophys. J.* **135,** 272 (1962).

180. P. Kepple and H. R. Griem, *Phys. Rev.* **173,** 317 (1968).

181. C. R. Vidal, J. Cooper, and E. W. Smith, *Astrophys. J. Suppl. Ser.* **214,** 37 (1973).

182. J. B. Shumaker and C. R. Yokley, *Appl. Opt.* **3,** 83 (1964).

183. J. B. Shumaker and C. H. Popenoe, *Phys. Rev. Lett.* **21,** 1046 (1968).

184. V. Helbig, in F. Rostas, ed., *Spectral Line Shapes*, vol. 3, Ch. 1, Walter de Gruyter, Berlin, New York (1985).

185. R. A. Hill, *J. Quant. Spectrosc. Radiat. Transfer* **4,** 857 (1964).

186. R. A. Hill, *J. Quant. Spectrosc. Radiat. Transfer* **7,** 401 (1967).

187. J. R. Greig, C. P. Lim, G. A. Moo-Yourig, G. Palumpo, and H. R. Griem, *Phys. Rev.* **172,** 148 (1968).

188. A. Czernikowski and J. Chapelle, *Acta Phys. Pol.* **63A,** 67 (1983).

189. D. R. Inglis and E. Teller, *Astrophys. J.* **90,** 434 (1939).

CHAPTER

11

EXCITATION MECHANISMS AND DISCHARGE CHARACTERISTICS—RECENT DEVELOPMENTS

M. W. BLADES

Department of Chemistry
University of British Columbia
Vancouver, British Columbia Canada

11.1. Introduction
11.2. Mechanisms: Collisional and Radiative Processes
11.3. Thermodynamic Equilibrium
11.4. Departures from LTE
- 11.4.1. LTE Populations
- 11.4.2. Partial LTE (p-LTE)
- 11.4.3. Further Deviations from LTE

11.5. Measurement of Departures from LTE
- 11.5.1. Introduction
- 11.5.2. Electron Density Measurements
- 11.5.3. Ion–Atom Emission Intensity Ratios
- 11.5.4. Analyte Ionization

11.6. Rate Models
11.7. Concluding Remarks
References

11.1. INTRODUCTION

Concurrent with the successful evolution of the inductively coupled plasma (ICP) as a powerful analytical instrument have been publications relating the basic physical, chemical, and spectral properties of the plasma discharge. Although a plasma can be simply described as an ionized gas, this belies the nature of this exceedingly complex state of matter. There are relatively few classes of particles in the atmospheric-pressure Ar plasma that is utilized for analytical spectrochemical applications, the most important being neutral Ar atoms, singly charged Ar ions, and unbound or "free" electrons. For a "pure" Ar ICP the complexity arises primarily as a result of three main factors:

1. The total energy of this "simple" three-component system is distributed in a multitude of bound, internal electronic states of the atoms and ions, and in translational degrees of freedom of the atoms, ions, and electrons.
2. The plasma is flowing and thus there is a net transport of all species away from the energy addition region.
3. The ICP discharge is spatially inhomogeneous and is therefore characterized by concentration and temperature gradients that give rise to convective and diffusional transport of species.

For many years physicists have devoted a great deal of effort to the study of gas plasmas both from a fundamental standpoint and because of the potential applications in fields such as astrophysics, energy production, and laser research.

For analytical spectroscopists a knowledge of the physical state of the Ar plasma is important because it is the environment in which atomized analyte is persuaded to emit the characteristic line spectra that are used for qualitative and quantitative analysis. Thus, discharge characterization in this context has been more concerned with the "analyte system" than the "Ar system." This introduces a further complexity, since in addition to analyte species, water is introduced along the discharge axis. Therefore the plasma must provide energy to desolvate, vaporize, and dissociate analyte species, and additionally, the water itself contributes OH, H, O, H^+, O^+, and OH^+ to the plasma environment.

It is generally acknowledged that the main aspects of ICP emission spectroscopy demanding a fundamental understanding are the apparent relatively high ionic line sensitivities and the relative absence of classical matrix effects [1]. It is an interesting dicotomy of the ICP that these problem areas, which have for many years eluded a complete, rational, fundamental description, are the very features that make ICP optical emission spectrometry analytically favorable.

The aim of fundamental studies is to identify the pertinent excitation, ionization, deexcitation, and recombination processes that are responsible for observed analyte behavior. The elucidation of these processes has been referred to as the study of "excitation mechanisms." The experimental techniques for studying these mechanisms are termed "plasma diagnostic" techniques and are basically a collection of well-developed experimental methods used to measure those physical properties of the plasma discharge that will yield significant information. An alternate and sometimes complementary approach to the study of excitation mechanisms is computer simulation of the plasma, usually through the use of a rate model constructed from an appropriate set of rate equations. Both of these approaches will be outlined and discussed in this chapter.

It has long been recognized that an important physical property of the ICP discharge that confers upon it desirable analytical characteristics is the relatively high temperature. In addition, the free electron concentration or electron number density (n_e) is important because the free electron is one of the most likely candidates causing analyte excitation and ionization. Over the past decade valuable contributions in the area of diagnostic temperature and electron density measurements have come from the work of Mermet and his colleagues [2–7], Fassel and his co-workers [8–10], Kornblum and de Galan [11, 12], and others [13–19].

In addition to these direct measurements of ICP physical properties, many researchers have attempted to deduce excitation mechanisms from the measurement of analyte emission intensity and species population distributions [11–13, 16, 17, 20–30].

The purpose of this chapter is to examine the current status of fundamental ICP studies and to bring recent developments in the field into perspective with the many measurements that have been made over the past decade.

11.2 MECHANISMS: COLLISIONAL AND RADIATIVE PROCESSES

Before we progress to a discussion of plasma diagnostics and rate models, it is necessary to identify the pertinent mechanisms through which energy is exchanged between various species in the plasma discharge. For a complete discussion of collisional and radiative processes in plasmas, the reader is urged to consult the excellent text by Mitchner and Kruger [31]. The relevant species for ICP are analyte atoms (M) in the ground (M) and excited (M_p) state (excited state p in this case); singly charged analyte ions (M^+) in the ground (M^+) and excited state (M_q^+) (excited state q in this case); free electrons (e^-); Ar atoms in the ground (Ar) and excited (Ar*) state; Ar ions (Ar^+); and photons, both continuum ($h\nu_{cont}$) and line ($h\nu_{line}$).

When two particles approach one another they may interact in some way, and this interaction is termed a collision. These collisions may be either elastic, in which the total kinetic energy of the colliding partners is conserved, or inelastic, in which the total energy is conserved, but kinetic energy is not. In the latter case kinetic energy may be lost to excite or ionize one of the particles, or conversely, energy "stored" in an excited state may be transferred to kinetic energy imparted to the colliding partners. Stored energy may also be released radiatively in the form of a photon. This discussion will be limited to inelastic collisions and radiative processes involving analyte species. The relevant processes are outlined in point form here:

1. Collisional excitation and deexcitation by electrons:

Analyte atoms $$M_p + e^- \rightleftarrows M_q + e^- \tag{11.1}$$

Analyte ions $$M_p^+ + e^- \rightleftarrows M_q^+ + e^- \tag{11.2}$$

In the forward process, kinetic energy is taken up from the electron by the analyte atom or ion that is in the process left in an excited state. In the reverse process, excitation energy is transferred from the excited atom or ion to the colliding electron. For the forward process, the energy of the electron must be equal to, or exceed the transition energy involved. For example, to excite Na I from the ground ($^2S_{1/2}$) to the first excited state ($^2P_{3/2}$) would require an energy of at least 2.1 eV.

For a process such as that described in Eq. (11.1), the overall rate can be expressed by the product of the concentration (population) of the colliding partners and the rate constant. The rate of the forward process (R_{CE})* is given by:

$$R_{\mathrm{CE}} = K_{\mathrm{CE}} n_p n_e \tag{11.3}$$

where K_{CE} is the collisional excitation rate constant, n_p is the number density of metal atoms in excited state p (or ions if n_p were replaced by n_p^+), and n_e is the number density of electrons. Details of the rate constants may be found in the excellent publication of Lovett [32].

The collisional deexcitation rate (R_{CD}) is given by:

$$R_{\mathrm{CD}} = K_{\mathrm{CD}} n_q n_e \tag{11.4}$$

where K_{CD} is the collisional decay rate constant and n_q is the number of density of atoms in state q. A similar expression can be written for the ion.

2. Collisional ionization and three-body recombination:

$$M_p + e^- \rightarrow M^+ + e^- + e^- \tag{11.5}$$

This process leads to the production of analyte ions normally in the ground state. The electron energy must match or exceed the ionization energy: For example, the ionization of Ca to yield Ca^+ requires 6.11 eV. The rate constant for this process (R_{CI}) is given by:

$$R_{\mathrm{CI}} = K_{\mathrm{CI}} n_p n_e \tag{11.6}$$

The reverse process, three-body recombination,

$$M^+ + e^- + e^- \rightarrow M_p + e^- \tag{11.7}$$

is an ion decay mechanism through which excited state analyte atoms may be produced. The rate for this process (R_{3B}) is given by:

$$R_{\mathrm{3B}} = K_{\mathrm{3B}}\, n_1^+ n_e^2 \tag{11.8}$$

where K_{3B} is the three-body recombination rate constant and n_1^+ is the population of ground-state ions.

*The R in this case is the total rate of the process rather than the partial rate constant used by Lovett [32].

3. Radiative recombination:

$$M^+ + e^- \rightarrow M_p + h\nu_{\text{cont}} \tag{11.9}$$

The products of this reaction are excited analyte atoms and a continuum photon ($h\nu_{\text{cont}}$). The photons produced in this reaction may have a continuous range of energies because the colliding electrons have a range of energies; from energy conservation

$$E_{h\nu} = E_{\text{electron}} + (E_{\text{ion}} - E_{M_p}) \tag{11.10}$$

The reverse process of reaction [Eq. (11.9)], photo ionization, is not as important in the ICP because, for the most part, the analyte lines are optically thin for reabsorption.

The rate for radiative recombination (R_{RR}) is

$$R_{\text{RR}} = K_{\text{RR}} n_1^+ n_e \tag{11.11}$$

where K_{RR} is the radiative-recombination rate constant.

4. Radiative deexcitation:

$$\text{Atoms} \quad M_p \rightarrow M_q + h\nu_{\text{line}} \tag{11.12}$$

$$\text{Ions} \quad M_p^+ \rightarrow M_q^+ + h\nu_{\text{line}} \tag{11.13}$$

where M_p (M_p^+) is an atom (ion) in an upper state (p) and M_q (M_q^+) is an atom (ion) in a lower state (q). These are, of course, the most significant processes in terms of the analytical aspects of ICP spectrometry since they create the characteristic line spectra used for chemical analysis. As was the case with radiative recombination, the reverse process is not favorable in an ICP due to its low optical density.

5. Penning ionization and excitation:

$$M + \text{Ar}_{\text{m}} \rightarrow M^+ + \text{Ar} + e^- \tag{11.14}$$

$$M + \text{Ar}_{\text{m}}^* \rightarrow M_p^+ + \text{Ar} + e^- \tag{11.15}$$

$$M + \text{Ar}_{\text{m}}^* \rightarrow M_p + \text{Ar} \tag{11.16}$$

where Ar_{m}^* represents an Ar atom excited in a metastable state.

Penning ionization reactions [33–36] produce ionized and/or ionized and excited analyte, depending on the energies involved with excess energy being carried away in the form of kinetic energy of the free electron (11.14) and (11.15). Thus, reaction (11.14) will produce analyte ions for any species whose ionization energy is less than the Ar metastable energies of 11.55 and 11.71 eV. Similarly in reaction (11.15), excited analyte ions are produced. Direct excitation of atoms and ions is also possible through reaction (11.16), although the energy restrictions are quite critical in that the excitation energy of the analyte and the Ar metastable energy should match closely. Beenakker [37] considers reaction (11.16) to be important in microwave-induced plasmas (MIP). In an atmospheric-pressure plasma, the lifetime of the metastable is very short [14, 38, 39] and hence a suprathermal population is unlikely; however, Penning reactions are still considered by some authors to be significant in the ICP [40].

6. Charge exchange with Ar:

$$Ar^+ + M \rightarrow Ar + M_p^+ + \Delta E \qquad (11.17)$$

In this process asymmetrical charge transfer takes place between Ar ions and analyte atoms, leading to the production of excited-state analyte ions. The difference in energy between M_p^+ and Ar^+ (ΔE) is dissipated as kinetic energy of the colliding partners. While this process need not be resonant ($\Delta E = 0$), a large energy mismatch is unfavorable.

This is not a complete list of all the possible excitation, deexcitation, ionization, and recombination processes. Lovett [32] also lists autoionization and the reverse process dielectronic recombination, Ar atom–analyte collisional processes, and stimulated emission. We shall return to a discussion of the significance of the rate equations in ICP fundamental studies in Section 11.6.

11.3. THERMODYNAMIC EQUILIBRIUM

For a partially ionized gas in a closed system, heated to temperature T, the macroscopic state of the hot gas may be described by only a few parameters such as pressure, temperature, concentration, and so forth. There does not have to be any detailed knowledge about the degree to which each of the mechanisms discussed in the previous section contributes to the state of the plasma. Rather, statistical methods enable one to describe the microscopic state of the plasma using macroscopic parameters (concentrations, temperature, pressure, etc.). This state can be completely characterized by a number of distribution functions, each of which are sensitive functions of temperature [41–43]. These distributions are the following:

1. Velocity of particles/Maxwell distribution. The Maxwell velocity distribution for a particle z is described by:

$$^zf(v) = 4\pi v^2 \left(\frac{^zm}{2\pi kT_K}\right)^{3/2} \exp\left(\frac{-\,^zmv^2}{2kT_K}\right) \qquad (11.18)$$

where ^zm is the mass of particle z; k is the Boltzmann constant (8.61706×10^{-5} eV K^{-1}); and T_K is the gas kinetic temperature of particle z. The electron kinetic temperature (T_e) is the temperature corresponding to the velocity distribution of free electrons in the plasma. In so-called thermal plasmas, the velocity and hence energy distribution for electrons is Maxwellian. In low-pressure, non-thermal discharges, such as hollow-cathode lamps, the electron temperature may be very high, and due to low collision rates the electron energy (velocity) distribution may often be non-Maxwellian [Eq. (11.18) not satisfied].

The number density $d^z n$ of particles of type z having velocities between v and $v + dv$ is given by

$$d^z n = {}^z n^z f(v) dv \tag{11.19}$$

where ${}^z n$ is the total number density of all particles of species z at all velocities.

2. Population distribution of bound states/Boltzmann distribution. The ratio of number densities of either atoms or ions in two energetically different bound states p and q is given by

$$\frac{n_p}{n_q} = \frac{g_p}{g_q} \exp\left[-(E_p - E_q)/kT_{\text{exc}}\right] \tag{11.20}$$

where n_p and n_q are the populations of particles in states p and q, respectively; g_p and g_q the statistical weights of the two states p and q, E_p and E_q the excitation energies of the two states p and q (with $E_p > E_q$); T_{exc} is the excitation (population) temperature of the species.

The number density of atoms or ions in state p (n_p) relative to the total number density of atoms or ions (n_{T}) is given by

$$\frac{n_p}{n_{\text{T}}} = \frac{g_p}{Q(T)} \exp(-E_p/kT_{\text{exc}}) \tag{11.21}$$

where $Q(T)$ is the partition function of the species. Methods for estimating the partition function for most elements have been published [43, 44].

3. Population distribution of ionization products/Saha distribution.

The ionization–recombination equilibrium

$$z_q + e^- \rightleftarrows z_p^+ + 2e^- \tag{11.22}$$

is governed by an equilibrium constant $S_{z,p}$ if the concentration of particles (z_q and z_p^+) in levels q and p of adjacent ionization stages are expressed as partial pressures. For $S_{z,p}$ we have

$$S_{z,p} = \frac{2g_p^+}{g_q} \frac{(2\pi m_e)^{3/2} (kT_i)^{5/2}}{h^3} \exp\left(\frac{-\Delta E_{p,q} - \Delta E}{kT_i}\right) \tag{11.23}$$

where m_e is electron mass; h, Planck's constant; $\Delta E_{p,q}$, energy difference between levels p and q; ΔE, lowering of ionization energy for species Z; and T_i, ionization temperature.

The lowering of ionization energy is normally negligible for temperatures at which ICPs operate. Equation (11.23) is one form of the Saha equation. The equilibrium constant can be expressed in terms of the number densities of states p and q by

$$S_z = \frac{n_p^+ n_e}{n_q} = \frac{2g_p^+}{g_q}\left(\frac{2\pi m_e kT_i}{h^2}\right)^{3/2} \exp\left(-\Delta E_{p,q}/kT_i\right) \qquad (11.24)$$

Alternatively, the Saha expression may be written for the total number density of ion and atom. If substitutions are made for the constants the net expression is [43]

$$\frac{n_T^+ n_e}{n_T} = 4.83 \times 10^{15}\, T_i^{3/2} \frac{Q^+(T)}{Q(T)} \exp\left(-E_i/kT_i\right) \qquad (11.25)$$

where E_i is the ionization energy of the atom and the Qs are the relevant partition functions.

4. Distribution of dissociation products/Guldberg–Waage distribution. For reactions of the type AB ⇄ A + B, the mass action law applies.

The number densities of the molecule (n_{AB}) and the dissociation products (n_A and n_B) are interrelated by:

$$\frac{n_A n_B}{n_{AB}} = \frac{{}^{A}Q(T)\,{}^{B}Q(T)}{{}^{AB}Q(T)}\left(\frac{m_A m_B}{m_A + m_B}\right)^{3/2}\left(\frac{2\pi kT_d}{h^2}\right)^{3/2} \exp\left(-E_{AB}/kT_d\right)$$

(11.26)

where ${}^{A}Q(T)$ is internal partition function for A; ${}^{B}Q(T)$, internal partition function for B; ${}^{AB}Q(T)$, internal partition function for AB; m_A, m_B, m_{AB}, the masses for species A, B, and AB; E_{AB}, the dissociation energy of the molecule AB; T_d, the dissociation temperature.

5. Distribution of radiation/Planck's law. The radiation density within a closed system is given by Planck's law for blackbody radiators. The blackbody temperature is called T_{rad}. Under the condition of complete thermodynamic equilibrium (CTE), each of the previously listed distributions is characterized by the same temperature; thus

$$T_K = T_{exc} = T_i = T_d = T_{rad} = T$$

In the many laboratory plasmas such as the analytical ICP, the low optical density and presence of high concentration and temperature gradients often prevents the establishment of CTE. The most prevalent form of deviation from CTE is due to loss of energy from the discharge in the form of radiation that is not reabsorbed within the plasma boundaries. Thus, Planck's law is rarely valid for laboratory plasmas. However, even in nonhomogeneous systems (systems in which density gradients are present), the temperature equivalence of the other distributions can remain locally valid if the velocity distribution of electrons satisfies Eq. (11.18). Generally, as long as the electron number density (n_e) is

$> 10^{11}$ cm^{-3}, this criterion is satisfied because of the rapid energy exchange between electrons [45]. Plasmas in which equilibrium is maintained for all distributions (except the Planck function) at any spatial point are said to be in local thermodynamic equilibrium (LTE). LTE is a special state for the plasma, and the presence of LTE simplifies plasma diagnostics since, as has been previously mentioned, the measurement of macroscopic parameters is sufficient to characterize the microscopic state (excitation mechanisms). Therefore it is a worthwhile endeavor to determine whether LTE is in existence for a particular plasma. The normal route for proof of the existence of LTE is to measure the temperature of the discharge using the distribution functions in Eqs. (11.20)–(11.26) [46, 47]. The identity of the numerical values of the temperatures is an indication that LTE exists. Alternatively, validity criteria may be established normally based on the magnitude of n_e [41, 48–51].

It is generally acknowledged that the analytical ICP is not in LTE [1–16]. This imposes a difficulty on those researchers who are attempting to elucidate excitation mechanisms, since when LTE is not established, diagnostic measurements that utilize the distribution functions expressed by Eqs. (11.20)–(11.26) cannot be used to infer the microscopic information. In other words, the validity of these distributions must be questioned when LTE does not exist; thus, the temperatures and densities measured using these distributions become difficult to interpret. There are basically two approaches that can be used to circumvent this difficulty. The first involves characterizing the discharge in terms of the extent and direction of deviation from LTE. The second involves computer simulations of excitation processes through the use of coupled rate equations. The remainder of this chapter will be devoted to outlining recent work using these approaches.

11.4. DEPARTURES FROM LTE

11.4.1. LTE Populations

Given that it has been amply demonstrated that the ICP is not in LTE, one approach to the study of excitation mechanisms is to quantify the nature of the deviation from LTE. First, let us consider an atmospheric pressure Ar plasma system, and the population distribution of excited states for analyte atoms and ions in this plasma. For the LTE case we can use the Saha (11.23) and Boltzmann (11.21) equations to provide the relative densities of excited states. If we assume that only singly charged ions and neutral atoms are significant, a Saha–Boltzmann plot would look like that provided in Fig. 11.1. This is a plot of the relative density per excited level ($\ln n_p/g_p$) as a function of the excited state energy of the level (E_p) for both the atom and ion of the analyte species. The slope of the lines ($-1/kT$) yields the population temperature for atom and ion,

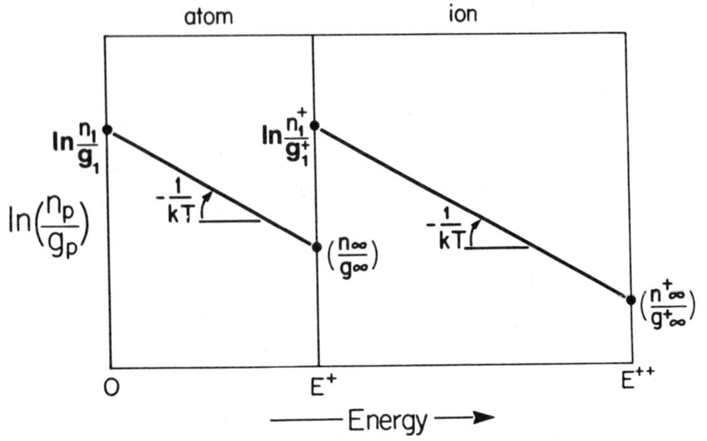

Figure 11.1. An LTE plot of a Boltzmann population distribution for analyte atom and ion. In the LTE case level populations (n_p/g_p) can be calculated from the Saha and Boltzmann equations. The slope of the lines ($-1/kT$) provides the excitation (population) temperature.

and since we are considering the LTE case, the temperature for atom and ion would be the same. In addition, the ratio (n_1^+/n_1), which is the population ratio for ground state ion (n_1^+) and ground state atom (n_1), would be given by the Saha equation as:

$$\frac{n_1^+}{n_1} = \frac{4.83 \times 10^{15}}{n_e} T^{3/2} \left(\frac{g_1^+}{g_1}\right) \exp\{-E_i/kT\} \qquad (11.27)$$

One thing that should be recognized is that the electron density is largely determined or controlled by the Ar support gas ionization process. Thus, the populations of analyte ions and atoms [the ground states of each in Eq. (11.27)] are determined by this "common ion" (n_e). An increase in the electron density alone would suppress ionization and a decrease would have the inverse effect. However, a further point to be stressed is that since we are considering the LTE case, the electron density cannot be changed in isolation from the temperature. The reason for this is the linking of temperature and electron density through the Saha equation for the Ar ionization equilibrium. For the LTE case, n_e and T are common for both Ar and analyte. The population ratio (n_1^+/n_1) is thus a complex function of n_e. Some examples of this calculation will be provided in Section 11.5.

With reference to Fig. 11.1, the equilibrium between excited levels and the ion ground level is called "Saha Equilibrium" and the equilibrium between the excited atom levels and the neutral ground level (n_1) is called "Boltzmann Equilibrium."

11.4.2. Partial LTE (p-LTE)

A frequently encountered departure for LTE is a population density distribution called partial LTE (p-LTE). For a analyte system in p-LTE a Saha equilibrium exists between the ion ground level (n_1^+) and highly excited levels of the atom [41]. In most cases this situation is created by radiative processes, the rates of which become significant enough to compete with collisional processes or by a particle transport rate that is so high that there is not sufficient time for the ion and excited-state populations to come into equilibrium with the surroundings. The point to be understood is that LTE is maintained by a condition of detailed balance for the collisional and radiative processes. Under this condition for each of the reactions outlined in Section 11.2, "the differential reaction rates for each microscopic process and for the corresponding inverse process are equal" [31]. The effect of radiative losses and transport is to upset this delicate balance so that the level populations must readjust to bring the system into balance.

Raaijmakers et al. [1] have outlined two approaches to describe the p-LTE regime. In spectrochemical work p-LTE may be described using two temperatures; the excitation temperature (T_{exc}) describes the excitation equilibrium between excited states, and the ionization temperature (T_{i}) describes the ionization equilibrium between the atom and ion ground levels. Alternatively, for a plasma physical description, in addition to the n_{e} or the electron temperature (T_{e}) an additional parameter, b_1, is introduced. This parameter is used to describe the overpopulated ($b_1 > 1$) or underpopulated ($b_1 < 1$) ground level relative to the LTE condition. The parameter b_1 is given by

$$b_1 = n_1/n_{1,\text{s}} \tag{11.28}$$

where n_1 is the actual ground state atom population and $n_{1,\text{s}}$ is the ground state population calculated from the Saha equation.

Most of the diagnostic work during the past decade has focused on the spectrochemical description, and the general finding has been that $T_{\text{i}} > T_{\text{exc}}$, leading to the notion of suprathermal ionization. More recent descriptions of the analytical ICP have focused on the plasma physical description [1, 52, 53]. The analogous population distribution to that in Fig. 11.1 for p-LTE is provided in Fig. 11.2 [1, 52]. Figure 11.2 is a plot of ln (n_p/g_p) as a function of excitation energy for analyte atoms. Figure 11.2*a* relates the overpopulated ground level case in which $b_1 > 1$. The triangles on the diagram are the actual level populations. We can see that for the higher energy levels these correspond to the Saha equilibrium curve. The extrapolation of this curve to ground state gives $n_{1,\text{s}}$, the Saha ground-state population.

The ratio b_1 (actually ln b_1) is indicated on the diagram. The Boltzmann plot relative to the atom ground level is indicated by the dotted line on Fig. 11.2*a*.

The underpopulated ground level situation is shown in Fig. 11.2*b*. This diagram may be understood by analogy with Fig. 11.2*a*. In this case $b_1 < 1$.

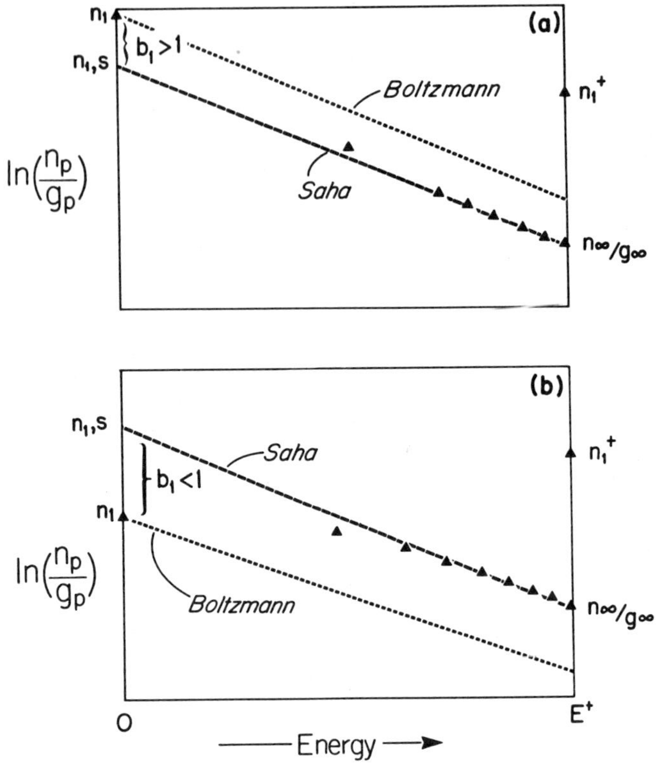

Figure 11.2. A schematic representation of a p-LTE population density distribution for analyte atoms. (*a*) A predominantly ionizing plasma, $b_1 > 1$; (*b*) a predominantly recombining plasma $b_1 < 1$. In both *a* and *b* the population of n_∞ remains coupled to the ground-state ion density (n_1^+) through electron collisions. The triangles in the diagrams represent actual populations, the dashed line is the "Saha" line, and parallel to this is the dotted "Boltzmann" line extrapolated from n_1.

Some further observations can be made about the plots provided in Fig. 11.2. The plot of ln (n_p/g_p) in Fig. 11.2*a* could be called an atom surplus case since there are more atoms for a given electron density value than are predicted by the Saha equation. This atom surplus case primarily manifests itself as an overpopulation of the ground atomic state and is characteristic of a predominantly ionizing plasma ($b_1 > 1$), which is normally created by a process of rapid heating [45]. Attempts to measure two-level excitation temperatures using a Boltzmann formula will result in temperatures that change with energy level and are lower for the lower levels and higher for the upper levels. The plot in Fig. 11.2*b* could be called an atom-deficient case, since there are fewer atoms for a given electron density value than are predicted by the Saha equation. This case is manifested by an underpopulation of the ground atomic state relative to LTE and is characteristic of a recombining plasma ($b_1 < 1$), which is normally created by a situation of rapid cooling. Excitation temperatures measured in this case will be higher for lower energy levels and lower for higher energy levels.

Alder et al. [14] have provided some interesting evidence that the aerosol channel can be characterized by the population distribution shown in Fig. 11.2*a*. These authors measured the relative emission intensities for 20 different Fe(I) lines, with the upper energy level varying from 26,000 cm^{-1} (3.22 eV) to 55,000 cm^{-1} (6.82 eV).

The emission intensity (I_{pq}) of a spectral line emanating from level p and terminating in level q is given by

$$I_{pq} = \frac{1}{4\pi} \frac{g_p n_T}{Q(T)} A_{pq} \frac{hc}{\lambda_{pq}} \exp\left(-E_p/kT_{\text{exc}}\right) \tag{11.29}$$

where g_p is the degeneracy or statistical weight of level p; A_{pq}, the transition probability for pq transition; λ_{pq}, the wavelength of pq transition; and n_{T}, the total number density of species in the ionization stage under consideration.

The excitation temperature (T_{exc}) may be determined by measuring the relative intensity of each spectral line and plotting ln $(I_{pq}\,\lambda_{pq}/g_p\,A_{pq})$ as a function of (E_p). The slope of the line $[-(1/kT_{\text{exc}})]$ then provides the temperature. Alternatively, A_{pq} may be replaced by f_{qp} (the transition oscillator strength) and ln $(I_{pq}\,\lambda_{pq}^3/g_q\,f_{qp})$ is plotted as a function of E_{p}. In this case the degeneracy g_q is the degeneracy of the lower level of the transition. ln $(I_{pq}\,\lambda_{pq}^3/g_q\,f_{qp})$ is a measure of the relative population of the upper level in question. The data collected by Alder et al. showed different temperatures for different levels. Kornblum and Smeyers-Verbeke [54] recast the original data for clarity. The population and temperature plots they obtained are shown in Fig. 11.3*a* and 11.3*b*, respectively. One can see in Fig. 11.3*a* that a Saha extrapolation from the upper energy levels (as in Fig. 11.2*a*) shows that the lower levels are overpopulated relative to the Saha curve. As expected from this type of distribution, the excitation temperature measured from the lower levels is lower than that measured from the higher levels (Fig. 11.3*b*).

Raaijmakers et al. have noted that the general domain of validity for p-LTE is approximately $10 > b_1 > 0.1$ [1]. When b_1 is outside these bounds, it is convenient to discuss non-LTE using other, more sophisticated models. These models dispense with Saha–Boltzmann analogies.

11.4.3. Further Deviations From LTE

Deviations from LTE for which p-LTE is no longer an adequate description ($10 > b_1 > 0.1$) are usually discussed using collisional-radiative models such as corona and saturation-phase models. These models have been extensively treated

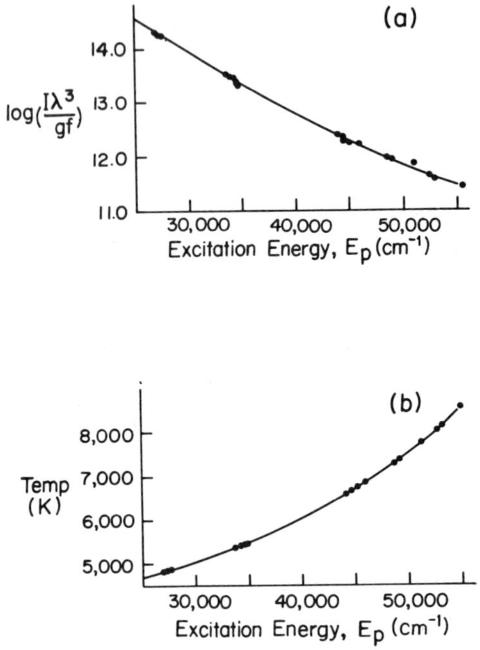

Figure 11.3. (*a*) A measured level population diagram for Fe (I). (*b*) The level-dependent excitation temperature derived from the population diagram in *A*. [Reprinted with permission from G. R. Kornblum and J. Smeyers-Verbeke, *Spectrochim. Acta* **37B**, 83 (1982). Copyright (1982), Pergamon Journals.]

by Fujimoto [55–59] and van der Mullen et al. [60], and discussed in relation to the ICP by Raaijmakers et al. [1]. In this treatment various regimes are considered. These are outlined briefly here.

1. Corona regime. The corona regime is characterized by radiation domination for deexcitation processes. Excitation takes place by electron collisions predominantly with the ground level, whereas for excited state deexcitation, radiative rates far exceed collisional rates so that the predominant deexcitation mechanism is radiative decay. This regime is typically found in plasmas in which the electron density is $<10^{12}$ cm^{-3}. Densities this low are not found in the spectrochemical ICP.

2. Saturation regime. As electron density increases, the transition between corona domain and saturation is marked by collisional processes becoming important for higher excited atom levels, while radiative processes still dominate at lower excited levels. One cause for this is that radiative transition rates generally decrease as one moves up in energy. This interfacial region is normally referred to as quasi-saturation. As electron density is further increased, collisional processes become more dominant; and as complete collision domination is reached the plasma can be described by p-LTE or LTE models. In saturation, the dominant population and depopulation processes for excited levels are collisional excitation and deexcitation between adjacent energy levels. This leads to so-called ''ladder-like'' excitation and deexcitation. If the net flow is up the ladder (is leading to ionization) the plasma can be considered to be net ionizing, and if the net flow is downward (that is, leading to recombination) the plasma can be considered to be recombining. In LTE the dominant mechanism is ionization to and recombination from the ion ground level. It is currently felt that for the range of electron density measured for the ICP (n_e = 10^{14} to 10^{16}), saturation and p-LTE are indicated.

To this point the discussion has been limited to what one might call theoretical considerations; that is, the potential physical characteristics analyte in a plasma discharge can assume. The actual population densities of excited states and of atoms and ions will be determined to a large degree by the physical environment created by the hot Ar gas. It has been suggested that the p-LTE domain is strongly indicated if one considers the ambient electron density [52]. From the previous discussion it is clear that a useful macroscopic description of the analyte can be obtained by comparing actual measured level populations to those that might be found in a discharge at LTE. The macroscopic state can then be interpreted in terms of the distribution provided in Fig. 11.2. For this description it is necessary to measure departures from LTE.

11.5. MEASUREMENT OF DEPARTURES FROM LTE

11.5.1. Introduction

If departures from LTE are to be determined, one is immediately faced with the problem of setting up an LTE model that can be used for comparison purposes. The approach currently considered to be the most useful and consistent is basing the LTE description on a single measured parameter, the measurement of which does not depend on the existence of LTE to be valid. The plasma electron density fulfills this criterion. The measurement of the full width at half-maximum (FWHM) of the H_β line at 486.13 nm is a convenient method of determining electron density. It is a strong line, sufficiently broadened for precise measurements, and for electron densities between 10^{14} and 10^{17} cm^{-3} the variation of the FWHM with temperature is small. Detailed calculations of the shape of the H_β line as a function of electron density and temperature are readily available in the literature [61, 62], and the agreement between experiment and theory indicates an accuracy in the range of 10% [63].

To compare measured analyte excited state and ion populations with hypothetical LTE populations, it is not necessary to assume that LTE is in existence

Table 11.1. LTE Temperatures Corresponding to Measured Electron Densities

Electron Density (cm^{-3})	LTE Temperature (K)
9×10^{13}	6450
1×10^{14}	6498
2×10^{14}	6827
3×10^{14}	7036
4×10^{14}	7192
5×10^{14}	7317
6×10^{14}	7423
7×10^{14}	7515
8×10^{14}	7596
9×10^{14}	7669
1×10^{15}	7736
2×10^{15}	8205
3×10^{15}	8506
4×10^{15}	8733
5×10^{15}	8919
6×10^{15}	9076
7×10^{15}	9212
8×10^{15}	9334
9×10^{15}	9443
1×10^{16}	9544

in the ICP for the experimental aspects, if the basis for this comparison is the plasma electron density. It has been noted [1, 52, 53, 64] that an appropriate approach to be used is to measure the electron density and use Dalton's law, charge neutrality, and the Saha formula to calculate an LTE temperature. It might be argued that this should be called an Ar ionization temperature, but it should be understood that the object of this exercise is to set up a hypothetical LTE model based on the measured n_e. Since at LTE all temperatures are the same, it is not necessary to make the distinction, T_i, for the LTE model. In this case the temperature derived from the calculation outlined is the plasma temperature, which will be referred to as T_{LTE}. Having established n_e and T_{LTE}, it is a straightforward matter to calculate relative excited state population densities and relative ion and atom densities. Table 11.1 is a listing of the T_{LTE} corresponding to the applicable electron density [53].

11.5.2. Electron Density Measurements

To establish the hypothetical LTE model, it is necessary to have a complete n_e description of the ICP itself. Many workers have measured electron density in the ICP [5, 8, 10–12, 14, 15, 17]; however, the most recent and comprehensive electron density characterization of the ICP has come from the measurements of Caughlin and Blades [65]. These authors measured the radial spatial distribution of electron density at 4, 8, 12, 16, and 20 mm above the load coil for radio frequency (rf) input powers of 1.0, 1.25, 1.5, 1.75, and 2.0 kW.

They also evaluated the precision of the measurement of n_e and found it to be between 2 and 10%. The spatial measurement was accomplished using a photodiode array spectrometer to detect the H_β line and by translating the ICP torch enclosure to 150 separate lateral spatial positions. The resultant lateral profiles were subjected to an Abel inversion, and isopleths of electron density were plotted as a function of both radial and vertical positions. The resultant electron density spatial profiles are provided in Fig. 11.4*a–e* [65]. It is clear that the magnitude of electron density increases with increasing power. It is also interesting to note that the electron density increases with height in the aerosol channel; at 1.5 kW, for example, the electron density is 0.5×10^{15} cm^{-3} at 4 mm and increases to a maximum of 2.5×10^{15} cm^{-3} between 12 and 17 mm above the load coil. For these profiles it was imperative that the basic lateral data were

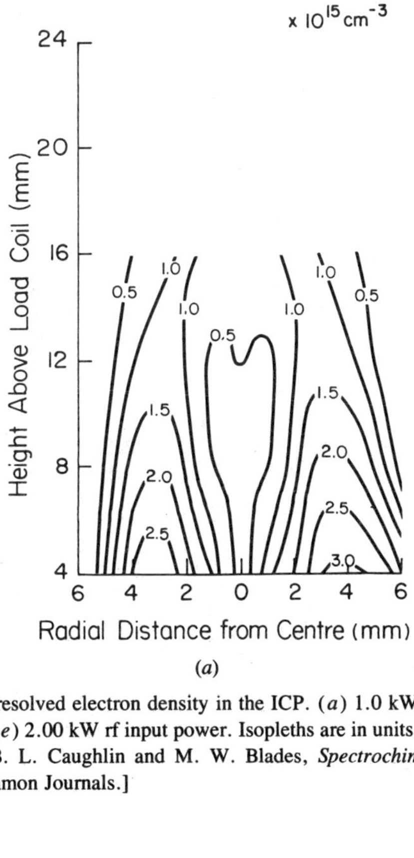

(*a*)

Figure 11.4. Spatially resolved electron density in the ICP. (*a*) 1.0 kW, (*b*) 1.25 kW, (*c*) 1.50 kW, (*d*) 1.75 kW, and (*e*) 2.00 kW rf input power. Isopleths are in units of 10^{15} cm^{-3}. [Reprinted with permission from B. L. Caughlin and M. W. Blades, *Spectrochim. Acta* **40B,** 98 (1985). Copyright (1985), Pergamon Journals.]

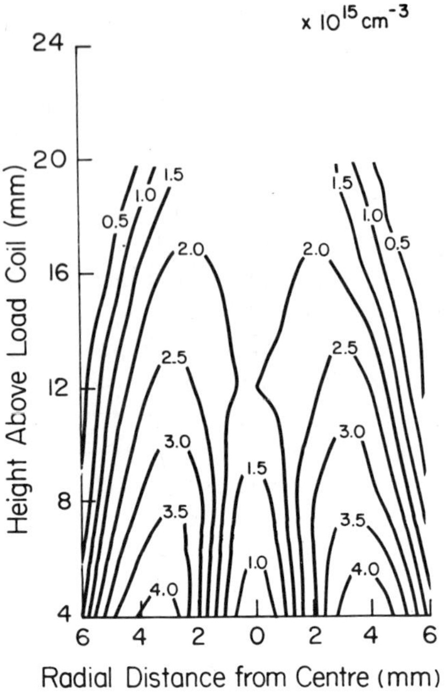

(*b*)

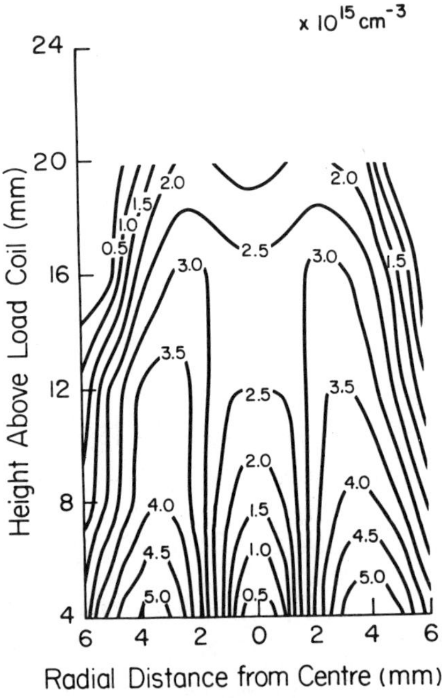

(*c*)

Figure 11.4. (*Continued*)

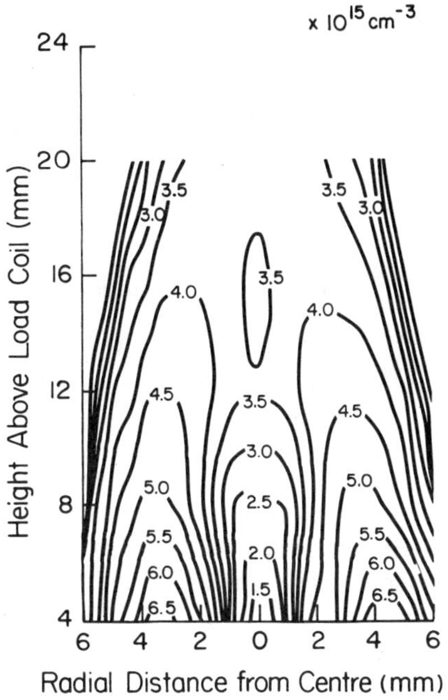

(*d*)

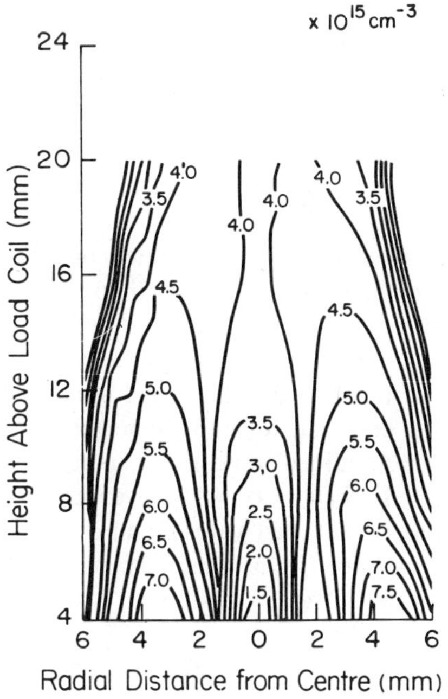

(*e*)

Figure 11.4. (*Continued*)

subjected to an Abel inversion. Without the Abel inversions, the high off-axis emission intensities from H_β mask the information emanating from the aerosol channel, particularly for lower observation heights. An increasng electron density in the aerosol channel with height would seem to indicate that between 4 and 16 mm above the load coil the aerosol channel is an ionizing plasma.

11.5.3. Ion–Atom Emission Intensity Ratios

The ratio of emission intensity of an ion line (I_i) to emission intensity of an atom line (I_a) (the ratio will be referred to as I_i/I_a) has previously been used as a measure of the energy characteristics of the ICP discharge [13, 66]. The finding by Boumans and de Boer [66] that experimental ion–atom emission intensities were from one to three orders of magnitude greater than those calculated from LTE expressions stimulated much discussion in the literature. More recently [53] ion–atom emission intensity ratios have been discussed in terms of a nonequilibrium parameter (b_r) analogous to the b_1 defined earlier. This parameter is also a measure of departure from LTE, based on measured and (calculated) LTE ion–atom emission intensity ratios and is given by:

$$(I_i/I_a)_{\text{exp}} = b_r\,(I_i/I_a)_{\text{LTE}} \tag{11.30}$$

where $(I_i/I_a)_{\text{exp}}$ is the experimentally measured ion–atom emission intensity ratio for a chosen pair of lines and $(I_i/I_a)_{\text{LTE}}$ is the analogous LTE ratio that is calculated from LTE formulas. Caughlin and Blades [53] have derived the relationship between b_1 and b_r. The resultant expression is

$$b_r = \frac{1}{b_1}\exp\left[\frac{E_q^+ - E_p}{kT_{\text{LTE}}} + \left(-\frac{E_q^+}{kT_{\text{exc}}^+} + \frac{E_p}{kT_{\text{exc}}}\right)\right] \tag{11.31}$$

where E_q^+ is the excitation energy of excited state q of the ion; E_p, the excitation energy of excited state p of the atom; T_{LTE}, the (hypothetical) LTE temperature; T_{exc}^+, the excitation temperature of the ion measured for level q; T_{exc}, the excitation temperature of the atom measured from level p.

For the derivation of this formula the actual ion density and the LTE ion density were assumed to be equal. It can be seen that if $T_{\text{exc}}^+ = T_{\text{exc}} = T_{\text{LTE}}$, $b_r = 1$ (given that $b_1 = 1$ as outlined earlier). In general, b_r varies inversely with b_1, leading to values < 1 when b_1 is >1; however, the exact value depends upon the level populations of excited states p and q.

The LTE ion–atom emission intensity ratio can be calculated from an expression given by Boumans [43]. The expression is

$$\left(\frac{I_i}{I_a}\right)_{\text{LTE}} = \frac{4.83\times 10^{15}}{n_e}\left(\frac{g_qA_{qp}}{\lambda_{qp}}\right)_{\text{ion}}\left(\frac{\lambda_{qp}}{g_qA_{qp}}\right)_{\text{atom}} T^{3/2}\,10^{-\{5040(E_i+E_q^+-E_p)/T_{\text{LTE}}\}} \tag{11.32}$$

Table 11.2. Wavelengths, Excitation Energies, Ionization Energies, and Transition Probabilities for Sr, Ca, Mg, Cd, and Zn

Species	λ (Å)	E_q (eV)	E_i (eV)	g_qA_{qp} (10^8 s^{-1})
Sr I	4607	2.692	5.629	6.03
II	4078	3.042		5.68
Ca I	4227	2.936	6.111	6.54
II	3936	3.152		5.88
Mg I	2852	4.35	7.644	14.4
II	2795	4.43		10.4
Cd I	2288	5.42	8.991	15.9
II	2265	5.47		6.0
Zn I	2139	5.80	9.391	21.27
II	2025	6.13		13.2

where g_q, is the degeneracy of excited states p and q, respectively; A_{qp}, the transition probability of qp transition; λ_{qp}, the wavelength of the qp atomic and ionic transitions. The other symbols have been previously defined.

It is important to note in connection with this formula that, since it is an LTE expression, the n_e and T should be internally consistent. Although the expression will be used to calculate ion–atom emission intensity ratios for analyte species, recall that n_e is dictated by Ar ionization and that the T_{LTE} used should be that provided in Table 11.1 for the corresponding n_e.

Caughlin and Blades have studied b_r values for a set of five elements: Sr, Ca, Mg, Cd, and Zn [67]. The ionization energies (E_i) for these elements cover a range from 5.7 to 9.4 eV, and the excitation energies (E_q^+ and E_q) of the lines studied cover a range from 2.7 to 6.1 eV. The relevant spectral data for these elements and the lines considered are listed in Table 11.2. Figure 11.5 shows plots of $(I_i/I_a)_{LTE}$ as a function of electron density over the range of interest for the ICP. It can be seen from the plot that the effect of increasing n_e (and hence T_{LTE}) on LTE ratios depends on the magnitude of the ionization energy. For elements with low ionization energy (Sr and Ca) the ion–atom ratio decreases as electron density increases. For Mg, an element with intermediate ionization energy, the variation is small, and for Cd and Zn the LTE ion–atom emission intensity ratio increases with increasing electron density.

A comparison of these LTE ion–atom ratios with those measured from an ICP is provided in Table 11.3 [67]. This table lists LTE and measured ion–atom ratios and also gives the value of b_r for each element for rf input powers of 1.0, 1.25, 1.50, 1.75, and 2.0 kW. The measured electron density corresponding to each power is also provided.

It is clear from the data contained in this table that, on the whole, the ICP is not characterized by LTE ion–atom emission intensity ratios. However, the experimental values are within a factor of five from LTE values. For all of the

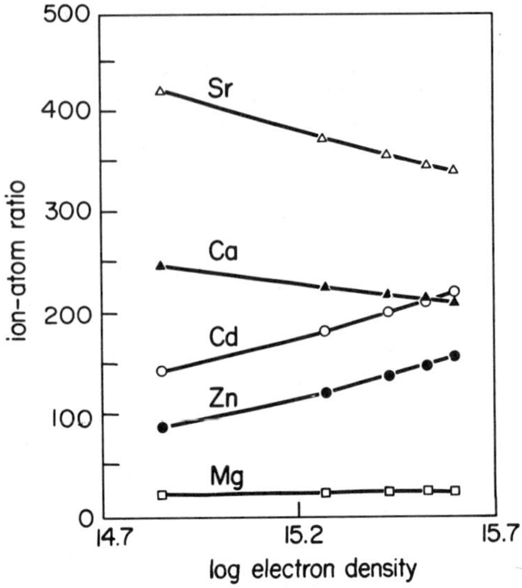

Figure 11.5. Ion–atom emission intensity ratios as a function of log of electron density for the lines listed in Table 11.2. The scale has been magnified × 100 for Cd and Zn.

Table 11.3. Ion–Atom Emission Intensity Ratios[a]

Power (kW)	1.00	1.25	0.50	1.75	2.00
n_e (cm^{-3})	7.1×10^{14}	1.8×10^{15}	2.7×10^{15}	3.4×10^{15}	4.0×10^{15}
Sr $(I_i/I_a)_{exp}$	82.7	147.1	372	463	518
$(I_i/I_a)_{LTE}$	423	375	358	348	342
b_r	0.196	0.392	1.04	1.33	1.51
Ca $(I_i/I_a)_{exp}$	43.4	95.6	163	212	260
$(I_i/I_a)_{LTE}$	249	228	221	217	213
b_r	0.174	0.419	0.738	0.977	1.22
Mg $(I_i/I_a)_{exp}$	4.93	6.62	10.3	12.3	13.6
$(I_i/I_a)_{LTE}$	21.8	23.6	24.4	24.9	25.2
b_r	0.226	0.280	0.422	0.494	0.540
Cd $(I_i/I_a)_{exp}$	—	1.08	1.34	1.63	1.89
$(I_i/I_a)_{LTE}$	1.47	1.85	2.03	2.15	2.24
b_r	—	0.584	0.660	0.758	0.844
Zn $(I_i/I_a)_{exp}$	—	0.37	0.55	0.72	0.82
$(I_i/I_a)_{LTE}$	0.90	1.23	1.40	1.51	1.59
b_r	—	0.301	0.393	0.477	0.516

[a][Reprinted with permission from B. L. Caughlin and M. W. Blades, *Spectrochim. Acta* **40B,** 1539 (1985). Copyright (1985), Pergamon Journals.]

elements, there is an increase in b_r as the rf power is increased, although b_r spans a much wider range for Sr and Ca as compared to Mg, Cd, and Zn. In addition, for Sr and Ca, b_r exceeds 1 at higher rf input powers (1.75 and 2.0 kW for Sr, and 2.0 kW for Ca).

A mechanistic interpretation of the data provided in Table 11.3 is difficult. It can be shown that b_r is a very sensitive measure of nonequilibrium, but its interpretation is complicated by the fact that the emission intensity information used to determine b_r comes from different ionization stages. The emission intensity for a spectral line depends on both the relative species abundance (ion or atom) and the populations of the relevant excited states. This gives several possibilities for non-LTE ion–atom emission intensity ratios. These are the following:

1. The actual degree of ionization and the LTE degree of ionization are the same, but either the atom or the ion, or both, have a non-Boltzmann excited state distribution. In this scheme the analyte ion ground state is in equilibrium with the electron density and the highly excited atom states through the Saha equilibrium.

2. The actual degree of ionization is different from the LTE degree of ionization, but the excited state populations are near-Boltzmann. In this scheme very small ionization equilibrium deviations can have a profound effect on b_r values. Consider the b_r value for Sr at 2.0 kW. The value of b_r from Table 11.3 indicates that the experimental ion–atom emission intensity differs from the LTE values by a factor of 1.5. If we interpret this in terms of a departure from Saha equilibrium while maintaining near-Boltzmann level populations, this would correspond to an experimental degree of ionization of 99.88% compared with the LTE value of 99.82%. Therefore Sr is only 0.06% overionized relative to LTE. However, this is enough to lead to a b_r value of 1.5. The reason is the relatively high absolute degree of ionization of Sr. Given a total of 10,000 Sr species, at LTE, 9982 would be in ion form, and 18 would be atoms. For the actual case, 9988 would be in ion form and 12 would be atoms. This difference in atom population would yield an experimental atom emission intensity that is a factor of one third lower than that that would be expected at LTE. Thus, even though the degree of ionization is very close to the LTE value, the ion–atom ratio departs significantly. This is a case in which b_r values are an extremely sensitive measure of non-LTE.

3. Both degree of ionization and excited-state populations are nonequilibrium values. This is the most complex case and requires detailed population measurements to verify. This is also the most probable state of analyte in the ICP. If lower levels of the atom are overpopulated with respect to the higher levels, for example, then the total population of atoms will exceed that predicted by LTE considerations. This implies that the ion must be underpopulated relative to LTE, since the excess population of the atom must come from somewhere. This would mean that the actual degree of ionization of analyte is different from that which would be expected at LTE; specifically, if b_r values are < 1, the analyte would be expected to be underionized relative to LTE.

Taken by itself the magnitude of b_r cannot be used to distinguish between these three possibilities. To understand the cause for non-LTE b_r values, actual level populations must be measured and a plot such as that in Fig. 11.2 constructed. Such detailed measurements are not available at this time, although two papers have appeared containing some level population data [14, 52].

The data reported by Alder et al. [14] and further discussed by Kornblum and Smeyers-Verbeke [54], some of which has been reproduced in Fig. 11.3, would seem to indicate that Scheme 1 or 3 is encountered in the ICP. Recall that in a p-LTE plasma highly excited states of the atoms are in equilibrium with the ground state ion population. The value of n_e reported by Alder et al. was 3.3×10^{15} cm^{-3} (at 10 mm above the load coil). The T_{LTE} corresponding to this n_e is about 8600 K. It is interesting to note that the Fe I excitation temperature measured by these authors for high-energy states was 8510 K, which is within experimental uncertainty of the LTE value. These values are consistent with Saha equilibrium between free electrons, the ion ground state, and highly excited atom levels. In addition, it has already been noted that the level populations for Fe (I) as measured by Alder et al. show an overpopulation of the lower levels relative to the higher levels. All of this evidence is consistent with an ionizing p-LTE plasma. Alder et al. also provided an ionization temperature calculated from the ion–atom emission intensity ratio for Fe I and Fe II. From the value of 8260 K they obtained, we conclude that the b_r value is less than one for this set of data. Since Fe has an ionization energy of 7.87 eV, we would expect that it would behave in a manner very similar to Mg, which has an ionization energy of 7.65 eV. This b_r value for Fe of less than one is also consistent with an ionizing, p-LTE plasma.

Schram et al. [52] have provided a "Boltzmann" plot for the level populations of Mg atom and ion species. These authors find near-Boltzmann behavior for excited levels of Mg atoms although the ground state appeared to be overpopulated by a factor of five. In addition, these authors found that the ion levels in the 16-eV range were overpopulated, whereas the remainder were near Boltzmann. They attributed this overpopulation to asymmetric charge transfer between the Ar ion and Mg atom.

Caughlin and Blades [66] have studied the spatial variation of b_r for Cd and Mg and found that the zone closest to LTE is the normal analytical zone (NAZ) [68] between 12 and 16 mm above the load coil. Below this zone b_r values increase with height, from 0.1 at 4 mm above the load coil to 0.55 at 12 mm above the load coil for Mg at 1.25 kW.

Although the measurement of non-LTE through the use of b_r, using n_e as a basis for determining the hypothetical LTE framework, is relatively new, Rezaaiyaan and Hieftje have recently reported b_r values that range from 0.08 for Mn to 0.35 for Zn in a low-flow, low-power ICP [69].

A significant aspect to this more consistent method of determining departures from LTE using ion–atom emission intensity ratios is the finding that, in general, ion–atom ratios are "infrathermal" rather than "suprathermal." Consequently the focus of ICP diagnostic and mechanistic work has recently shifted away from trying to identify potential causes for suprathermal ion–atom emission intensity ratios and toward trying to identify the cause of excitation temperatures, which are lower than the temperature derived from electron density (Ar ionization temperatures). Non-Boltzmann level populations are strongly indicated, although detailed level population determinations are required to verify this postulate. These determinations will, no doubt, be complicated by the often-encountered inaccuracy in published transition probability data.

11.5.4. Analyte Ionization

There is little doubt that the ICP is a very effective ionization source. The recent evolution of its use as an ion source for mass spectrometry is a testament to this fact. In a recent paper, Caughlin and Blades [67] have studied analyte ionization in the ICP. In this study the experimental degree of ionization was estimated from measured ion–atom emission intensity ratios. The approach used was to assume that both ion and atom have near-Boltzmann level populations and to use the Saha–Boltzmann relationship [43]:

$$\frac{\alpha}{1-\alpha} = \frac{(I_{qp})_{\text{ion}}}{(I_{qp})_{\text{atom}}} \left(\frac{g_q A_{qp}}{\lambda_{qp}}\right)_{\text{atom}} \left(\frac{\lambda_{qp}}{g_q A_{qp}}\right)_{\text{ion}} \frac{Q_{\text{ion}}}{Q_{\text{atom}}} 10^{-(5040(E_q - E_q^+)/T)} \tag{11.33}$$

to evaluate α, the degree of ionization. The temperature used for this calculation was the temperature derived from the measurement of electron density (see Table 11.1). The argument used was that, since the measured ion–atom emission intensity ratios were within a factor of five of LTE values, deviations from LTE are not too pronounced to preclude the use of Eq. (11.33). In addition, the inaccuracy in α is further reduced by the fact that, if the difference between E_q and E_q^+ is small, the effect of temperature on the power term is minimized. This was largely the case for the elements they studied, which are those listed in Table 11.2. For example, an uncertainty of ± 2000 K in the temperature introduces only a 0.03% uncertainty in the value of α for Ca.

The results obtained are listed in Table 11.4. The table lists the experimental and hypothetical LTE degree of ionization for Sr, Ca, Mg, Cd, and Zn. Two trends are evident in the table. First, as rf power is increased, the degree of ionization also increases. Second, the degree of ionization decreases with in-

Table 11.4. $\alpha \times 100$ **Degree of Ionization in Percent**[a]

Power (KW)	1.0	1.25	1.50	1.75	2.0	Uncertainty
Sr expt	99.40	99.62	99.84	99.87	99.88	±0.02
LTE	99.88	99.85	99.84	99.83	99.82	±0.04
Ca expt	98.90	99.44	99.65	99.73	99.77	±0.03
LTE	99.81	99.76	99.74	99.73	99.72	±0.05
Mg expt	93.00	94.39	96.22	96.77	97.03	±0.26
LTE	98.33	98.36	98.37	98.37	98.37	±0.42
Cd expt	—	85.99	88.43	90.29	91.46	±0.24
LTE	89.39	91.35	92.06	92.45	92.76	±1.35
Zn expt	—	63.89	72.30	77.39	79.40	±3.23
LTE	81.96	85.66	86.98	87.71	88.20	±2.12

[a][Reprinted with permission from B. L. Caughlin and M. W. Blades, *Spectrochim. Acta* **40B**, 1539 (1985). Copyright (1985), Pergamon Journals.]

creasing ionization energy of the element under consideration. The relationship between α and ionization energy is provided in Fig. 11.6, in which both experimental and LTE degrees of ionization are plotted as a function of ionization energy for an rf power of 1.25 kW. It is clear that below about 6 eV analyte is essentially 100% ionized in the ICP. In addition, even below 9 eV analyte is more than 60% ionized. The data in Table 11.4 and Fig. 11.6 are consistent with degrees of ionization measured using an ICP–mass spectrometer [70].

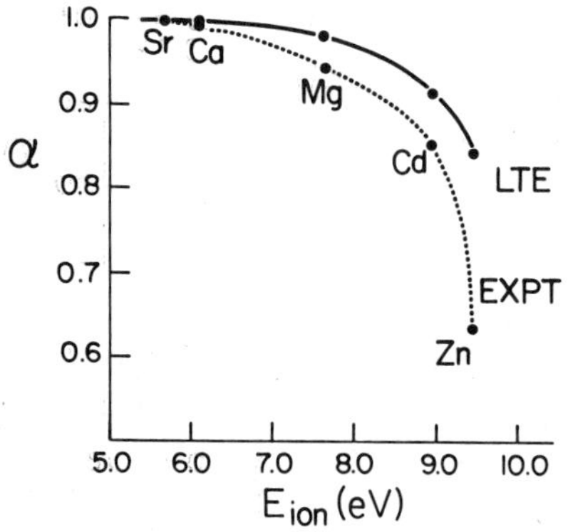

Figure 11.6. Degree of ionization as a function of ionization energy for analyte in the ICP at 1.25 kW, 16 mm above the load coil. The solid line is an LTE value and the dashed line is an experimental value. [Reproduced with permission from B. L. Caughlin and M. W. Blades, *Spectrochim. Acta,* **40B,** 1539 (1985). Copyright (1985), Pergamon Journals.

In terms of discharge characterization, it is apparent that on the whole, given the assumptions that went into the calculation of α, analyte is underionized relative to LTE. The degree of underionization varies from 25% for Cd to 0.23% for Sr at 1.25 kW. It should be noted that underionization is consistent with an ionizing p-LTE plasma in which atoms are overpopulated relative to the LTE case. The cause of underionization is probably the influence of a radiative recombination rate, which is significant compared with three-body recombination. This aspect will be discussed further in the next section.

11.6. RATE MODELS

Studies relating to the measurement of departure from LTE have been very useful for elucidating the physical state of analyte in the plasma, but such studies are unable to provide detailed mechanistic evidence. However, from the previous discussion, p-LTE is strongly indicated with the highly excited atom levels in equilibrium with the ion ground state, and lower energy levels overpopulated with respect to higher energy levels. Since LTE is a condition in which collisional processes dominate excitation and deexcitation, and radiative contributions are considered to be relatively insignificant, one might surmise that the cause of the observed departure from LTE is significant radiative contributions to excited-state depopulation. To understand the degree to which the various processes contribute to analyte excitation and deexcitation, several workers have recently adopted a rate model approach to the study of ICP phenomena [32, 40, 71–74].

The application of rate models to the study of excitation mechanisms in plasmas can be traced to the treatment of Bates et al. [75–77] and McWhirter and Hearn [78]. The formulation is based on what is generally known as the CR model, which has been used by many workers to study excitation mechanisms and population distributions in plasmas. The work of Srivastava and Ghosh [79] applied to a He plasma is a good example of a rate model approach.

The collisional-radiative (CR) model is based on the principle that in a homogeneous plasma the number density of atoms in a given excited state is a balance between both the collisional and radiative excitation and deexcitation of the state, and the rate of change of the population of that state. In addition, the ion population is a balance between the rates of ionization and recombination. In a plasma in which there are density and temperature gradients, transport of species by convection and diffusion must also be included. For the most part the latter have not been considered to date for the ICP except by Schram et al. [52], who have included convection and diffusion in a continuity equation for Ar.

To understand how the rate model approach works, let us consider the simple

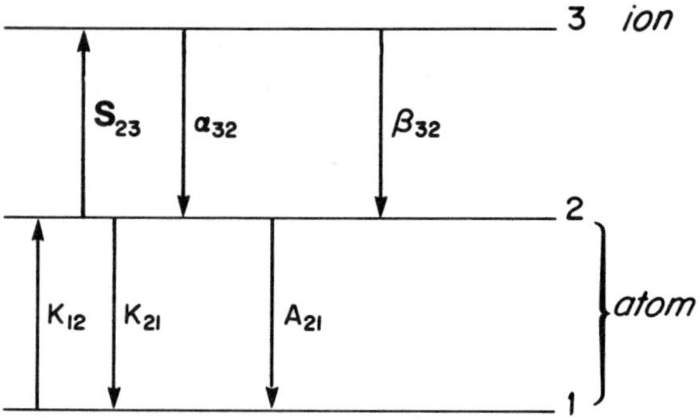

Figure 11.7. Simple hypothetical three-level system. Levels 1 and 2 are ground state and excited state of the atom, and level 3 is ground state for the ion. Populating and depopulating processes for level 2 are:
K_{12}, collisional excitation rate constant; K_{21}, collisional deexcitation rate constant; A_{21}, radiative deexcitation rate constant; S_{23}, collisional ionization rate constant; α_{32}, three-body recombination rate constant; β_{32}, radiative recombination rate constant.

three-level system depicted in Fig. 11.7 consisting of an atom ground state (level 1), an excited atom state (level 2), and an ion ground state (level 3). The rate model approach considers all population and depopulation processes for a particular level. For example, for level 2 in this simple system consider the following processes:

1. Electron excitation and deexcitation from level 2 to and from level 1:

$$M(1) + e^- \underset{K_{21}}{\overset{K_{12}}{\rightleftharpoons}} M(2) + e^-$$

2. Electron impact ionization and three-body recombination to and from level 3:

$$M(2) + e^- \underset{\alpha_{32}}{\overset{S_{23}}{\rightleftharpoons}} M^+(3) + e^- + e^-$$

3. Spontaneous emission from level 2 to level 1:

$$M(2) \xrightarrow{A_{21}} M(1) + h\nu_{\text{line}}$$

4. Radiative recombination from level 3 to level 1:

$$M^+(3) + e^- \xrightarrow{\beta_{32}} M(2) + h\nu_{\text{cont}}$$

The rate equations for these processes have already been discussed in Section

11.2. The more usual rate constant symbols are used in 1 to 4 rather than the symbols introduced in Section 11.2.

To continue with the establishment of the model, we recognize that the rate of change of the population of level 2 will be the difference between the rates of all populating processes and depopulating processes. Thus, we can write for the change of the population of level 2 with time (dn_2/dt):

$$\frac{dn_2}{dt} = K_{12}n_1n_e + \alpha_{32}n_3n_e^2 + \beta_{32}n_3n_e - K_{21}n_2n_e - S_{23}n_2n_e - n_2A_{21} \qquad (11.34)$$

For the simple three-level system, this rate equation expresses the time-dependent population of level 2. To solve this equation for the population of level 2 we make the steady-state approximation; that is, that under homogeneous conditions $dn_2/dt = 0$. This is not unrealistic for a small spatial volume of the ICP since relatively time-invariant emission intensities (which are a measure of the population of level 2) are observed. The solution to the equation then is given by

$$n_2 = \frac{n_e\left[K_{12}n_1 + n_3\left(\alpha_{32}n_e + \beta_{32}\right)\right]}{A_{21} + n_e\left(S_{23} + K_{21}\right)} \qquad (11.35)$$

The reader will notice that to solve for the population of level 2 the populations of levels 1 and 3 must be known, in addition to the electron density. Normally one can set up analogous equations for the populations of levels 1 and 3 and use a measured electron density to enable the solution of the three simultaneous equations. This solution will explicitly provide the populations of levels 1–3.

For a real analyte system the situation is much more complex because of the number of excited levels involved for both atom and ion. The most comprehensive rate model treatment of the analytical ICP has been carried out by Lovett [32]. Lovett set up and solved the rate equations for several elements. He used 80 levels for Be, 112 for Mg, 170 for Ca, 141 for Sr, and 148 for Ba. The mechanisms he considered were collisional excitation and deexcitation, collisional ionization and three-body recombination, radiative recombination, autoionization and dielectronic recombination, Ar atom collisional processes, radiative emission and absorption, and stimulated emission. The relevant rate constants were taken from a variety of sources.

Lovett used an electron density of 3.34×10^{15} cm^{-3}, a value that would be encountered in the ICP operating between 1.5 and 1.75 kW at 12–16 mm above the load coil [65]. His findings are very interesting, particularly when compared with those that have been obtained from non-LTE diagnostic studies. Lovett compared rate model populations with LTE populations using a nonequilibrium parameter b_n, which is the ratio of the rate model population to the LTE population. For the ground state this would correspond to the b_1 discussed in preceding sections of this chapter. In general Lovett found "... an increase in the ground-state atom population for elements whose ion ground state is heavily populated'' (that is, for elements with a high degree of ionization) compared to LTE populations. The relative rates for the processes indicated that, in general, these departures from LTE were caused by radiative rates that were competitive with collisional rates for both ion recombination and excited-state depopulation. The most significant excitation processes were found to be electron collisions, although Lovett felt that Penning processes were significant for Be ionization. For all of the elements studied, infrathermal ion–atom emission intensities are indicated ($b_r < 1$). Interestingly, he also found Mg to be slightly underionized relative to LTE, but Ba was found to be slightly overionized relative to LTE. This finding correlates very well with the experimental evidence that has been presented in Table 11.4 and discussed in Section 11.5.4. Ba with a low ionization energy would be expected to behave much like Sr, which was found to be overionized relative to LTE at a similar electron density [67].

More recently Hieftje et al. [40] have used a steady-state rate approach to the study of excitation mechanisms in the ICP. In this paper the authors generate ''product maps'' of expected population distributions by multiplying the appropriate reaction partners to produce product population spatial maps. For example, to test whether electron impact is important for analyte atom excitation Hieftje et al. multiply the concentration profile of ground-state atom (n_1) with that of electrons to provide an expected excited state spatial map due to electron collisions. In a similar manner they tested for radiative and three-body recombination and Penning ionization. Their findings were that atom excitation occurs by ion–electron recombination, whereas ion generation proceeds via Penning ionization in hotter positions of the discharge and by electron impact in the cooler aerosol channel. These results are considered to be speculative at this point.

The use of a CR rate model shows great promise for ICP fundamental studies. The chief difficulties encountered are their complexity, the computational difficulties associated with finding solutions, and the uncertainty in the values of rate constants.

11.7. CONCLUDING REMARKS

During the past few years there has been much progress in the area of fundamental characterization of analytical ICP discharges. The cumulative effort of many different research groups scattered around the world has enabled a rational description of excitation mechanisms that can be used to understand most of the spectral and physical properties of the ICP. It is clear that the ICP is not in LTE. Most workers agree that diagnostic temperature and electron density measurements lead to the conclusion that all of the relevant LTE expressions cannot be satisfied by a single temperature. In addition, the data of Alder et al. [14]

provide evidence that even for a single analyte [Fe I in this case] the excitation temperature varies with the energy level used to measure it. There appears to be a lack of Boltzmann equilibrium within the excitation manifold of atom species. It should be pointed out that there is a general lack of data on level populations, in particular, data that have been subjected to an Abel inversion. It is not certain at this time that non-Boltzmann populations are a general phenomenon, and detailed work on this aspect is warranted.

Recently the use of ion–atom emission intensity ratios as a method of quantifying departures from LTE has come under review [53, 67]. Earlier reports that the experimental ion–atom ratios were from one to three orders of magnitude greater than the corresponding LTE ratios [66] have been questioned, since the electron density values and temperatures used for this calculation were not consistent with use in LTE expressions. The more appropriate use of a temperature derived from electron density measurements yields a different result. Several research groups [1, 52, 53] now advocate the use of either an accurately determined electron temperature or Ar-ionization temperature derived from the measurement of electron density for setting up a hypothetical LTE model. In general, this approach leads to ion–atom emission intensity ratios that are lower than the corresponding LTE values. There are some exceptions to this general finding, specifically for elements with low ionization energies (<6.5 eV) observed at relatively high rf input powers (>1.75 kW). However, while ion–atom emission intensity ratios depart from LTE, the magnitude of this departure is, on the whole, less than a factor of five.

The synthesis of all of the current diagnostic information into a single generalized description of analyte behavior in the ICP strongly suggests that the p-LTE model is adequate to rationalize level population data, ion–atom emission intensity ratios, and the results of rate model calculations. In this model, highly excited states of atom species are in Saha equilibrium with ground state ion species through collisional ionization and recombination with electrons. The lower energy levels of the atom may be overpopulated relative to the Saha level populations, leading to excitation temperatures that are less than ionization temperatures and that vary with energy level. These departures from LTE are most likely caused by lack of collision domination through the influence of radiative processes—both radiative decay of excited states and radiative recombination of ions and electrons.

Lovett [32] has pointed out that the relative populations of any two adjacent levels in any ionization stage are determined by the rates of the processes coupling these levels. The actual population of an excited level (n^*) relative to the ground level (n_1) is given by the ratio of rates of excitation and deexcitation processes. This ratio is given by

$$\left(\frac{n^*}{n_1}\right) = \frac{\mathrm{CE}}{\mathrm{CD} + \mathrm{RD}}$$

where CE is the total rate of collisional excitation, CD is the total collisional decay rate, and RD is the radiative decay rate. At LTE the value would be given by:

$$\left(\frac{n^*}{n_1}\right)_{\text{LTE}} = \frac{\text{CE}}{\text{CD}}$$

Lovett calculates that the ratio of RD to CD for Mg is 0.054, leading to an overpopulated ground state for Mg. Lovett concludes that "deviations from LTE can be expected at normally accepted ICAP temperatures due solely to the contribution of radiative decay."

Overall, then, it would appear that the dominant excitation and deexcitation mechanisms are electron collisions and radiative decay. At this time there is no overwhelming evidence to suggest that Penning ionization or asymmetric charge transfer are dominant excitation–ionization mechanisms.

Further work at this and other laboratories will, no doubt, refine our understanding of this complex source and provide the complete physical understanding necessary to ensure optimized analytical application.

REFERENCES

1. I. J. M. Raaijmakers, P. W. J. M. Boumans, B. van der Sijde, and D. C. Schram, *Spectrochim. Acta* **38B,** 697 (1983).
2. J. Jarosz, J. M. Mermet, and J. Robin, *C. R. Acad. Sci. Ser. B* **27,** 885 (1974).
3. J. M. Mermet, *Spectrochim. Acta* **30B,** 383 (1975).
4. J. M. Mermet and J. Jarosz, *J. Quant. Spectrosc. Radiat. Transfer* **17,** 237 (1977).
5. J. Jarosz, J. M. Mermet, and J. P. Robin, *Spectrochim. Acta* **33B,** 55 (1978).
6. M. H. Abdallah and J. M. Mermet, *J. Quant. Spectrosc. Radiat. Transfer* **19,** 83 (1978).
7. M. H. Abdallah and J. M. Mermet, *Spectrochim. Acta* **37B,** 391 (1982).
8. D. J. Kalnicky, R. N. Kniseley, and V. A. Fassel, *Spectrochim. Acta* **30B,** 511 (1975).
9. D. J. Kalnicky, V. A. Fassel, and R. N. Kniseley, *Appl. Spectrosc.* **31,** 137 (1977).
10. A. Montaser, V. A. Fassel, and G. Larsen, *Appl. Spectrosc.* **35,** 385 (1981).
11. G. R. Kornblum and L. de Galan, *Spectrochim. Acta* **29B,** 249 (1974).
12. G. R. Kornblum and L. de Galan, *Spectrochim. Acta* **32B,** 71 (1977).
13. N. Furuta and G. Horlick, *Spectrochim. Acta* **37B,** 53 (1982).
14. J. F. Alder, R. M. Bombelka, and G. F Kirkbright, *Spectrochim. Acta* **35B,** 163 (1980).
15. H. Uchida, K. Tanabe, Y. Nojiri, H. Haraguchi, and K. Fuwa, *Spectrochim. Acta* **36B,** 711 (1981).

16. H. Kawaguchi, T. Ito, and A. Mizuike, *Spectrochim. Acta* **36B,** 615 (1981).
17. W. H. Gunter, K. Visser, and P. B. Zeeman, *Spectrochim. Acta* **38B,** 949 (1983).
18. L. M. Faires, B. A. Palmer, R. Engleman, and T. M. Niemczyk, *Spectrochim. Acta* **39B,** 819 (1984).
19. M. W. Blades and B. L. Caughlin, *Spectrochim. Acta* **40B,** 579 (1985).
20. G. F. Larson, V. A. Fassel, R. H. Scott, and R. N. Kniseley, *Anal. Chem.* **47,** 238 (1975).
21. T. Edmonds and G. Horlick, *Appl. Spectrosc.* **31,** 536 (1977).
22. G. Dubé and M. I. Boulos, *Can. J. Spectrosc.* **22,** 68 (1977).
23. G. R. Kornblum and L. de Galan, *Spectrochim. Acta* **32B,** 455 (1977).
24. J. A. C. Broekaert, F. Leis, and K. Laqua, *Spectrochim. Acta* **34B,** 167 (1979).
25. G. Horlick and M. W. Blades, *Appl. Spectrosc.* **34,** 229 (1980).
26. H. Kawaguchi, T. Ito, K. Ota, and A. Mizuike, *Spectrochim. Acta* **35B,** 199 (1980).
27. N. Omenetto, S. Nikdel, R. D. Reeves, J. B. Bradshaw, J. N. Bower, and J. D. Winefordner, *Spectrochim. Acta* **36B,** 507 (1981).
28. M. W. Blades and G. Horlick, *Spectrochim. Acta* **36B,** 861 (1981).
29. M. W. Blades and G. Horlick, *Spectrochim. Acta* **36B,** 881 (1981).
30. Y. Nojiri, K. Tanabe, H. Uchida, H. Haraguchi, K. Fuwa, and J. D. Winefordner, *Spectrochim. Acta* **38B,** 61 (1983).
31. M. Mitchner and C. H. Kruger, *Partially Ionized Gases,* Wiley-Intersicence, New York (1973).
32. R. J. Lovett, *Spectrochim. Acta* **37B,** 969 (1982).
33. R. S. Berry, *Radiato. Res.* **59,** 367 (1974).
34. H. Hotop, *Radiato. Res.* **59,** 379 (1974).
35. M. J. Shaw, *Contemp. Phys.* **15,** 445 (1974).
36. J. M. Mermet, *C. R. Hebd. Seances Acad. Sci. Paris Ser. B* **281,** 273 (1975).
37. C. I. M. Beenakker, *Spectrochim. Acta* **32B,** 173 (1977).
38. J. W. Mills and G. M. Hieftje, *Spectrochim. Acta* **39B,** 859 (1984).
39. M. W. Blades and G. M. Hieftje, *Spectrochim. Acta* **37B,** 191 (1982).
40. G. M. Hieftje, G. D. Rayson, and J. W. Olesik, *Spectrochim. Acta* **40B,** 167 (1985).
41. H. W. Drawin, *High Pressure-High Temperatures* **2,** 359 (1970).
42. H. R. Griem, *Plasma Spectroscopy,* McGraw-Hill, New York (1964).
43. P. W. J. M. Boumans, *Theory of Spectrochemical Excitation,* Hilger and Watts, London (1966).
44. L. de Galan, R. Smith, and J. D. Winefordner, *Spectrochim. Acta* **23B,** 521 (1968).
45. C. Park, *J. Quant. Spectrosc. Radiat. Transfer* **22,** 113 (1979).
46. W. Lochte-Holtgreven (Ed.), *Plasma Diagnostics.* North-Holland, Amsterdam (1968).

47. R. Huddlestone and S. L. Leonard, ed., *Plasma Diagnostic Techniques,* Academic Press, New York (1965).
48. H. R. Griem, *Phys. Rev.* **131,** 1170 (1973).
49. H. W. Drawin, *Z. Naturforsch.* **24A,** 1492 (1969).
50. H. W. Drawin, *Z. Physik* **228,** 99 (1969).
51. H. W. Drawin, *J. Quant. Spectrosc. Radiat. Transfer* **10,** 33 (1970).
52. D. C. Schram, I. J. M. M. Raaijmakers, B. van der Sijde, H. J. W. Schenkelaars, and P. W. J. M. Boumans, *Spectrochim. Acta* **38B,** 1545 (1983).
53. B. L. Caughlin and M. W. Blades, *Spectrochim. Acta* **39B,** 1583 (1984).
54. G. R. Kornblum and J. Smeyers-Verbeke, *Spectrochim. Acta* **37B,** 83 (1982).
55. T. Fujimoto, *J. Phys. Soc. Japan* **47,** 265 (1979).
56. T. Fujimoto, *J. Phys. Soc. Japan* **47,** 273 (1979).
57. T. Fujimoto, *J. Phys. Soc. Japan* **49,** 1561 (1980).
58. T. Fujimoto, *J. Phys. Soc. Japan* **49,** 1569 (1980).
59. T. Fujimoto, Y. Ogata, I. Sugiyama, K. Tachibane, and K. Fukuda, *Jpn. J. Appl. Phys.* **11,** 718 (1972).
60. J. J. A. M. van der Mullen, B. Van der Sijde, and D. C. Schram, *Phys. Lett.* **79A,** 51 (1980).
61. H. R. Griem, *Spectral Line Broadening in Plasmas*, Academic, New York (1974).
62. C. R. Videl, J. Cooper, and E. W. Smith, *Astrophys. J. Suppl. Series #214*, **25,** 37 (1973).
63. W. L. Wiese, D. E. Kelleher, and D. R. Paquette, *Phys. Rev.* **A6,** 1132 (1972).
64. M. W. Blades and N. Lee, *Spectrochim. Acta* **39B,** 879 (1984).
65. B. L. Caughlin and M. W. Blades, *Spectrochim. Acta* **40B,** 987 (1985).
66. P. W. J. M. Boumans and F. J. de Boer, *Spectrochim. Acta* **32B,** 365 (1977).
67. B. L. Caughlin and M. W. Blades, *Spectrochim. Acta* **40B,** 1539 (1985).
68. S. R. Koirtyohann, J. S. Jones, and D. A. Yates, *Anal. Chem.* **52,** 1966 (1980).
69. R. Rezaaiyaan and G. M. Hieftje, *Anal. Chem.* **57,** 412 (1985).
70. R. S. Houk and J. A. Olivares, *Spectrochim. Acta* **39B,** 575 (1984).
71. T. Hasegawa, K. Fuwa, and H. Haraguchi, *Chem. Lett.* 2027 (1984).
72. T. Hasegawa and H. Haraguchi, *Spectrochim. Acta* **40B,** 1067 (1985).
73. T. Hasegawa and H. Haraguchi, *Spectrochim. Acta* **40B,** 1505 (1985).
74. G. D. Rayson and G. M. Hieftje, *Spectrochim. Acta* **41B,** 683 (1986).
75. D. R. Bates, A. E. Kingston, and R. W. P. McWhirter, *Proc. R. Soc.* London *Ser. A* **267,** 297 (1962).
76. D. R. Bates, A. E. Kingston, and R. W. P. McWhirter, *Proc. R. Soc. London Ser. A* **270,** 155 (1962).
77. D. R. Bates and A. E. Kingston, *Planetary Space Sci.* **11,** 1 (1963).
78. R. W. P. McWhirter and A. G. Hearn, *Proc. Phys. Soc. London* **82,** 641 (1963).
79. H. C. Srivastava and P. K. Ghosh, *J. Quant. Spectrosc. Radiat. Transfer* **25,** 59 (1981).

CHAPTER

12

STATUS AND TRENDS OF DEVELOPMENT OF ATOMIC SPECTROMETRIC METHODS FOR ELEMENTAL TRACE DETERMINATIONS

J. A. C. BROEKAERT

and

G. TÖLG

Institut für Spektrochemie und angewandte Spektroskopie, Dortmund, Federal Republic of Germany
Laboratorium für Reinststoffanalytik, Max-Planck-Institut für Metallforschung (Stuttgart), Dortmund, Federal Republic of Germany

12.1. Introduction
12.2. Atomic Absorption Spectroscopy
12.2.1. Flame Atomic Absorption Spectrometry
12.2.2. Graphite Furnace Atomic Absorption Spectrometry
12.2.3. Hydride and Cold-Vapor Techniques
12.3. Atomic Emission Spectrometry
12.3.1. Introduction
12.3.2. Classical Sources
12.3.3. Plasma Sources
12.3.3.1. Direct Current Plasmas
12.3.3.2. Capacitively Coupled Microwave Plasmas
12.3.3.3. Inductively Coupled Plasmas
12.3.3.4. Microwave-Induced Plasmas
12.3.3.5. Afterglow Systems
12.3.3.6. Reduced Pressure Discharges
12.3.3.7. Conclusion
12.4. Atomic Fluorescence Spectroscopy
12.5. Laser-Enhanced Ionization
12.6. Coherent Forward Scattering
12.7. Laser Evaporation Methods
12.8. X-Ray Spectrometry

This work has been supported by the Ministerium für Wissenschaft und Forschung des Landes Nordrhein-Westfalen and by the Bundesministerium für Forschung und Technologie.

12.9. Mass Spectrometry for Elemental Analysis
12.10. An Assessment in Comparison to Other Analytical Methods
12.10.1. Introduction
12.10.2. Power of Detection
12.10.3. Multielement Capability
12.10.4. Matrix Effects and Calibration
12.10.5. Distribution (Local) and Microanalysis
12.10.6. The Physical State of Samples
12.10.7. Analytical Precision
12.10.8. Capabilities for Speciation
References

12.1. INTRODUCTION

In this chapter, innovation in atomic spectrometric methods is discussed. First, developments in optical atomic spectrometry and in related areas are treated. This involves a presentation of the figures of merit of (1) new atomic absorption spectrometric techniques using flames or furnaces, (2) atomic emission spectrometry (AES) with plasma sources or glow discharges, and (3) atomic fluorescence spectrometry using lasers. The possibilities of related techniques such as laser-enhanced ionization spectrometry and coherent forward scattering are also evaluated. In the field of X-ray spectrometry, progress is shown to lie on the excitation side. The capabilities of total reflection X-ray spectrometry (TRXRF), proton-induced X-ray emission (PIXE), and synchrotron radiation-excited X-ray spectrometry (SYNFXRF) are discussed with respect to progress in power of detection. The revival of mass spectrometry for inorganic analysis is the result of the availability of reasonably low-priced quadrupole mass spectrometers and of work with new ion sources such as the ICP or glow discharges. However, also related techniques such as laser mass spectrometry and Rutherford backscattering spectrometry must be treated in this connection. The aim of this chapter is to assess the state of the art of the methods listed above, but also to show how, by suitable combinations of existing techniques or by using other types of information, new techniques are being or will be developed. Finally, the methods of atomic spectrometry are compared with other instrumental methods for inorganic analysis, so as to enable the analyst to make critical use of them when developing strategies for solving complex analytical problems.

An entirely objective assessment of the methods on the basis of literature data is hardly possible because of differences in experimental conditions. The analytical figures of merit depend on the very elements studied, the compromise conditions selected for a certain combination of elements in the case of multielement determinations, and, last but not least, on the matrix and the concomitants. Consequently, methods can only be compared objectively on the basis of well-defined analytical problems. Because of the complexity and multitude of methods, these limitations have to be accepted, however, since otherwise it would be impossible to assess the present status in the field.

12.2. ATOMIC ABSORPTION SPECTROSCOPY

Flame atomic absorption techniques are known to be rather free from interferences and it is shown that by appropriate means they can be considerably improved with respect to power of detection. Graphite furnace atomic absorption spectrometry (AAS) still is the most sensitive of the established methods of atomic spectrometry. Indeed, detection limits are in the sub-nanogram per milliliter region [1], while the sample size is at the 10 to 50-μL level. However, matrix effects and interferences, especially at this low level of concentration, are high, and current innovation lies in this field.

12.2.1. Flame Atomic Absorption Spectrometry

In "classical" flame atomic absorption spectrometry (FAAS), the detection limits are at a level of 1–100 ng mL^{-1} [1], being two to three orders of magnitude higher than in furnace AAS. However, several paths can be followed to improve the power of detection.

It was found, for example, that by using slotted-tube atomic trap techniques, the analyte density in the absorption volume of the flame and, accordingly, the power of detection can be considerably increased [2]. The technique has been successfully applied to water analyses [3] and to the analysis of biological samples [4].

Further, the required volumes, which are normally about 2 mL, can be made considerably smaller. This can be achieved with the injection technique as introduced by Sebastiani and Ohls [5] and Berndt and Jackwerth [6]. Here, with sample aliquots down to 50 μL, signals approaching the steady-state values of continuous pneumatic nebulization can be reached. Accordingly, sample consumption and absolute detection limits, but also nebulization effects and risks of clogging, because of the short contact between nebulizer and analyte solution, are lowered. The technique, in fact, is only a special form of flow injection analysis (FIA). The latter will generally improve the analytical performance of all methods where liquid samples are fed to the signal generation unit. FIA facilitates (1) automated analysis including sampling and dilution, (2) work with small-volume samples, and (3) on-line combination of preenrichment and determination. FIA is useful not only for FAAS [7], but also in combination with hydride generation. Furthermore, as for AAS, it also is of interest for plasma AES [8].

Another improvement of the power of detection in FAAS can be attained with wire loop techniques. Their principle goes back as far as the Delves cup technique [9]. However, the application of external heating made them attractive for in situ sample treatment. For the Pt loop, Berndt and Messerschmidt [10] reported an increase by a factor of 10 in power of detection. As recently reviewed by Langmyhr and Wibetoe [11], direct solids sampling in AAS is useful to increase the power of detection but also to avoid a wet decomposition that is labor-intensive and introduces risks of contamination. In FAAS, heating of the sample by irradiation with high-intensity lamps together with decomposition in an oxygen atmosphere has been successfully used for evaporating volatile elements such as Cd, Zn, Pb, and Tl from biological samples [12].

These developments indicate that in an easy and a widespread method of analysis, considerable gain in analytical figures of merit can be achieved by relatively simple approaches.

12.2.2. Graphite Furnace Atomic Absorption Spectrometry

Graphite furnace AAS (GFAAS) is the most sensitive established atomic spectrometric method for a majority of elements except those forming thermally stable oxides and carbides (Ti, Hf, B, rare earths) and nonmetals (for example, C, S, N, O, P, halogens). The detection limits are in the subnanogram per milliliter range. Only for a few elements do other methods yield better values, for example, neutron activation analysis, inverse voltammetry, and chelate gas chromatography. Also because of its relatively simple instrumentation, it should be the ''ideal method'' for elemental trace analysis. However, the euphoria of the early years died down when a lot of interferences and thus systematic errors were found. Such errors arise from the thermochemical processes in the furnace, which are time dependent and overlap in time with the evaporation, dissociation, and radiation absorption phenomena. Accordingly, progress in GFAAS now lies in the domain of analytical accuracy rather than in the domain of power of detection.

A considerable reduction of systematic errors and interferences could be achieved by improving the surface of the graphite furnace, for example, by pyrolytically coating or by introducing some metal oxides and also by using tubes entirely made of pyrolytic graphite [13]. This is mainly due to the reduction of the diffusion of the analyte into the tube material [14].

The reduction of interferences was also achieved by controlling the temperature of the atomizer and especially by realizing a constant temperature across the atomizer. The use of a wire loop [15], a graphite platform, or a cup-in-tube technique [16] makes analyte vapor enter a hot gas and this results in a well-defined atomization, even for direct solids sampling [16]. In this respect double-furnace techniques as proposed by Frech et al. (see, for example, [17]) are now promising because they allow the separate optimization of the analyte volatilization and atomization. Also by using matrix modifiers, such as quarternary ammonium bases, a certain isoformation of samples and a like reduction of matrix interferences can be achieved. When applying such measures, thermochemical processes and temperature profiles in the furnace used have to be well known, as can be concluded from studies by Sturgeon et al. [18] and Chang and Chakrabarti [19], for instance.

Interferences, as far as they originate from background absorption (unspecific for the analyte), can be decreased by physical means rather than by chemical treatment. Background correction systems using continuous radiation sources such as deuterium discharge lamps were introduced by Koirtyohann and Pickett [20] in 1965. They are now standard accessories in GFAAS equipment. However, as discussed by Massmann et al. [21], their application required much experience because they may lead to false corrections in the case of a fine-structured background arising from molecular bands. Moreover, they cannot be applied in the whole wavelength range, and their low radiance necessitates the operation of the primary radiation source at a correspondingly low spectral output, which may limit the power of detection.

The use of the Zeeman effect for background correction in AAS was already proposed by Prugger and Torge [22] in 1969. Zeeman AAS (ZAAS) is based on the splitting of atomic spectral lines into several components under the influence of a magnetic field. Because of the wavelength shift, they are either influenced or not influenced by background absorption. As the σ and π components are polarized in perpendicular planes, they can easily be discriminated by using a polarization system. Several practical realizations are possible and, as discussed by de Galan and de Loos-Vollebregt [23], the differences in realization have not only instrumental, but also analytical implications. Three different setups are commercially available [24] (Fig. 12.1). When the magnet is placed around the primary radiation source, there are no limitations for the dimensions and the shape of the graphite furnace, which is a considerable advantage, for example, for direct solids sampling (case B). However, conventional hollow cathode lamps cannot be used. Since the σ components are shifted away from the original wavelength and the π components remain at the original wavelength, turning the polarizer, as in version A, allows the alternate measurement of the sum of line and background absorption and background absorption alone; then a permanent magnet can be used. In the case of an ac magnet, the background and the line plus background signals can be obtained by measuring the σ component at zero and at maximum field strength (case C). The latter technique yields the highest sensitivity, but might be somewhat more prone to errors in the case of molecular bands that display the Zeeman effect as

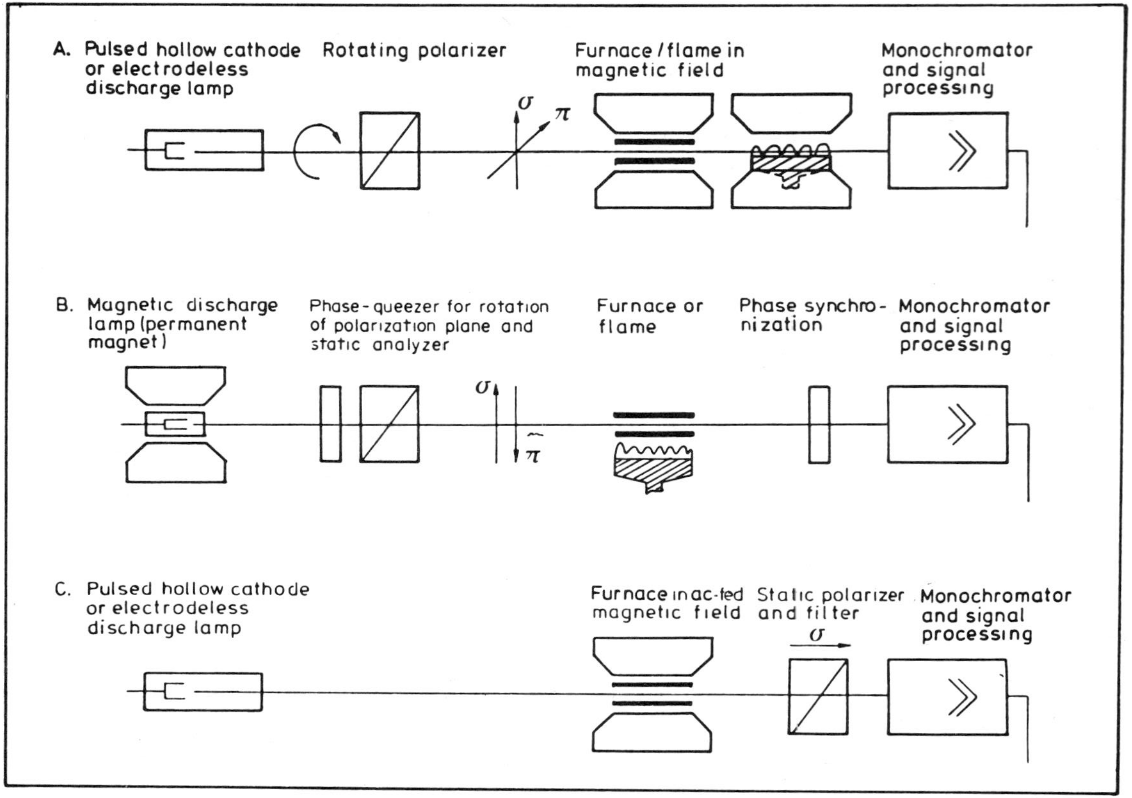

Figure 12.1. Principle of commercial ZAA spectrometers. (*A*) Hitachi, Model 180 Series; (*B*) Erdmann + Grün; and (*C*) Perkin-Elmer, Zeeman 5000 [24]. [Reprinted with permission from J. A. C. Broekaert, *Spectrochim. Acta* **37B,** 65 (1982). Copyright (1982), Pergamon Journals.]

well. Whereas ZAAS does not limit the power of detection in the way D_2-lamp background correction does, and it enables the elimination of a series of background interferences, calibration because of rollover may become difficult [25, 26].

Direct solids sampling in GFAAS first made a breakthrough by the use of ZAAS, since the latter could correct high background interferences. Although direct solid sampling has advantages, it should be borne in mind that difficulties in sample homogenization and limitations in power of detection due to the small amounts of sample used may be disadvantageous. For various elements with simple thermochemical behavior (As, Cd, Cr, Cu, Pb, and Mn), reliable direct determinations in biological materials and soils, for example, have already been reported (see, for example [16]).

Apart from instruments using ZAAS, the technique proposed by Smith and Hieftje, or modifications (see [27]), is now available. They make use of line reversal induced by pulsing the hollow cathode lamp up to high currents; however, they are hampered by losses in analytical sensitivity.

12.2.3. Hydride and Cold Vapor Techniques

Both represent a special combination of chemical separation and preenrichment with a determination by AAS. Higher power of detection can be achieved for elements that form volatile hydrides (As, Bi, Se, Sb, Ge, Pb, and Sn) and Hg. However, high accuracy can only be attained with properly optimized working conditions. Meyer et al. [28], for instance, already reported on the chemical interferences in the determination of Se by hydride AAS. As shown by Piwonka et al. [29], the power of detection can be increased and the interferences can be reduced by properly isolating the hydride (Fig. 12.2). In the case of Se, the hydride was dried, frozen, and subsequently released by heating and swept with He, Ar, or N_2 into the AAS measuring cell. This technique allows an accurate determination of picogram amounts of Se. Thus, the Se content of fragments of a single human hair could be determined with a relative standard deviation (RSD) of 5% and with high accuracy.

Hydride techniques are still in evolution. On the one hand, this leads to a better understanding of the hydride technique, which is helpful for avoiding interferences. Both the kinetics of the hydride formation and the breakdown of the hydride in the atomizer have been studied. Welz and Melcher [30], for instance, concluded that traces of hydrogen catalyze the breakdown. On the other hand, hydride systems will evolve; for instance, continuous hydride systems (see, for example [31]) similar to those used for plasma source atomic emission spectrometry are applied in AAS.

From the achievements discussed it may be concluded that AAS is a powerful method also for extreme trace analysis. However, many problems with

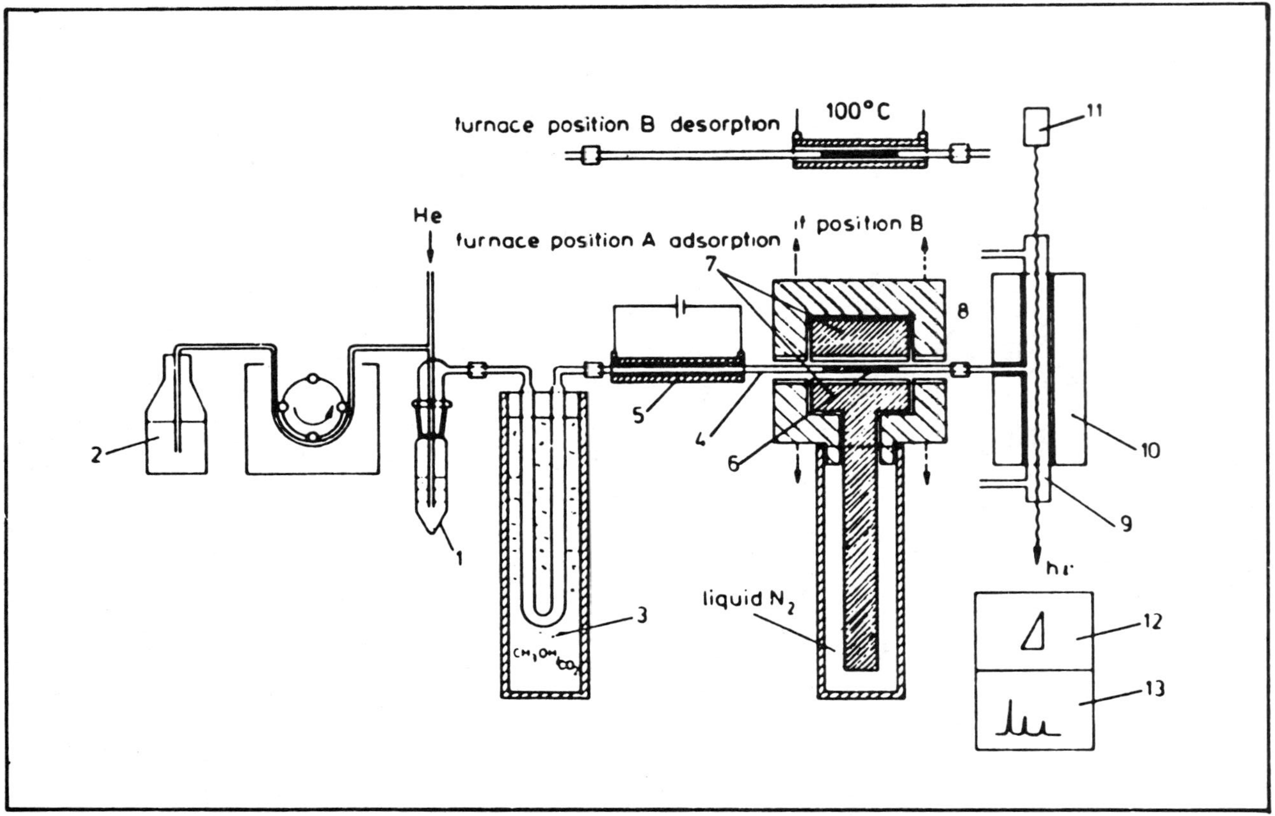

Figure 12.2. Hydride generation/AAS for the determination of Se in biotic matrices after decomposition in $HNO_3/HClO_4$ and preconcentration of H_2Se. (1) Quartz vessel; (2) reduction solution; (3) cold trap (−70°C); (4) quartz adsorption tube; (5) heating system; (6) adsorption region Chromosorb W 30/60; (7) liquid N_2-cooled Al block; (8) insulation; (9) quartz cuvette; (10) furnace; (11) EDL; (12) AA-Spectrometer; (13) recorder [29]. [Reprinted with permission from J. Piwonka, G. Kaiser, and G. Tölg, *Fresenius Z. Anal. Chem.* **321,** 225 (1985). Copyright (1985), Springer-Verlag, Berlin.]

respect to systematic errors remain unsolved. They are not due to imperfection of the instruments, but to the complex thermochemical reactions associated with the signal generation. For avoiding the resulting chemical interferences, the results of investigations of the analyte losses during thermal matrix decomposition are particularly informative (see Krivan and Schmid [32]).

12.3. ATOMIC EMISSION SPECTROMETRY

12.3.1. Introduction

Atomic emission spectrometry (AES) is discussed in Part 1, Chapter 1, and in particular Section 1.7. This section briefly discussed arcs, sparks, glow discharges, hollow cathode discharges, lasers, furnaces, exploding conductors, flames, plasma sources, and combined sources. Also the key literature was covered in that context. For these reasons, the scope of the present discussion is limited.

12.3.2. Classical Sources

Atomic emission spectrometric methods using flames, arcs, and sparks are the oldest atomic spectrometric techniques for multielement analysis. At present, flame emission is used only for the determination of the alkalis.

Direct-current (dc) arc emission spectroscopy developed in the 1950s into a routine method for multielement determinations in solids, in particular nonconducting powders. Universal procedures have been described by Addink [33], Kroonen and Vader [34], and Harvey [35] and are used for (semi-)quantitative analysis of a broad variety of materials. As shown in a review by Avni [36], dc arc spectrography permits trace determinations in the microgram per gram range, even in difficult matrices such as U_3O_8.

Spark atomic emission spectrometry is widely used for rapid multielement analyses, as required for process control. In the steel industry [37], for example, it is very important for both process control and product analysis. Nevertheless, for elements with an atomic number ≥ 10, X-ray fluorescence spectrometry has become at least equally important. Recent developments are the high-repetition rate spark and high-energy prespark techniques, which provide for large sampling areas and, thus far, improved accuracy. Since also large amounts of material enter the spark, a better power of detection is attained. Fundamental work in this field is performed by Walter's group [38, 39]). Also pulse distribution spark analysis, which was found useful for speciation of Al in steel [40], is of topical interest.

12.3.3. Plasma Sources

Since the mid-1970s AES experienced a remarkable revival. This was mainly due to the development of plasma sources for multielement determinations in liquid samples. The inductively coupled plasma (ICP), the direct current plasma (DCP), the microwave-induced plasma (MIP), and the capacitively coupled microwave plasma (CMP) now are of paramount importance for trace analysis. The principles of these sources have been discussed in detail in Part 1, Chapters 2 and 3. In the present context, the four sources and their main analytical characteristics are featured in Fig. 12.3.

12.3.3.1. Direct Current Plasmas

The development of the DCP, which has been reviewed in Part 1, Section 2.2, began with the first efforts of Margoshes and Scribner [42] and Korolev and Vainshtein [43]; it reached a final stage in the commercial three-electrode DCP. In this version, predominantly atom lines are excited and the detection limits using pneumatic nebulization are below 10–100 ng · mL^{-1}. This applies to the

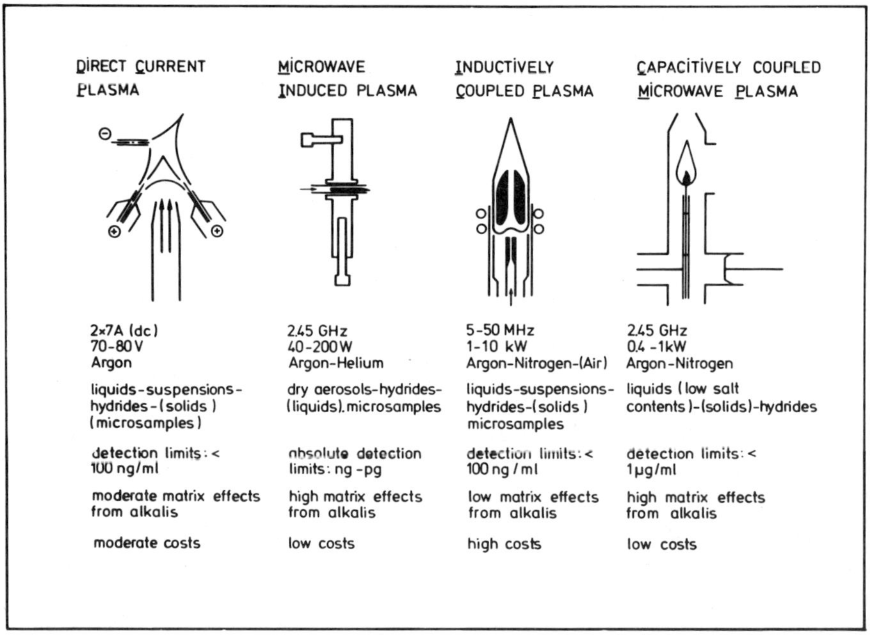

Figure 12.3. Principle and analytical figures of merit of plasma sources for optical emission spectrometry [41]. [Reprinted with permission from *Trends Anal. Chem.* **1**, 249 (1982). Copyright (1982), Elsevier, Amsterdam.]

DCP in combination with a high-resolution echelle spectrometer. Here, due to high radiant densities, detector noise limitations at short UV-wavelengths may be somewhat lower than in the case of an ICP–echelle spectrometer [44], and accordingly, the DCP detection limits become comparable to those of ICP–AES. Varying amounts of alkalis in the samples cause high matrix effects as compared to the ICP. DCP–AES is used as a routine technique for the analysis of widely differing samples such as water (for example [45]) and various industrial products [46].

12.3.3.2. Capacitively Coupled Microwave Plasmas

The principle and history of the single electrode CMP have been discussed in Part 1, Section 2.3. The detection limits attained with this source are rather poor because the aerosol does not penetrate the plasma efficiently (see for example [47, 48]). Matrix effects from alkalis are severe [49–51]. However, these interference effects can be circumvented by the use of excess alkali as buffer [52, 53]. The use of a Cs buffer also leads to better detection limits. At present, however, the CMP is hardly used for practical analysis.

12.3.3.3. Inductively Coupled Plasmas

Since the ICP forms the main subject of the two parts of this work, the present discussion primarily serves as a key to topics treated in the previous chapters and points out some features of direct interest in the present context.

The history and principles of conventional Ar and N_2–Ar ICPs was considered in Part 1, Section 3.2, while the performance was outlined in Section 3.3. The performance characteristics were considered in a broader context in Part 1, Chapter 4, where concepts such as detection limits, precision, limits of determination, accuracy, dynamic range, and multielement capability were treated and numerical values of figures of merit were stated. Chapter 4 also covers ICP optimization and includes a discussion on the significance of the solvent load for when organic solvents are fed into an Ar ICP (Section 4.7.8). That section deals with recent insights derived from the work of Maessen et al. [54]. Section 4.9 summarizes the results of new systematic studies of ICPs in which Ar is partly or wholly replaced by molecular gases. One of the targets of such studies is the reduction of the Ar consumption to reduce the running costs. In the majority of the efforts aiming at this target, however, Ar was maintained for all gas flows, but the outer flow was decreased by redesigning the torch. This subject is detailed in Part 1, Chapter 5. More recently, Chan and Montaser [55] reported the successful operation of an annular He ICP, which shows promises for the determination of halogens.

ICP–AES using pneumatic nebulization has become a routine analytical technique in many fields. Chapters 1–6 of this book, dealing with applications in the various fields, testify to this.

Efforts have been made to improve the performance of ICP–AES by (1) designing devices for microvolume samples, (2) using alternative methods of aerosol production to raise the detection power, and (3) combining the ICP with devices that permit direct solids sampling. (Fig. 12.4). The various approaches have been essentially discussed in Part 1, Chapter 6, and Part 2, Chapter 7.

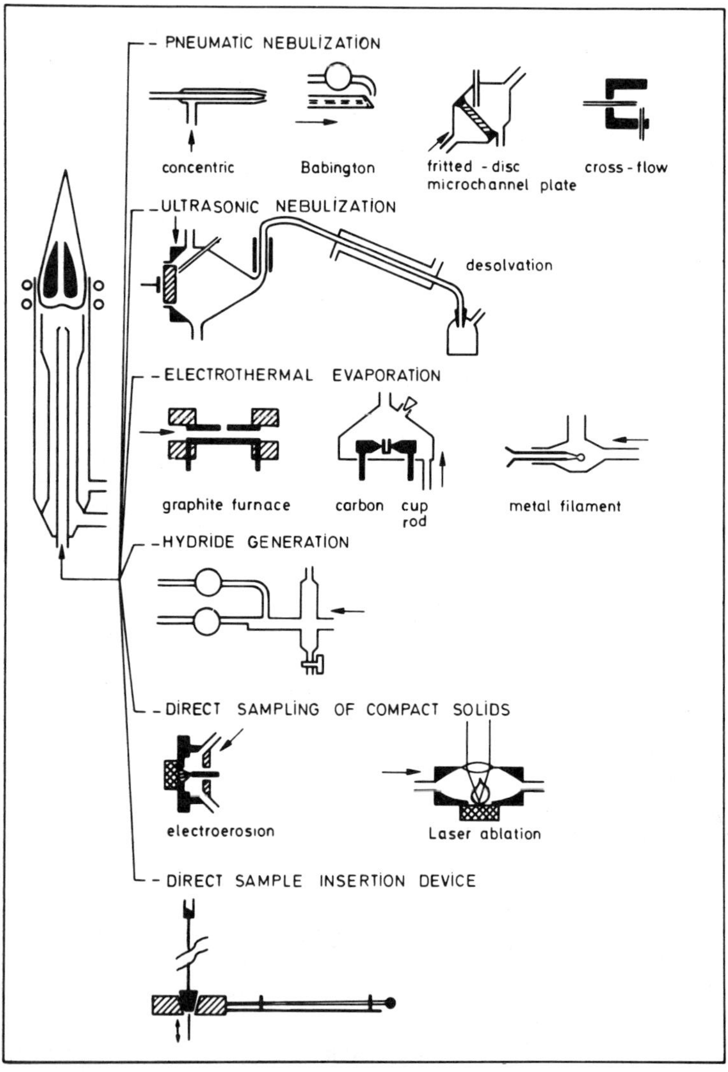

Figure 12.4. Different techniques for sample introduction in ICP-OES.

Those chapters cover, inter alia, the injection technique, electrothermal nebulization devices, hydride techniques, and direct solids sampling. Tables 6.1–6.10 in Part 1 provide illustrative examples of the results achived. The various developments indicate that appreciable improvements in analytical performance can be achieved by the use of special devices and/or techniques.

Further improvements in ICP–AES can be expected from progress in the acquisition and handling of spectral information. Part 1, Chapter 7, dealt with spectral interferences and line selection. In that context, the benefits of high spectral resolution were covered (Section 7.7). The prime benefit stems from the increase in selectivity, as has been shown by Botto [56] and Boumans and Vrakking [57, 58]. Improved selectivity manifests in lower limits of determination and also in lower limits of detection, if the latter are defined in such a way that the errors due to interfering signals other than smooth background are correctly covered. This point is explicitly treated in Boumans' recent review article: "A century of spectral interferences in atomic emission spectroscopy—Can we master them with modern instruments and approaches?" [59].

Conventionally, high-resolution spectra are obtained with wavelength-dispersive spectrometers using gratings with a high groove density or echelles (see Part 1, Section 8.6). As an alternative, high-resolution Fourier transform spectrometry has been proposed by Faires et al. [60, 61]. This approach has many potentials, but it requires further investigations before a more definitive assessment can be made.

Further progress in signal acquisition can be achieved by the use of multichannel detectors, so that the spectral environment of a line can be recorded at various wavelengths simultaneously. In particular, this may improve the quality of background correction, as can be judged from the work done with a silicon intensified target (SIT) vidicon by Furuta et al. [62] as well as from records of complex ICP spectra in Fig. 12.5. We can expect here a new type of sequential ICP spectrometry (cf. Part 1, Section 8.11), provided that the limitations of detector noise are overcome, as seems to be the case with a new type of a silicon photodiode array incorporating a microchannel plate [63]. The simultaneous intensity measurements at different wavelengths achieved with array detectors potentially provide for fully exploiting the advantages of internal standards, as discussed in Part 1, Section 4.2.4.

Finally, quantitative spectral analysis, in the sense proposed by Taylor and Schutyser [64] (cf. Part 1, Section 7.5), has the potential to bring substantial improvements in accuracy, since it makes background correction far less dependent on the sample composition and the operator's choice of the wavelengths at which the background is measured.

12.3.3.4. Microwave-Induced Plasmas

The MIP is covered in Part 1, Section 2.4. Among the recently described applications, we mention the combination of the MIP with a device for electro-

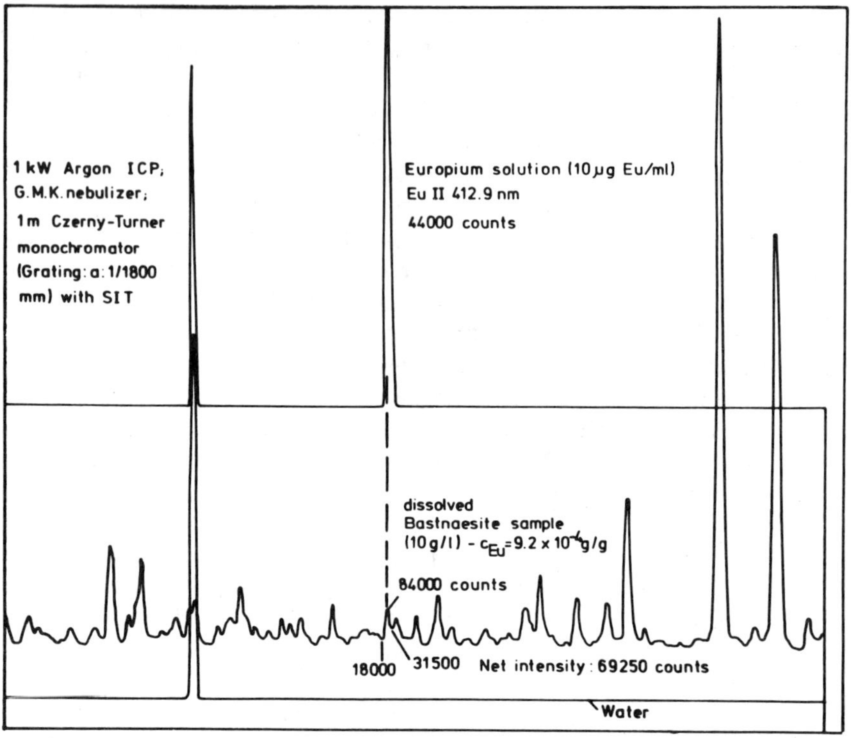

Figure 12.5. Record of ICP spectra obtained with a SIT spectrometer [63].

thermal evaporation, such as a graphite furnace (Fig. 12.6) [65] or a wire loop [66]. With such systems multielement determinations are possible also in real samples, in the same way as with systems incorporating an ICP (see Part 1, Section 6.4). With the aid of electrodeposition on graphite electrodes, ultratrace analyses are possible [67]. In the case of Hg, picogram amounts can be detected by reduction to metallic Hg, collection on a gold wire, and subsequent sweeping into the MIP [68]. Since He can be used as working gas, S, P, and the halogens can also be detected by MIP–AES [69], which makes it attractive for element-specific detection in gas chromatography, as indicated in Part 1, Section 2.4, where the relevant key literature is also stated.

With the aid of new tube configurations, for example, those described by Bollo-Kamara and Codding [70] or Kollotzek et al. [71, 72], multifilament and toroidal MIPs could be realized. The latter accept wet aerosols, but the detection limits are poorer and the matrix effects higher than for ICPs [72]. However, they are useful as element-specific detectors in liquid chromatography, as illustrated by the determination of various Hg species in soils (Fig. 12.7) [73]. Coupling these new MIPs with other aerosol generation devices, such as de-

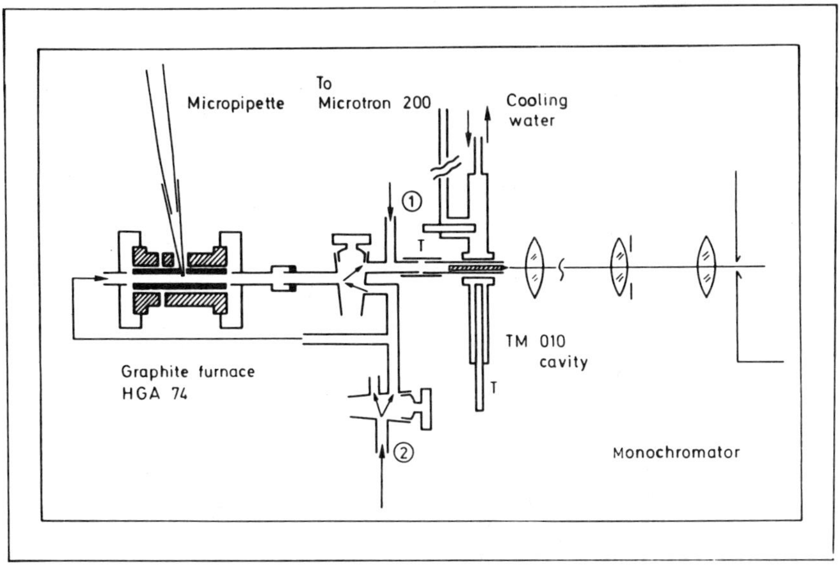

Figure 12.6. Graphite furnace MIP instrumentation. (1) Plasma gas; (2) carrier gas; (*T*) tuning [65]. [Reprinted with permission from A. Aziz, J. A. C. Brockaert, and F. Leis, *Spectrochim. Acta* **37B,** 381 (1982). Copyright (1982), Pergamon Journals.]

vices for hydride generation, electrothermal evaporation, or electroerosion, could lead to further improvements in power of detection, because of possible improvements in aerosol sampling efficiency compared to earlier MIPS.

12.3.3.5. Afterglow Systems

Other plasma discharges at atmospheric pressure have been described in the literature and were applied for special purposes. In this respect afterglow systems [74], which may be useful for element-specific detection in gas chromatography, should be mentioned.

12.3.3.6. Reduced Pressure Discharges

Apart from discharges at atmospheric pressure, discharges at reduced pressure have gained considerable interest for spectrochemical analysis. As already discussed by Mandelstam and Nedler [75] and later by Falk [76], these sources benefit from the absence of local thermal equilibrium (LTE). It is known that the electron temperatures (≥ 5000 K) by far exceed the gas temperatures (≤ 1000 K). Perhaps also because of a non-Maxwellian energy distribution, electrons with high energies are available; therefore lines with high excitation energies

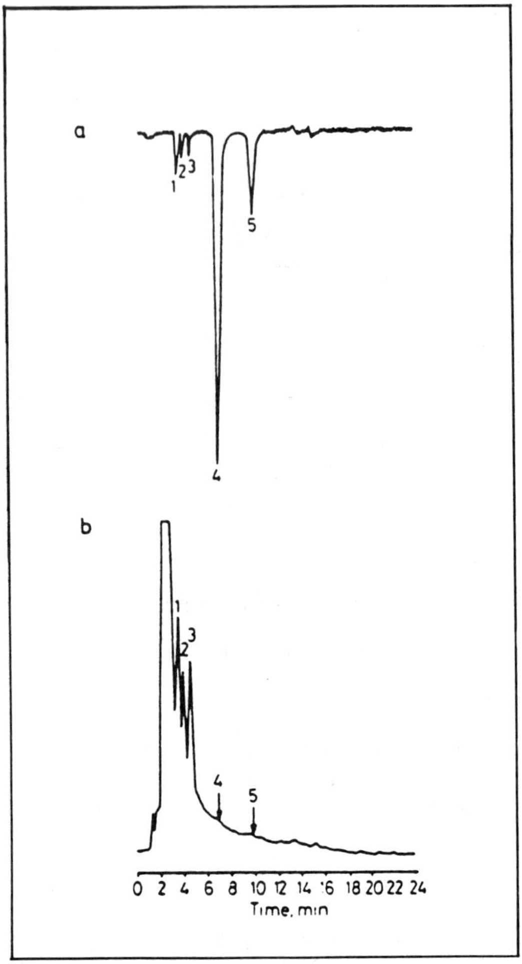

Figure 12.7. Element-specific detection in HPLC by MIP–OES. Chromatographic separation of different Hg species after their extraction with ethanol from Hg-contaminated highland peat bog soil. Column: RP6, 5 μm 250 mm × 4.6 mm i.d. Mobile phase: water/methanol (40/60 v/v), flow rate 1 mL·min. (*a*) MIP detection; Hg I 253.6 nm. (*b*) UV–vis detection: absorption at 250 nm, reference wavelength 430 nm [73]. [Reprinted with permission from D. Kollotzek, D. Oechsle, G. Kaiser, P. Tschöpel, and G. Tölg, "Application of a Mixed-Gas Microwave Induced Plasma as On-Line Element-Specific Detector in High-Performance Liquid Chromatography," *Fresenius Z. Anal. Chem.* **318,** 485 (1984). Copyright (1984), Springer-Verlag, Heidelberg.]

are well excited while the spectral background remains low. Accordingly, better power of detection compared to sources in LTE may be achieved.

For this reason hollow cathode sources have been used. For the determinations of volatile elements with high excitation energies (P, S, Se, As, Tl ...) at sub-microgram per gram concentrations in high-temperature alloys, the hot hollow cathode is very useful, as shown by Berglund and Thelin [77].

In the glow discharge lamp introduced by Grimm [78] for the analysis of compact solid samples, the sample is volatilized by cathodic sputtering and the analyte material is excited in the negative glow. As shown by Dogan et al. [79, 80] and many subsequent investigations in Laqua's group (see, for example [81, 82]), the technique has low matrix effects and in some cases is a viable alternative to X-ray fluorescence spectrometry (XRFS). The state-of-the-art detection limits may be higher than for the spark; however, improvements by the use of additional analyte excitation by superimposed dc, high-frequency or microwave discharges can be expected. As shown by Berneron [83], the glow discharge lamp may be of use for in-depth profiling. In contrast to Auger and secondary ion mass spectrometry, the in-depth resolution is poorer but the penetration rate is higher (50 nm $\cdot$ s^{-1}). The technique is suitable for the control of coatings [84]. The power of detection, the precision, and the in-depth resolution are interrelated [85].

Graphite furnace emission spectrometry, as explored by Littlejohn and Ottaway [86] in the 1970s, first became of real use after the furnace was operated under reduced pressure, while an additional discharge was applied for analyte excitation, as realized by Falk et al. in furnace atomic nonthermal excitation spectrometry (FANES) (see [87]). Graphite furnace atomic emission spectrometry at atmospheric pressure permits simultaneous multielement determinations, unlike GFAAS, but, since the excitation of atomic lines at 3000 K is not very efficient and the background intensities are high, the power of detection is poor. Moreover, a separate optimization of volatilization and excitation conditions is not possible. In FANES, the volatilization takes place in the furnace, as in GFAAS, and the analyte is excited in a non-LTE plasma, which is built up between the furnace and a supplementary electrode. Owing to the nonthermal excitation, detection limits down to the picogram range are obtained [88]. However, up to now it is not known to what extent easily ionizable elements will disturb the nonthermal excitation and influence the power of detection or cause high matrix effects.

12.3.3.7. Conclusion

As a result of the development of new excitation sources, the concomitant improvement of the spectrometers and the use of advanced systems for sample introduction, the power of detection, the capability for microanalysis, and the accuracy in AES has improved considerably since the early 1970s. This trend can be expected to continue for some time, while the multielement capability of the method is maintained or even improved.

12.4. ATOMIC FLUORESCENCE SPECTROSCOPY

The fluorescence techniques combine the advantages of the large dynamic range of emission techniques with zero-background or high selectivity of absorption techniques. Therefore, their potential use has long been propagated for analytical purposes; their actual use is rather limited, however.

Flames have been extensively used as atom reservoirs. For elements with refractory oxides, however, the ICP was found to be a better atom reservoir for atomic fluorescence spectroscopy (AFS).

A system for hollow cathode lamp excited ICP–AFS, as proposed by Demers and Allemand [89], is now commercially available as a modular simultaneous multielement ICP system. By using ultrasonic nebulization and adding propane gas to the Ar carrier gas, its analytical performance could be considerably improved [90].

It has also been proposed to use an ICP in which a concentrated solution of the elements to be determined is introduced as primary radiation source and a second ICP as atom reservoir in which the sample solution is introduced [91]. The so-called ASIA (atomizer, source, ICPs in AFS) is shown by Greenfield [92] to be a viable alternative to emission spectrometry to alleviate spectral interferences that may occur in complex sample matrices, without the need for an expensive, high-resolution monchromator.

As discussed by Omenetto and Winefordner [93], however, saturation of the fluorescence level is required to reach a maximum radiance of fluorescence and to realize the full power of detection of the method. This can only be achieved with monochromatic primary sources with high radiant output like the laser. As tunable lasers have become available, it was possible to improve the power of detection compared with that obtained with continuous sources (for example, Xe lamps) or line sources such as hollow cathode or electrodeless discharge lamps [94] by orders of magnitude. To cover the whole analytical wavelength range, tunable dye lasers with different dyes have to be used. They are pumped in some cases with excimer lasers. Frequency doubling and wavelength shifting have to be applied to get access to short UV wavelengths [95]. The drawbacks of laser AFS lie in the facts that up to now a random access to any wavelength is still troublesome and that the use of lasers is at the expense of the multielement capability.

In addition to flames and ICPs, furnaces can also be applied as atom reservoirs. Then, extremely low detection limits can be obtained. Bolshov et al.

[96], reported values in the picogram per milliliter range. Matrix interferences may be expected to be much higher than with an ICP as atom reservoir. However, recent work shows that atomization in vacuum appreciably reduces the matrix effects [97].

The lowest detection limits are obtained in the case of nonresonance fluorescence [98]. Then, stray radiation from the exciting source is no longer limitative. The detection limits of laser-excited atomic or ionic nonresonance ICP–AFS for elements such as Ga, Pb, Si, Sn, and Tl already are considerably lower than those of ICP–AES [99]. As wavelengths below 230 nm become accessible, a considerable gain in power of detection for elements such as As, Se, and Cd, which have their resonance lines in this range, becomes possible. Laser induced double-resonance ionic fluorescence provides the highest spectral selectivity [100]. At the present state of the art, however, such techniques serve for basic studies rather than for analytical work.

12.5. LASER-ENHANCED IONIZATION

Laser enhanced ionization (LEI) techniques are based on the detection of analyte ions produced by ionization from laser-excited states. The power of detection of LEI is extremely high. The excitation of analyte atoms is highly selective and, furthermore, the ions produced can be detected with a higher efficiency than radiation because there is no solid-angle limitation. Flames [101, 102], furnaces [103], and plasmas [104] have been used as atom reservoirs. In flame LEI, the detection limits of alkalis, for example, are at the picogram per milliliter level [101]. In recently published applications, it was shown that by flame LEI submicrogram per gram concentrations of Tl in Pb can be directly and accurately determined [105]. It has also been shown that ionization is more effective with continuous wave than with pulsed lasers. Further, as shown by an example of biological samples [106], matrix effects caused by alkalis are lower for continuous-wave lasers. The method may become more attractive when the expensive tunable lasers can be replaced by semiconductor lasers with a sufficient radiant output. The selectivity of LEI can still be considerably improved by Doppler-free spectroscopy. Then, not only the power of detection is further increased, but it also becomes possible to detect the single isotopes of an element, as shown by Niemax and co-workers (see [107]) (Fig. 12.8). This offers potentials for calibration by isotopic dilution in optical atomic spectrometry.

A further increase in sensitivity may be expected from the use of a thermionic diode instead of optogalvanics for detection. The thermionic diode was described 60 years ago by Hertz [108] and Kingdon [109]. Basically it consists of a heated tungsten wire acting as a cathode surrounded by a metallic cylinder.

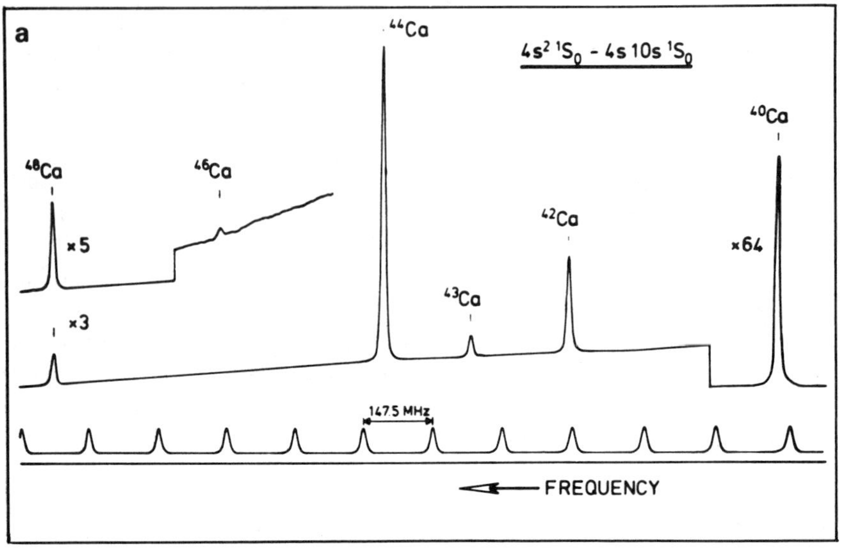

Figure 12.8. Doppler-free 2-photon spectrum of the $4s^2\ {}^1S_0 - 4s\ 10s\,{}^1S_0$ transition in the natural abundant isotopes of Ca [107]. [Reprinted with permission from K. Niemax, "Spectroscopy Using Thermionic Diode Detectors," *Appl. Phys. B* **38,** 147 (1985). Copyright (1985) Springer-Verlag, Heidelberg.]

Ions produced by laser action enter the negative space charge region of the cathode and recombine with electrons, lowering the potential wall around the cathode and inducing a diode current. This method of detection, as investigated by Niemax and co-workers (literature is included in [107]) is far more sensitive than optogalvanic detection. In experiments where analyte material was evaporated electrothermally and selectively excited by laser radiation, one was able to detect 10^3–10^4 atoms cm^{-3} in the vapor cloud. The practical realization of this principle for ultratrace analysis now centers on suitable analyte introduction systems and atom reservoirs for thermionic detection. Also, a thorough characterization of noise sources and a study of the effects of concomitants are necessary.

12.6. COHERENT FORWARD SCATTERING

Coherent forward scattering (CFS), as explored for atomic spectrometry [110, 111], is based on the measurement of stimulated emission of resonance radiation. The set-up consists of a primary radiation source and an atom reservoir placed in a strong magnetic field. When the magnetic field direction is perpendicular to the optical axis, the scattered radiation is linearly polarized (Voigt effect); when the field is parallel, circular polarization takes place (Faraday effect). Owing to this magnetic-optical effect, which is strong at resonance wavelengths, the scattered radiation can be isolated by placing the atom reservoir between crossed polarization filters.

The technique, just as AFS and LEI, has the high dynamic range of emission techniques and the zero-background of AAS methods. The spectra show virtually only resonance lines. This is a great advantage compared with AES, where spectra may be very line rich. Accordingly, a spectral apparatus of low practical resolving power can be used. A flame, a furnace, and, in principle, also a plasma can serve as atom reservoir and hollow cathode lamps or a Xe lamp as primary source. With a Xe lamp, simultaneous multielement determinations are possible, as shown by Debus et al. [111], who used a SIT vidicon in a CFS experiment. The detection limits are in the nanogram per milliliter range if a furnace is the atom reservoir. The limiting source of noise has not yet been established and consequently it is not clear whether the detection power can be increased by using more intensive primary sources such as tunable dye lasers. The technique has high analytical sensitivity because the signal grows quadratically with concentration. Moreover, it has been postulated that molecular species do not produce a measurable CFS signal. If this can be substantiated, then, matrix effects would be lower than in GFAAS. Because CFS can be done with the same setup as used for ZAAS, a combined instrument could provide for a highly reliable system for trace determinations.

12.7. LASER EVAPORATION METHODS

As shown in Laqua's review [112], solid-state lasers (e.g., Nd–YAG) are suitable for the evaporation of both conducting and nonconducting samples. As the diameters of the craters produced lie in a range between 20 and 80 μm for laser pulse energies between 0.1 and 2 J, they can be used for local analysis. Conventionally, the radiation from the vapor plume produced by the laser is unsuitable for analysis because of its high optical density (cf. Part 1, Section 1.7.6). This situation could be different for high-power electronically switched lasers. Laser ablation might become more attractive when combined with a low-cost MIP for additional excitation of the vapor cloud or in combination with atomic fluorescence or LEI detection. A considerable gain in power of detection for atomic spectrometric analysis of solids can be expected here. The laser not only has promises as an atomizer in optical spectrometry, but also as an expedient for ion production in mass spectrometry [113, 114].

12.8. X-RAY SPECTROMETRY

Apart from developments in the instrumentation of the well-known wavelength-dispersive [115–118] and energy-dispersive XRFS, there is no revolutionary innovation in this field. The lowest detection limits are obtained for elements with an atomic number in the vicinity of 30. Here detection limits of about 0.1 μg are normally obtained. However, as shown for energy-dispersive XRFS with direct excitation, it is possible to reach the picogram level [119]. The detection limits worsen toward both higher and lower atomic numbers. In the upward direction, it is the decrease in energy of the characteristic X-ray radiation that is responsible for this; in the downward direction, it is due to the transition from the K series to the L series. The main limitation of the power of detection, however, lies in the Compton background. The latter increases with the penetration depth of the exciting radiation in the sample. Furthermore, Compton background radiation not only limits the power of detection, but also contributes to matrix effects. Therefore, calibration in XRFS methods requires calibration samples having a matrix matched to that of the analysis samples or extensive software corrections for eliminating matrix effects (for key literature see, [116]).

The limitations of the power of detection by Compton background radiation can be reduced by mainly three new developments. Total reflection X-ray fluorescence spectrometry (TRXRF) has been introduced by Aiginger and Wobrauschek [120] and has been refined, chiefly by Schwenke and Knoth [121]. In this technique, the incident angle of the exciting beam is made small and total reflection takes place. When thin sample films are deposited on a target with very low sample roughness, one succeeds in lowering the absolute detection limits of energy-dispersive XRFS by a factor of up to 1000. This is possible by the drastic reduction of the Compton background, since the exciting beam hardly penetrates into the target. An extension of this approach is multiple total reflection of the exciting beam (Fig. 12.9) [122].

The main difficulty of the technique at present is the optimal preparation of microsamples. It was found that the problem already begins with rinsing the quartz targets used. The latter could be successfully covered homogeneously by evaporation of the solution or by electrodeposition [123]; however, histological slices can also be analyzed directly [124].

The power of detection of TRXRF spectrometry is at least 1 or 2 orders of magnitude better than that of proton-induced X-ray emission (PIXE) spectrometry (see, [125]). The detection limits of PIXE are lower than those of normal energy dispersive XRFS; this is so because the penetration depth is smaller while the proton beam can be focussed to cross sections as small as 1 μm^2. The most important advantage of PIXE is its capability for microdistribution analyses. Thus, for example, element distributions across the cross section of hair segments can be measured [126].

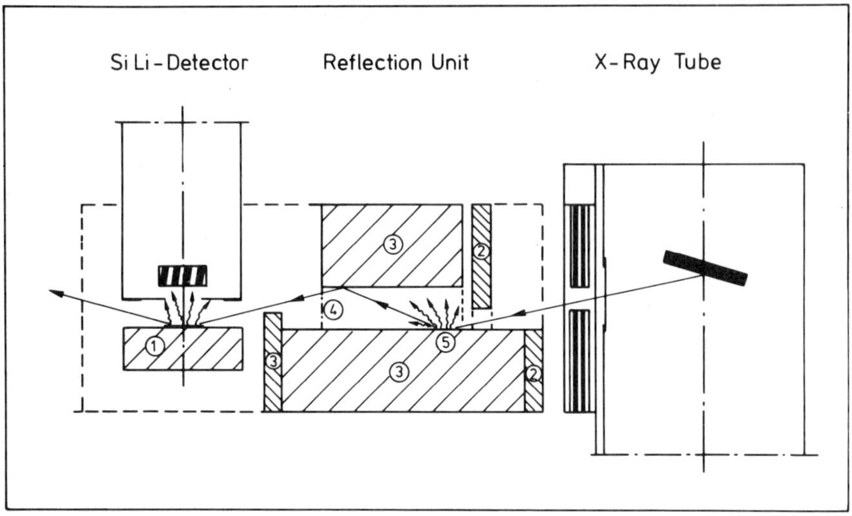

Figure 12.9. Multiple reflection X-ray fluorescence spectrometer. (1) sample, (2) diaphragms, (3) reflectors, (4) spacer (20 μm) [122]. [Reprinted with permission from H. Schwenke and J. Knoth, "High-Sensitivity Multielement Trace Analysis Using EDXRF Spectrometry with Multiple Total Reflection of the Exciting Beam," P. Brätter, and P. Schramel, eds., in *Trace Element Analytical Chemistry in Medicine and Biology*, p. 307. Copyright (1980), Walter de Gruyter, Berlin, New York.]

Synchrotron-charged particle beams can also be used for the excitation of X-ray spectra. The absolute detection limits obtained are 10^{-12}–10^{-10} g [127, 128] whereas beam diameters down to 10 μm can be produced.

The three techniques, however, still have to be more thoroughly studied to permit an assessment. Nevertheless, examples in the literature indicate that the power of detection, the multielement capability, and the accuracy are high.

12.9. MASS SPECTROMETRY FOR ELEMENTAL ANALYSIS

Mass spectrometry (MS) for inorganic analysis has the following interesting features.

1. It is the only universal multielement method, since it permits the sequential or simultaneous determination of all elements and their isotopes in both solids and liquids.
2. The detection limits for virtually all elements are low.
3. It can be more easily applied as an absolute method compared with other

spectroscopic methods, because the analyte atoms produce the analytical signal themselves; their presence is thus not indirectly derived from their emitted radiation.

4. The spectra are simple compared to the line-rich spectra often found, for instance, in AES.
5. The resolving power provided by conventional instruments is sufficient to separate all isotope signals and, therefore, high-precision results can be achieved by isotope dilution techniques.

In conclusion, inorganic mass spectrometry can be used for a broad variety of tracc analysis problems, as is evident from recent reviews [129, 130]. However, conventional double-focussing mass spectrometers with arc, spark, or laser sources and photographic detection are complex and expensive. Moreover, for many analytical problems the analytical precision and accuracy are insufficient. The technique has proved useful for the analysis of high-purity materials, for instance, where very low impurity concentrations in relatively simple matrices of metals or semiconductors must be determined. For some years, there has been a considerable development in mass spectrometry for inorganic trace analysis. On the one hand, this is due to the fact that powerful but less expensive mass spectrometers have become available. On the other hand, excitation sources from AES, such as the ICP and the glow discharge, have been successfully applied as ion sources for mass spectrometry.

Thermal ionization MS, as applied by Heumann and co-workers [131, 132], and field desorption techniques, as investigated by Schulten et al. [133], have a very high absolute power of detection.

For analysis of solutions, ICP–MS (see Part 1, Section 3.5) is very promising. Detection limits for most elements are in the 0.1 to 1 $ng \cdot mL^{-1}$ range, which is two orders of magnitude lower than for ICP–AES, where in addition, there are differences of up to five orders of magnitude from element to element. Furthermore, high-precision analyses are possible with the aid of isotope dilution. ICP–MS, though still troublesome for various types of samples, is maturing as an analysis technique [134]. However, less expensive plasma sources might become attractive for MS. For a number of years, MIP has been used as ion source for MS [135] and the improved analytical figures of merit of new MIP types justify its reevaluation. Also, just as in plasma source AES, special techniques for work with microsamples will be developed for plasma source MS.

For the direct analysis of solid samples, glow discharge mass spectrometry (GD–MS) [136] and mass spectrometry of sputtered neutral atoms (SNMS) [137] are of interest.

Hollow cathode glow discharges have since long been proposed by Mattson et al. [138] as ion sources for MS. Really steady-state sample volatilization conditions can be more easily realized in glow discharges with a plane cathode such as that developed for AES (Section 12.3.3.6). If suitable interfacing is employed, the mass spectra obtained show little background from clusters [139]. It may be expected that detection limits in the microgram per gram range will be obtained for most elements.

Mass spectrometry of sputtered neutral particles has been developed by Oechsner et al. [137]. A first system makes use of an ion source and an RF plasma for ionization of extracted sputtered neutral particles. This version is especially suited for in-depth analysis (resolution, 1 nm) and seems to be more sensitive than secondary ion mass spectrometry (SIMS). In another version, both sputtering and ionization of the sputtered material is achieved with the rf plasma. This version is suitable for bulk analysis; then, detection limits are below 1 $\mu g \cdot g^{-1}$ for most elements and matrix effects seem to be very low.

Laser mass spectrometry [113] as a technique for direct analysis of solids has already been mentioned in Section 12.7. Commercially available is the laser micro mass analyzer (LAMMA). This instrument permits the determination of absolute amounts down to 10^{-18} g [114,130]. To a certain degree, even speciation analysis at a microscale is possible. This technique is certainly one of the most sensitive methods of atomic spectrometry. However, as holds for other microprobe surface analysis methods, quantitative analyses are still difficult.

Techniques such as Rutherford backscattering spectrometry and Auger electron spectroscopy are also based on atomic spectrometric principles and are of use for inorganic analysis. However, they are typically surface analysis techniques, which are beyond the scope of this chapter (for a review, see [140]).

12.10. AN ASSESSMENT IN COMPARISON TO OTHER ANALYTICAL METHODS

12.10.1. Introduction

It is relevant to compare the analytical figures of merit of the atomic spectroscopic methods discussed here with those of other methods presently available to the analyst. A rigorous, unambiguous, and unbiased comparison is virtually impossible. However, it may be useful to list the order of magnitude of the power of detection, to give an indication about the matrix effects, assess the multielement capability, and make some statements about the possibilities for local, micro-, and in-depth analysis, high-precision analyses, and speciation, and about the state of the sample (liquid, solid, gas). Such a listing is provided in Table 12.1, which is commented upon in the following sections.

Table 12.1 Evaluation of the Analytical Capabilities of Some Important Methods for Inorganic Analysis

Method	c_L (Order of Magnitude)			Matrix Effects	Multielement Capability	Micro(M), Local(L) in-Depth(D) analysis	Solid(S), Liquid(L) High$^+$, Medium^0, Low$^-$ Precision	Speciation	Reference
	$\mu g \cdot g^{-1}$	$ng \cdot mL^{-1}$	pg						
ICP–AES									
Pneumatic nebulizer	1	5	1000	Low	High	M(50 μL)	L +	Yes +C	[152]
Ultrasonic nebulizer	1	0.5	100	Medium	High	M(50 μL)	L 0	Yes +C	[153]
ETV	0.1	1	10	High	High	M(50 μL, 1–2 mg)	S L 0	Yes	[154]
DSID						M(1–10 μL,			
Graphite rod	1	5	1	High	High	1–2 mg)	S L 0	Yes	[155,156]
Wire loop	0.1	0.1	1	High	High	M(10 μL)	L 0	Yes	[157]
Hydride generation					Limited				
	1	1	1000	Medium	(As,Se,..)	M(50 μL)	L +	Yes	[158,159]
Spark sampling	1			Low	High	L(5 mm)			[160,161]
						D(0.2 mm)	S +	No	
MIP–AES									
ETV	1	1	100	High	High	M(50 μL, 1–2 mg)	S L 0	Yes	[65]
Toroidal MIP	10	10	1000	High	High	M(50 μL)	L +	Yes +C	[72]
DCP–AES									
Pneumatic nebulizer	0.5	5	1000	High	High	M(50 μL)	L +	Yes +C	[46]
DC arc AES	1	5	100	High	High	M(50 μL, 1–2 mg)	S L 0	Yes +C	[36,162]
Spark AES	1	10	1000	Medium	High	M(50 μL) L(5 mm) D(0.2 mm)	S L +	Yes +C	[163]
Glow discharge AES	5			Low	High	L(2 mm) D(1 mm)	S L +	No	[79–82]

Hollow cathode AES	1	1	100	High	High	M(1–50 μL, 1–2 mg)	S	L	0	Yes	[77,162]
FANES	0.1	0.1	10	High	High	M(1–50 μL)	S	L	0	Yes	[87,88]
AAS flame											
Pneumatic nebulizer	0.5	10	1000	Low	Absent	M(10 μL)		L	+	Yes +C	[1]
Wire loop	0.5	0.5	10	Medium	Absent	M(10 μL)		L	+	No	[10]
Hydride generation and cold valpor	0.5	0.1	100	High	Absent (As,Se,..)	M(10 μL)		L	+	No	[164]
AAS furnace	0.1	0.1	10	High	Absent	M(1–50 μL, 1–2 mg)	S	L	0	Yes	[1]
AFS											
Laser, ICP/ICP	0.1	1	100	Low	Limited	M(50 μL)		L	+	Yes +C	[98,91]
Hollow cathode/ICP	1	5	100	Low	High	M(50 μL)		L	+	Yes +C	[89]
Laser/furnace	0.01	0.1	1	High	Limited	M(1–50 μL)	S	L	0	Yes	[96]
CFS furnace	1	1	100	High	High	M(1–50 μL)	S	L	0	Yes +C	[110,111]
XRS											
XRFS	1	100		Medium	High ($Z \geq 6$)	D(100 μm)	S	L	+	Yes	[115,118]
Electron microprobe	10		0.001	Medium	High ($Z \geq 5$)	L(0.5 μm) D(10 μm)	S		+	Yes +C	[139]
PIXE	10	10	10^5	Medium	High ($Z \geq 14$)	D(10 μm)	S	L	+	Yes	[125,140]
Optogalvanic methods	0.01	0.01	1	Low	Absent	M(50 μL, 1–2 mg)	S	L	+	No	[165]

Table 12.1 *(Continued)*

Method	c_L (Order of Magnitude)			Matrix Effects	Multielement Capability	Micro(M), Local(L) in-Depth(D) analysis	Solid(S), Liquid(L) High$^+$, Medium^0, Low$^-$ Precision		Speciation	Reference
	$\mu g \cdot g^{-1}$	$ng \cdot mL^{-1}$	pg							
Auger electron spectroscopy	1000			Medium	High ($Z \geq 3$)	M(50 μL) L(0.2 μm) D(10 Å)	S	+	Yes	[140]
Mass spectroscopy										
Spark	0.01			Medium	High	M(1–50 μL, 1 mg) D(0.2 nm)	S L	0	No	[129]
Thermionic		0.01	1	Low	Limited	M($\leq \mu$L)	L	+	No	[131,132]
Field desorption		0.01	1	Low	Limited	M($\leq \mu$L)	L	+	No	[133]
Glow discharge	0.1			Low	High	M(1–5 μL,1 mg)	S L	+	No	[166]
LAMMA			10^{-6}	Medium	High	M L(1–3 μm)	S	–	Yes	[114,130, 167]
SIMS	0.1			Low	High	M L(0.5 μm) D(1–10 Å)	S	0	Yes	[130]
SNMS	0.1			Low	High	M D(10 Å)	S	0	Yes	[139]
ICP/MS	1	0.1	100	Low	High	M(1–50 μL)	L	+	No	[168,169]
Activation analysis										
NAA	0.1	0.1	0.1	Low	High	M($\leq$mg, μL)	S L	0	Yes +C	[144]
INAA	1	1	10	Low	High	M($\leq$mg, μL)	S L	0	Yes +C	[144]
Electrochemistry (DPASV)		0.1		High	Limited(Bi,Cd, Cu, Pb,Tl,..)	M($\leq$mL)	L	+	Yes	[141]

Spectrophotometry	1		High	Limited	M(≤mL)	L	+	Yes	
Spectrofluorimetry	0.01		High	Limited	M(≤mL)	L	+	Yes	
Chromatography									
Chelate, GC								Yes	
EDC,									
Plasma AES,		1,							
MS*		≤ 1*	High	Limited	M(≤μL)	L	+		[142,143]
HPLC		1, ≤1*	High	Limited	M(≤mL)	L	+	Yes	[151]
Ion chromatography	10		Medium	Anions/cations	M(≤mL)	L	+	Yes	[145]

ETV, electrothermal vaporization; DSID, direct sample insertion device; DPASV, differential pulse anodic stripping voltammetry; +C; speciation possible in combination with chromatography only;*, with mass spectrometric detection; for MS and NAA matrix effects can be decreased and precision increased by using ID.

12.10.2. Power of Detection

Here, the statements clearly differ from one element to another. The detection limits of plasma source AES with pneumatic nebulization lie between those of furnace and flame AAS. For elements with a high ionization potential and prominent lines with high excitation potential (for example, Pb, Cd, T1), furnace AAS and even advanced flame AAS techniques, such as the wire loop technique [10], yield lower values than plasma source AES. For the latter elements, electrochemical techniques [141] and chelate gas chromatography (GC) [142,143] are also superior. With optogalvanic spectroscopy and novel mass spectrometric methods, the gap in power of detection existing between atomic spectrometry and the expensive, time-consuming, labor-intensive, but very sensitive neutron activation analysis (NAA) [144] or instrumental neutron activation analysis (INAA), has become smaller. Atomic spectrometry will maintain a higher power of detection than ion chromatography [145]. However, the latter can be also used for the determination of anions.

12.10.3. Multielement Capability

Emission spectrometry inherently has a high multielement capability, whereas AAS is fundamentally a single-element method. In laser-excited AFS and LEI, the multielement capability is limited by the wavelength presetting of the tunable lasers. The fact that solid-state lasers operating at analytically interesting wavelengths have now become available could change the situation and open the door for applications of these techniques to practical analytical problems.

X-ray spectrometry and mass spectrometry have excellent multielement capabilities. Whereas the former is limited to elements with an atomic number higher than 10, mass spectrometry can be applied to all elements. The multielement capability of other methods such as electrochemical methods and neutron activation analysis is often limited by the nature of the sample; this contrasts with atomic spectrometry.

12.10.4. Matrix Effects and Calibration

It has been shown that by appropriate measures, matrix effects in atomic spectrometry methods can be efficiently controlled. To this aim, isotopic dilution (ID) is very useful. This approach is possible in mass spectrometry and in new optical techniques using Doppler-free laser spectroscopy.

Systematic errors in atomic spectrometry for trace analysis still arise mainly from contamination. They may be recognized by analyzing certified reference samples. Contamination can be avoided to some extent by appropriate purification of all reagents, appropriate rinsing of vessels, and performing the sample preparation and also the eventual determinations under clean conditions, as realized in clean benches or clean rooms. These topics are extensively treated in [146] and [147] and related papers by Tölg et al.

12.10.5. Distribution (Local) and Microanalysis

For local analyses, electron microprobe [148] and ion microprobe techniques [130] permitting high lateral resolution and area-mapping facilities are the appropriate tools. Surface and in-depth information may be obtained by Auger electron spectroscopy, SIMS [149], and related techniques [140], also, however, by GD–AES [84] (cf. Section 12.3.3.6).

With the aid of suitable preparation techniques, the material present in an inclusion or a layer can be isolated from the sample and thus the problem of local analysis can often be transformed into a microanalytical problem. Electropolishing procedures that enable layer-by-layer ablation, for instance, can be used for such purposes. For the microdetermination step, graphite furnace AAS, ETV/ICP–AES, or TRXRF are useful. For instance, with ETV/ICP–AES, absolute amounts of material down to 10^{-11} g can be determined and direct sample insertion (DSID) in ICP–AES improves this level by one order of magnitude. These combined microanalytical techniques may be used for solving a wide variety of analytical problems in biology, toxicology, environmental sciences, phase studies in metallurgy, and other fields. They fill a gap between macroanalysis and ultramicroanalysis methods such as LAMMA.

12.10.6. The Physical State of the Samples

In atomic spectrometry, techniques for each physical state of the samples are available. Although the presentation of the sample to the instrument may be easy, implications for the ease of calibration must be taken into consideration, in particular, in the direct analysis of solids.

12.10.7. Analytical Precision

Analytical precision in AES, AAS, or XRFS can be considerably improved mainly by using an internal standard (cf. Part 1, Section 4.2.4) and further by the method of binary ratios [150] where concentration ratios are determined as a function of intensity ratios. However, the precision of the above methods at trace element level is lower than in some MS techniques, where by the use of isotopic dilution at the microgram per gram level, RSDs of 0.1% can be obtained.

12.10.8. Capabilities for Speciation

Speciation may be based on the coupling of the effluent of a chromatographic system for separation of elemental species with an atomic spectrometric element-specific detector. However, one may also make use of the difference in volatility of compounds in electrothermal evaporation. Of the new plasma discharges, MIP especially is suitable for speciation work. It is compact, the investment and operation costs are low, and since He can be used as working gas, the halogens can also be detected. Element-specific detection in gas and liquid chromatography by MIP–AES, however, still requires optimization with respect to interfacing. Here, the experience gained in chelate GC using electron capture detection (ECD) [142,143] or high-performance liquid chromatography (HPLC) using UV-VIS detection [151] is certainly useful.

REFERENCES

1. B. Welz, *Atomic Absorption Spectrometry*, Verlag Chemie, Weinheim, New York (1976).
2. A. A. Brown, B. A. Milner, and A. Taylor, *Analyst* **110,** 501 (1985).
3. C. Hallam and K. C. Thompson, *Analyst* **110,** 497 (1985).
4. A. A. Brown and A. Taylor, *Analyst* **110,** 579 (1985).
5. E. Sebastiani, K. Ohls, and G. Riemer, *Fresenius Z. Anal. Chem.* **264,** 105 (1973).
6. H. Berndt and E. Jackwerth, *Spectrochim. Acta* **30B,** 169 (1975).
7. J. F. Tyson, *Analyst* **110,** 419 (1985).
8. S. Greenfield, *Spectrochim. Acta* **38B,** 93 (1982).
9. H. T. Delves, *Analyst* **95,** 431 (1970).
10. H. Berndt and J. Messerschmidt, *Anal. Chim. Acta* **136,** 407 (1982).
11. F. J. Langmyhr and G. Wibetoe, *Prog. Anal. Atom. Spectrosc.* **8,** 193 (1985).
12. H. Berndt, *Spectrochim. Acta* **39B,** 1121 (1984).
13. T. C. Dymott, M. P. Wassell, and P. J. Whiteside, *Analyst* **110,** 467 (1985).
14. H. M. Ortner, G. Schlemmer, B. Welz, and W. Wegscheider, *Spectrochim. Acta* **40B,** 959 (1985).
15. H. Berndt and J. Messerschmidt, *Fresenius Z. Anal. Chem.* **316,** 2 (1983).
16. U. Völlkopf, Z. Grobenski, R. Tamm, and B. Welz, *Analyst* **110,** 57 (1985).
17. D. C. Baxter, W. Frech, and E. Lundberg, *Analyst* **110,** 475 (1985).
18. R. E. Sturgeon and C. L. Chakrabarti, *Prog. Anal. Atom. Spectrosc.* **1,** 5 (1978).
19. S. B. Chang and C. L. Chakrabarti, *Prog. Anal. Atom. Spectrosc.* **8,** 83 (1985).
20. R. S. Koirtyohann and E. E. Pickett, *Anal. Chem.* **37,** 601 (1965).
21. H. Massmann, Z. El Gohary, and S. Gücer, *Spectrochim. Acta* **31B,** 399 (1976).

22. M. Prugger and R. Torge, German Patent 1964469, filed December 23, 1969.
23. L. de Galan and M. T. C. de Loos-Vollebregt, in *Proc. 21th Coll. Spectrosc. Int. and 8th Int. Conf. Atomic Spectrosc.*, Cambridge 1979, Keynote Lectures, Heyden, London (1979) p. 49.
24. J. A. C. Broekaert, *Spectrochim. Acta* **37B,** 65 (1982).
25. M. T. C. de Loos-Vollebregt and L. de Galan, *Prog. Anal. Atom. Spectrosc.* **8,** 47 (1985).
26. W. B. Barnett, W. Bohler, G. R. Carnick, and W. Slavin, *Spectrochim. Acta* **40B,** 1689 (1985).
27. D. D. Siemer, *Appl. Spectrosc.* **37,** 552 (1983).
28. A. Meyer, C. Hofer, G. Tölg, S. Raptis, and G. Knapp, *Fresenius Z. Anal. Chem.* **296,** 337 (1979).
29. J. Piwonka, G. Kaiser, and G. Tölg, *Fresenius Z. Anal. Chem.* **321,** 225 (1985).
30. B. Welz and M. Melcher, *Analyst* **108,** 213 (1983).
31. B. T. Sturman, *Appl. Spectrosc.* **39,** 48 (1985).
32. W. Schmid and V. Krivan, *Anal. Chem.* **57,** 30 (1985).
33. N. W. H. Addink, *DC Arc Analysis*, Macmillan, London (1971).
34. J. Kroonen and D. Vader, *Line Interference in Emission Spectrographic Analysis*, Elsevier, Amsterdam (1963).
35. C. E. Harvey, *Semiquantitative Spectrochemistry*, Applied Research Laboratories, Glendale/California (1964).
36. R. Avni, in E. L. Grove, ed., *Applied Atomic Spectroscopy*, vol. 1, Ch. 4. Plenum, New York (1978).
37. K. H. Koch, *Spectrochim. Acta* **39B,** 1067 (1984).
38. J. P. Walters and W. S. Eaton, *Anal. Chem.* **55,** 57 (1983).
39. A. Scheeline, Prog. *Anal. Atom. Spectrosc.* **7,** 21 (1984).
40. K. Tohyama, J. Ono, M. Onodera, and M. Saeki, *Pulse Distribution Analysis Methods for Emission Spectrochemical Analysis*, Okochi Memorial Foundation (1978).
41. J. A. C. Broekaert, *Trends Anal. Chem.* **1,** 249 (1982).
42. M. Margoshes and B. F. Scribner, *Spectrochim. Acta* **15,** 138 (1959).
43. V. V. Korolov and E. E. Vainshtein, *Zh. Anal. Khim.* **14,** 658 (1959).
44. L. A. Fernando, *Spectrochim. Acta* **37B,** 859 (1982).
45. G. W. Johnson, H. E. Taylor, and R. K. Skogerboe, *Spectrochim. Acta* **34B,** 197 (1979).
46. J. Reednick, *Am. Lab.*, May, 127 (1979).
47. A. Disam, P. Tschöpel, and G. Tölg, *Fresenius Z. Anal. Chem.* **310,** 131 (1982).
48. G. Wünsch, N. Czech, and G. Hegenberg, *Fresenius Z. Anal. Chem.* **310,** 62 (1982).
49. P. W. J. M. Boumans, F. J. De Boer, F. J. Dahmen, H. Hoelzel, and A. Meier, *Spectrochim. Acta* **30B,** 449 (1975).

50. G. F. Larson and V. A. Fassel, *Anal. Chem.* **48,** 1161 (1976).
51. I. O. Burman and K. Bostrom, *Anal. Chem.* **51,** 516 (1979).
52. K. Govindaraju, G. Mevelle, and C. Chouard, *Anal. Chem.* **48,** 1325 (1976).
53. J. Dahmen, *ICP Information Newslett.* **6,** 576 (1981).
54. F. J. M. J. Maessen, G. Kreunig, and J. Balke, *Proc. 1985 European Winter Conf. Plasma Spectrochemistry, Leysin, Spectrochim. Acta* **41B,** 3 (1986).
55. S. Chan and A. Montaser, *Spectrochim. Acta* **40B,** 1467 (1985).
56. R. I. Botto, *Spectrochim. Acta* **38B,** 129 (1983).
57. P. W. J. M. Boumans and J. J. A. M. Vrakking, *Spectrochim. Acta* **40B,** 1085 (1985).
58. P. W. J. M. Boumans and J. J. A. M. Vrakking, *Spectrochim. Acta* **40B,** 1107 (1985).
59. P. W. J. M. Boumans, *Fresenius Z. Anal. Chem.* **324,** 397 (1986).
60. L. M. Faires, B. A. Palmer, R. Engelmann Jr., and T. M. Niemczyk, *Spectrochim. Acta* **39B,** 891 (1984).
61. L. M. Faires, *Spectrochim. Acta* **40B,** 1473 (1985).
62. N. Furuta, C. W. McLeod, H. Haraguchi, and K. Fuwa, *Appl. Spectrosc.* **34,** 211 (1980).
63. J. A. C. Broekaert, F. Leis, and K. Laqua, unpublished work.
64. P. Taylor and P. Schutyser, *Proc. 1985 European Winter Conf. Plasma Spectrochemistry, Leysin, Spectrochim. Acta* **41B,** 81 (1986).
65. A. Aziz, J. A. C. Broekaert, and F. Leis, *Spectrochim. Acta* **37B,** 381 (1982).
66. E. I. Brooks and K. J. Timmins, *Analyst* **110,** 557 (1985).
67. G. Volland, P. Tschöpel, and G. Tölg, *Spectrochim. Acta* **36B,** 901 (1981).
68. G. Kaiser, D. Götz, P. Schoch, and G. Tölg, *Talanta* **22,** 889 (1975).
69. J. P. J. van Dalen, P. A. de Lezenne-Coulander, and L. de Galan, *Anal. Chim. Acta* **94,** 1 (1977).
70. A. Bollo-Kamara and E. G. Codding, *Spectrochim. Acta* **36B,** 973 (1981).
71. D. Kollotzek, P. Tschöpel, and G. Tölg, *Spectrochim. Acta* **37B,** 91 (1982).
72. D. Kollotzek, P. Tschöpel, and G. Tölg, *Spectrochim. Acta* **39B,** 625 (1984).
73. D. Kollotzek, D. Oechsle, G. Kaiser, P. Tschöpel, and G. Tölg, *Fresenius Z. Anal. Chem.* **318,** 485 (1984).
74. G. W. Rice, A. P. D'Silva, and V. A. Fassel, *Spectrochim. Acta* **40B,** 1573 (1985).
75. S. L. Mandelstam and V. V. Nedler, *Spectrochim. Acta* **17,** 885 (1961).
76. H. Falk, *Spectrochim. Acta* **32B,** 437 (1977).
77. B. Berglund and B. Thelin, *Analyst* **107,** 867 (1982).
78. W. Grimm, *Spectrochim. Acta* **23B,** 443 (1968).
79. M. Dogan, K. Laqua, and H. Massmann, *Spectrochim. Acta* **26B,** 631 (1971).
80. M. Dogan, K. Laqua, and H. Massmann, *Spectrochim. Acta* **27B,** 65 (1972).

81. R. Klockenkämper, K. Laqua, and M. Dogan, *Spectrochim. Acta* **35B,** 527 (1980).
82. J. B. Ko, *Spectrochim. Acta* **39B,** 1405 (1984).
83. R. Berneron, *Spectrochim. Acta* **33B,** 665 (1978).
84. K. H. Koch, M. Kretschmer, and D. Grunenberg, *Mikrochim. Acta* **II,** 225 (1983).
85. A. Quentmeier and K. Laqua, in K. H. Koch and H. Massmann, eds., *13. Spektrometertagung*, de Gruyter, Berlin (1981) p. 37.
86. D. Littlejohn and J. M. Ottaway, *Analyst* **104,** 208 (1979).
87. H. Falk, E. Hoffmann, and C. Lüdke, *Spectrochim. Acta* **39B,** 283 (1984).
88. H. Falk, E. Hoffmann, C. Lüdke, J. M. Ottaway, and S. K. Giri, *Analyst* **108,** 1459 (1983).
89. D. R. Demers and C. H. Allemand, *Anal. Chem.* **53,** 1915 (1981).
90. E. B. M. Jansen and D. R. Demers, *Analyst* **110,** 541 (1985).
91. R. J. Krupa, G. L. Long, and J. D. Winefordner, *Spectrochim. Acta* **40B,** 1485 (1985).
92. S. Greenfield and M. Thomsen, *Spectrochim. Acta* **40B,** 1369 (1985).
93. N. Omenetto and J. D. Winefordner, *Prog. Anal. Atom. Spectrosc.* **2,** (1/2) (1979).
94. A. Montaser and V. A. Fassel, *Anal. Chem.* **48,** 1490 (1976).
95. N. Omenetto and H. G. C. Human, *Spectrochim. Acta* **39B,** 1333 (1984).
96. M. A. Bolshov, A. V. Zybin, and I. I. Smirenkins, *Spectrochim. Acta* **36B,** 1143 (1981).
97. M. A. Bolshov, A. V. Zybin, V. G. Koloshnikov, I. A. Mayorov, and I. I. Smirenkina, *Spectrochim. Acta* **41B,** 487 (1986).
98. H. G. C. Human, N. Omenetto, P. Cavalli, and G. Rossi, *Spectrochim. Acta* **39B,** 1345 (1984).
99. N. Omenetto, H. G. C. Human, P. Cavalli, and G. Rossi, *Spectrochim. Acta* **39B,** 115 (1984).
100. N. Omenetto, B. W. Smith, L. P. Hart, P. Cavalli, and G. Rossi, *Spectrochim. Acta* **40B,** 1411 (1985).
101. J. C. Travis, G. C. Turk, and R. B. Green, *Anal. Chem.* **44,** 1006A (1974).
102. G. C. Turk, J. C. Travis, J. R. De Voe, and T. C. O'Haver, *Anal. Chem.* **51,** 1890 (1979).
103. I. Magnusson, O. Axner, I. Lundgren, and H. Rubinstein-Dunlop, *Appl. Spectrosc.* **40,** 968 (1986).
104. G. C. Turk and R. L. Watters, *Anal. Chem.* **57,** 1979 (1985).
105. N. Omenetto, T. Berthoud, P. Cavalli, and G. Rossi, *Anal. Chem.* **57,** 1256 (1985).
106. G. J. Havrilla, S. J. Weeks, and J. C. Travis, *Anal. Chem.* **54,** 2566 (1982).
107. K. Niemax, *Appl. Phys. B* **38,** 147 (1985).

108. G. Hertz, *Z. Physik* **18,** 307 (1923).
109. K. H. Kingdon, *Phys. Rev.* **21,** 408 (1923).
110. M. Yamamoto, M. Ito, and M. Yasuda, *Spectrochim. Acta* **35B,** 43 (1980).
111. P. Wirz, H. Debus, W. Hanle, and A. Scharmann, *Spectrochim. Acta* **37B,** 1013 (1982).
112. K. Laqua, "Analytical Spectroscopy using Laser Atomizers," in N. Omenetto, ed., *Analytical Laser Spectroscopy*, Ch. 2, Wiley-Interscience, New York (1979) p. 47.
113. J. A. J. Jansen and A. W. Witmer, *Spectrochim. Acta* **37B,** 483 (1982).
114. E. Denoyer, R. Van Grieken, F. Adams, and D. F. S. Natusch, *Anal. Chem.* **54,** 26A (1982).
115. L. S. Birks, *X-Ray Spectrochemical Analysis*, Wiley-Interscience, New York (1969).
116. J. N. Kikkert, Proc. 23rd Coll. Spectrosc. Int., Amsterdam 1983, *Spectrochim. Acta* **38B,** 1497 (1983).
117. L. S. Birks, *Spectrochim. Acta* **40B,** 1161 (1985).
118. R. Jenkins, *Einführung in die Röntgenspektralanalyse*, Heyden, London (1977).
119. C. Ruch, F. Rastegar, R. Heimburger, E. Maier, and M. J. F. Leroy, *Anal. Chem.* **57,** 1691 (1985).
120. P. Wobrauschek and H. Aiginger, *Anal. Chem.* **47,** 852 (1975).
121. J. Knoth and H. Schwenke, *Fresenius Z. Anal. Chem.* **291,** 200 (1978).
122. H. Schwenke and J. Knoth, in P. Brätter and P. Schramel, eds., *Trace Element Analytical Chemistry in Medicine and Biology*, de Gruyter, Berlin, (1980) p. 307.
123. R. Eller, R. Klockenkämper, and G. Tölg, XXIV Coll. Spectrosc. Intern., Garmisch-Partenkirchen 1985, book of abstracts, p. 736.
124. B. Baumgardt, R. Klockenkämper, and G. Tölg, *XXIV Coll. Spectrosc. Intern., Garmisch-Partenkirchen 1985*, book of abstracts, p. 738.
125. R. P. H. Garten, "Protoneninduzierte Röntgen-Emissionsspektrometrie (PIXE). Analytische Anwendungen," in *Analytiker Taschenbuch*, vol. 4. Springer, Berlin (1984).
126. A. J. J. Bos, C. C. A. H. Van Der Stap, R. D. Vis, and V. Valkovic, *Spectrochim. Acta* **38B,** 1209 (1983).
127. J. V. Gilfrich, E. F. Skelton, S. B. Quadri, J. P. Kirkland, and D. J. Nagel, *Anal. Chem.* **55,** 187 (1983).
128. A. Knöchel, W. Petersen, and G. Tolkiehn, *Anal. Chim. Acta* **173,** 105 (1985).
129. J. R. Bacon and A. Ure, *Analyst* **109,** 1229 (1984).
130. F. Adams, Proc. 23rd Coll. Spectr. Int., Amsterdam 1983, *Spectrochim. Acta* **38B,** 1379 (1983).
131. K. G. Heumann, *Toxicol. Environm. Chem. Rev.* **3,** 111 (1980).
132. K. G. Heumann, F. Beer, and H. Weiss, *Mikrochim. Acta,* **I,** 95 (1983).
133. H. R. Schulten, P. B. Monkhouse, and R. Müller, *Anal. Chem.* **54,** 654 (1982).

134. A. L. Gray, *Spectrochim. Acta* **40B,** 1525 (1985).

135. D. J. Douglas and J. B. French, *Anal. Chem.* **53,** 37 (1981).

136. W. W. Harrison, K. R. Hess, R. K. Markus, and F. L. King, *Anal. Chem.* **58,** 341 A (1986).

137. H. Oechsner and E. Stumpe, *Appl. Phys.* **14,** 43 (1983).

138. W. A. Mattson, B. L. Bentz, and W. W. Harrison, *Anal. Chem.* **48,** 489 (1976).

139. N. Jakubowski, D. Stuewer, and G. Toelg, *Int. J. Mass Spectrosc. Ion Proc.* **71,** 183 (1986).

140. H. W. Werner and R. P. H. Garten, A comparative study of methods for thin-film and surface analysis, *Rep. Prog. Phys.* **47,** 221 (1984).

141. H. W. Nürnberg, *Pure Appl. Chem.* **54,** 853 (1982).

142. P. C. Uden, in R. M. Barnes, ed., *Developments in Atomic Plasma Spectrochemical Analysis*, Heyden, Philadelphia (1981) p. 302.

143. A. Tavlaridis and R. Neeb, *Fresenius Z. Anal. Chem.* **292,** 199 (1978).

144. V. Krivan, ''Neutronenaktivierungsanalyse,'' in *Analytiker Taschenbuch*, vol. 5, Springer, Berlin (1985) p. 36.

145. G. Schwedt, B. Rössner, and D. Yan, in B. Sansoni, ed., *Instrumentelle Multielementanalyse*, p. 445. Verlag Chemie, Weinheim (1985).

146. P. Tschöpel, L. Kotz, W. Schulze, M. Weber, and G. Tölg, *Fresenius Z. Anal. Chem.* **302,** 1 (1980).

147. P. Tschöpel and G. Tölg, *J. Trace Microprobe Techniques* **1,** 1 (1982).

148. L. S. Birks, *Electron Probe Microanalysis, Chemical Analysis Series*, vol. 17, Wiley-Interscience, New York (1971).

149. M. Grasserbauer, *Fresenius Z. Anal. Chem.* **322,** 105 (1985).

150. 1969 ASTM Book of ASTM Standards, *Recommended Practices for Spectrochemical Computations* (E 158–66), American Society for Testing and Materials, Philadelphia (1971).

151. B. Steinbrech and K. H. König, *Fresenius Z. Anal. Chem.* **316,** 68 (1983).

152. R. K. Winge, V. J. Peterson, and V. A. Fassel, *Appl. Spectrosc.* **33,** 206 (1979).

153. P. D. Goulden and D. H. J. Anthony, *Anal. Chem.* **56,** 2327 (1984).

154. A. Aziz, J. A. C. Broekaert, and F. Leis, *Spectrochim. Acta* **37B,** 369 (1982).

155. G. F. Kirkbright and S. J. Walton, *Analyst* **107,** 276 (1982).

156. Z. Li-Xing, G. F. Kirkbright, M. J. Cope, and J. M. Walton, *Appl. Spectrosc.* **37,** 250 (1983).

157. E. D. Salin and R. L. A. Sing, *Anal. Chem.* **56,** 2596 (1984).

158. M. Thompson, B. Pahlavanpour, and S. J. Walton, *Analyst* **103,** 568 (1978).

159. J. A. C. Broekaert and F. Leis, *Fresenius Z. Anal. Chem.* **300,** 22 (1980).

160. A. Aziz, J. A. C. Broekaert, K. Laqua, and F. Leis, *Spectrochim. Acta* **39B,** 1091 (1984).

161. J. A. C. Broekaert, F. Leis, and K. Laqua, in B. Sansoni, ed., *Instrumentelle Multielementanalyse*, Verlag Chemie, Weinheim (1985), p. 359.

162. K. I. Zil'bershtein, *Spectrochemical Analysis of Pure Substances*, Adam Hilger, London (1977).

163. K. Slickers, *Die automatische Emissionsspektralanalyse*, Brühldruck und Pressehaus Giessen, P.O. Box 5506, D-6300 Giessen, Fed. Rep. Germany (1977).

164. B. Welz and M. Melcher, *Spectrochim. Acta* **36B,** 439 (1981).

165. J. C. Travis, G. C. Turk, J. R. De Voe, and P. K. Schenk, *Prog. Anal. Atom. Spectrosc.* **7,** 199 (1984).

166. J. W. Coburn and W. W. Harrison, *Appl. Spectrosc. Rev.* **17,** 95 (1981).

167. H. Vogt, H. J. Heinen, and S. Meier, *Laser and Optoelektronik* **1,** 23 (1983).

168. A. L. Gray and A. R. Date, *Analyst* **108,** 1033 (1983).

169. R. S. Houk, V. A. Fassel, G. D. Flesch, H. J. Svec, A. L. Gray, and C. E. Taylor, *Anal. Chem.* **52,** 2283 (1980).

INDEX

AAS:
analytical characterization, 445f, 447
atomic absorption spectrometry, 2
detection limits, biological samples, 88f
direct solids analysis, 221, 427
hydride generation, 427, 447. *See also* Hydride generation
interferences, 427, 447
kinetics, 427
geological materials, 28
see also ETA–AAS; FAAS; GFAAS; ZAAS (Zeem atomic absorption spectrometry)
AAS *vs.* ICP–AES:
see ICP–AES *vs.* AAS, steel and metals; ICP–AES *vs.* FAAS
Abel inversion, 356
H_β line, 403f
Accumulator tissues:
elemental concentrations, table, 73
types and composition, 71
Accuracy, geological analysis, 38f
Acid digestion:
airborne particulates, 50
environmental samples, 50
vs. fusion, 50
losses, 50
see also Digestion, acid-; Dissolution; Wet-ashing
Acid dissolution:
closed systems, geological samples, 31f
open systems, geological samples, 31
see also Acid digestion
Active carbon, for trace enrichment, 17
Aerosol(s):
aerosol *vs.* vapor load, 277f
bimodal distribution, 252
characterization, 245f, 248
drop size, 246f
aerodynamic diameter, 254f
cutoff diameter, 271f, 275
distribution curves, 250f
equilibrium diameter, 276f
equivalent diameter, 252
lognormal distribution, 251f
median diameter, 249
USN, 270f
normal distribution, 250f
Nukiyama-Tanasawa equation, 155, 253, 267
numerical count distribution, 254
role of impact bead, 271f
role of spray chamber, 272f
Sauter mean diameter, 249, 267
tertiary processes, 272f
drop size distribution, 274, 248f, 250f, 252f
curves, 250f
and evaporation, 255f, 258
PN, 267f
various solvents, 268f
drop sizing:
sampling, 258f
using:
aerodynamic properties, 252, 254f
Andersen impactor, 255f
cascade impaction, 252, 255f
Fraunhofer scattering, 252, 259f
gravitational settling, 252, 255f
isokinetic sampling, 259
microscopic collection, 252, 253f
Mie scattering, 252, 259f
optical scattering, 252, 259f
equilibrium drop diameter, 276f
equivalent diameter, 252
evaporation factor, 159f
table, 160
interferences, *see* Nebulizer effects
ionic redistribution, 282f
mass *vs.* numerical distribution, 249
mass transport properties, 246f
mass transport rate and SBR, 284f

Aerosol(s), drop sizing (*Continued*)
mass transport rate W, 246f
measurement, 261f
mass transport terms, 247
median diameter, USN, 270f
median droplet diameter, 249
numerical count distribution, 254
particle size, *see* Aerosols, drop size
primary, 267f, 272f
properties, effect of evaporation, 275f
role of solvent transport, 248, 277f
role of solvent vapors, 248, 277f
role of spray chamber temperature, 277f
Sauter mean diameter, 249, 267
Stokes number, 253
techniques for particle sizing, 252
tertiary, 249, 272f
trajectories, in ICP, 312f
transport efficiency, 124f, 246f
measurement, 261f
direct, 262f
fiber glass filter, 263
indirect, 264
silica gel, 263f
organic solvents, 162
table, 163
vapor introduction, 125f
transport and ICP excitation characteristics, 248, 277f
USN, 268f
W-parameter, 246f
see also Nebulization; Organic solvents
Aerosol characteristics:
additional literature, 285
optimization for ICP, 284f
Aerosol cooling, *see* Aerosol vapor thermostatting
Aerosol desolvation, 176f, 277f
Aerosol generation, 264f
"bursting bag" model, PN, 266
"cavitation" model, USN, 270
centrifugal loss, 274f
droplet cutoff diameter, 271f
fundamentals, 244f
"geyser formation" model, USN, 270
gravitational loss, 274f
high-velocity liquid jets, 264f
high-voltage electrification, 264f
impact bead, 271f
impaction loss, 271f, 273f
additional literature, 285
models, 265f
PN, 266f
USN, 270
"oscillating ligand" model, PN, 266
secondary processes, 271f
secondary shattering, 272
"surface stripping" model, PN, 267
tertiary modifying processes, 272f
turbulence loss, 273f
see also Nebulization
Aerosol ionic redistribution, 282f
Aerosol transport, fundamentals, 244f
Aerosol vapor thermostatting:
organic solvents, 177f
detection limits, 177f
maximum temperature, 178
AES (atomic emission spectrometry):
arc, *see* Arc AES
atomic emission spectrometry, 1
background correction, 433
DC arc, detection limits, biological samples, 88f
graphite furnace-, 437. *See also* FANES
multichannel detection, 433
selectivity, 433
spark, *see* Spark AES
AFS (atomic fluorescence spectrometry):
analytical characterization, 445f, 447
atomic fluorescence spectrometry, 438
direct solids analysis, 221
flame, 438
furnace, 438f
laser excited, 438
selectivity, 438
Afterglow systems, 435
Agricultural analysis:
arguments for use of ICP–AES, 87
ICP–AES, assessment, 93
interferences, 92f
Agricultural materials:
analysis of, 65f
analyte loss in drying, 75
dry-ashing, 75f
drying, 75
extraction procedures, 81f, 83
hydride generation, *see* Hydride generation
ion exchange extraction, 82f

preconcentration techniques, 81f
sample preparation, 73f
literature, 74
speciation, 85f
types and element concentrations, 66f
see also Animal tissues; Biological materials; Feedstuffs, composition; Fertilizers, types and composition; Food(s); Plant materials; Sediments; Soil(s); Tissues
AIR (aerosol ionic redistribution), 282f
Airborne particulates:
acid digestion, 50
analysis of, 43, 60f, 230f
ICP–AES, spectral interferences, 60
Air ICP, 294
Air particulates, *see* Airborne particulates
Air quality monitoring, 60f
Albacore tuna:
analysis of, ICP–AES, 121f, 135f
SRM, digestion and analysis of, 121f
Alumina, physical properties, 318
Alumina powder, particle trajectories, in ICP, 317
Aluminates, polymerization in solution, 29
Aluminum determination:
blood, 74
urine, 74
Analysis:
of agricultural samples, 65f
of airborne particulates, 43, 50, 60f, 230f
of animal bone, 123f, 129f, 135f
of biological materials, 100f
chronological review, 134, 135f
of biological SRMs, 121f, 130f, 132f
of biological-clinical materials, 100f
of blood, *see* Blood
of blood plasma, 135f
of blood serum, *see* Serum
of bovine liver, 115, 121f, 127, 129, 132f, 135f. *See also* Bovine liver
of chemicals, 17f
of chlorides, MIP, 234
of coal, 15, 61
of environmental samples, 48f
of faeces, 135f
of ferrous alloys, 11f
of fish, 59
of fly ash, 15, 61
of food, 65f
of fresh waters, 53f
of geological materials, 28f
direct, 230
of hair, 135f
of human kidney and liver, 128f, 135f
of human milk, Ni determination, ETV–ICP–AES, 133
of human stones, 135f
of industrial chemicals, 17f
of industrial materials, 1f
of iron ores, 15
of lubricating oil, 16, 154, 189f, 195
of metal chelates:
ICP, 127, 195, 211, 235f
MIP, 234
of metallic thin layers, 2
of metals, 1f
of milk, determination of P, 74
of monitor organisms, water, 58f
of muscle tissue, 135f
of oil, 16, 154, 189f, 195
of organics, 151f
assessment of ICP–AES, 213f
of organic vapors, 154
of oxidic materials, steel industry, 14f
of oyster tissue, 121f, 135f
of petroleum products, 153f
of plankton, 43
of platinum group metals, 2
of powders, 222f, 230f
of seawater, 55f, 119f, 123
of sediments, 59
of serum, *see* Serum
of silicates, 28, 30
of skin, 135f
of slags, 14, 15, 28
of slurry, 92, 227f
of soil, *see* Soil(s)
of soil extracts, 80f
of solids, direct, *see* Direct solids analysis
of solution residues, 222, 236
of steels, 2. *See also* Steel(s)
of teeth, 135f
of urine, 57f, 74, 76, 84, 123f, 130f, 135f. *See also* Urine
of waste waters, 57
of water, 53f. *See also* Water analysis
of wear metals, 16, 154, 189f, 195
see also Trace analysis

Analysis lines:
iron and ferrous alloys, 9f
metals, 13f
nonferrous alloys, 13f
polychromator, steel industry, 7f
Analytical characterization, various methods, 445f
Analytical curves, via plasma modeling, 336
Anderson impactor:
schematic diagram, 257
see also Aerosol(s)
Animal bone, complete analysis procedure, 129f
Animal tissues:
analysis of, ICP–AES, 135f
types and composition, 71f
Arc, capillary, and ICP, 230
Arc ablation, and ICP, 229
Arc AES:
detection limits, 446
direct solids analysis, 219
geological materials, 28
universal procedures, 429
Arc AES *vs*. ICP–AES:
biological materials, 141
geological analysis, 36
Ar ICP:
analytical performance, 431, 445, 446, 450f
detection limits, steel, table, 7f
electron density, *see* Electron density
electron temperature, *see* Electron temperature
intermediate power, signal changes, organic solvents, 186f
nebulizer effects, 280
see also ICP
Ar ICP *vs*. N_2–Ar ICP:
analytical performance, 3
detection limits, polychromator, table, 10
Ar lines:
Stark effect, 378f
data, 378
Stark *vs*. Doppler effect, 378f
Array detector, 433
Ashing, *see* Digestion, acid-; Dry-ashing; Wet-ashing
Aspiration rates, limiting, organic solvents, table, 160
Atomic absorption spectrometry, *see* AAS
Atomic emission spectrometry, *see* AES (atomic emission spectrometry)
Atomic fluorescence spectrometry, *see* AFS (atomic fluorescence spectrometry)
Atomic level, *see* Level
Atomic spectrometry, comparison with other methods, 445f
Atomic spectrometry, trace analysis, 421f
Auger electron spectroscopy, 445
analytical characterization, 445f, 448
Autoionization, 392
Automation, fusion of silicates, 30
A values, *see gA* values

Babington nebulizer:
fresh water analysis, 54
see also Nebulizer; V-groove nebulizer
Back extraction, into aqueous phase, 12, 81f, 121
Background:
continuum, 285, 354, 360, 391
from line wings, 285
organic solvents, 168, 179f
oxygen addition, 170f
Background absorption, GFAAS, 425
Background correction, 52, 54
AES, 433
blank subtraction, 91
GFAAS, deuterium lamp, 425
quantitative spectral analysis, 433
ZAAS, 425f
Background shift, 52
Background spectra, organics, 168, 179f, 182f
Balmer series, H lines, *gA* values, 368
Band emission:
organic solvents, 168, 179f, 182f
band heads, table, 182f
Bimodal distribution, 252
Biogeochemical prospecting, 43f
Biological function, trace elements with, 69, 71, 101f
Biological materials:
analysis of, 100f
analysis by TRXRF, 442
assessment of ICP–AES, 140f

chelation and vapor introduction, 127
chronological review of analyses, 134, 135f
combustion, 110, 112
  Berthelot oxidation, 112
  contamination, 112
  Schöeniger oxidation, 112
detection limits, various techniques, table, 88f
digestion:
  ashing, 109f
  comparison of methods, 116f
  high-temperature ashing, *see* Dry-ashing
  mineralization, 109f
  procedures, survey, 108f
  see also Dry-ashing; Wet-ashing
dry-ashing, *see* Dry-ashing
FAAS, decomposition in O_2, 424
high-temperature ashing, *see* Dry-ashing
hydride generation, *see* Hydride generation
ICP–AES, comparison with other techniques, 140f
microsample technique(s), 124f, 131f
  ETV, 127f, 133
  FIA, 125, 132
pyrolysis, 110
rotating disk technique, 141
sample homogenization, 106f
  fragmentation, 107
  instruments/methods, 106f
  refrigerants/grinding, 107
  segregation, 107
sample pretreatment, 107f
  anticoagulants, 107
  chemical preservatives, 107f
  contamination, 108
  washing procedures, 108
sample storage:
  cleaning procedures, 104f
  contact time, 105f
  container materials, 104f
  containers, impurity levels, 105f
  defrosting, 106
  freezing, 106f
  homogenization, 106f
  lyophilization, 106
  pH, 105f
  sorption losses, 105f
  specific container surface, 105f
  storage time, 105f
sample storage and transport, 104f
sampling, 102f
  and contamination, 103f, 106f
  instruments and tools, 103f
  using laser beam, 104
solubilization/leaching, *see* Solubilization
SRMs, digestion and analysis of, 121f, 130f
trace analysis, 102
  enrichment procedures, 119f
  extraction, procedures, 120f. *See also* Extraction
  ion exchange procedures, *see* Ion exchange, procedures, biological fluids
  matrix separation, 119f
  preconcentration, 119f. *See also* Preconcentration
  SRMs, 121f
trace elements:
  classification, 101f
  clinical relevance, 101f
  concentrations, 118f
  essential, 101f
  functions of, 69, 71, 101f
  hydride forming, 102
  nonessential, 101f
  speciation, 102
  toxic, 101f
vapor introduction techniques, *see* Chelate vapor introduction; Hydride generation
wet-ashing, *see* Wet-ashing
see also Agricultural materials; Body fluids; Tissues
Biological SRMs, digestion and analysis, 121f, 130f
Biological tissues, *see* Tissues
Blood, 741
  analysis of, ICP–AES, 135f
  chelation and vapor introduction, 127
  determination of Al, 74
  extraction, 115
  microliter sample introduction, 125
  rf plasma ashing, 76
  SRM, digestion and analysis of, 122f
Blood plasma, analysis of, ICP–AES, 135f
Blood serum, *see* Serum

Body fluids:
human, composition, major elements and Fe, 109
trace element concentrations, 118f
sampling, 103f
types and composition, 71f
Boltzmann distribution, 329, 358, 393
Boltzmann equation, 329, 358, 393
Boltzmann equilibrium, 396f
lack of, 417
Boltzmann plot, 395f. *See also* T measurement
Bomb digestion, 61, 79, 113f, 121f, 133
biological SRMs, 121f
coal and fly ash, 61
Bone:
analysis of, ICP–AES, 135f
animal, complete analysis procedure, 129f
preconcentration and ICP analysis, 123f
Boron determination, waste waters, 57
Bovine liver:
analysis of, 115, 121f, 127, 129, 132f, 135f
chelation and vapor introduction, 127
ETV–ICP–AES, 128
extraction, 115
ICP–AES, FIA, 132
ICP–AES, hydride generation, Se, 132f
microlitersample introduction, 132
SRM, digestion and analysis of, 121f
Brain, analysis of, ICP–AES, 135f

Ca-Al interference, 281f
Calibration:
environmental samples, 50f
organic standards, 189, 195f
table, 196f
Calibration equation, linear *vs.* logarithmic, 51
Capacitively coupled microwave plasma, *see* CMP (capacitively coupled microwave plasma)
Ca-phosphate interference, 281f
Capillary arc, and ICP, 230
Capsule-in-flame atomizer, direct solids analysis, 220f
Carbon:
active, for trace enrichment, 17
deposits in torch, organics, 172f
Carbonatite, fluid inclusions, 44
Carbon rod:
and ICP, direct solids analysis, 236, 238f
interferences, 239f
Carrier gas:
saturation with solvent, 29, 32f, 92
wetting, 29, 32f, 92
Carrier gas flow:
effect on physical ICP characteristics, 305f
Carrier gas tube:
clogging, 92
use of sheathing gas, 92
Cascade impactor:
schematic diagram, 257
segration of mass distribution, 256
see also Aerosol(s)
C_2 bands, organic solvents, 168f, 179f, 182, 184
CCl bands, organic solvents, 179, 183
Cellulose, rf plasma ashing, 76
Cerebrospinal fluid, analysis of, ICP–AES, 135f
CFS:
analytical characterization, 445f, 447
coherent forward scattering, 440
Faraday effect, 441
magneto-optical effect, 441
Voigt effect, 441
Charge exchange, *see* Charge transfer
Charge neutrality, 402
Charge neutrality equation, 328
Charge transfer, 392, 410, 418
CH bands, organic solvents, 179, 182
Chelate(s), metal, analysis of:
ICP, 127, 195, 211, 235f
MIP, 234
Chelate vapor introduction, 127, 195
detection limits, 195
table, 211
see also ICP–GC; ICP–HPLC; ICP–LC
Chelating ion exchange, *see* Ion exchange
Chelation, for sample introduction, *see* Chelate vapor introduction
Chemicals, industrial, trace analysis, 17f
Chlorides, analysis of, MIP, 234f
Chromatography:
analytical characterization, 445f
see also GC; LC; HPLC

Classical analysis methods, geological materials, 28
C lines, organic solvents, 179, 183f
CMP (capacitively coupled microwave plasma), 29, 430
  detection limits, 431
  interferences, 431
CMP–AES, geological materials, 29
CMP–AES *vs.* ICP–AES, geological analysis, 36f
CN bands, organic solvents, 179f, 182, 183
Coal:
  analysis of, 15, 61
  digestion, 61
CO bands, organic solvents, 179
Coherent forward scattering, *see* CFS
Cold vapor technique, Hg, 21f, 427
Collisional deexcitation, *see* Deexcitation
Collisional excitation, *see* Excitation
Collisional ionization, *see* Ionization, collisional
Collisional-radiative models, 399f, 413f
Colorimetric procedures, geological materials, 28
Combustion, biological materials, 110, 112
  Berthelot oxidation, 112
  contamination, 112
  Schöeniger oxidation, 112
Composition:
  accumulator tissues, 73
  animal tissues, 71f
  diet composites, 72
  feedstuffs, 71
  fertilizers, 69f
  food(s), 71f
  human body fluids:
    major elements and Fe, 109
    trace elements, 118f
  human tissues, elemental, 71, 73, 110, 118f
  humus, 67
  oxidic materials, steel industry, 7f
  peat, 67
  plant materials, 70
  soil, 67f
  steel, 7f
Compost, *see* Fertilizers, types and composition
Compton background, XRFS, 442
Computer simulation, plasma modeling, *see* Plasma modeling
Concentration profile, analyte, ICP, 334
Configuration factor, torch, 294
Contamination, 450f
  biological samples:
    pretreatment, 71, 74f, 108f
    sampling, 71, 74f, 103f, 108f
    soil, sampling, 74
Continuity equation, 295
Continuous background, 285, 354, 360, 391
Continuum, 285, 354, 360, 391
Corona regime, 400
CR models, *see* Collisional-radiative models
CS bands, organic solvents, 179, 182f
CTE (complete thermal equilibrium), 357, 394
Cyclone spray chamber, *see* Spray chamber

Dalton's law, 328, 402
DC arc AES:
  analytical characterization, 445f
  detection limits, biological samples, 88f
  see also Arc AES
DCP:
  direct current plasma, 430
  and graphite cup, direct solids analysis, 224f
  three electrode, 430
    detection limits, 430, 446
DCP–AES, analytical characterization, 445f
DCP–GC, interfacing, 22
Decomposition, solid particles in ICP, 318
Deconvolution, line broadening, 374
Decrepitation, geological samples, fluid inclusions, 44
Deexcitation:
  collisional, 389, 414
  ladder-like, 401
  radiative, 391
Degeneracy, of excited state, *see* Statistical weight
Degree of ionization, 409, 411f
Density:
  organic solvents:
    table, 156
    and transport efficiency, PN, 162
  solvents, effect on nebulizer performance, 158

Depopulating processes, 389, 415
Detailed balance, 389, 397, 414f
Detection limit(s):
arc AES, 446
Ar ICP, steel, table, 7f
FAAS, 423, 447, 450
GFAAS, 424, 447, 450
hydride generation, 125f
Se, 132f
ICP–AES, 431, 446
with chelate vapor introduction, 195
table, 211
in organic solvents, table, 198f
SET *vs.* injection technique, table, 20
toluene solvent, aerosol cooling, 177f
table, 178
ICP–GC, 211
table, 212
N_2–Ar ICP *vs.* Ar ICP, 9f
polychromator, table, 10
organics, effect of spray chamber cooling, 177f
and selectivity, 433
spark AES, 446
and spectral bandwidth, 91
and spectral interferences, 91
three-electrode DCP, 430, 446
various analytical techniques, table, 88f, 446f
Detection power:
various analytical techniques, table, 88f, 446f
see also Detection limit(s)
Diagnostics:
plasma, non-LTE studies, 416
spectroscopic, 353f, 388
Dialysis fluids:
analysis of, ICP–AES, 135f
preconcentration and ICP analysis, 123f
Dielectronic recombination, *see* Recombination
Diet composites:
definition, 71
elemental concentrations, 71
table, 72
Diffusion coefficients, Chapman–Enskog kinetic theory, 328
Digestion:
acid:
closed systems, geological samples, 31f
environmental samples, 50
vs. fusion, 50
losses, 50
open systems, geological samples, 31
see also Dry-ashing; Wet-ashing
biological materials:
comparison of methods, 116f
survey of methods, 108f. *See also* Biological materials
combustion, *see* Combustion, biological materials
dry-ashing, *see* Dry-ashing
wet-ashing, *see* Wet-ashing
see also Bomb digestion; Dissolution; Pressure digestion; rf plasma ashing
Digestion and analysis:
animal bone, 129f
biological SRMs, 121f
human kidney and liver, 129
Direct current plasma, *see* DCP
Direct reader, *see* Polychromator, line programs, steel industry
Direct solids analysis, 217f
AAS, 220f
AAS or AFS, 221
arc AES, 219
assessment, 240f
capsule-in-flame atomizer, 220f
carbon rod, 236, 238f
interferences, 239f
circular cavity furnace, 220f
DCP, graphite cup, 224f
ETA–AAS, 219
FAAS, 424
graphite furnace, 220f, 425, 427
ICP, 19f, 432f, 446, 451
agitation of powders, 22f
and capillary arc, 230
fluidized bed chamber, 223f
graphite cupped graphite rod, 225f
graphite platform, 228
interrupted arc ablation, 229
laser vaporization, 43, 230f
SET, 19f, 225f
spark ablation, 19f, 228f
spark elutriation, 230f
tantalum filament, 236f
MIP, and induction furnace, 233f
not involving plasmas, 220f
plasma methods, 221f

slurry, 92, 227f
solution residues, 222
graphite braid, 236
spark AES, 219
traces, various approaches, 218, 221f
Direct solids sampling, *see* Direct solids analysis
Discharge characteristics, 387f
Discrete nebulization, *see* FIA; Injection technique(s); Microsample technique(s); PN
Dissociation equilibrium, 394
Dissolution:
acid:
airborne particulates, 50
vs. fusion, 50
geological samples:
closed systems, 31f
open systems, 31
losses, 50
acid digestion, *see* Wet-ashing
bomb digestion, *see* Wet-ashing
dry-ashing, *see* Dry-ashing
environmental samples, 49f
fusion, *see* Fusion
geological samples, 28f
residues, geological samples, 29
sediments, 59
sewage sludges, 78f
soil, residues, 78
steel, 12
wet-ashing, *see* Wet-ashing
see also Digestion; rf plasma ashing
Distillation, subboiling, 75
Distribution analysis, comparison, various methods, 445f
Doppler effect:
equations, 357f
vs. Stark effect, Ar lines, 378f
T measurement, 373f
Doppler-free spectroscopy, *see* LEI
Doppler temperature, ICP, 374
Doppler width, equation, 358
DPASV (differential pulse anodic stripping voltammetry), 448f
Drag coefficient, 315
Drop diameter, *see* Aerosol(s); Nebulization; PN; USN
Drop size, *see* Aerosol(s); Nebulization; PN; USN
Drop sizing, *see* Aerosol(s)
Dry-ashing:
agricultural materials, 75f
biological materials, 75f, 110f
ash dissolution, 75f, 111
crucible materials, 111f
low-temperature, 111f
parameters, 110f
reduced pressure, 111f
retention losses, 111
loss of analyte, 76, 110f
plant materials, 75f
see also rf plasma ashing
DSID (direct sample insertion device), 449. *See also* SET

Easily ionizable element, *see* EIE (easily ionizable element); Interference(s)
Echelle monochromator, line width measurements, 374
Echelle spectrometer, 431
with DCP, 431
Effluents, *see* Waste waters
Einstein transition probability, 329, 359
EIE (Easily ionizable element), 285. *See also* interference(s)
Electrochemical methods, analytical characterization, 445f, 448
Electron concentration, *see* Electron density
Electron density, 294, 389, 395f
Ar ICP, with organic solvents, 171
radial profiles, 332, 402f
see also n_e (electron number density)
Electronic level, *see* Level
Electron microprobe, analytical characterization, 445f, 447
Electron number density, *see* Electron density
Electron temperature, 294, 360, 374, 379, 397, 417
reduced pressure discharge, 435
Electrothermal:
atomization, *see* ETA (electrothermal atomization)
nebulization, *see* ETV
vaporization, *see* ETV
volatilization, *see* ETV
Element-selective detector, *see* Element-specific detector

Element-specific detector, 22f, 86f, 134, 154, 170, 177, 211f, 233f
ICP, 5, 22f, 86f, 134, 154, 170, 177, 211f
MIP, 22f, 233f, 434
see also ICP–GC; ICP–HPLC; ICP-LC (interfacing)
Elenbaas–Heller equation, 291
Emission, spontaneous, 414
Emission coefficient, 356
Emission pattern, *see* Emission profiles, ICP
Emission profiles, ICP, 326f, 332f, 335f
Emission spectrometry, *see* AES (atomic emission spectrometry)
Energy transfer equation, 295
Engine oil, *see* Lubricating oil, analysis of, wear metals
Enrichment procedures, *see* Preconcentration
Environmental analysis:
advantages of ICP–AES, 48f
analysis of, 48f
calibration, 50f
dissolution, 49f
hydride generation, 58
spectral interferences, 52
ETA (electrothermal atomization), 2
ETA–AAS:
analytical characterization, 445f, 447
detection limits, biological samples, 88f
direct solids analysis, 219
ETA–AAS *vs.* ICP–AES,
biological materials, 140
geological analysis, 35
Ethanol, plasma tolerance, Ar ICP, 169
ETV:
biological samples, 127f, 133
with fluoridation using freon, 240f
human milk, Ni determination, 133
see also Carbon rod; Graphite braid, and ICP; Tantalum filament, and ICP, direct solids analysis
ETV–ICP–AES, 432f, 446
biological samples, 127f
human milk, Ni determination, 133
ETV–ICP–AES *vs.* ETA–AAS, 127f
Evaporation curves, nebulization, organic samples, 159, 161
Evaporation factor, nebulization, organic samples, 159f
table, 160
Excitation:
charge transfer, 392
collisional, 389, 414
ladder-like, 401
nonthermal, reduced pressure discharge, 437
Penning, *see* Penning excitation
Excitation conditions, ICP:
and aerosol transport, 248, 277f
with organic solvents, 171, 189
Excitation mechanisms, 387f
collisional processes, 389
definition, 388
radiative processes, 389
Excitation temperature, 358, 373, 397, 417. *See also* *T* measurement
Excited level, 356
Extraction, 4f, 152f
agricultural samples, 81f, 83
back extraction into aqueous phase, 12, 81f, 121
biological samples, 120f
recovery, 120f
biological SRMs, 121f
chelating agents, 120f, 152f
complexing agents, 120f, 152f
extractants, 120f, 153
and injection technique, 17f
organic solvents, and ICP–AES, 121
precious metals, 121
reasons for using ICP–AES, 153
seawater, 55f
solvents for, 153
steel industry, 17
see also Ion exchange; Preconcentration

FAAS:
analytical characterization, 445f, 447
biological samples, decomposition in O_2, 424
detection limits, 423, 447, 450
direct solids sampling, 424
and FIA, 423
flame atomic absorption spectrometry, 28
geological materials, 28
injection technique, 423
slotted-tube atom trap techniques, 423
wire loop technique, 424, 447
FAAS *vs.* ICP–AES,
biological materials, 141
geological analysis, 34f

Faeces, analysis of, ICP–AES, 135f
FANES:
  analytical characterization, 445f, 447
  furnace atomic nonthermal excitation spectrometry, 437
Faraday effect, *see* CFS
Feedstuffs, composition, 71
Ferrous alloys:
  analysis lines, 11f
  ICP procedures, 11f
Fertilizers, types and composition, 69f
FIA (flow injection analysis)
  bovine liver, ICP–AES, 132
  flow injection analysis, 125, 132, 175, 423
Filters:
  for airborne particulate analysis, 60
  glass fiber *vs.* other types, 60
Fish analysis, hydride generation, 59
Fish meal, *see* Fertilizers, types and composition
Flame AFS, 438
Flame atomic absorption spectrometry, *see* FAAS
Flame ionization detector, for GC, 22
Flow injection, *see* FIA (flow injection analysis)
Flow injection analysis, *see* FIA (flow injection analysis)
Fluid inclusions, geological samples, 44
Fluidized bed chamber, direct powder analysis, 223f
Fluorescence spectrometry, *see* AFS (atomic fluorescence spectrometry)
Fluoridation, using freon, 240f
Flux(es):
  agricultural samples, 80
  coal or fly ash, 61
  for fusion, 14f, 29f, 50, 61, 80
  geological samples, 29f
  oxidic materials, steel industry, 14f
  see also Fusion
Fly ash:
  analysis of, 15, 61
  digestion, 61
Food(s):
  analysis of, 65f
  composition, 71f
  diet composites, 71
Fourier transform spectrometry, 433
  line width measurement, 374
Free-bound transitions, 360
Free-free transitions, 360
Freon, for fluoridation, 240f
Frequency:
  angular, 355
  effect on physical ICP characteristics, 303f
  and SBR, 293f
  and wavelength, 354
Fresh water analysis, 53f
  Babington nebulizer, 54
  PN *vs.* USN nebulizer, 54
Fritted disk nebulizer:
  organic solvents, 176
    plasma tolerance, 170f
Furnace:
  direct solids analysis, 220f
  induction, direct solids analysis, 233f
  see also ETA (electrothermal atomization); ETV; GFAAS, Graphite furnace AAS; ZAAS (Zeeman atomic absorption spectrometry)
Furnace AFS, 438f
Furnace atomic nonthermal excitation spectrometry, *see* FANES
Fusion, 79
  vs. acid digestion, 50
  agricultural samples, 80
  coal and fly ash, 61
  environmental samples, 50
  fluxing agents, *see* Flux(es)
  geological samples, 29f
  microfusion, 15
  nebulizer performance, 79
  oxidic materials, steel industry, 14f
  silicate analysis, 30
  silicates, automatic procedure, 30
FWHM (full width at half maximum), 355

Gas chromatography, *see* GC
Gas kinetic temperature, *see* Temperature, kinetic
Gas supplies, ICP, routine laboratory, 3
Gas temperature, *see* Temperature, kinetic
gA values:
  Ar lines for *T* measurement, 363, 365
    references, 364
  Balmer series, H lines, 368

GC:
  detector, 22f
  flame ionization detector, 22
  gas chromatography, 22
  see also ICP–GC; MIP–GC, interfacing
GC separation:
  metallo-organic compounds, 5
  see also ICP–GC
GD–MS:
  analytical characterization, 445f
  glow discharge MS, 444
Geochemical prospecting, 230f
  selective sample preparation, 32
Geochemistry, marine, 43
Geological analysis:
  accuracy, 38f
  biogeochemical prospecting, 43f
  collaborative study, 39f
  fluid inclusions, 44
  ICP–AES *vs.* arc AES, 36
  ICP–AES *vs.* CMP–AES, 36f
  ICP–AES *vs.* ETA–AAS, 35
  ICP–AES *vs.* FAAS, 34f
  ICP–AES *vs.* spark AES, 35f
  ICP–AES *vs.* XRFS, 38
  marine geochemistry, 43
  rare earths, 42
  types, 34
Geological materials:
  AAS, 28, 34f
  acid dissolution:
    closed systems, 31f
    open systems, 31
  analysis of, 28f
  arc AES, 28, 36
  classical analysis methods, 28
  colorimetric procedures, 28
  direct analysis of, 230
  dissolution, 28f
  FAAS, 28
  fusion methods, 29f
  halogen determination, XRFS, 38
  laser vaporization, 43, 230f
  preconcentration, 30f
  selective sample preparation, 32
  spark AES, 28, 35f
  trace elements, 30f
  XRFS, 28, 38
Geological prospecting, 42
Geological reference sample, *see* GRS
gf values:
  Fe lines for *T* measurement, 363, 366
  Ti lines for *T* measurement, 363, 367
GFAAS:
  accuracy, 424
  analytical characterization, 445f, 447
  background absorption, 425
  background correction, 425
  detection limits, 424, 447, 450
  direct solids sampling, 425, 427. *See also* Direct solids sampling
  interferences, 239f, 424
  matrix modifier, 425
  Zeeman AAS, *see* ZAAS (Zeeman atomic absorption spectrometry)
  see also ETA–AAS
Glow discharge AES, analytical characterization, 445f, 446
Glow discharge lamp, 437
  bulk analysis, 437
  in-depth profiling, 437
Glow discharge mass spectrometry, *see* GD–MS
Graphite braid, and ICP, 236
Graphite cup, direct solids analysis:
  DCP, 224f
  ICP, 225f
Graphite filament, and ICP, 236
Graphite furnace AAS, *see* GFAAS
Graphite furnace AES, 437. *See also* FANES
Graphite furnace AFS, 438f
Graphite rod atomizer, ICP–AES, 133
  direct solids analysis, 236, 238f
  interferences, 239f
Grimm glow discharge lamp, *see* Glow discharge lamp
GRS:
  French GRSs, major elements, 40
    traces, 41
  geological reference sample, 39f
Guldberg–Waage distribution, 394

Hagen–Poiseuille equation, 278f
Hair, analysis of, ICP–AES, 135f
Halfwidth, *see* FWHM (full width at half maximum); Line width(s)
Halides, volatilization into ICP, 85
Halogens determination, directly in solids, MIP, 233f

He ICP, 431
Height profiles, ICP, organic solvents, 181, 183f
He MIP, 434
Hg, cold vapor technique, 21f, 427
Hg detection, MIP, 434
High-efficiency ICP, *see* ICP, high-efficiency
High-performance liquid chromatography, *see* HPLC
H_β line, n_e measurement, 376f, 378, 401f, 403f
H lines:
  Balmer series, *gA* values, 368
  organic solvents, 179
  Stark effect, 368, 376f, 378
  see also H_β line, n_e measurement
Hollow cathode discharge, 437
Holtsmark broadening, 375
HPLC:
  analytical characterization, 449
  high-performance liquid chromatography, 23
  see also ICP-HPLC
Human blood, *see* Blood
Human body fluids:
  composition, major elements and Fe, 109
  trace element concentrations, 118f
Human kidney, complete analysis procedure, 128f
Human liver, complete analysis procedure, 129
Human milk, Ni determination, ETV–ICP–AES, 133
Human stones, analysis of, ICP–AES, 135f
Human tissues:
  accumulator, 73
  composition, major elements and Fe, 110
  elemental concentrations, 71, 73, 110, 118f
  trace element concentrations, 118f
Humus, composition, 67
Hybrid techniques, trace analysis, 218, 222
Hydride generation, 6, 125f
  AAS, 427, 447
    interferences, 427, 447
    kinetics, 427
  agricultural samples, 84f
  arsine trapping, 85
  biological samples, 102, 125f, 132f
  detection limits, 126
  environmental samples, 58
  fish analysis, 59
  ICP, 433, 446
  interferences, 21, 84f, 126, 132f
  MIP, 435
  plasma stability, 84f
  preconcentration, 85
  Se determination, bovine liver, 132f
  sediments, 59
  steel, 21
  see also Vapor introduction techniques
Hydride technique, *see* Hydride generation
Hydrofluoric acid, complexing with boric acid, 31

ICP:
  air–Ar, tolerance for organic solvents, 168
  air ICP, 294
  analyte concentration profiles, 334
  analytical performance, 431, 445, 446, 450f
  annular He ICP, 431
  Ar ICP:
    analytical performance, 431, 445, 446, 450f
    intermediate power, tolerance for organic solvents, 168f
    organic solvents, excitation conditions, 189
  axial temperature profiles, 333
  computer simulation, *see* Plasma modeling
  concentration fields, 306
  detection limits, *see* Detection limit(s)
  deviation from LTE, 290, 331, 349
  direct solids sampling, *see* Direct solids analysis
  discharge characteristics, 387f
  Doppler temperature, 374
  effect of carrier gas, *see* Carrier gas
  effect of frequency, *see* Frequency
  effect of power, *see* Power, effect on physical ICP characteristics
  effect of swirl of outer gas, 298f
  electron density, 294, 378, 402f
    measurement, *see* n_e measurement
    radial profiles, 332, 403f

ICP (*Continued*)
electron temperature, 294, 360, 374, 379, 397, 417
element-specific detector, *see* Element-specific detector
emission profiles, 335f
ETV, *see* ETV; ETV–ICP–AES
excitation characteristics, and aerosol transport, 248, 277f
excitation mechanisms, 387f
flow and temperature fields, 300f, 304, 306, 342
gas supplies, routine laboratory, 3
height profiles, background, organic solvents, 181, 183f
high-efficiency, 431
high-efficiency, plasma tolerance, organic solvents, 171f
high-power, tolerance for organic solvents, 164, 168
hydride generation, *see* Hydride generation
inductively coupled plasma, 1
injection technique, 432f, 446, 451
ion-atom intensity ratio, 406f, 408, 417
ionic line advantage, 388
low-flow, 431
matrix effects, classical, 388. *See also* Interference(s); Matrix effects
microvolume sampling, *see* Microsample technique(s); Microsampling, ICP
in molecular gas, *see* Molecular gas ICP
N_2 ICP, 294
tolerance for organic solvents, 164, 168
non-LTE effects, 290, 331, 349, 395f, 400f, 406f, 409f, 415f, 417
O_2 ICP, 294
operating conditions, organic solvents, 164
table, 165f
organic solvents, *see* Organic solvents
particle trajectories, alumina powder, 317
physical characterization, 401, 406f, 416f
plasma modeling, *see* Plasma modeling
p-LTE, 413f, 417
power distribution, 312
power loss:
by conduction, 313f
by radiation, 314f
signal changes, organic solvents, 186f
and spark ablation, 19f, 228f, 446
synopsis, 431
temperature, *see* Temperature
tolerance for organic solvents, *see* Plasma tolerance, organic solvents
torch geometry, 294, 296
velocity fields, 309
see also Air ICP; Ar ICP; N_2–Ar ICP; O_2 ICP; Plasma

ICP–AES:
advantages in environmental analysis, 48f
agricultural analysis, assessment, 93
agricultural samples, 65f
analysis of organics, assessment, 213f
analytical characterization, 431, 445f
and arc ablation, 229
arguments for use in agricultural analysis, 87
biological materials, 100f, 121f
analyzed, 134, 135f
assessment, 140f
ETV, 127f, 133
microsample technique(s), 124f, 131f
needs, 142
and carbon rod, direct solids analysis, 236, 238f
interferences, 239f
detection limits, *see* Detection limit(s)
determination of molecular species, 134
direct powder analysis, 222f, 230, 231
direct solids analysis, *see* Direct solids analysis
environmental samples, 48f
and FIA, bovine liver, 132
geological materials, 28f
and graphite braid, 236
and graphite filament, 236
graphite rod atomizer, 133
hydride generation, *see* Hydride generation
industrial materials, 1f
injection technique, *see* Injection technique(s)
and laser vaporization, 43, 230f
metals, 1f
organics, sample introduction, 152, 213f. *See also* Organic solvents
organic samples analyzed, listing, 190f
organics standards, table, 196f
reasons for use with solvent extraction, 153

silicate analysis, 28
solids analysis, *see* Direct solids analysis
and spark ablation, 19f, 228f
and spark elutriation, 230, 231
see also Analysis
ICP–AES *vs.* AAS, steel and metals, 2
ICP–AES *vs.* arc AES:
biological materials, 141
geological analysis, 36
ICP–AES *vs.* CMP–AES, geological analysis, 35, 36f
biological materials, 140f
trace analysis, 6
ICP–AES *vs.* FAAS:
biological materials, 141
geological analysis, 34f
soil extracts, 80f
ICP–AES *vs.* rotating disk spark AES, biological materials, 141
ICP–AES *vs.* spark AES, geological analysis, 35f
ICP–AES *vs.* XRFS:
biological materials, 141
geological analysis, 38
ICP–AFS, 438
analytical characterization, 438, 445f
hollow cathode lamp excited, 438
laser excited, 438
ICP characteristics, physical, effect of:
carrier gas flow, 305f
frequency, 303f
plasma confinement, 299f
power, 311
swirl of outer gas, 298f
total gas flow, 310f
ICP characterization, physical, 401, 406f, 416f
ICP-chromatography, interfacing, *see* ICP–GC; ICP–HPLC; ICP–LC, interfacing
ICP–ETV, *see* ETV–ICP–AES
ICP–GC:
chelation, biological samples, 127
detection limits, 211f
table, 212
interfacing, 5, 22f, 86f, 154, 177, 211f
ICP–HPLC:
interfacing, 23, 86f, 134, 154
protein separation, 86f
speciation of phosphate compounds, 134
ICP–LC, interfacing, 170, 212f
ICP–MS, 444
analytical characterization, 445f, 448
ICP parameters, optimization, 93
INAA:
analytical characterization, 445f
instrumental neutron activation analysis, 450
In-depth profiling, glow discharge lamp, 437
Inductively coupled plasma, *see* ICP
Industrial materials, analysis of, 1f
Infrathermal intensity, 416
Infrathermal populations, 411
Injection technique(s), 5, 17f, 20, 57, 432f, 446, 451
blood serum, 131
FAAS, 423
vs. SET, detection limits, table, 20
see also FIA
Injector tube, *see* Carrier gas tube
Instrumental neutron activation analysis, *see* INAA
Intensity:
continuous background, 285, 354, 360, 391
emission coefficient, 356
infrathermal, 416
lateral, 355
radial, 355f
total line, 354f
see also Line intensity
Interelement corrections, for spectral interferences, 91
Interference(s):
aerosol ionic redistribution, 282f
aerosol-related, *see* Nebulizer effects
Ca–Al, 281f
Ca–phosphate, 281f
carbon rod, and ICP, 239f
classical, and ICP, 388
CMP, 431
comparison, various methods, 445f
from easily ionizable elements, 93, 285
GFAAS, 424
hydride generation, 21, 84f, 126, 132f
spectral, *see* Spectral interference(s)
USN, 284f
volatilization, 281f, 284f
see also Matrix effect(s)
Interferometry, line width measurements, 374

Internal standard, 433
Inverse voltammetry, detection limits, biological samples, 88f
Ion-atom intensity ratio, ICP, 406f, 408, 417
Ion chromatography, analytical characterization, 449
Ion decay, 390
Ion exchange, 17
  agricultural samples, 82f
  biological samples, 119f, 123f, 130f
  polyacrylamide-oxine resin, 57f, 123f
  polydithiocarbamate resin, 57f, 84, 123f, 130f
  procedures, biological fluids, 57f, 84, 119f, 123f
  resins, 56, 57f, 82f, 120, 123f
  seawater, 56
  water analysis, 57f
Ionic level, *see* Level
Ionic line advantage, ICP, 388
Ionization:
  analyte, 411f
  auto-, *see* Autoionization
  collisional, 390
  degree of, 409, 411f
  Penning, *see* Penning ionization
  photo, *see* Photo ionization
  suprathermal, 397
Ionization energy, lowering of, 393
  Griem's theory, 329
Ionization temperature, 359f, 397, 402, 417
Ionization-recombination equilibrium, 359f, 393
Ionization-recombination temperature, 359f, 397, 402, 417
Ionizing plasma, 398
Ions:
  molecular, 356
  negative, 356
  positive, 356
Iron analysis:
  analysis lines, 11f
  ICP procedures, 11f
Iron ores, analysis of, 15
Isotope dilution:
  Doppler-free laser spectroscopy, 450
  MS, 444, 450

Kidney:
  analysis of, ICP–AES, 135f
  see also Human kidney

Ladder-like (de)excitation, 401
LAMMA:
  analytical characterization, 445f, 448
  laser micro mass analyzer, 445
Laser, ion source for MS, 441
Laser ablation:
  local analysis, 441
  with MIP, 441
  see also Laser vaporization, and ICP
Laser-enhanced ionization, *see* LEI
Laser evaporation, *see* Laser ablation
Laser excited AFS, *see* LEAFS
Laser excited atomic fluorescence spectrometry, *see* LEAFS
Laser micro mass analyzer, *see* LAMMA
Laser microprobe, 231f
Laser vaporization, and ICP, 43, 230f
LC:
  element-specific detection, MIP, 434
  liquid chromatography, 170
  see also HPLC
LC–ICP, *see* ICP–LC, interfacing
Leaching, *see* Solubilization/leaching
LEAFS:
  analytical characterization, 445f
  laser excited atomic fluorescence spectrometry, 438f
LEI:
  analytical characterization, 445f
  Doppler-free spectroscopy, 439, 450
    isotope dilution, 450
  laser-enhanced ionization, 439
  optogalvanic detection, 439f
  selectivity, 439
  thermoionic diode detection, 439f
  see also Optogalvanic methods, analytical characterization
Level:
  excited, 356
  metastable, 356
  resonant, 356
Limit of detection, *see* Detection limit(s)
Limit of determination, and selectivity, 433
Limit(s) of determination:
  Ar ICP, steel, table, 7f
  vs. detection limit(s), steel, 6f
Line, *see also* Analysis
Line broadening:
  Doppler effect, 357f
  Holtsmark broadening, 375
  Lorentz broadening, 375
  Stark effect, 375

Line coincidences, typical, steel industry, 6
Line intensity:
  equation, 359, 399
  peak, 355
  total, 354f
Line overlap, correction for, 52
Line profile, 354f
Line selection, AES, 433
Line width measurement:
  echelle monochromator, 374
  Fourier transform spectrometry, 374
  interferometry, 374
Line width(s):
  physical, prominent ICP lines, 374
  see also FWHM (full width at half maximum)
Line wing, 355
Liquid chromatography, *see* HPLC; ICP–HPLC; ICP–LC interfacing; LC
Liquid uptake, *see* Solvent uptake, Control of
Liquid-liquid extraction, *see* Extraction
Lithium metaborate, as flux, 29f
Liver:
  analysis of, ICP–AES, 135f
  see also Bovine; Human
Local analysis:
  comparison, various methods, 445f
  laser ablation, 441
Local thermal equilibrium, *see* LTE
Lognormal distribution, 251f
Lorentz broadening, 375
Low-flow ICP, 431
LTA:
  biological SRMs, 121f
  low-temperature ashing, 111f
LTE:
  departures from, 290, 331, 349, 400f, 409f, 417
    classification, 395f
    diagnostic studies, 415f
    measurement, 401f
  local thermal equilibrium, 290, 395f
LTE temperature, 402, 406f
  numerical data, 402
Lubricating oil, analysis of, wear metals, 16, 154, 189f, 195

Magneto-optical effect, *see* CFS
Manure, *see* Fertilizers, types and composition
Marine geochemistry, 43
Marine sediments, *see* Sediments
Mass action law, 394
Mass spectrometry, *see* ICP–MS; MS; SIMS; SNMS; SSMS
Mass transfer equation, 296
Matrix effects:
  animal bone, ICP–AES, 129f
  classical, and ICP, 388
  comparison, various methods, 445f
  ETV–ICP–AES, human milk, 133
  serum analysis, injection technique, 131
  see also Interference(s)
Matrix modifier, GFAAS, 425
Matrix separation, biological samples, chelating ion exchange, 119f, 123f
Maxwell distribution, 357, 392f
Maxwell equations, 291
Maxwell velocity, 392f
Metal chelate, *see* Chelate(s), metal, analysis of
Metallic thin layers, analysis of, 2
Metallo-organic standards, 189, 195f
  table 196f
Metals analysis:
  analysis lines, 1f, 13f
  ICP procedures, 13f
  sample preparation, 2
    trace analysis, 4f
  trace, 17f
  wear, 16, 154, 189f, 195
Metastable level, 356
Methanol, plasma tolerance, Ar ICP, 169
Microanalysis, comparison, various methods, 445f. *See also* Microsample technique(s); Microsampling, ICP
Microfusion, 15
Microsample technique(s), 17f
  biological samples, 124f, 127f, 131f
    ETV, 127f, 133
    FIA, 125, 132, 175
  bovine liver, 132
  serum, 131
Microsampling, ICP, 432, 446, 451
  graphite rod, 236, 238f
  tantalum filament, 236f
  see also Microsample technique(s)
Microwave induced plasma, *see* MIP
Milk:
  determination of P, 74
  human, analysis of, ICP–AES, 135f

Mineralization, *see* Digestion
MIP:
  analysis of, chlorides, 234f
    metal chelates, 234
  and electrodeposition, 434
  element-specific detection, *see* Element-specific detector
  and ETV, 434
  He–MIP, 434
  Hg detection, 434
  and induction furnace, direct solids analysis, 233f
  with laser ablation, 441
  microwave induced plasma, 392, 430, 434
  Penning processes, 392
  speciation, 233f
MIP–AES, analytical characterization, 445f
MIP–ETV, 435
MIP–GC, interfacing, 22, 233f
MIP–LC, interfacing, 434
Molecular bands, organics, *see* Band emission
Molecular gas ICP, 431. *See also* Air ICP; N_2 ICP; O_2 ICP
Molecular species, determination using ICP–AES, 134
Momentum transfer equation, 295
Monitoring:
  air quality, 60f
  water quality, 53f
    organisms, 58f
  wear metal concentrations, 16, 154, 189f, 195
MS:
  analytical characterization, 445f, 448
  double focussing mass spectrometers, 444
  field desorption, analytical characterization, 445f, 448
  GD–MS, 444, 448
  ICP–MS, *see* ICP–MS
  isotope dilution, 444, 450
  mass spectrometry, 87, 88f, 443f
  secondary ion MS, *see* SIMS
  of sputtered neutral atoms, *see* SNMS
  thermal ionization, 444
  thermoionic, analytical characterization, 445f, 448
Multichannel detection, 433
Multielement capability, comparison, various methods, 445f
Muscle tissue, analysis of, ICP–AES, 135f
NAA:
  analytical characterization, 445f, 448
  neutron activation analysis, 87, 88f, 450
N_2–Ar ICP:
  signal changes, organic solvents, 186
  nebulizer effects, 280
N_2–Ar ICP *vs.* Ar ICP:
  analytical performance, 3
  detection limits, polychromator, 9f
    table, 10
National Bureau of Standards, *see* NBS
N_2^+ bands:
  T measurement, 368f, 370
NBS (National Bureau of Standards), 227
NBS standard reference material, *see* SRM(s)
n_e (electron number density), 294, 374f, 389, 395f. *See also* Electron density
Nebulization:
  aerosol properties, effect of evaporation, 275f
  carrier gas wetting, 29, 32f
  centrifugal loss, 274f
  control of sample flow to nebulizer, 158
  discrete, *see* FIA (flow injection analysis); Injection technique(s); PN
  drop size, *see* Aerosol(s)
  evaporation factor, 159f
    table 160
  impact bead, 271f
  impaction loss, 273f
  organic solvents, 154f, 155f, 157, 159f, 162
    evaporation curves, 159, 161
    evaporation factor, table, 160
    limiting aspiration rates, table, 160
    transport efficiency, 162
      table, 163
  pneumatic, *see* PN
  preatomization drop size, 158f
  primary aerosol, 267f
  primary drop size, 155f
  primary *vs.* tertiary drop diameter, 158f
  secondary processes, 271f
  secondary shattering, 272
  slurry, 92
  solution with high solids content, 92
  spray chamber, *see* Spray chamber
  tertiary drop size, 158f

transport efficiency, *see* Aerosol(s)
turbulence loss, 273f
ultrasonic, *see* USN
see also Aerosol generation
Nebulizer design, organic solvents, 176f
Nebulizer effects:
aerosol ionic redistribution, 282f
aerosol washout, 284
carry-over problems, 284
drop size-related, 281f, 284f
high-power N_2–Ar ICP, 280
intermediate power Ar ICP, 280
organic solvents, 280f
solution viscosity effects, 278f, 281
see also Interference(s); Spray chamber
Nebulizer(s), 91f
Babington, 92
blockage, 23, 54, 91f
concentric *vs.* cross-flow, 92
clogging problems, 29f, 54, 91f
prevention, 92
concentric, 265f
cross-flow, 265f
fritted disk, *see* Fritted disk,
ICP vs. AAS, 158
performance, 91f
pneumatic, *see* PN
V-groove, 92
n_e measurement:
from absolute continuum intensity, 379
classification, 374f
H_β line, 401f, 403f
ICP, experimental results, 332, 378, 402f
Inglis–Teller method, 379f
Stark effect, 375
Ar lines, 378f
data, 378
H lines, 376f
results, 332, 378, 402f
Neutron activation analysis, *see* NAA
NH bands, organic solvents, 179, 183
N_2 ICP, 294
Nitrogen determination, organic compounds, 212
NO bands, organic solvents, 179, 183
Nonequilibrium parameter, *b*, 397, 406f
Nonferrous alloys:
analysis lines, 13f
ICP procedures, 13f
Non-LTE excitation, 437. *See also* FANES
Non-LTE parameter *b*, 397, 406f
Non-LTE phenomena, 400f, 409f, 417
classification, 395f
corona regime, 400
diagnostic studies, 415f
ICP, 290, 331, 349
ladder-like (de)excitation, 401
measurement, 401f
saturation regime, 400f
two-temperature models, 337f
see also LTE
Normal distribution, 250f
Norm temperature, 360
Nukiyama–Tanasawa equation, 155, 253, 267

O_2 bands:
use for O determination, 173
organic solvents, 179
O_2^+ bands, organic solvents, 179
OH bands:
organic solvents, 179, 183
T measurement, 370f
data, 372
O_2 ICP, 294
Oil analysis, 154, 189f, 195
organic standards, 189, 195f
rotating disk technique, 16f
Operating conditions, ICP, organic solvents, 164
table, 165f
Optimization:
aerosol characteristics, ICP, 284f
ICP parameters, 93
Optogalvanic detection, *see* LEI
Optogalvanic methods, analytical characterization, 445f, 447
Ores, iron, analysis of, 15
Organics, analysis of, 151f
with ICP–AES, assessment, 213f
Organic sample introduction, *see* Organic solvents
Organic samples:
analysis of, 151f
analyzed by ICP–AES, 189
listing, 190f
N determination, 212
nebulization, 155f
O determination, 212
types, 150
see also Agricultural materials; Biological materials; Organic solvents

Organic solvents:
  aerosol properties, effect of evaporation, 275f
  aerosols, drop size distribution, 268
  aerosol vapor thermostatting, 177f
  Ar ICP, excitation conditions, 189
  background emission, 168, 179
    band heads, table, 182f
    height profiles, 181, 183f
    oxygen addition, 170f
  carbon deposits in torch, 172f
  control of uptake, 173
  detection limits in, table, 198f
  as diluents for organic samples, listing, 196f
  evaporation characteristics, 159f
  evaporation curves, 159, 161
  evaporation factors, table, 160
  fritted disk nebulizer, 176
  high-power N_2–Ar ICP, signal changes, 186
  and ICP excitation characteristics, 171, 189, 248, 277f
  ICP operating conditions, 164
    table, 165f
  ICP temperature, 185
  intermediate power Ar ICP, signal changes, 186f
  introduction into ICP, 81, 154f
  limiting aspiration rates, table, 160
  nebulization, 155f, 157, 159f
    primary drop size, 155f, 157
    transport efficiency, table, 163
  nebulizer design, 176f
  nebulizer effects, 280f
  physical properties, table, 156
  plasma tolerance, 164, 168f, 277f, 285
    air-Ar ICP, 168
    fritted disk nebulizer, 170f
    intermediate power, Ar ICP, 168f
    low-power, low-flow Ar ICP, 171f
    N_2–Ar ICP, 164, 168
    oxygen addition, 170f
  signal changes, 185
  torch design, 92, 172f, 174
  transport efficiency, PN, 162
  USN, 176f
  see also Solvent uptake, control of
Organic standards, for ICP–AES, 189, 195f
  table, 196f
Organic vapor introduction, see Chelate vapor introduction
Organic vapors, analysis of, 154
Organometallic standards, 189, 195f
  table, 196f
Oscillator strength, 359, 399. *See also gA* values; *gf* values; Transition probability
Outer gas, effect of swirl, 298f
Overpopulation, 417
  relative to LTE, 397
Oxidic materials:
  composition range, steel industry, 7f
  sample preparation, 2, 4f
  and spark ablation, 19f
  standardized analysis methods, 14f
  steel industry, analysis lines, 14f
    fusion, 14f
    ICP procedures, 14f
  trace analysis, 17f
Oxygen determination, in organics, 173, 212
Oyster tissue:
  analysis of, ICP–AES, 121f, 135f
  SRM, digestion and analysis of, 121f

Partial LTE, *see* p–LTE (partial LTE)
Particle decomposition, in ICP, 318
Particle size, *see* Aerosols, drop size
Particle trajectories:
  governing equations, 315f
  ICP, alumina powder, 317
Particle-plasma interactions, 318f
  governing equations, 319f
Particulate analysis, *see* Airborne particulates
Partition function, 329, 358, 393
Peat, composition, 67
Pegmatite, fluid inclusions, 44
Penning excitation, 391
Penning ionization, 391, 418
Penning processes, MIP, 392
Performance, analytical, various methods, 445f
Petroleum products, ICP–AES, 153f
Phosphate status, soil, 81
Phosphorus determination:
  milk, 74
  spectral interference from Cu, 57

steel, 9
waste waters, 57
Photodiode array, 433
Photo ionization, 391
Physical line widths, prominent ICP lines, 374
Physical properties, organic solvents, table, 156
PIXE, 442
analytical characterization, 447
proton-induced X-ray emission, 422
Planck's law, 394
Plankton, analysis of, 43
Plant materials:
composition, 69
table, 70
essential elements:
for animal life, 69
for plant growth, 69
toxic elements, 69
Plasma:
definition, 387
ionizing, 398
recombining, 398
species in, 388f
Plasma ashing, *see* rf plasma ashing
Plasma confinement, effect on ICP characteristics, 299f
Plasma diagnostics, 353f, 388
non-LTE studies, 415f
Plasma modeling, 289f
aerosol trajectories in ICP, 312f
analytical curves, 336
Ar ICP, 293
channel model, 291f
collisional-radiative models, 399f, 413f
computer simulation, 388
concentration fields, 291f, 330f
continuity equation, 295
corona regime, 400
detailed balance, *see* Detailed balance
effect of:
carrier gas flow, 305f
frequency, 303f
plasma confinement, 299f
power, 311
swirl in outer gas, 298f
total gas flow, 310f
Elenbaas–Heller equation, 291
emission pattern, *see* Emission profiles, ICP
energy transfer equation, 295
flow fields, 291f, 330f
governing equations, boundary conditions, 295f
mass transfer equation, 296
Maxwell equations, 291
momentum transfer equation, 295
N_2 ICP, 293
non-LTE, two-temperature models, 337f
one-dimensional models, 291f
O_2 ICP, 293
particle trajectories in ICP, 312f
governing equations, 315
prospects, 349
rate models, *see* Rate models, plasma modeling
saturation regime, 400f
temperature fields, 291f, 330f
temperature histories, 312f
two-dimensional models, 292f
Plasma solvent load, *see* Organic solvents, plasma tolerance
Plasma sources, synopsis, 430f
Plasma stability, and hydride generation, 84f
Plasma tolerance, organic solvents, *see* Organic solvents
Platform, graphite, ICP, direct solids analysis, 228
Platinum group metals, analysis of, 2
p-LTE (partial LTE), 396f, 413f, 417
PN, 124f, 264f
clogging, 29f, 54, 91f
discrete nebulization, *see* FIA (flow injection analysis); Injection technique(s); Microsample technique(s)
drop size distribution, 155
heated spray chamber, 54
limitations, 91f
median drop diameter, 155f
nebulizer design, organic solvents, 176
Nukiyama–Tanasawa equation, 155, 253, 267
organic solvents, 155f
pneumatic nebulization, 54
transport efficiency, 124f
see also Aerosol(s); Nebulization

Pneumatic nebulizer, *see* Nebulizer; PN
Poiseuille equation, 278f
Polyacrylamide-oxine resin:
  urine analysis, 123f
  water analysis, 57f
Polychromator, line programs, steel industry, 6
Polydithiocarbamate resin, urine analysis, 57, 84, 123f, 130f
Polyethylene containers, for sample storage, 105
Polymerisation:
  aluminates, in solution, 29
  silicates, in solution, 29
Polytetrafluoroethylene, *see* PTFE
Populating processes, 389, 415
Powder injection, ICP:
  apparatus, ICP, 223
  fluidized bed chamber, 223f
Powders:
  analysis of, ICP, 222f, 230, 231
  thermal treatment in plasmas, modeling, 313f
Power, effect on physical ICP characteristics, 311
Power of detection, *see* Detection limit(s)
Power distribution, Ar and N_2 ICPs, 312
Power loss, ICP:
  by conduction, 313f
  by radiation, 314f
Precision, comparison, various methods, 451f
Preconcentration, 4f, 54f
  using active carbon, 17
  adsorption, seawater, 56
  agricultural samples, 81f
  biological SRMs, 121f
  chelating ion exchange, 119f, 123f
  coprecipitation, seawater, 56
  ion exchange, *see* Ion exchange
  precipitate exchange, 17
  seawater analysis, 55f
  solvent extraction, *see* Extraction
  trace analysis:
    biological samples, 119f
    geological samples, 30f
  urine, *see* Urine
  water analysis, 57f
Pressure digestion, 61, 79, 113f, 121f, 133
  biological SRMs, 121f
  coal and fly ash, 61
Prospecting:
  biogeochemical, 43f
  geochemical, 230f
    selective sample preparation, 32
  geological, 42
Proton-induced X-ray emission, *see* PIXE
PTFE:
  containers for sample storage, 105
  crucibles for digestion, 114
  polytetrafluoroethylene, 105
Pump, control of sample flow to nebulizer, 158, 173
Pyrolysis, biological materials, 110

QAH (quarternary ammonium hydroxides), 115
Quantitative spectral analysis, background correction, 433
Quarternary ammonium hydroxides, *see* QAH (quarternary ammonium hydroxides)

Radiation temperature, 360
Radiative decay, 418
Radiative deexcitation, *see* Deexcitation
Radiative recombination, 374, 391, 414
Rare earth elements, *see* REE
Rate constant, 390
Rate equations, 414f
Rate models, plasma modeling, 388f, 413f
Recombination:
  dielectronic, 392
  radiative, 374, 391, 414
  three-body, 390
Recombining plasma, 398
Reduced pressure discharges, 435
REE:
  determination, geological samples, 42
  rare earth elements, 42
Reference material, *see* SRM(s)
Reference solutions:
  multielement, 51
  see also Calibration
Residues, solution, analysis of, 222
  graphite braid, 236
Resin, *see* Ion exchange
Resonant level, 356
Reynolds number, 316
rf (radiofrequency), 268
rf plasma ashing, 111f
  biological samples, 76

coal or fly ash, 61
Rotating disk technique:
biological materials, 141
oil analysis, 16f
Rotational temperature, 359, 368
Rotrode technique, *see* Rotating disk technique
Rutherford backscattering, 445

Saha–Boltzmann plot, 395f
Saha distribution, 393, 396f
Saha–Eggert equation, *see* Saha equation
Saha equation, 328, 359f, 396
Saha equilibrium, 393, 396f
Sample elevator technique, *see* SET
Sample handling:
analyte loss in drying, 75
contamination, 75
drying, agricultural samples, 75
plant materials, 74f
reagent purity, 75
soils, 74f
see also Biological materials; Sample preparation
Sample introduction, 91f, 128
biological materials, 124f
microliter introduction, 124f, 131f
ETV, 127f, 133
hydride generation, 125f
survey of modes, 128
using syringe, organics, 175
vapor introduction, 125f
volatile chelates, *see* Chelate(s), metal, analysis of
see also Aerosols; Direct solids analysis; Microsample technique(s); Nebulization
Sample preparation:
agricultural samples, *see* Agricultural materials
biological samples, *see* Biological materials
environmental samples, 49f
geological samples, *see* Geological materials
metals and oxides, 2, 4f
Sample storage, biological materials, *see* Biological materials
Sampling:
biological materials, 102f
contamination, 103f
instruments and tools, 103f
using laser beam, 103f
plant materials, 74f
sample collection, 102f
sample homogenization, 102f, 106f
fragmentation, 107
instruments/methods, 106f
refrigerants and grinding, 107
sample pretreatment, 102f
segregation, 107
sample storage and transport, 104f
soils, 74f
see also Biological materials
Saturation regime, plasmas, 400f
Sauter mean diameter, aerosols, 249, 267
SBR:
and aerosol mass transport rate, 284f
and ICP frequency, 293f
signal-to-background ratio, 284f
Scott spray chamber, *see* Spray chamber
Seawater analysis, 55f, 119f, 123
preconcentration techniques, 55f, 119f
see also Water analysis; Water quality monitoring
Seaweed, *see* Fertilizers, types and composition
Secondary ion MS, *see* SIMS
Se determination, bovine liver, hydride generation, 132f
Sediments:
analysis of, 59
dissolution, 59
hydride generation, 59
spectral interferences, 59
Selectivity:
AES, 433
and detection limit, 433
AFS, 438
LEI, 439
Semi-halfwidth, 355
Serum:
analysis of, ICP–AES, 135f
chelation and vapor introduction, 127
ETV–ICP–AES, 128
injection technique, 131
microliter sample introduction, 125, 131
preconcentration and ICP analysis, 123f
rf plasma ashing, 76
SET, 225f
detection limits, 20
sample elevator technique, 18f

SET (*Continued*)
see also DSID (direct sample insertion device)
Sewage sludge(s):
extraction of, 79
see also Fertilizers, types and composition
Sheathing gas, carrier gas flow, 92
Silicate analysis:
fusion methods, 30
ICP–AES, 28
Silicates:
fusion, automatic procedure, 30
methods, 29f
polymerisation in solution, 29
SIMS:
analytical characterization, 445f, 448
secondary ion mass spectrometry, 445
Skin, analysis of, ICP–AES, 135f
Slags, analysis of, 14, 15, 28
Slurry, analysis of , 227f
Slurry nebulization, 92
SNMS:
analytical characterization, 445f, 448
sputtered neutral atoms mass spectrometry, 444f
Sodium peroxide, as flux, 29f
Soil extracts, analysis of, 80f
ICP–AES *vs.* FAAS, 80f
Soil(s):
composition, 67
table, 68
dissolution, 78
extractable contents, 80
extractants, 67, 80
extraction procedures, 82
minerals, 67
organic content, 67
phosphate status, 81
sampling, 74f
Solid particles:
decomposition in ICP, 318
heating in plasma, 313f
thermal treatment in plasmas, modeling, 313f
trajectories in ICP, 313f
see also Particle decomposition; Particle size; Particle trajectories; Particle-plasma interactions
Solids, direct analysis, *see* Direct solids analysis
Solubilization:
calibration, 116
capacity, 115
QAHs, 115
TMAH, 115f
see also Solubilizer(s)
Solubilization/leaching, 114f
aqueous leachants, 115
body fluids, 114f
calibration leachants, 114, 115f
deproteinization, 115
dilution, 114f
extraction:
of bovine liver, 115
of whole blood, 115
matrix modification, 115
multielement extraction, 115
Solubilizer(s):
QAHs, quaternary ammoniumhydroxides, 115
TMAH, tetramethylammoniumhydroxide, 114, 115f
trade names, 115
Solution residues, *see* Residues, solution, analysis of
Solution techniques, advantages and disadvantages, 73
Solvent, *see* Organic solvents
Solvent extraction, *see* Extraction
Solvent load, *see* Organic solvents, plasma tolerance
using positive displacement, 175
using syringe, 175
Spark ablation, and ICP, 19f, 228f, 446
Spark AES:
analytical characterization, 445f, 446
detection limits, 446
direct solids analysis, 219
geological materials, 28
high-energy prespark techniques, 429
high-repetition rate, 429
pulse distribution analysis, 429
rotating disk technique:
biological materials, 141
oil, 16f
Spark AES *vs.* ICP–AES, geological analysis, 35f
Spark elutration, and ICP, 230, 231

Spark erosion, and ICP, 19f, 228f
Spark source mass spectrometry, *see* SSMS
Speciation:
  agricultural samples, 85f
  As(III) and As(V), 126, 139
  biological materials, 102, 134, 139
  comparison, various methods, 445f
  Cr(III) and Cr(VI), ion exchange, 123f
  Fe(II) and Fe(III), ion exchange, 123f
  inorganic compounds, 140
  MIP, 233f
  oxidation state of Se, 139
  phosphate compounds, using ICP–HPLC, 134
Spectral interference(s), 52, 433
  airborne particulates, 60
  animal bone, ICP–AES, 129f
  continua, 53f
  environmental samples, 52
  human kidney and liver, 129
  interelement corrections, 91
  line broadening, 53f
  P from Cu, 57
  single-channel *vs.* multichannel spectrometer, 90f
  steel, 6, 9
  stray light, 53f, 92f
  typical, steel industry, 6
Spectral line, *see* Line
Spectral resolution, benefits, 433
Spectrofluorimetry, analytical characterization, 445f, 449
Spectrometer, single-channel *vs.* multichannel, 87f
Spectrophotometry, analytical characterization, 445f, 449
Spectroscopic diagnostics, 353f, 388
  non-LTE studies, 415f
Spectroscopic interference, *see* Spectral interference(s)
Spontaneous emission, 414
Spray chamber:
  cyclone, 275
    droplet cutoff diameter, 275
  design, 285
  dual concentric, 273f
  nebulizer effects, conical *vs.* Scott type, 280
  role of temperature, 277f
  Scott type, 273f
  temperature control, 177f
  vapor condensation device, 177f
  water-cooled, organics, effect on detection limits, 177f
SRM(s):
  biological, 121f, 130f
  standard reference material, 195
SSMS:
  analytical characterization, 445f, 448
  detection limits, biological samples, 87, 88f
  spark source mass spectrometry, 87, 88f
Standard reference material, *see* SRM(s); GRS
Stark broadening, 375
  see also Stark effect
Stark effect:
  Ar lines, n_e measurement, 378f
  H lines, 368, 376f
  n_e measurement, 375f, 378
Statistical weight, 329, 358, 393
Steel(s):
  analysis of, 2
  composition range, 7f
  dissolution procedure, 12
  limits of determination, Ar ICP, table, 7f
  spectral interferences, 6, 9
  typical line coincidences, 6
  see also Oxidic materials
Stokes number, aerosols, 253
Storage, of samples, *see* Sample storage, biological materials
Stray light:
  Ca and Mg, 53f
  water analysis, 53f
Subboiling distillation, 75
Sugar, rf plasma ashing, 76
Sulfur determination, steel, 9
Suprathermal ionization, 397
Suprathermal populations, 411
Surface tension:
  effect on nebulizer performance, 158
  organic solvents, table, 156
    and transport efficiency, PN, 162
Swirl, outer gas, effect of, 298f
Swirl velocity, torch, 294
Synchrotron radiation-excited X-ray spectrometry, *see* SYNFXRF (synchrotron radiation-excited X-ray spectrometry)

SYNFXRF (synchrotron radiation-excited X-ray spectrometry), 422, 443
Syringe sample injection, organics, 175

T (temperature), 356
Tantalum filament, and ICP, direct solids analysis, 236f
Teeth, analysis of, ICP–AES, 135f
Teflon, *see* PTFE
Temperature:
  definition, 356
  Doppler, 374
  electron, 294, 360, 374, 379, 397, 417
  excitation, 358, 373, 397, 417
  ICP:
    axial profiles, 333
    histories, 312f
    organic solvents, 171, 185
    profiles, 310f, 331, 333
    radial profiles, 331
    and USN, 277f
  ionization, 359f, 397, 402
  ionization-recombination, 359f, 397, 402
  kinetic, 357, 392
  LTE, 402, 406f
    numerical data, 402
  norm, 360
  radiation, 360
  rotational, 359, 368
  translational, 357
  see also *T* (temperature)
Temperature measurement, *see* *T* measurement
Tetramethylammoniumhydroxide, *see* TMAH (tetramethylammoniumhydroxide)
Thermal equilibrium, 392f
  complete, *see* CTE (complete thermal equilibrium)
  local, *see* LTE
  partial local, *see* p-LTE (partial LTE)
Thermodynamic equilibrium, *see* Thermal equilibrium
Thermoionic diode detection, *see* LEI
Tissues:
  accumulator, 73
  animal, analysis of, ICP–AES, 135f
  biological, analysis by TRXRF, 442
  dry-ashing, 58f
  human, composition, major elements and Fe, 110
    trace element concentrations, 118f
  rf plasma ashing, 76
  for water quality monitoring, 58f
  wet-ashing, 58f
  see also Animal tissues; Human tissues; Plant materials
TMAH (tetramethylammoniumhydroxide), 114, 115f
T measurement:
  from absolute intensity, 361
  from Boltzmann plot, 361, 396
    Ar lines, *gA* and λ values, 363, 365
      references, 364
    Cd, Mn, V and W lines, 368
    Fe lines, 363, 366, 400
      gf and λ values, 363, 366
    H lines, 368
    Ti lines, *gf* and λ values, 363, 367
  classification, 361
  from electron density, 402
  excitation temperature, 361, 399
  line pair method, 361, 373
  line width method, 373f
  from line-to-continuum intensity ratio, 374
  from rotational bands:
    N_2^+ bands, 368
      data, 370
    OH bands, 368
      data, 372
Tolerance, plasma, *see* Plasma tolerance, organic solvents
Torch:
  carbon deposits, organics, 172f
  demountable, 173f
  streamlined, 92
Torch design:
  for high salt content, 92
  for organic solvents, 92, 172f, 174
Torch geometry, ICP, 296
  configuration factor, 294
  swirl velocity, 294
Trace analysis:
  atomic spectrometric methods, 421f
  biological samples:
    matrix separation, 119f
    preconcentration, 119f
  directly in solids, various approaches, 218, 222
  geological samples, 30f
  ICP–AES *vs.* ETA–AAS, 6

industrial chemicals, 17f
metals, 17f
metals and oxides, sample preparation, 4f
oxidic materials, 17f
see also Analysis
Trace elements, biological materials:
concentrations, human body fluids, 118f
human tissues, 118f
SRMs, 121f
see also Biological materials
Transition probability, 329, 359
Ar I 430.0 nm, various values, 361f
see also gA values; Oscillator strength
Transport efficiency, nebulization, *see* Aerosol(s)
TRXRF (total reflection X-ray spectrometry), 422, 442

Ultrasonic nebulizer, *see* USN; Nebulizer(s)
Underpopulation, relative to LTE, 397
Urine:
determination of Al, 74
groupwise analyte enrichment, 130f
preconcentration and ICP analysis, 57f, 74, 76, 84, 123f, 130f, 135f
rf plasma ashing, 76
Urine analysis, ICP–AES, 57f, 84, 123f, 130f, 135f
polyacrylamide-oxine resin, 57f, 123f, 130f, 135f
polydithiocarbamate resin, 57f, 84, 123f, 130f, 135f
USN:
and aerosol desolvation, 176f, 277f
aerosols, median diameter, 270f
and background from line wings, 285
"cavitation" model, 269
"geyser formation" model, 269f
interferences from, 284f
organic solvents, 176f
and plasma temperature, 277f
ultrasonic nebulization, 54, 124f, 264f, 268f
see also Aerosol generation; Aerosol(s); Nebulization
UV (ultraviolet), 354

Vapor introduction techniques, *see* Chelate vapor introduction; Hydride generation
V-groove nebulizer, *see* Babington nebulizer; Nebulizer(s)
Viscosity:
and nebulizer effects, 158, 278f, 281
Poiseuille equation, 278f
organic solvents, control of uptake, 173
table, 156
and transport efficiency, PN, 162
Voigt effect, *see* CFS
Voltammetry, *see* DPASV (differential pulse anodic stripping voltammetry); Inverse;
VUV (vacuum ultraviolet), 9

Waste waters:
analysis of, 57
boron determination, 57
phosphorus determination, 57
Water analysis, 53f
preconcentration techniques, 57f
stray light, 53f
see also Freshwater analysis; Seawater analysis; Wastewaters
Water quality monitoring, 53f
organisms, 58f
Wavelength, and frequency, 354
Wear metals, concentration monitoring, 16, 154, 189f, 195
Wet-ashing, 77f, 113f
cautions required, 78, 113
closed systems:
autoclaves, 113f
losses, 113f
teflon bombs, 61, 79, 113f, 121f, 133
vessel materials, 113f
formation of nitroglycerine, 77
hydrogen peroxide, 78
loss of analyte, 78
nitric/perchloric acid, 77
nitric/perchloric/sulfuric acid, 77
open systems:
Betge apparatus, 113
charring, 113f
equipment, 113f
oxydizing agents, 113f
reagents, 113f
perchloric acid, risks, 78, 113
SRMs, 121f
trace analysis, literature review, 114
various acid mixtures, 113f
Whole blood, *see* Blood

Wing, of spectral line, 355
Wire loop technique, FAAS, 424, 447

X-ray emission, proton-induced, *see* PIXE
  synchrotron radiation-excited, *see* SYNFXRF (synchrotron radiation-excited X-ray spectrometry)
X-ray fluorescence spectrometry, *see* XRFS
X-ray spectrometry, *see* PIXE; SYNFXRF (synchrotron radiation-excited X-ray spectrometry); XRFS; XRS
XRFS, 422, 437, 441f
  analytical characterization, 445f, 447
  Compton background, 442
  detection limits, biological samples, 87, 88f
  geological materials, 28
    halogens, 38
  X-ray fluorescence spectrometry, 1
XRFS *vs.* ICP–AES:
  biological materials, 141
  geological analysis, 30, 38
XRS, 422, 437, 441f
  total reflection, *see* TRXRS

ZAAS (Zeeman atomic absorption spectrometry), 425f
Zeeman AAS, 425f

(continued from front)

Vol. 67. **An Introduction to Photoelectron Spectroscopy.** By Pradip K. Ghosh

Vol. 68. **Room Temperature Phosphorimetry for Chemical Analysis.** By Tuan Vo-Dinh

Vol. 69. **Potentiometry and Potentiometric Titrations.** By E. P. Serjeant

Vol. 70. **Design and Application of Process Analyzer Systems.** By Paul E. Mix

Vol. 71. **Analysis of Organic and Biological Surfaces.** Edited by Patrick Echlin

Vol. 72. **Small Bore Liquid Chromatography Columns: Their Properties and Uses.** Edited by Raymond P. W. Scott

Vol. 73. **Modern Methods of Particle Size Analysis.** Edited by Howard G. Barth

Vol. 74. **Auger Electron Spectroscopy.** By Michael Thompson, M. D. Baker, Alec Christie, and J. F. Tyson

Vol. 75. **Spot Test Analysis: Clinical, Environmental, Forensic and Geochemical Applications.** By Ervin Jungreis

Vol. 76. **Receptor Modeling in Environmental Chemistry.** By Philip K. Hopke

Vol. 77. **Molecular Luminescence Spectroscopy—Part 1: Methods and Applications.** Edited by Stephen G. Schulman

Vol. 78. **Inorganic Chromatographic Analysis.** Edited by John C. MacDonald

Vol. 79. **Analytical Solution Calorimetry.** Edited by J. K. Grime

Vol. 80. **Selected Methods of Trace Metal Analysis: Biological and Environmental Samples.** By Jon C. VanLoon

Vol. 81. **The Analysis of Extraterrestrial Materials.** By Isidore Adler

Vol. 82. **Chemometrics.** By Muhammad A. Sharaf, Deborah L. Illman, and Bruce R. Kowalski

Vol. 83. **Fourier Transform Infrared Spectrometry.** By Peter R. Griffiths and James A. de Haseth

Vol. 84. **Trace Analysis: Spectroscopic Methods for Molecules.** Edited by Gary Christian and James B. Callis

Vol. 85. **Ultratrace Analysis of Pharmaceuticals and Other Compounds of Interest.** Edited by S. Ahuja

Vol. 86. **Secondary Ion Mass Spectrometry: Basic Concepts, Instrumental Aspects, Applications and Trends.** By A. Benninghoven, F. G. Rüdenauer, and H. W. Werner

Vol. 87. **Analytical Applications of Lasers.** Edited by Edward H. Piepmeier

Vol. 88. **Applied Geochemical Analysis.** by C. O. Ingamells and F. F. Pitard

Vol. 89. **Detectors for Liquid Chromatography.** Edited by Edward S. Yeung

Vol. 90. **Inductively Coupled Plasma Emission Spectroscopy—Part 1: Methodology, Instrumentation, and Performance, Part 2: Applications and Fundamentals.** Edited by P. W. J. M. Boumans